Microprocessors and Microcontrollers

for Anna University

EEE | E&I | ICE Courses

Microprocessors and Microcontrollers

for Anna University

EEE | E&I | ICE Courses

A Nagoor Kani

Founder, RBA Educational Group
Chennai

CBS

CBS Publishers & Distributors Pvt Ltd

New Delhi • Bengaluru • Chennai • Kochi • Kolkata • Lucknow • Mumbai
Hyderabad • Jharkhand • Nagpur • Patna • Pune • Uttarakhand

Microprocessors and **Microcontrollers**

for Anna University
EEE | E&I ICE Courses

ISBN: 978-93-5466-156-3

Copyright © Author and Publisher

First Edition: 2022

Published by Satish Kumar Jain and produced by Varun Jain for
CBS Publishers & Distributors Pvt Ltd
4819/XI Prahlad Street, 24 Ansari Road, Daryaganj, New Delhi 110 002, India
Ph: 011-23289259, 23266861, 23266867   Fax: 011-23243014
Website: www.cbspd.com   e-mail: delhi@cbspd.com; cbspubs@airtelmail.in

Corporate Office: 204 FIE, Industrial Area, Patparganj, Delhi 110 092, India
Ph: 011-49344934   Fax: 011-49344935   e-mail: publishing@cbspd.com; publicity@cbspd.com

Branches

- **Bengaluru:** Seema House 2975, 17th Cross, K.R. Road, Banasankari 2nd Stage, Bengaluru 560 070, Karnataka, India
  Ph: +91-80-26771678/79   Fax: +91-80-26771680   e-mail: bangalore@cbspd.com
- **Chennai:** 7, Subbaraya Street, Shenoy Nagar, Chennai 600 030, Tamil Nadu, India
  Ph: +91-44-26680620, 26681266   Fax: +91-44-42032115   e-mail: chennai@cbspd.com
- **Kochi:** 42/1325, 1326, Power House Road, Opposite KSEB, Power House, Ernakulum-682018, Kochi, Kerala, India
  Ph: +91-484-4059061-67   Fax: +91-484-4059065   e-mail: kochi@cbspd.com
- **Kolkata:** 147, Hind Ceramics Compound, 1st Floor, Nigunj Road, Belghoria, Kolkata-700056, West Bengal, India
  Ph: +91-33-22891126, 22891127, 22891128   e-mail: kolkata@cbspd.com
- **Lucknow:** Basement, Khushuma Complex, 7 Meerabai Marg (Behind Jawahar Bhawan), Lucknow-226001, UP, India
  Ph: +522-40000032   e-mail: tiwari.lucknow@cbspd.com
- **Mumbai:** PWD Shed, Gala No. 25/26, Ramchandra Bhatt Marg, Next JJ Hospital Gate No. 2
  Opp. Union Bank of India, Noorbaug, Mumbai-400009, Maharashtra, India
  Ph: +91-22-66661880/89   Mob: 0-8424005858   e-mail: mumbai@cbspd.com

Representatives

• **Hyderabad**	0-9885175004	• **Jharkhand**	0-9811541605	• **Nagpur**	0-9421945513
• **Patna**	0-9334159340	• **Pune**	0-9623451994	• **Uttarakhand**	0-9716462459

Printed at: Mudrak, Noida, UP, India.

to
_____

My Father-in-law Capt R Chandrasekaran and
Mother-in-law Mrs Amala Chandra

PREFACE

This text on microprocessor and microcontroller has been crafted and designed for the latest syllabus of Anna university EEE/E&I/ICE courses. The various concepts of 8085 microprocessor are discussed in Chapters 1 and 2 and the concepts of 8051 microcontroller are discussed in Chapters 3 and 5. Interfacing of peripherals with 8085 and 8051 are discussed in Chapter 4.

Chapter 1 briefs about evolution of microprocessor and the basic terms involved in a microprocessor literature. The architecture, memory organisation and timing diagram of machine cycles of 8085 microprocessor are discussed in detail in Chapter 1. A brief discussion about semiconductor memory and peripheral devices and their interfacing with an 8085 microprocessor are presented in Chapter 1. Design examples are also included for better understanding of the concept of the memory and IO interfacing with an 8085 microprocessor. The importance of interrupts and the various interrupts of an 8085 processor are also discussed in Chapter 1.

The instruction set and assembly language programming concepts of 8085 microprocessor are discussed in Chapter 2. The timing diagram of some instructions are presented clearly for better understanding of execution of instructions. A number of 8085 microprocessor assembly language programming examples are included in Chapter 2 which are useful for microprocessor practical laboratory exercises.

Chapter 3 briefs about the basic concepts of 8051 microcontroller, architecture and memory organisation. The machine cycles of 8051 microcontroller with timing diagrams and instruction set of 8051 are also explained in detail in Chapter 3. The various interrupts of 8051 are also discussed in Chapter 3.

The popular peripheral devices 8255, 8259, 8254, 8279, ADC0809 and DAC0800 and their interfacing with 8085 and 8051 are discussed in detail in Chapter 4. Programming examples of ADC and DAC interface with 8051 are also presented in Chapter 4.

A number of 8051 microcontroller assembly language programming examples are included in Chapter 5 which are useful for microcontroller practical laboratory exercises. Detailed discussion about the timers of 8051, serial communication using serial port of 8051, interrupts of 8051 and keyboard and LCD interface with ports of 8051 along with programming examples are included in Chapter 5. Some application case studies of 8051 microcontroller-based system and programming are also presented in Chapter 5.

The instruction set of 8085 and 8051 are listed in Appendices II to IV, for the use of assembly language programmers. Anna university question papers are given at the end of the book. I have taken care to present the concepts of 8085 microprocessor and 8051 microcontroller in a simple manner and hope that the teaching and student community will welcome the book. The readers can feel free to convey their criticism and suggestions to nagoorkani65@yahoo.com for further improvement of the book.

A Nagoor Kani

ACKNOWLEDGEMENTS

I express my heartfelt thanks to my wife Ms C Gnanaparanjothi Nagoor Kani and my sons N Bharath Raj Alias Chandrakani Allaudeen and N Vikram Raj for the support, encouragement and cooperation they have extended to me throughout my career. I thank Ms TA Benazir, Manager, RBA Group, and all my office-staff for their cooperation in carrying out my day-to-day activities.

It is my pleasure to acknowledge the contributions of our technical editors, Ms S Saranya and Ms P Kanimozhi and Ms K G Sathyapriya for editing, proof-reading and type-setting of the manuscript and preparing the layout of the book.

My sincere thanks to all reviewers for their valuable suggestions and comments which helped me to explore the subject to a greater depth.

I am also grateful to Mr Satish K Jain, CMD, CBS Publishers & Distributors, for his keen interest in publishing this work in CBS banner. My sincere thanks to all team members of CBS Publishers & Distributors, for their concern and care in publishing this work.

Finally, a special note of appreciation is due to my sisters, brothers, relatives, friends, students and the entire teaching community for their overwhelming support and encouragement to my writing.

A Nagoor Kani

CONTENTS

CHAPTER 2: Programming of 8085 Processor   2.1–2.124

CHAPTER 4: Peripheral Interfacing   4.1–4.76

APPENDICES   A1–A10

ANNA UNIVERSITY QUESTION PAPER   Q1–Q4

GENERAL INDEX   I.1–I.5

CHIP INDEX   I.6

LIST OF ABBREVIATIONS

Abbreviations

ADC	-	Analog to Digital Converter
AF	-	Auxiliary carry Flag
ALE	-	Address Latch Enable
ALU	-	Arithmetic Logic Unit
ASCII	-	American Standard Code for Information Interchange
BCD	-	Binary Coded decimal
CF	-	Carry Flag
CMOS	-	Complementary Metal Oxide Semiconductor
CPU	-	Central Processing Unit
CRT	-	Cathode Ray Tube
DAC	-	Digital to Analog Converter
DIP	-	Dual In-line Package
DMA	-	Direct Memory Access
DRAM	-	Dynamic Random Access Memory
EEPROM	-	Electrically Erasable Programmable Read Only Memory
EOI	-	End of Interrupt
EPROM	-	Erasable-Programmable Read Only Memory
FIFO	-	First In First Out
HMOS	-	High density Metal Oxide Semiconductor
IC	-	Integrated Circuit
IO	-	Input-Output
IP	-	Instruction Pointer
IR	-	Instruction Register
IRR	-	Interrupt Request Register
ISP	-	In-System Programmable
ISR	-	Interrupt Service Routine/ In-Service Register
ISS	-	Interrupt Service Subroutine

LCD	-	Liquid Crystal Display
LDT	-	Local Descriptor Table
LED	-	Light Emitting Diode
LIFO	-	Last In First Out
LRU	-	Least Recently Used
LSB	-	Least Significant Byte
LW	-	Lower Word
MSB	-	Most Significant Byte
MSW	-	Machine Status Word
NDRO	-	Non-Destructive Read Out Memory
NMI	-	Non Maskable Interrupt
NMOS	-	N-type Metal Oxide Semiconductor
NVRAM	-	Non Volatile Random Access Memory
OCW	-	Operational Common Word
OF	-	Overflow Flag
PC	-	Program Counter/ Personal Computer
PCB	-	Printed Circuit Board
PF	-	Parity Flag
PGA	-	Pin Grid Array
PMOS	-	P-type Metal Oxide Semiconductor
PROM	-	Programmable Read Only Memory
PS	-	Priority Resolver
PSW	-	Program Status Word
PWM	-	Pulse Width Modulation
RAM	-	Random Access Memory
RI	-	Receive Interrupt
SEC	-	Single Edge Connector
SF	-	Sign Flag
SFR	-	Special Function Registers
SI	-	Source Index
SIMD	-	Single Instruction Multiple Data
SP	-	Stack Pointer
SPP	-	Speed Power Product

TI	-	Transmit Interrupt
TLB	-	Translation Look Aside Buffer
TTL	-	Transistor Transistor Logic
UART	-	Universal Asynchronous Receiver Transmitter
USART	-	Universal Synchronous Asynchronous Receiver Transmitter
UV	-	Ultra Violet
VLSI	-	Very Large Scale Integration

WDT	-	Watchdog Timer
ZF	-	Zero Flag

Symbols

&	-	Logical AND
|	-	Logical OR
^	-	Logical Exclusive OR
~	-	Logical NOT
$	-	Denote the end of a String

CHAPTER 1

8085 PROCESSOR

1.1 INTRODUCTION TO MICROPROCESSORS

1.1.1 TERMS USED IN MICROPROCESSOR LITERATURE

Bit : A digit of the binary number or code is called a bit.

Nibble : The 4-bit (4-digit) binary number or code is called a nibble.

Byte : The 8-bit (8-digit) binary number or code is called a byte.

Word : The 16-bit (16-digit) binary number or code is called a word.

Double Word : The 32-bit (32-digit) binary number or code is called a double word.

Multiple Word : The 64, 128, ... bit / digit binary numbers or codes are called multiple words.

Data : The quantity (binary number/code) operated by an instruction of a program is called data. The size of data is specified as bit, byte, word,etc.

Address : Address is an identification number (in binary) for memory locations. The 8085 processor uses a 16-bit address for memory.

Memory Word Size : The memory word size or addressability is the size of binary information
(or Addressability) that can be stored in a memory location. The memory word size for an 8085 processor-based system is 8-bit.

[Address and program codes in a microprocessor system are given in binary (i.e., as a combination of "0" and "1"). With n-bit binary we can generate 2^n different binary codes or addresses.]

Microprocessor : The microprocessor is a program-controlled semiconductor device (IC), which fetches instruction and data (from memory), decodes and executes instructions. It is used as CPU (**Central Processing Unit**) in computers. The basic functional blocks of a microprocessor are ALU (**Arithmetic Logic Unit**), an array of registers and a control unit. The microprocessor is identified with the size of data the ALU of the processor can work with at a time. The 8085 processor has a 8-bit ALU; hence, it is called a 8-bit processor. The 80486 processor has a 32-bit ALU; hence, it is called a 32-bit processor.

Bus : A bus is a group of conducting lines that carries data, address and control signals. Buses can be classified into Data bus, Address bus and Control bus.

The group of conducting lines that carries data is called a data bus.

The group of conducting lines that carries address is called an address bus.

The group of conducting lines that carries control signals is called a control bus.

CPU Bus : The group of conducting lines that are directly connected to the microprocessor is called a CPU bus. In a CPU bus, the signals are multiplexed, i.e., more than one signal is passed through the same line but at different timings.

System Bus : The group of conducting lines that carries data, address and control signals in a microcomputer system is called System bus. Multiplexing is not allowed in a system bus.

[In microprocessor-based systems, each bit of information (data/address/control signal) is sent through a separate conducting line. Due to practical limitations, the manufacturers of microprocessors may provide multiplexed pins, i.e., one pin is used for more than one purpose. This leads to a multiplexed CPU bus. For example, in an 8086 processor, the address and data are sent through the same pins but at different timings. But when the system is formed, the multiplexed bus lines should be demultiplexed by using latches, ports, transceivers, etc. The demultiplexed bus lines are called system bus. In a system bus, separate conducting lines will be provided for each bit of data, address and control signals.]

Clock : A clock is a square wave used to synchronize various devices in the microprocessor and in the system. Every microprocessor system requires a clock for its functioning. The time taken for the microprocessor and the system to execute an instruction or program are measured only in terms of the time period of its clock.

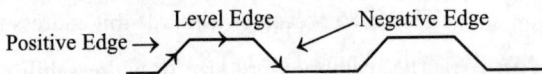

A clock has three edges: rising edge (positive edge), level edge and falling edge (negative edge). The device is made sensitive to any one of the edges for better functioning (it means that the device will recognize the clock only when the edge is asserted or arrived).

Tristate Logic : Almost all the devices used in a microprocessor-based system use tristate logic. In devices with tristate logic, three logic levels will be available: **High** state, **Low** state and **High impedance** state.

The **high** and **low** level states are normal logic levels for data, address or control signals. The **high impedance** state is an electrical open-circuit condition. The **high impedance** state is provided to keep the device electrically isolated from the system. The tristate devices will normally remain in the **high impedance** state and their pins are physically connected in the system bus but electrically isolated. In the **high impedance** state, they cannot receive or send any signal or information. These devices are provided with chip enable/chip select pins. When the signal at this pin is asserted to the right level, they come out from the **high impedance** state to normal levels.

1.1.2 EVOLUTION OF MICROPROCESSORS

History tells us that it was the ancient Babylonians who first began using the abacus (a primitive calculator made of beads) in about 500 BC. This simple calculating machine eventually sparked the human mind into the development of calculating machines that use gears and wheels (Blaise Pascal in 1642). The giant computing machines of the 1940s and 1950s were constructed with relays and vacuum tubes. Next, the transistor and solid-state electronics were used to build the mighty computers of the 1960s. Finally, the advent of the Integrated Circuit (IC) led to the development of the microprocessor and microprocessor-based computer systems.

In 1971, INTEL Corporation released the world's first microprocessor the INTEL 4004, a 4-bit microprocessor. It addresses 4096 memory locations of 4-bit word size. The instruction set consists of 45 different instructions. It is a monolithic IC employing large-scale integration in PMOS technology. The INTEL 4004 was soon followed by a variety of microprocessors, with most of the major semiconductor manufacturers producing one or more types.

First-Generation Microprocessors

The microprocessors introduced between 1971 and 1973 were the first-generation processors. They were designed using PMOS technology. This technology provided low cost, slow speed and low output currents and was not compatible with TTL (Transistor Transistor Logic) levels.

The first-generation processors required a lot of additional support ICs to form a system, sometimes as high as 30 ICs. The 4-bit processors are provided with only 16 pins, but 8-bit and 16-bit processors are provided with 40 pins. Due to limitations of pins, the signals are multiplexed. A list of first-generation microprocessors are as follows:

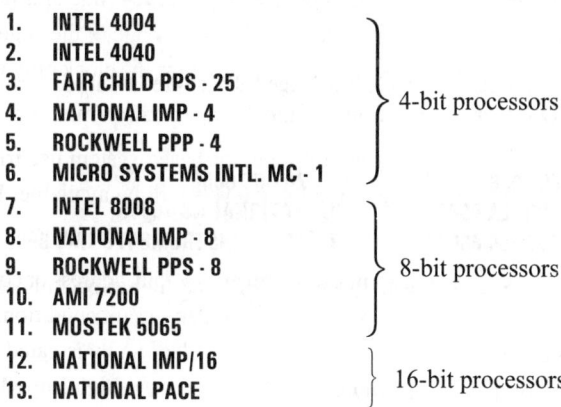

1. INTEL 4004
2. INTEL 4040
3. FAIR CHILD PPS - 25
4. NATIONAL IMP - 4   } 4-bit processors
5. ROCKWELL PPP - 4
6. MICRO SYSTEMS INTL. MC - 1
7. INTEL 8008
8. NATIONAL IMP - 8
9. ROCKWELL PPS - 8   } 8-bit processors
10. AMI 7200
11. MOSTEK 5065
12. NATIONAL IMP/16   } 16-bit processors
13. NATIONAL PACE

Second-Generation Microprocessors

The second-generation microprocessors appeared in 1973 and were manufactured using the NMOS technology. The NMOS technology offers faster speed and higher density than PMOS and it is TTL compatible. Some of the second-generation processors are as follows:

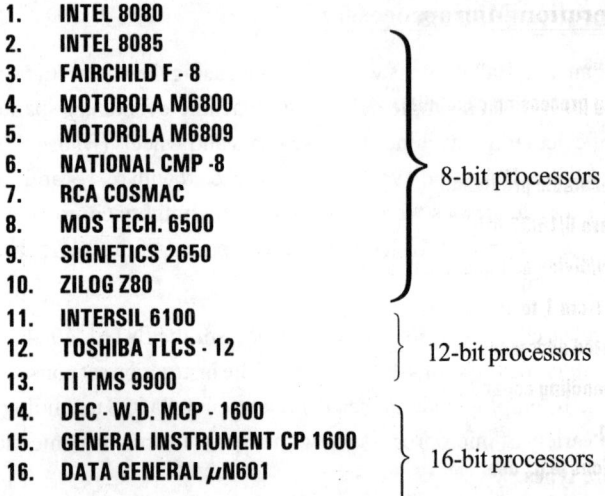

1. **INTEL 8080**	
2. **INTEL 8085**	
3. **FAIRCHILD F - 8**	
4. **MOTOROLA M6800**	
5. **MOTOROLA M6809**	
6. **NATIONAL CMP -8**	8-bit processors
7. **RCA COSMAC**	
8. **MOS TECH. 6500**	
9. **SIGNETICS 2650**	
10. **ZILOG Z80**	
11. **INTERSIL 6100**	
12. **TOSHIBA TLCS - 12**	12-bit processors
13. **TI TMS 9900**	
14. **DEC - W.D. MCP - 1600**	
15. **GENERAL INSTRUMENT CP 1600**	16-bit processors
16. **DATA GENERAL μN601**	

Characteristics of Second-Generation Microprocessors

1. **Larger chip size (170 × 200 mil). [1mil = 10^{-3} inch]**
2. **40 pins.**
3. **More numbers of on-chip decoded timing signals.**
4. **The ability to address large memory spaces.**
5. **The ability to address more IO ports.**
6. **Faster operation.**
7. **More powerful instruction set.**
8. **A greater number of levels of subroutine nesting.**
9. **Better interrupt-handling capabilities.**

Third-Generation Microprocessors

After 1978, the third-generation microprocessors were introduced. These are 16-bit processors and designed using HMOS (**H**igh density **MOS**) technology. Some of the third generation microprocessor are given below:

1. **INTEL 8086**	4. **INTEL 80286**	7. **ZILOG Z8000**
2. **INTEL 8088**	5. **MOTOROLA 68000**	8. **NATIONAL NS 16016**
3. **INTEL 80186**	6. **MOTOROLA 68010**	9. **TEXAS INSTRUMENTS TMS 99000**

The HMOS technology offers better Speed Power Product (SPP) and higher packing density than NMOS.

Speed Power Product (SPP) = Speed × Power

Unit of SPP = Nanoseconds × Milliwatts

= Picojoules

1. **Speed Power Product of HMOS is four times better than NMOS.**
   SPP of NMOS = 4 picojoules (pJ)
   SPP of HMOS = 1 picojoules (pJ)
2. **Circuit densities provided by HMOS are approximately twice those of NMOS.**
   Packing density of NMOS = 1852.5 gates/mm^2
   Packing density of HMOS = 4128 gates/mm^2 (1 mm = 10^{-6} metre)

Characteristics of Third-Generation Microprocessors

- **Provided with 40/48/64 pins.**
- **High speed and very strong processing capability.**
- **Easier to program.**
- **Allow for dynamically relocatable programs.**
- **Size of internal registers are 8/16/32 bits.**
- **The processor has multiply/divide arithmetic hardware.**
- **Physical memory space is from 1 to 16 megabytes.**
- **The processor has segmented addresses and virtual memory features.**
- **More powerful interrupt-handling capabilities.**
- **Flexible IO port addressing.**
- **Different modes of operations (e.g., user and supervisor modes of M68000).**

Fourth-Generation Microprocessors

The fourth-generation microprocessors were introduced in the year 1980. These generation processors are 32-bit processors and are fabricated using the low-power version of the HMOS technology called HCMOS. These 32-bit microprocessors have increased sophistications that compete strongly with mainframes. Some of the fourth-generation microprocessors are given below:

1.	**INTEL 80386**	4.	**MOTOROLA M68020**	7.	**MOTOROLA MC88100**
2.	**INTEL 80486**	5.	**BELLMAC - 32**		
3.	**NATIONAL NS16032**	6.	**MOTOROLA M68030**		

Characteristics of Fourth-Generation Microprocessors

1. **Physical memory space of 2^{24} bytes = 16 MB (megabytes).**
2. **Virtual memory space of 2^{40} bytes = 1 TB (terabytes).**
3. **Floating-point hardware is incorporated.**
4. **Supports increased number of addressing modes.**

Fifth-Generation Microprocessors

In microprocessor technology, INTEL has taken a leading edge and is developing more and more new processors. The latest processor by INTEL is the **pentium** which is considered a fifth-generation processor. The pentium is a 32-bit processor with 64-bit data bus and is available in a wide range of clock speeds from 60 MHz to 3.2 GHz. With improvement in semiconductor technology, the processing speed of microprocessors has increased tremendously. The 8085 released in the year 1976 executes 0.5 Million Instructions Per Second (0.5 MIPS). The 80486 executes 54 Million Instructions Per Second. The pentium is optimized to execute two instructions in one clock period. Therefore, a pentium processor working at 1 GHz clock can execute 2000 Million Instructions Per Second (2000 MIPS). The various processors released by INTEL are listed in Appendix I.

1.1.3 FUNCTIONAL BUILDING BLOCKS OF A MICROPROCESSOR

(AU, Nov/Dec' 19, 13 Marks)

A microprocessor is a programmable IC which is capable of performing arithmetic and logical operations. The basic functional block diagram of a microprocessor is shown in Fig. 1.1.

The basic functional blocks of a microprocessor are ALU, Flag register, Register array, Program Counter (PC)/Instruction Pointer (IP), Instruction decoding unit, and the Timing and Control unit.

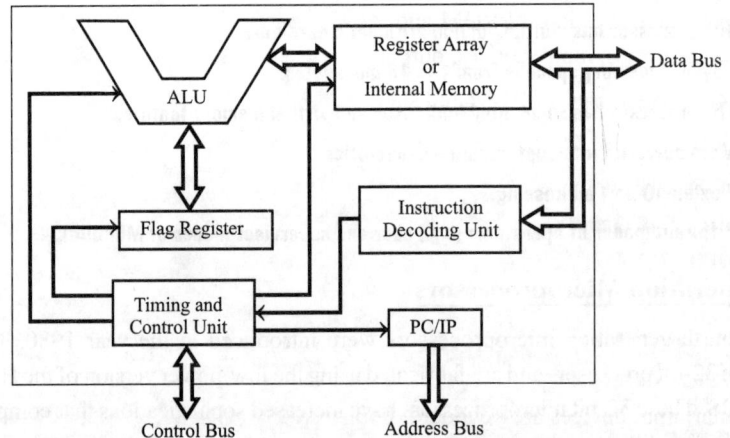

Fig. 1.1: Block diagram showing functional blocks of a microprocessor.

ALU is the computational unit of the microprocessor which performs arithmetic and logical operations on binary data. The various conditions of the result are stored as status bits called flags in the flag register. For example, consider sign flag. One of the bit positions of the flag register is called sign flag and it is used to store the status of sign of the result of the ALU operation (output data of ALU). If the result is negative then "1" is stored in the sign flag and if the result is positive then "0" is stored in the sign flag.

The register array is the internal storage device and so it is also called internal memory. The input data for ALU, the output data of ALU (result of computations) and any other binary information needed for processing are stored in the register array.

For any microprocessor, there will be a set of instructions given by its manufacturer. For doing any useful work with the microprocessor, we have to first write a program using these instructions, and store them in a memory device external to the microprocessor.

The instruction pointer generates the address of the instructions to be fetched from the memory and sends it through the address bus to the memory. The memory will send the instruction codes and data through the data bus. The instruction codes are decoded by the decoding unit and it sends information to the timing and control unit. The data is stored in the register array for processing by the ALU.

The control unit will generate the necessary control signals for internal and external operations of the microprocessor.

1.1.4 MICROPROCESSOR-BASED SYSTEM (ORGANIZATION OF A MICROCOMPUTER)

A microprocessor is a semiconductor device (or integrated circuit) manufactured by using the VLSI (**V**ery **L**arge **S**cale **I**ntegration) technique. It includes the ALU, the register arrays and the control circuit on a single chip. To perform a function or useful task we have to form a system by using the microprocessor as a CPU (**C**entral **P**rocessing **U**nit) and interfacing the memory, input and output devices to it. A system designed by using a microprocessor as its CPU is called a microcomputer or a single board microcomputer. A microprocessor-based system consists of a microprocessor as the CPU, semiconductor memories like EPROM and RAM, an input device, an output device and interfacing devices. The memories, input devices, output devices and interfacing devices are called peripherals.

The commonly used EPROM and static RAM in microcomputers are given below:

EPROM	Static RAM
INTEL 2708 (1 kB)	MOTOROLA 6208 (1kB)
INTEL 2716 (2 kB)	MOTOROLA 6216 (2kB)
INTEL 2732 (4kB)	MOTOROLA 6232 (4kB)
INTEL 2764 (8kB)	MOTOROLA 6264 (8kB)

Note: kB refers to kilobytes.

Popular input devices are keyboard, floppy disk, etc., and output devices are printer, LED and LCD displays, CRT monitor, etc.

The block diagram of a microprocessor-based system (or organization of microcomputer) is shown in Fig. 1.2. In this system the microprocessor is the master and all other peripherals are slaves. The master controls all the peripherals and initiates all operations.

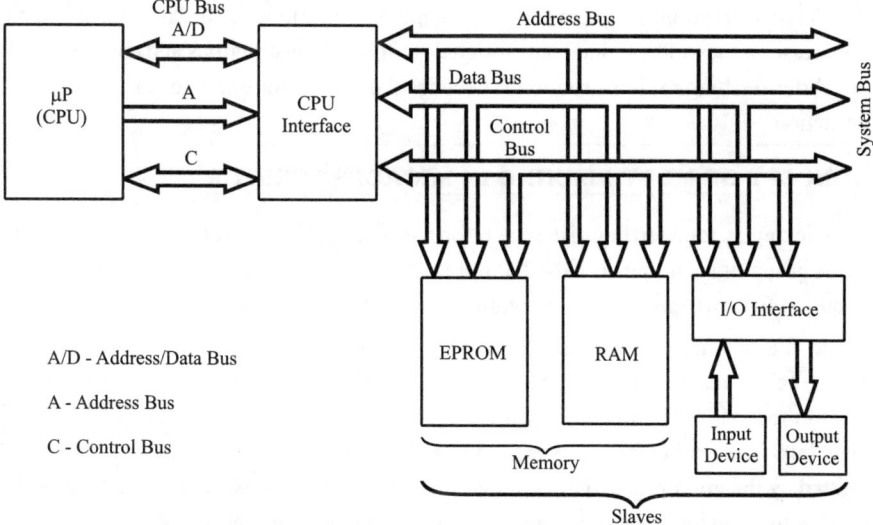

Fig. 1.2: Microprocessor-based system (organization of microcomputer).

Buses are groups of lines that carry data, addresses or control signals. The CPU bus has multiplexed lines, i.e., the same line is used to carry different signals. The CPU interface is provided to demultiplex the multiplexed lines to generate chip select signals and additional control signals. The system bus has separate lines for each signal.

All the slaves in the system are connected to the same system bus. But communication takes place between the master and one of the slaves at any one time. All the slaves have tristate logic and hence normally remain in **high impedance** state. The processor selects a slave by sending an address. When a slave is selected, it comes to the normal logic and communicates with the processor.

The EPROM memory is used to store permanent programs and data. The RAM memory is used to store temporary programs and data. The input device is used to enter the program, data and to operate the system. The output device is used for examining the results. Since the speed of IO devices does not match with the speed of the microprocessor, an interface device is provided between the system bus and the IO devices. Generally IO devices are slow devices.

The work done by the processor can be classified into the following three groups:

    1. Work done internal to the processor.

    2. Work done external to the processor.

    3. Operations initiated by the slaves or peripherals.

The work done internal to the processor are additions, subtractions, logical operations, data transfer within registers, etc. The work done external to the processor are reading/writing the memory and reading/writing the IO devices or the peripherals. If the peripheral requires the attention of the master, then it can interrupt the master and initiate an operation.

The microprocessor is the master which controls all the activities of the system. To perform a specific job or task, the microprocessor has to execute a program stored in the memory. The program consists of a set of instructions stored in consecutive memory locations. In order to execute the program, the microprocessor issues address and control signals to fetch the instructions and data from the memory one by one. After fetching each instruction it decodes the instructions and performs the task specified by the instruction.

1.1.5 CONCEPT OF MULTIPLEXING IN MICROPROCESSOR

Multiplexing is transferring different information at different well-defined times through the same lines. A group of such lines is called a multiplexed bus. The result of multiplexing is that fewer pins are required for microprocessors to communicate with the outside world.

Due to the pin number limitations, most microprocessors cannot provide simultaneously similar lines (such as address, data, status signals, etc.). Hence multiplexing of one or more of these buses is performed. Most often data lines are multiplexed with some or all address lines to form an address/data bus. (e.g., In 8085 the lower 8-address lines are multiplexed with data lines.) The status signals emitted by the microprocessor are sometimes multiplexed either with the data lines (as done in the INTEL 8080A) or with some of the address lines (as done in the INTEL 8086).

Whenever multiplexing is used, the CPU interface of the system must include the necessary hardware to demultiplex those lines to produce separate address, data and control buses required for the system. Demultiplexing of a multiplexed bus can be handled either at the CPU interface or locally at appropriate points in the system. Besides a slower system operation, a multiplexed bus also results in additional interface hardware requirements.

Demultiplexing of Address/Data Lines in 8085 Processor

In order to demultiplex the address/data lines (of the processor), the processor provides a signal called the ALE (**Address Latch Enable**). The ALE is asserted **high** and then **low** by the processor at the beginning of every machine cycle. At the same time the low byte address is given out through the AD_0 - AD_7 lines. The demultiplexing of address/data lines using an 8-bit D-latch 74LS373 is shown in Fig. 1.3.

The ALE is connected to the enable pin (EN) of an external 8-bit latch. When the ALE is asserted **high** and then **low**, the addresses are latched into the output lines of the latch. It holds the low byte of the address until the next machine cycle. After latching the address, the AD_0 - AD_7 lines are free for data transfer. The first T-state of every machine cycle is used for address latching in 8085 and the remaining T-states are used for reading or writing operation.

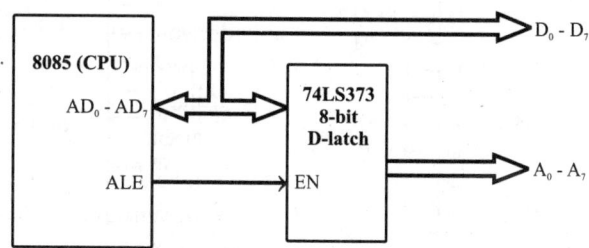

Fig. 1.3: Demultiplexing of address and data lines in an 8085 processor.

1.2 HARDWARE ARCHITECTURE AND PINOUTS

The INTEL 8085 is an 8-bit microprocessor released in the year 1976. The 8085 was originally designed using NMOS technology but now it is manufactured using HMOS technology and contains approximately 6500 transistors. The 8085 is packed in a 40-pin DIP (**D**ual **I**n-line **P**ackage) and requires a single 5V supply.

The 8085 has an internal clock oscillator. It generates a clock signal internally and divides by two for use as internal clock. This internal clock is also given out through the CLK pin for the clock requirement of peripheral devices.

The NMOS 8085 is available in two versions: 8085A and 8085A-2, with a maximum internal clock frequency of 3.03 MHz and 5 MHz respectively. The enhanced version of the 8085 is designed with HMOS transistors. It is available in three versions: 8085AH, 8085AH-2 and 8085AH-1 with maximum internal clock of 3 MHz, 5 MHz and 6 MHz respectively.

The basic data size of an 8085 is 8-bit. Therefore the memory word size of the memories interfaced with a 8085 processor is also 8-bit or byte. The 8085 uses a 16-bit address to access memory and hence it can address up to $2^{16} = 65,536_{10} = 64$k memory locations. Since, one byte of information can be stored in one memory location, the maximum memory capacity of an 8085-based system is 64 kilobytes. For accessing IO-mapped devices, the 8085 uses a separate 8-bit address and so it can generate $2^8 = 256_{10}$ IO addresses.

1.2.1 PIN CONFIGURATION OF 8085

The pin configuration of an 8085 microprocessor is shown in Fig. 1.4. The signals of the 8085 are listed in Table 1.1. The 8085 has 8 pins AD_0 to AD_7 for data transfer, which are multiplexed with low byte of address. The 8085 provides a signal ALE (**A**ddress **L**atch **E**nable) to demultiplex the low byte address and data using an external latch. The demultiplexing of address and data lines in an 8085 is shown in Fig. 1.3 in Section 1.1.5.

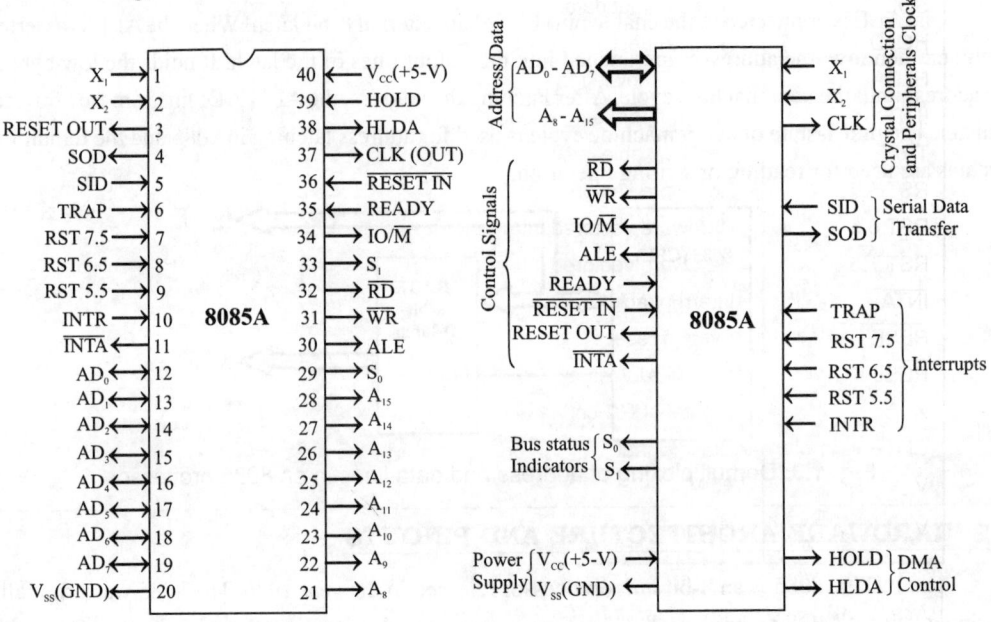

Fig. 1.4: 8085 microprocessor signals and pin assignment.

During memory access, the 16-bit memory address are output on AD_0 to AD_7 and A_8 to A_{15} lines. During IO access of IO-mapped devices the 8-bit IO address are output on both AD_0 to AD_7 and A_8 to A_{15} lines. The 8085 processor differentiates the memory and IO address using the signal $IO/\overline{M}$. When the processor outputs a memory address, the $IO/\overline{M}$ is asserted **low** and when the processor outputs an IO address, the $IO/\overline{M}$ is asserted **high**.

The $\overline{RD}$ signal is asserted **low** by the processor during a memory or IO read operation. The $\overline{WR}$ signal is asserted **low** by the processor during a memory or IO write operation. The S_0 and S_1 are bus status indicators. The output signals on these lines during various bus activity (or machine cycles) are listed in Table 1.2.

Table 1.1: 8085 Signal Description Summary

Pin Name	Description	Type
AD_0 - AD_7	Address/Data	Bidirectional, Tristate
A_8 - A_{15}	Address	Output, Tristate
ALE	Address latch enable	Output, Tristate
$\overline{RD}$	Read control	Output, Tristate
$\overline{WR}$	Write control	Output, Tristate
IO/$\overline{M}$	IO or memory indicator	Output, Tristate
S_0, S_1	Bus state indicators	Output
READY	Wait state request	Input
SID	Serial input data	Input
SOD	Serial output data	Output
HOLD	Hold request	Input
HLDA	Hold acknowledge	Output
INTR	Interrupt request	Input
TRAP	Nonmaskable interrupt request	Input
RST 5.5	Hardware vectored interrupt request	Input
RST 6.5	Hardware vectored interrupt request	Input
RST 7.5	Hardware vectored interrupt request	Input
$\overline{INTA}$	Interrupt acknowledge	Output
$\overline{RESET\ IN}$	System reset	Input
$\overline{RESET\ OUT}$	Peripherals reset	Output
X_1, X_2	Crystal or RC connection	Input
CLK (OUT)	Clock signal	Output
V_{cc}	+5 V	Power supply
V_{ss}	Ground	Power supply

Note: A overbar on the signal, indicates that it is active low. (i.e., the signal is normally high and when the signal is activated it is low).

Table 1.2: Bus Status Signals

IO/$\overline{M}$	S_1	S_0	Operation performed by the 8085
0	0	1	Memory write
0	1	0	Memory read
1	0	1	IO write
1	1	0	IO read
0	1	1	Opcode fetch
1	1	1	Interrupt acknowledge

READY is an input signal that can be used by slow peripherals to get extra time in order to communicate with the 8085. The 8085 will work only when READY is tied to logic **high**. Whenever READY is tied to logic **low**, the 8085 will enter a wait state. When the system has slow peripheral devices, additional hardware is provided in the system to make the READY input **low** during the required extra time while executing a machine cycle, so that the processor will remain in wait state during this extra time.

The HOLD and HLDA signals are used for **D**irect **M**emory **A**ccess (DMA) type of data transfer. This type of data transfers are achieved by employing a DMA controller in the system. When DMA is required, the DMA controller will place a **high** signal on the HOLD pin of the 8085. When the HOLD input is asserted **high**, the processor will enter a wait state and drive all its tristate pins to a **high impedance** state and send an acknowledgement signal to the DMA controller through the HLDA pin. Upon receiving the acknowledgement signal, the DMA controller will take control of the bus and perform DMA transfer and at the end it asserts HOLD signal **low**. When HOLD is asserted **low** the processor will resume its execution.

The 8085 has five interrupt pins. The order of priority of the interrupts is TRAP, RST 7.5, RST 6.5, RST 5.5 and INTR. The interrupts TRAP, RST 7.5, RST 6.5 and RST 5.5 are hardware vectored interrupts and are enabled by appropriate signals at the appropriate pins of the 8085. When a vectored interrupt is enabled and if it is accepted then the program execution branches to the vector addresses specified by INTEL. The interrupts RST 7.5, RST 6.5 and RST 5.5 are maskable interrupts by software.

The INTR is enabled by appropriate signals at its pin. In order to service the INTR, one of the eight opcodes (RST 0 to RST 7) has to be provided on the AD_0 - AD_7 bus by external logic. The 8085 then executes this instruction and vectors to the appropriate address to service the interrupt.The vector address for an interrupt RST n is given by $(08 \times n)_H$. The vector addresses of the interrrupts of 8085 are listed in Table 1.3. (The interrupt TRAP is RST 4.5.)

Table 1.3: Vector Addresses of Interrupts

Interrupt	Vector address	Interrupt	Vector address
RST 0	0000_H	RST 5	0028_H
RST 1	0008_H	RST 5.5	$002C_H$
RST 2	0010_H	RST 6	0030_H
RST 3	0018_H	RST 6.5	0034_H
RST 4	0020_H	RST 7	0038_H
TRAP	0024_H	RST 7.5	$003C_H$

The 8085 has the clock generation circuit on the chip but an external quartz crystal or LC circuit or RC circuit should be connected at the pins X_1 and X_2. The frequency at X_1 and X_2 is divided by two internally and used as an internal clock. The frequency of the output clock signal at the CLK(OUT) pin is same as that of the internal clock.

$\overline{\text{RESET IN}}$ is the system reset input signal and it is used to bring the processor to a known state. For proper reset, the $\overline{\text{RESET IN}}$ pin should be held **low** for at least three clock periods. When pin is asserted **low**, the program counter, instruction register, interrupt mask bits and all internal registers are cleared/reset. Also the RESET OUT signal is asserted **high** to clear/reset all the peripheral devices in the system. After a reset, the content of the program counter will be 0000_H and so the processor will start executing the program stored at 0000_H.

The pins SID and SOD can be used for serial data communication between the 8085 and any serial device under software control.

Driving X_1 and X_2 Inputs

The X_1 and X_2 pins of an 8085 processor are provided to connect an external quartz crystal or LC circuit. It can also be driven by an RC circuit or an external clock source. This connection is necessary for the internal oscillator to generate the clock signal for the processor. An oscillator consists of an amplifier and a feedback circuit. The feedback circuit of an oscillator can be of RC type, LC type or quartz crystal (a quartz crystal is electrically equivalent to an RLC circuit.) Also the feedback circuit decides the frequency of the signal generated by the oscillator.

In an 8085 processor, the oscillator circuit is provided internally except the feedback circuit. This feature facilitates the system designer to choose his own frequency for clock signals. But this frequency should not exceed the maximum clock frequency specified by the manufacturer. Another reason for keeping feedback circuit external to the processor is that the high Q circuits (quartz crystal or large values of L) cannot be fabricated by IC technology.

In an 8085, the frequency generated by the oscillator circuit will be double that of the internal clock frequency. (The maximum clock frequencies specified by the manufacturer are internal clock frequencies.) In other words, the frequency at X_1 - X_2 pins of an 8085 is divided by two internally. This means that in order to obtain an internal clock of 3.03 MHz, a clock source of 6.06 MHz must be connected to X_1 - X_2. (Crystal/LC/RC should be designed for double the internal frequency.)

Quartz crystals are the best choice for connecting at X_1 - X_2, because they are less expensive, highly stable, have a large Q, occupy a very small space and frequencies do not drift with ageing. For crystals with less than 4 MHz, a capacitor of 20 pF should be connected between X_2 and ground to ensure the starting up of the crystal at the right frequency.

When an LC circuit is used, the value of L_{ext} and C_{ext} can be chosen using the formula,

$$f = \frac{1}{2\pi L_{ext}(C_{ext} + C_{int})}$$

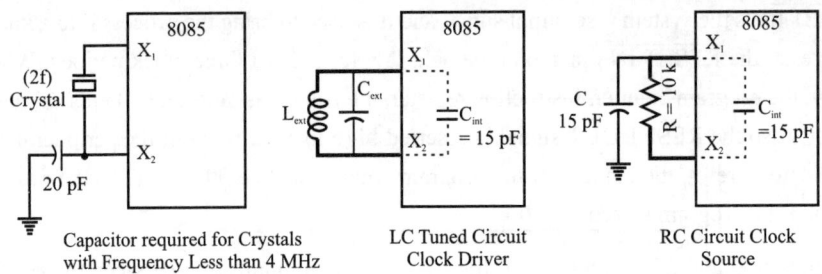

Fig. 1.5: Clock driver circuits for an 8085.

To minimize the variations in frequency, it is recommended that the value for C_{ext} should be chosen which is twice that of C_{int} or 30 pF. The use of LC circuit is not recommended for external frequencies higher than 5 MHz.

An RC circuit may also be used as the clock source for the 8085A if an accurate clock frequency is of no concern. Its advantage is the low component cost. The values shown in Fig. 1.5 are for generating an approximate external frequency of 3 MHz. Note that frequencies higher or lower than 3 MHz should not be attempted on this circuit.

1.2.2 HARDWARE ARCHITECTURE OF 8085

The architecture of an 8085 is shown in Fig. 1.7. The 8085 includes an ALU, a timing and control unit, a instruction register and a decoder, a register array, an interrupt control and a serial IO control.

The ALU performs the arithmetic and logical operations. The operations performed by the ALU of an 8085 are **addition, subtraction, increment, decrement, logical AND, OR, EXCLUSIVE-OR, compare, complement and left/right shift.** The accumulator and temporary register are used to hold the data during an arithmetic/logical operation. After an operation the result is stored in the accumulator and the flags are set or reset according to the result of the operation. The accumulator and flag register together is called the **Program Status Word (PSW).**

There are five flags in an 8085: **Sign Flag (SF), Zero Flag (ZF), Auxiliary Carry Flag (AF), Parity Flag (PF) and Carry Flag (CF).** The bit positions reserved for these flags in the flag register are shown in Fig. 1.6.

D_7	D_6	D_5	D_4	D_3	D_2	D_1	D_0
SF	ZF		AF		PF		CF

Fig. 1.6: Bit positions of various flags in the flag register of 8085.

After an ALU operation if the most significant bit of the result is 1, the sign flag is set. The zero flag is set if the ALU operation results in zero and it is reset if the result is nonzero. In an arithmetic operation, when a carry is generated by the lower nibble, the auxiliary carry flag is set. After an arithmetic or logical operation if the result has an even number of 1's, the parity flag is set, otherwise it is reset.

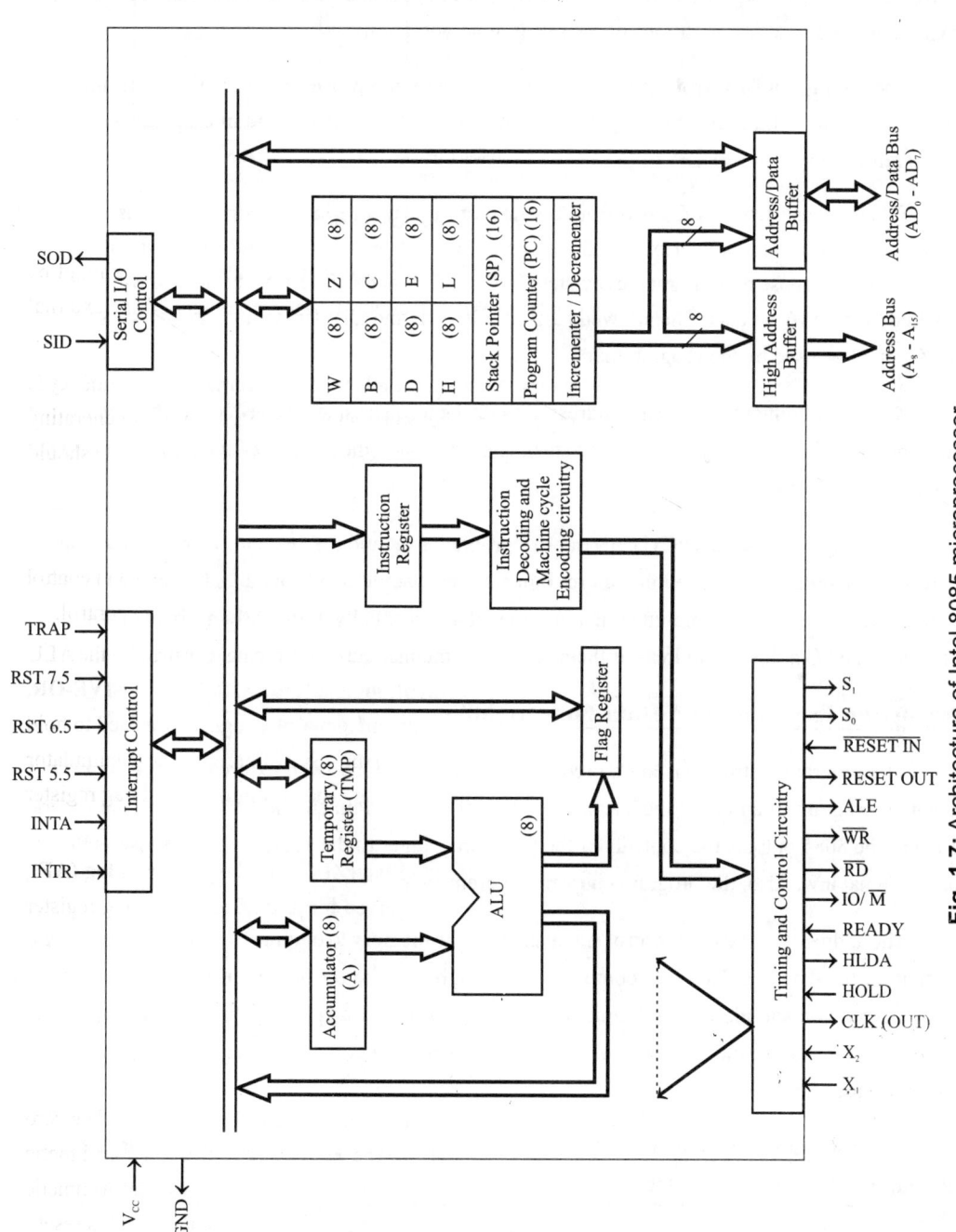

Fig. 1.7: Architecture of Intel 8085 microprocessor.

If an arithmetic operation results in a carry, the carry flag is set, otherwise it is reset. Among the five flags, the AF flag is used internally for BCD arithmetic and other four flags can be used by the programmer to check the conditions of the result of an operation.

The **timing and control unit** synchronizes all the microprocessor operations with the clock, and generates the control signals necessary for communication between the microprocessor and the peripherals.

When an instruction is fetched from the memory it is placed in the instruction register. Then it is decoded and encoded into various machine cycles. A part from the **Accumulator** (A-register) there are six general purpose programmable registers B, C, D, E, H and L. They can be used as 8-bit registers or paired to store 16-bit data. The allowed pairs are BC, DE and HL. The temporary registers TMP, W and Z cannot be used by the programmer.

The Stack Pointer SP holds the address of the stack top. The stack is a sequence of RAM memory locations defined by the programmer. The stack is used to save the content of the registers during the execution of a program.

The Program Counter (PC) keeps track of program execution. To execute a program the starting address of the program is loaded in the program counter. The PC sends out an address to fetch a byte of instruction from memory and increment its content automatically. Hence when a byte of instruction is fetched, the PC holds the address of the next byte of the instruction or the next instruction.

Instruction Execution and Data Flow in 8085

The program instructions are stored in the memory, which is an external device. In order to execute a program in an 8085, the starting address of the program should be loaded in the program counter. The 8085 outputs the contents of the program counter to the address bus and asserts the read control signal **low**. Also, the program counter is incremented.

The address and the read control signal enables the memory to output the content of the memory location on the data bus. Now the content of the data bus is the opcode of an instruction. The read control signal is made **high** by the timing and control unit after a specified time. At the rising edge of read control signals, the opcode is latched into the microprocessor internal bus and placed in the instruction register.

The instruction decoding unit decodes the instructions and provides information to the timing and control unit to take further action.

1.3 MEMORY ORGANIZATION

A memory unit is an integral part of any microcomputer system and its primary purpose is to store programs and data. In a broad sense, a microcomputer memory system can be logically divided into three groups. They are as follows:

- **Processor memory**
- **Primary or main memory**
- **Secondary memory**

Processor memory refers to registers inside the microprocessor. These registers are used to hold data and results temporarily when computation is in progress. Since the registers of the processor are fabricated using the same technology as that of a microprocessor, there is no speed disparity between these registers and a microprocessor. However, the cost involved in this approach forces a manufacturer to include only a few registers in the microprocessor.

Primary or main memory refers to the storage area which can be directly accessed by the microprocessor. Therefore, all programs and data must be stored only in primary memory prior to execution. In primary memory the access time should be compatible with the read/write time of the processor. Therefore, only semiconductor memories are used as primary memories and they (the latest versions) are fabricated using CMOS technology. Primary memory normally includes ROM, EPROM, static RAM, DRAM and NVRAM.

Secondary memory refers to the storage medium which comprises of slow devices such as magnetic tapes and disks (hard disk, floppy disc and Compact Disc (CD)). They are called as auxiliary or backup storage devices. These devices are used to hold large data files and huge programs such as operating systems, compilers, data bases, permanent programs, etc. The microcomputer system copies the required programs and data from secondary memory to main memory and work directly with main memory only.

1.3.1 INTERFACING STATIC RAM AND EPROM

The primary function of memory interfacing is that the microprocessor should be able to read from and write into a set of semiconductor memory IC chips. Generally, EPROM is interfaced for read operations and RAM is interfaced for read and write operations. The procedure for interfacing SRAM for read/write operation and EPROM for read operation are similar. So, they are dealt commonly in this section.

In order to perform the read/write operation the memory access time should be less than the read/write time of processor, chip select signals should be generated for selecting a particular memory IC, suitable control signals have to be generated for read/write operation and a specific address should be allotted to each memory location.

Hence, memory interfacing deals with choosing memories with suitable access time, designing address decoding circuit to generate chip select signals, generating control signals for read/write operation and allocation of addresses to various memory ICs and their locations.

Typical EPROM and Static RAM

A typical semiconductor memory IC will have **n** address pins, **m** data pins (or output pins) and a minimum of two power supply pins (one for connecting required supply voltage (V_{cc}) and the other for connecting ground). The control signals needed for static RAM are chip select (chip enable), read control (output enable) and write control (write enable). The control signals needed for read operation in EPROM are chip select (chip enable) and read control (output enable). A typical static RAM and EPROM are shown in Fig. 1.8 and Fig. 1.9 respectively.

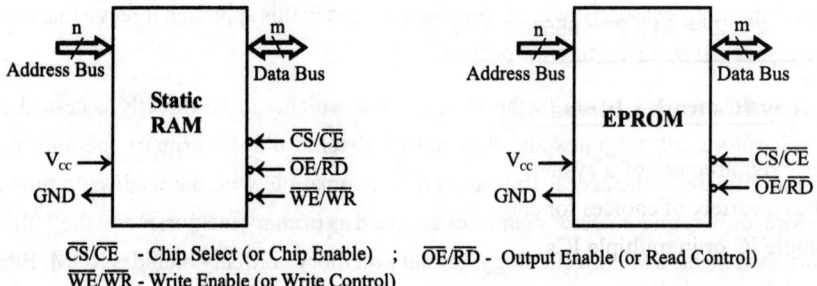

$\overline{CS}/\overline{CE}$ - Chip Select (or Chip Enable)   ; $\overline{OE}/\overline{RD}$ - Output Enable (or Read Control)
$\overline{WE}/\overline{WR}$ - Write Enable (or Write Control)

Fig. 1.8: A typical static RAM IC.   **Fig. 1.9:** A typical EPROM IC in read mode.

> *Note:* *The pins of EPROM are redefined for write operation. An EPROM requires a different hardware setup and high supply voltage for write operation.*

Memory Capacity

A semiconductor memory IC will have **n** address pins and **m** data pins. Such a memory has 2^n locations and each location can store **m**-bit data. The size of data stored in each memory location is called memory word size. In INTEL 8085-based systems normally memories with word size of 1-byte are used. (But we can even interface memories with word size 1-bit, 2-bit and 4-bit.)

The memory capacity is specified in kilobytes. If the memory IC has **m** data pins and **n** address pins, then the memory IC will have a capacity of $2^n \times m$ bits. When m = 8, the memory capacity is 2^n bytes. One kilobyte is $1024_{10} (= 400_H)$ bytes. The relation between address pins and capacity of memory ICs are listed in Table 1.4.

Table 1.4: Relation between Number of Address Pins and Memory Capacity

Number of address pins	Memory capacity			Range of address in hexa
	in decimal	in kilo	in hexa	
10	$2^{10} = 1024$	1k	400	000 to 3FF
11	$2^{11} = 2 \times 2^{10} = 2048$	2k	800	000 to 7FF
12	$2^{12} = 2^2 \times 2^{10} = 4 \times 2^{10} = 4096$	4k	1000	000 to FFF
13	$2^{13} = 2^3 \times 2^{10} = 8 \times 2^{10} = 8192$	8k	2000	0000 to 1FFF
14	$2^{14} = 2^4 \times 2^{10} = 16 \times 2^{10} = 16384$	16k	4000	0000 to 3FFF
15	$2^{15} = 2^5 \times 2^{10} = 32 \times 2^{10} = 32768$	32k	8000	0000 to 7FFF
16	$2^{16} = 2^6 \times 2^{10} = 64 \times 2^{10} = 65536$	64k	10000	0000 to FFFF

Choice of Memory ICs and Address Allocation

The memory requirement of a system depends on the application for which it is designed. A system designer has a variety of choices for choosing memory ICs. The total memory requirement can be realized in a single IC or in multiple ICs.

The total memory requirement of the system will be split between EPROM and RAM memories. The EPROM memories are used for storing monitor programs, other permanent programs and data. The RAM memories are used for stack operations, temporary program and data storage.

Popular EPROM and static RAM ICs with 8085 systems and their capacity are listed here. Table 1.5 shows the number of address pins and data pins available on these ICs.

EPROM

2708 (1k × 8 = 8 kilobits/1kB)

2716 (2 k × 8 = 16 kilobits/2 kB)

2732 (4 k × 8 = 32 kilobits/4 kB)

2764 (8 k × 8 = 64 kilobits/8 kB)

27256 (32 k × 8 = 256 kilobits/32 kB)

27512 (64 k × 8 = 512 kilobits/64 kB)

Static RAM

6208 (1k × 8 = 8 kilobits/1kB)

6216 (2 k × 8 = 16 kilobits/2 kB)

6232 (4 k × 8 = 32 kilobits/4 kB)

6264 (8 k × 8 = 64 kilobits/8 kB)

62256 (32 k × 8 = 256 kilobits/32 kB)

62512 (64 k × 8 = 512 kilobits/64 kB)

Note: In this book kB refers to kilobytes.

Table 1.5: Number of Address and Data Pins in Memory ICs

Memory IC EPROM/RAM	Capacity	Number of address pins	Number of data pins
2708/6208	1 kB	10	8
2716/6216	2 kB	11	8
2732/6232	4 kB	12	8
2764/6264	8 kB	13	8
27256/62256	32 kB	15	8
27512/62512	64 kB	16	8

Note: 16 kB memory is not available as a standard product.

Generation of Chip Select Signals

Decoders are used to generate chip select signals. The 2-to-4 decoder will give four chip select signals. The 3-to-8 decoder will give eight chip select signals. The 4-to-16 decoder will give sixteen chip select signals.

Decoder is a logic circuit that identifies each combination of the signals present at its input. Decoders have **n** input lines and 2^n output lines. In logic **low** decoder, at any one time one of the 2^n outputs will remain **low** and all other outputs will remain **high**.

The output which remains **low** depends on the input signal. Hence if the decoder outputs are connected to chip select pins of ICs in the microprocessor system at any one time, only one chip will be selected. The input to the decoders are unused address lines or high order address lines.

While interfacing memories, low order address lines are connected to memory ICs. The remaining unused address lines (or high order address lines) are connected to the input of the decoder. The outputs of the decoder are connected to CS or CE pins of memory ICs.

In a microprocessor-based system, all the memory ICs and peripheral ICs are connected to a common system bus. Therefore, the data, address and control lines are connected to all the slaves (memory/peripheral ICs). But all the slaves remain in **high impedance** state. So, they cannot communicate with the master (processor) through bus (i.e., they are physically connected but electrically isolated).

When the address is given out by the processor for read/write operation, only one of the memory ICs is selected and the selected memory IC will come to normal logic. The selection logic depends on address decoding logic. All other memory ICs will remain in **high impedance** state. So, they are electrically isolated from the system. The read/write operation is performed by the processor with the selected memory IC.

Decoder

Popular decoders used in the microprocessor-based system are 74LS138 and 74LS139. The 74LS138 is a 3-to-8 decoder and 74LS139 is dual 2-to-4 decoder.

The 74LS138 decoder consists of 3-input lines, 8-output lines (logic **low**) and three enables or ground. In the three enables, two are logic **low** and one is a logic **high** enable. The pin configuration of 3-to-8 decoder (74LS138) is shown in Fig. 1.10. The truth table of the decoder is given in Table 1.6.

The 74LS139 decoder consists of two numbers of 2-to-4 decoder packed in a single IC package. Each decoder has two input pins, four output lines and a logic **low** enable. The pin configuration of 74LS139 is shown in Fig. 1.11. The truth table of 2-to-4 decoder is given in Table 1.7. In the 74LS139 each decoder can work independently.

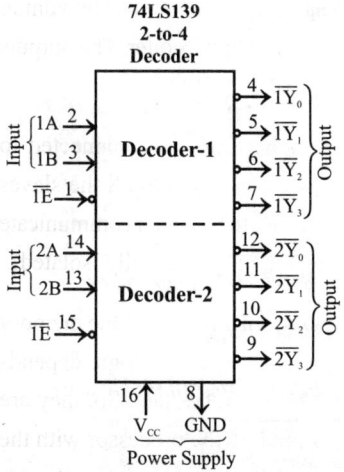

Fig. 1.10: Signals of 74LS138.

Fig. 1.11: Signals of 74LS139.

Table 1.6: Truth Table of 3-to-8 Decoder

Enables			Input			Output							
G_1	$\overline{G}_{2A}$	$\overline{G}_{2B}$	C	B	A	$\overline{Y}_7$	$\overline{Y}_6$	$\overline{Y}_5$	$\overline{Y}_4$	$\overline{Y}_3$	$\overline{Y}_2$	$\overline{Y}_1$	$\overline{Y}_0$
1	0	0	0	0	0	1	1	1	1	1	1	1	0
1	0	0	0	0	1	1	1	1	1	1	1	0	1
1	0	0	0	1	0	1	1	1	1	1	0	1	1
1	0	0	0	1	1	1	1	1	1	0	1	1	1
1	0	0	1	0	0	1	1	1	0	1	1	1	1
1	0	0	1	0	1	1	1	0	1	1	1	1	1
1	0	0	1	1	0	1	0	1	1	1	1	1	1
1	0	0	1	1	1	0	1	1	1	1	1	1	1
0	1	1	X	X	X	H	H	H	H	H	H	H	H

Table 1.7: Truth Table of The 2-to-4 Decoder

Enable	Input		Output			
$\overline{E}$	B	A	$\overline{Y}_3$	$\overline{Y}_2$	$\overline{Y}_1$	$\overline{Y}_0$
0	0	0	1	1	1	0
0	0	1	1	1	0	1
0	1	0	1	0	1	1
0	1	1	0	1	1	1
1	X	X	H	H	H	H

1.3.2 MEMORY ORGANIZATION IN 8085-BASED SYSTEM

A microprocessor-based system requires both EPROM and RAM. Hence the available memory space has to be divided between EPROM and RAM. The 8085 has 64 kB of addressable memory space and allotting this address space for EPROM and RAM depends on the system designer as well as the application for which the system is designed.

Some systems may require large memory space. So, the full memory space is utilized. But in some systems the memory requirement may be less and in this case the full memory space will not be utilized. When the full memory space is not utilized, the unused memory addresses can be used for addressing IO devices. Such IO devices are called memory-mapped IO devices and they can be accessed similar to that of a memory device.

In 8085 system, the EPROM is mapped at the beginning of memory space (i.e., 0000_H address is allotted to EPROM memory location). Whenever the power supply is switched ON, the microprocessor chip will be reset. This power-on reset will be implemented by the system designer. When the processor is reset all the internal registers, flag register and program counter will be cleared. Hence, after a reset, the program counter will have an address 0000_H and so the processor starts fetching and executing the instruction stored at 0000_H.

The system designer will store the monitor program starting from the address 0000_H. The monitor program should be executed to initialize system peripherals whenever the system is switched ON. To enable automatic execution of monitor program, whenever the system is switched ON, the EPROM should be mapped from 0000_H location in 8085-based system. Monitor program is a permanent program written by the system designer to take care of system initializations. System initializations includes the following :

    (a) **Programming 8279 for keyboard scanning and display refreshing.**

    (b) **Programming peripheral ICs 8259, 8257, 8255, 8251, 8254, etc.**

    (c) **Initializing stack.**

    (d) **Display a message on display (output) device.**

    (e) **Initializing interrupt vector table.**

Note: *8279 - Programmable keyboard/display controller.*	*8257 - DMA controller.*
8259 - Programmable interrupt controller.	*8251 USAR*
8255 - Programmable peripheral interface.	*8254 - Programmable timer.*

The total address space and its allocation is shown in Fig. 1.12.

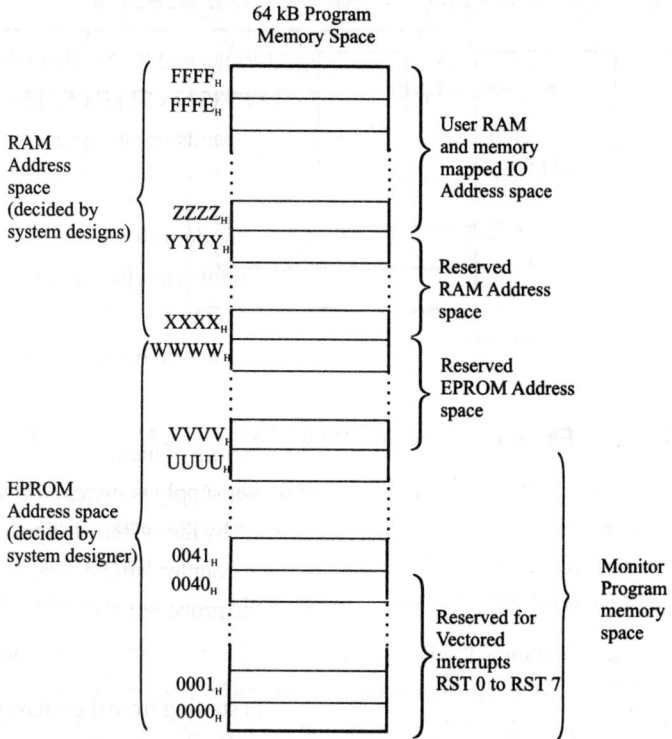

Fig. 1.12: Memory Address Space in 8085 Microprocessor.

The required EPROM memory capacity of the system can be implemented in one IC or in multiple ICs. Similarly the RAM capacity of the system can be implemented in one IC or in multiple ICs. This choice depends on the availability of memory IC and the system designer. Some examples of memory organizations for 8085 microprocessor-based system are discussed in this section.

Consider a system in which the full memory space 64 kB is utilized for EPROM memory. In this system the entire 16 address lines of the processor are connected to address input pins of memory IC in order to address the internal locations of memory and Chip Select (CS) pin of EPROM is permanently tied to logic **low** (i.e., tied to ground) as shown in Fig. 1.13. Now the range of address for EPROM is 0000_H to $FFFF_H$.

Consider a system in which the available 64 kB memory space is equally divided between EPROM and RAM. Let us implement 32 kB memory capacity of EPROM using single IC 27256. Similarly, 32 kB RAM capacity is implemented using single IC 62256. The 32 kB memory requires 15 address lines and so the address lines A_0 - A_{14} of the processor are connected to 15 address pins of both EPROM and RAM as shown in Fig. 1.13.

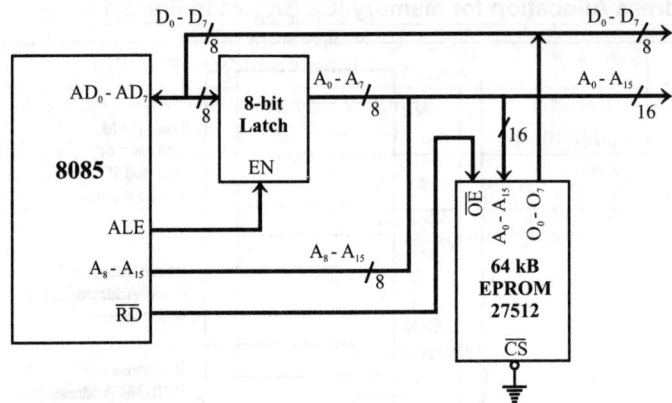

Fig. 1.13: Example of implementing 64 kB EPROM in the 8085 system.

The unused address line A_{15} is used as a chip select signal for selecting either EPROM or RAM. The A_{15} line is directly connected to the CS pin of EPROM and it is inverted and connected to CS pin of RAM. Therefore, the EPROM is selected when $A_{15} = 0$ and RAM is selected when $A_{15} = 1$. The address range of EPROM will be 0000_H to $7FFF_H$ and that of RAM will be 8000_H to $FFFF_H$.

Consider a system in which 32 kB memory space is implemented using four 8 kB memory. Let two 8 kB memory be EPROM and the remaining two be RAM. Each 8 kB memory requires 13 address lines. So, the address lines $A_0 - A_{12}$ of the processor are connected to 13 address pins of all the memory ICs. The address lines A_{13} and A_{14} can be decoded using a 2-to-4 decoder to generate four chip select signals. These four chip select signals can be used to select one of the four memory IC at any one time. The address line A_{15} is used as an enable for the decoder. The simplified schematic of this memory organization is shown in Fig. 1.15 and address allotted to each memory IC is shown in Table 1.8.

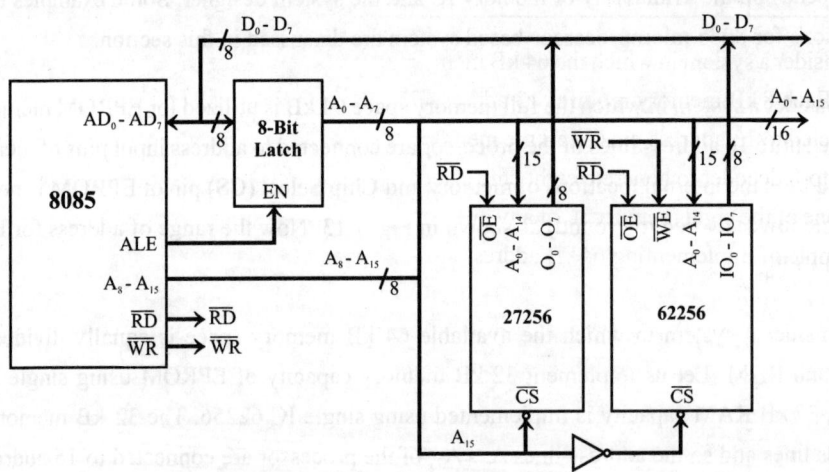

Fig. 1.14 : Example of implementing 32 kB EPROM and 32 kB RAM in an 8085 system.

Table 1.8: Address Allocation for Memory ICs Shown in Fig. 1.15

Device	Decoder enable/input A_{15} A_{14} A_{13} A_{12}	Input to address pins of memory IC A_{11} A_{10} A_9 A_8	A_7 A_6 A_5 A_4	A_3 A_2 A_1 A_0	Hexa address
8 kB EPROM - I	0 0 0 0	0 0 0 0	0 0 0 0	0 0 0 0	0000
	0 0 0 0	0 0 0 0	0 0 0 0	0 0 0 1	0001
	0 0 0 0	0 0 0 0	0 0 0 0	0 0 1 0	0002
	. . . .	. . . .	. . . .	. . . .	.
	0 0 0 1	1 1 1 1	1 1 1 1	1 1 1 1	1FFF
8 kB EPROM - II	0 0 1 0	0 0 0 0	0 0 0 0	0 0 0 0	2000
	0 0 1 0	0 0 0 0	0 0 0 0	0 0 0 1	2001
	0 0 1 0	0 0 0 0	0 0 0 0	0 0 1 0	2002
	. . . .	. . . .	. . . .	. . . .	.
	0 0 1 1	1 1 1 1	1 1 1 1	1 1 1 1	3FFF
8 kB RAM - I	0 1 0 0	0 0 0 0	0 0 0 0	0 0 0 0	4000
	0 1 0 0	0 0 0 0	0 0 0 0	0 0 0 1	4001
	0 1 0 0	0 0 0 0	0 0 0 0	0 0 1 0	4002
	. . . .	. . . .	. . . .	. . . .	.
	0 1 0 1	1 1 1 1	1 1 1 1	1 1 1 1	5FFF
8 kB RAM -II	0 1 1 0	0 0 0 0	0 0 0 0	0 0 0 0	6000
	0 1 1 0	0 0 0 0	0 0 0 0	0 0 0 1	6001
	0 1 1 0	0 0 0 0	0 0 0 0	0 0 1 0	6002
	. . . .	. . . .	. . . .	. . . .	.
	0 1 1 1	1 1 1 1	1 1 1 1	1 1 1 1	7FFF

Consider a system in which the 64 kB memory space is implemented using eight numbers of 8 kB memory. Each 8 kB memory requires 13 address lines and so the address line A_0-A_{12} of the processor are connected to 13 address pins of all the memory ICs. The address lines A_{13}, A_{14} and A_{15} are decoded using a 3-to-8 decoder to generate eight chip select signals. These eight chip select signals can be used to select one of the eight memory IC at any one time. Design example-2 given at the end of this chapter is an example of implementing 64 kB address space using 8 numbers of 8 kB memory.

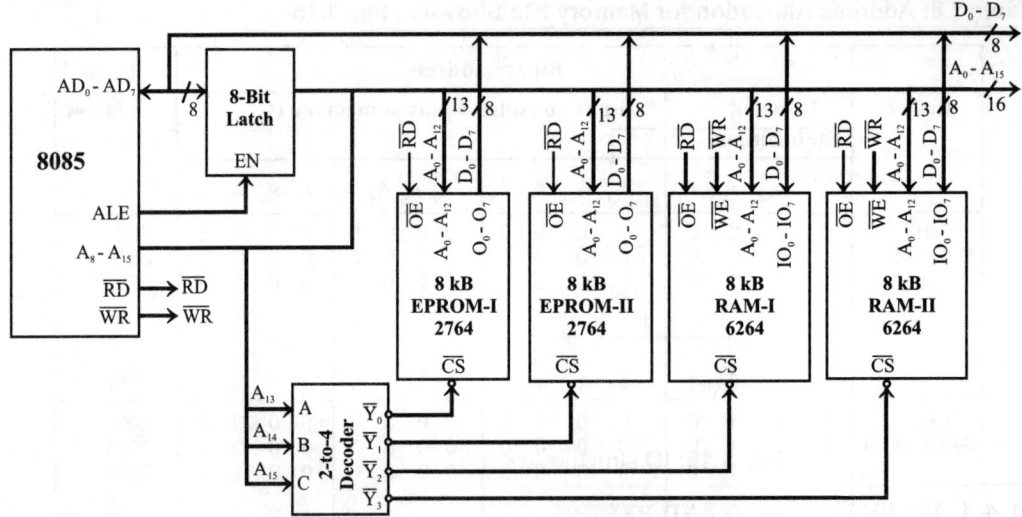

Fig. 1.15: Example of implementing 16 kB EPROM and 16 kB RAM in an 8085 system.

1.4 IO PORTS AND DATA TRANSFER CONCEPTS

The IO devices connected to a microcomputer system provides an efficient means of communication between the microcomputer system and the outside world. These IO devices are commonly called peripherals and include keyboards, CRT displays, printers and disks (floppy disk, hard disk and Compact Disc (CD)).

The characteristics of the IO devices are normally different from the characteristics of the microprocessor. Since the characteristics of the IO devices are not compatible with that of the microprocessor, interface hardware circuitry between the microprocessor and IO device are necessary.

There are three major types of data transfer between the microcomputer and an IO device. They are as follows:

- **Programmed IO**
- **Interrrupt driven IO**
- **Direct memory access (DMA)**

(AU, Nov/Dec' 19, 2 Marks)

In programmed IO the data transfer is accomplished through an IO port and controlled by software. In interrupt driven IO, the IO device will interrupt the processor and initiate data transfer. In DMA, the data transfer between memory and IO can be performed by bypassing the microprocessor. Each type of data transfer scheme mentioned above, includes different methods of data transfer schemes. Fig. 1.16 shows all the types of data transfer schemes in a microcomputer and it can also be called **IO structure of a microcomputer**.

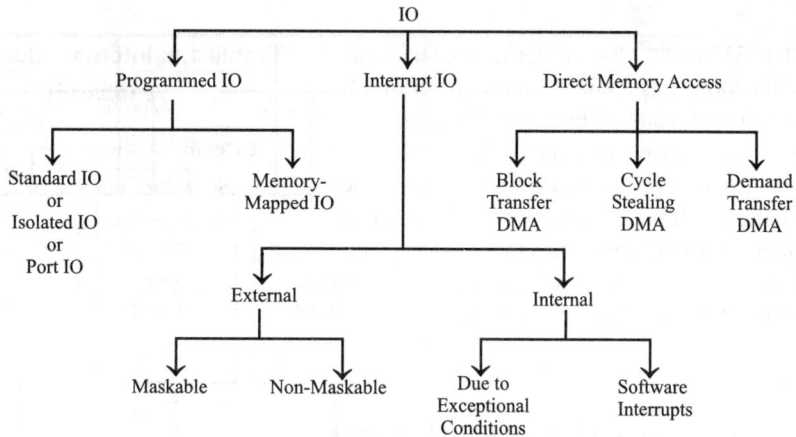

Fig. 1.16: IO structure of a typical microcomputer.

1.4.1 INTERFACING IO AND PERIPHERAL DEVICES

The IO devices are generally slow devices. So, they are connected to the system bus through ports. The ports are buffer IC which is used to temporarily hold the data transmitted from the microprocessor to IO device or to hold the data transmitted from IO device to the microprocessor.

To data transfer from the input device to the processor the following operations are performed:

- **The input device will load the data to the port.**
- **When the port receives the data, it sends message to the processor to read the data.**
- **The processor will read the data from the port.**
- **After the data has been read by the processor the input device will load the next data into the port.**

To data transfer from the processor to the output device the following operations are performed:

- **The processor will load the data to the port.**
- **The port will send a message to the output device to read the data.**
- **The output device will read the data from the port.**
- **After the data has been read by the output device the processor can load the next data to the port.**

INTEL IO Port Devices

The various INTEL IO port devices are 8212, 8155/8156, 8255, 8355 and 8755.

INTEL 8212

The 8212 is a 24-pin IC. It consists of eight number of D-type latches, each followed by a tristate buffer. It has 8-input lines DI_1 to DI_8 and 8-output lines DO_1 to DO_8. The 8212 can be used as an input or output device and the function is determined by the mode pin. However, it cannot be used simultaneously for input and output in the same circuit, since its mode pin is hardwired. It has 2-device select signals DS_1 and DS_2. The port is selected by the processor by sending appropriate address to device select pins.

Output Port : When $MD = 1$, $DS_1 = 0$ and $DS_2 = 1$

Input Port   : When $MD = 0$, $DS_1 = 0$ and $DS_2 = 1$

INTEL 8155

INTEL 8155 has 256×8 static RAM, two numbers of 8-bit parallel IO port (ports A and B), one number of 6-bit parallel IO port (port-C) and 14-bit timer. The ports A and B can be programmed to work as simple or handshake input or output port. If port-A and port-B are simple ports then port-C can be used as input or output port. The timer can be programmed to operate in four different modes. INTEL 8155 requires six internal addresses and has one logic **low** Chip Select pin (CS). The addresses of internal devices of 8155 are listed in Table 1.9.

INTEL 8156

INTEL 8156 is same as 8155, but it has logic **high** Chip Select (CS), i.e., the chip is selected when CS = 1.

Table 1.9: Internal Address of 8155/8156

Internal device	A_2	A_1	A_0
Control Register/ Status Register	0	0	0
Port-A	0	0	1
Port-B	0	1	0
Port-C	0	1	1
LSB of Timer	1	0	0
MSB of Timer	1	0	1

INTEL 8255

It has 3 numbers of 8-bit parallel IO ports (ports A, B and C).

Port-A can be programmed in mode 0, mode1 or mode-2 as input or output port. Port-B can be programmed in mode-1 and mode-2 as IO port. When ports A and B are in mode-0, port-C can be used as IO port. The individual pins of port-C can be set or reset. INTEL 8255 requires four internal addresses and has one logic **low** Chip Select (CS) pin. The address of internal devices of 8255 are listed in Table 1.10.

Table 1.10: Internal Address OF 8255

Internal device	A_1	A_0
Port-A	0	0
Port-B	0	1
Port-C	1	0
Control Register	1	1

INTEL 8355

It has $2\,k \times 8$ ROM and two numbers of 8-bit port (Ports A and B). The individual pins of ports A and B can be programmed as input or output lines by sending a control word to DDR (**Data Direction Register**). The address of internal devices of 8355 are listed in Table 1.11. The 8355 requires four internal addresses and has one logic **low** Chip Select (CS) pin.

INTEL 8755

Same as 8355 but has $2\,k \times 8$ EPROM.

INTEL peripheral devices

Apart from port ICs, dedicated programmable controller/peripheral ICs are used in the system for various activities. Some of the controller/peripheral devices used in the 8085 system and their functions and internal addresses are listed in Table 1.11.

Table 1.11: Internal Address of 8355/8755

Internal device	A_1	A_0
Port-A	0	0
Port-B	0	1
DDR A	1	0
DDR B	1	1

Table 1.12: Functions and Internal Addresses of Peripheral Devices

Device	Function	Internal addresses
INTEL 8279	Keyboard/display controller. Used for keyboard scanning and display refreshing.	**Two-internal addresses** $A_0 = 0 \rightarrow$ Data register $A_0 = 1 \rightarrow$ Control register
INTEL 8257 or INTEL 8237	DMA controller. Used for supporting DMA access to the IO device. It acts as a master during the DMA mode. It is a slave device during programming mode.	**Sixteen-internal addresses** $\quad A_3 \quad A_2 \quad A_1 \quad A_0$ $\quad 0 \quad\; 0 \quad\; 0 \quad\; 0$ $\quad 0 \quad\; 0 \quad\; 0 \quad\; 1$ $\quad .\quad\;\; .\quad\;\; .\quad\;\; .$ $\quad 1 \quad\; 1 \quad\; 1 \quad\; 1$
INTEL 8259	Interrupt controller. Used to expand the hardware interrupt INTR to eight interrupts in an 8085-based system and 256 interrupts in an 8086-based system.	**Two-internal addresses** $A_0 = 0$ $A_0 = 1$
INTEL 8253/ 8254	Programmable timer. Used in the system to produce various timing signals. It has three independent counters and can be programmed in six operating modes.	**Four-internal addresses** $\qquad\qquad\qquad A_1 \quad A_0$ Counter-0 $\quad\; 0 \quad\;\; 0$ Counter-1 $\quad\; 0 \quad\;\; 1$ Counter-2 $\quad\; 1 \quad\;\; 0$ Control Register $\; 1 \quad\;\; 1$
INTEL 8251 (USART)	Universal **S**ynchronous/**A**synchronous **R**eceiver **T**ransmitter. Used for serial data communication.	**Two-internal addresses** $C/D = 0 \rightarrow$ Data register $C/D = 1 \rightarrow$ Control register

IO Mapping

The port and peripheral devices will have one logic **low/high** chip select pin. The processor can access the port/peripheral device by supplying internal address and chip select signals. Therefore, the port and peripheral device interfacing (IO interfacing) deals with allocation of various internal addresses and generation of chip select signals.

There are two ways of interfacing IO devices in 8085-based system.

- **Memory-mapped IO device.**
- **Standard IO-mapped IO device or Isolated IO mapping.**

> *Note:* *The interfacing of IO ports and controller/peripheral ICs are commonly referred as IO device mapping.*

In memory mapping of IO devices the ports are allotted a 16-bit address like that of the memory location. Some of the chip select signals generated to select memory ICs are used for selecting the IO port devices. Hence, the processor treats the IO ports as memory locations for reading and writing (i.e., the devices which are mapped by memory mapping are accessed by executing memory read cycle or memory write cycle).

In standard IO mapping or isolated IO mapping, a separate 8-bit address is allotted for the IO ports and the peripheral ICs. The processor differentiates the IO-mapped devices, from the memory-mapped devices in the following ways:

1. For accessing the IO-mapped devices the processor executes IO read or write cycle.

2. During IO read or write cycle, the 8-bit address is placed on both low order address lines and the high order address lines.

3. IO/$\overline{M}$ is asserted high to indicate the IO operation (for read as well as write).

A 8085 processor does not provide separate read ($\overline{RD}$) and write ($\overline{WR}$) signals for memory and IO devices. But it differentiates the memory and IO device accessed by IO/$\overline{M}$ signal. The three signals $\overline{RD}$, $\overline{WR}$ and IO/$\overline{M}$ can be decoded as shown in Fig. 1.17 to provide separate read and write control signals for IO devices and memory devices.

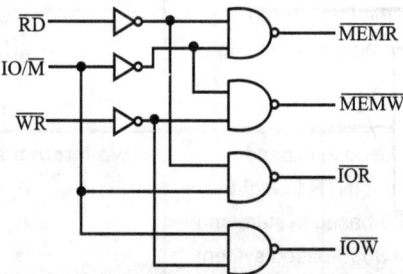

Fig. 1.17: Circuit to generate separate read and write signals for memory and IO devices in an 8085-based system.

When the devices are IO-mapped, then only IN and OUT instructions have to be used for data transfer between the device and the processor. For the IO-mapped devices a separate decoder should be used to generate the required chip select signals.

Table 1.13: Comparison of Memory Mapping and IO Mapping of IO Device

Memory mapping of IO device	IO mapping of IO device
1. 16-bit addresses are provided for IO devices.	1. 8-bit addresses are provided for IO devices.
2. The devices are accessed by memory read or memory write cycles.	2. The devices are accessed by IO read or IO write cycle. During these cycles, the 8-bit address is available on both low order address lines and high order address lines.
3. The IO ports or peripherals can be treated like memory locations and so all instructions related to memory can be used for data transfer between the IO device and the processor.	3. Only IN and OUT instructions can be used for data transfer between the IO device and the processor.
4. In memory-mapped ports, the data can be moved from any register to the ports and vice versa.	4. In IO-mapped ports, the data transfer can take place only between the accumulator and the ports.
5. When memory mapping is used for IO devices, the full memory address space cannot be used for addressing memory. Hence memory mapping is useful only for small systems, where the memory requirement is less.	5. When IO mapping is used for IO devices, then the full memory address space can be used for addressing the memory. Hence it is suitable for systems which requires a large memory capacity.
6. In memory-mapped IO devices, a large number of IO ports can be interfaced.	6. In IO mapping, only 256 ports (2^8 = 256) can be interfaced.
7. For accessing memory-mapped devices, the processor executes the memory read or write cycle. During this cycle, IO/$\overline{M}$ is asserted **low** (IO/$\overline{M}$= 0)	7. For accessing the IO-mapped devices, the processor executes the IO read or write cycle. During this cycle, IO/$\overline{M}$ is asserted **high** (IO/$\overline{M}$=1).

DESIGN EXAMPLE - 1

Interface two numbers of 4 kB EPROM and one number of 8 kB RAM with 8085 processor. Explain the interface diagram and allocate binary addresses to memory ICs.

Solution

The IC 2732 is selected for EPROM memory and the IC 6264 is selected for RAM memory. Both the memory IC's have time compatibility with 8085 processor.

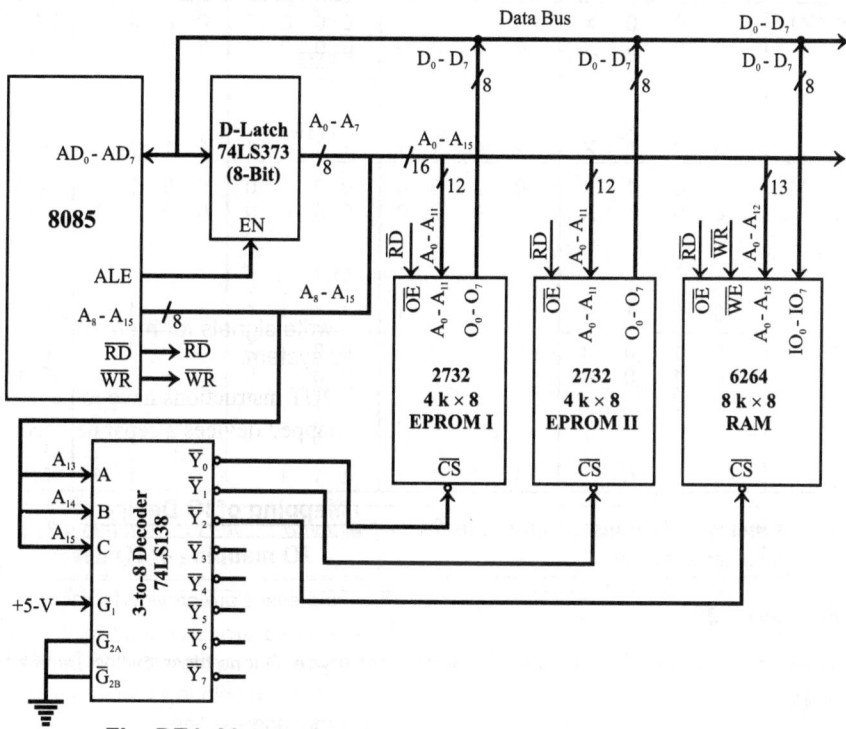

Fig. DE1: Memory interface diagram for Design Example - 1.

The $4\,kB$ EPROM IC requires 12 address lines ($2^{12} = 4$ k). The 8 kB RAM IC requires 13 address lines ($2^{13} = 8$ k). The address lines A_0 - A_{11} are connected to both EPROM and RAM address input pins. The address lines A_{13}, A_{14} and A_{15} are not used for memory address. Hence by decoding these address lines we can generate chip select signals.

The 3-to-8 decoder, 74LS138 is employed to produce the chip select signals for the system. The decoder has 8-output lines which can be used as 8-chip select signals. In this, three chip select signals are used for selecting memory ICs and the remaining five can be used for selecting other peripheral ICs in the system or for future expansion of the memory capacity. The interface diagram is shown in Fig. DE1. Address allotted to memory ICs are shown in Table DE1.

The EPROM's are mapped in the beginning of memory space. The remaining addresses can be allotted to RAM's. The EPROM memory is mapped from 0000_H to $0FFF_H$ and 2000_H to $2FFF_H$. The RAM memory is mapped from 4000_H to $5FFF_H$.

Table DE1: Address Allocation Table for Design Example - 1

Memory IC	Decoder input				Input to memory address pins																Hexa address
	A_{15}	A_{14}	A_{13}	A_{12}	A_{11}	A_{10}	A_9	A_8	A_7	A_6	A_5	A_4	A_3	A_2	A_1	A_0					
EPROM I 2732	0	0	0	X	0	0	0	0	0	0	0	0	0	0	0	0					0000
	0	0	0	X	0	0	0	0	0	0	0	0	0	0	0	1					0001
	0	0	0	X	1	1	1	1	1	1	1	1	1	1	1	1					0FFF
EPROM II 2732	0	0	1	X	0	0	0	0	0	0	0	0	0	0	0	0					2000
	0	0	1	X	0	0	0	0	0	0	0	0	0	0	0	1					2001
	0	0	1	X	1	1	1	1	1	1	1	1	1	1	1	1					2FFF
RAM 6264	0	1	0	0	0	0	0	0	0	0	0	0	0	0	0	0					4000
	0	1	0	0	0	0	0	0	0	0	0	0	0	0	0	1					4001
	0	1	0	1	1	1	1	1	1	1	1	1	1	1	1	1					5FFF

> *Note: X indicates the unused address line for the particular memory IC and they are considered as zero.*

DESIGN EXAMPLE - 2

Interface three numbers of 8 kB EPROM and 5 numbers of 8 kB static RAM to microprocessor 8085 to have a total memory capacity of 64 kB.

Solution

The IC 2764 is selected for EPROM memory and the IC 6264 is selected for RAM memory. Both the memory ICs have time compatibility with the 8085 processor.

The 8 kB EPROM IC requires 13 address lines ($2^{13} = 8$ k). The 8 kB RAM IC also requires 13 address lines ($2^{13} = 8$ k). The address lines $A_0 - A_{12}$ are connected to all the EPROM's and RAMs. Hence $A_0 - A_{12}$ will select the required memory location. The address lines A_{13}, A_{14} and A_{15} are not used for memory address. Hence by decoding these address lines we can generate chip select signals.

The 3-to-8 decoder, 74LS138 is employed to produce the chip select signals for the system. The decoder has 8-output lines which can be used as 8-chip select signals. All the 8-chip select signals are used to select memory ICs. EPROM's are mapped at the beginning of memory space. The decoder will select a memory IC by decoding the address lines A_{13}, A_{14} and A_{15}. The address lines $A_0 - A_{12}$ will select a particular memory location in the selected IC. The interface diagram is shown in Fig. DE2 and address allocation table is shown in Table DE2.

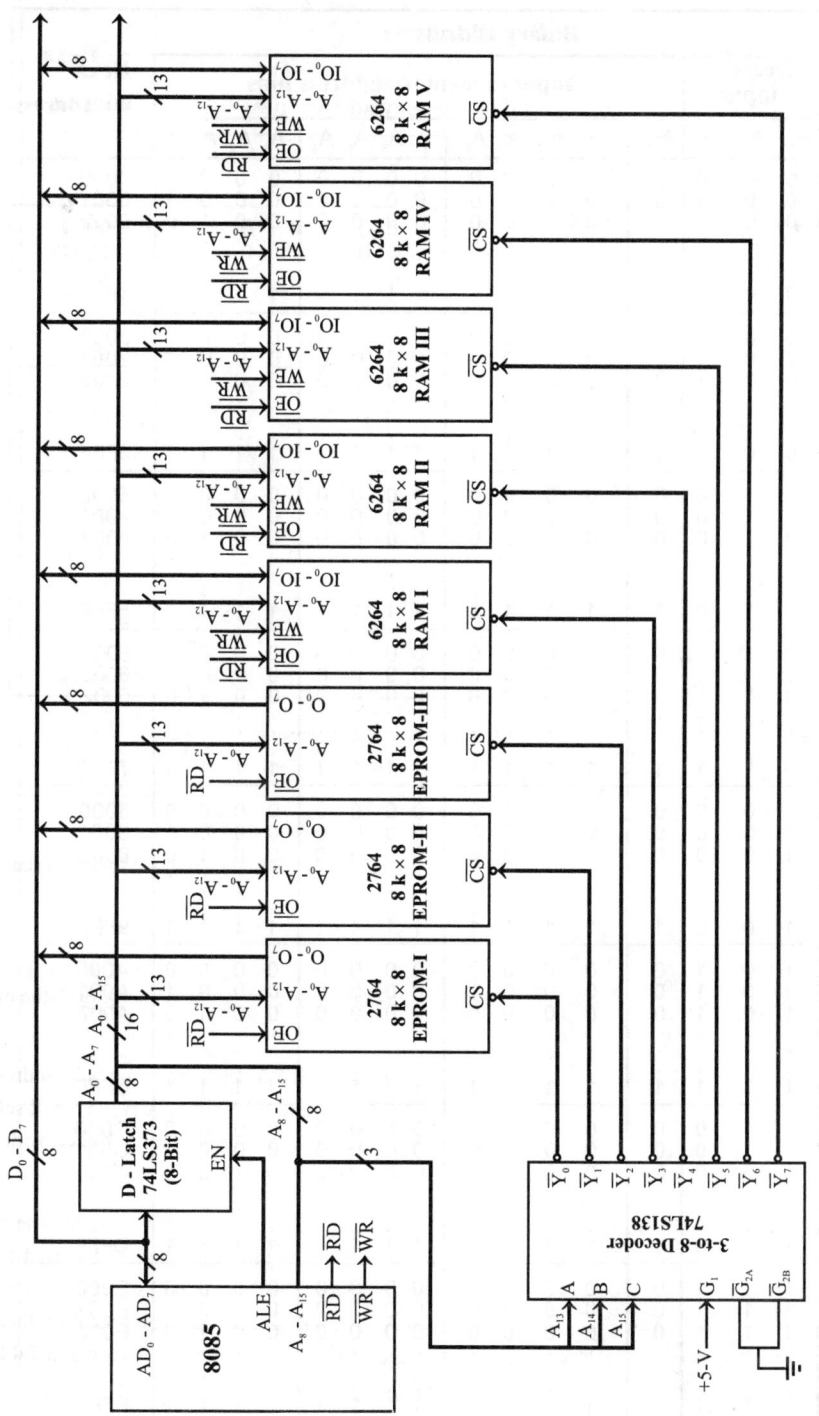

Fig. DE2: Memory interface diagram for Design Example - 2.

Table DE2: Address Allocation Table for Design Example- 2

Memory IC chip	Binary address								Hexa address
	Decoder input			Input to memory address pins					
	A_{15} A_{14} A_{13}	A_{12}	A_{11} A_{10} A_9 A_8	A_7 A_6 A_5 A_4	A_3 A_2 A_1 A_0				
EPROM I	0 0 0	0	0 0 0 0	0 0 0 0	0 0 0 0				0000
	0 0 0	0	0 0 0 0	0 0 0 0	0 0 0 1				0001
	0 0 0	0	0 0 0 0	0 0 0 0	0 0 1 0				0002
	: : :	:	: : : :	: : : :	: : : :				:
	0 0 0	1	1 1 1 1	1 1 1 1	1 1 1 1				1FFF
EPROM II	0 0 1	0	0 0 0 0	0 0 0 0	0 0 0 0				2000
	0 0 1	0	0 0 0 0	0 0 0 0	0 0 0 1				2001
	0 0 1	0	0 0 0 0	0 0 0 0	0 0 1 0				2002
	: : :	:	: : : :	: : : :	: : : :				:
	0 0 1	1	1 1 1 1	1 1 1 1	1 1 1 1				3FFF
EPROM III	0 1 0	0	0 0 0 0	0 0 0 0	0 0 0 0				4000
	0 1 0	0	0 0 0 0	0 0 0 0	0 0 0 1				4001
	0 1 0	0	0 0 0 0	0 0 0 0	0 0 1 0				4002
	: : :	:	: : : :	: : : :	: : : :				:
	0 1 0	1	1 1 1 1	1 1 1 1	1 1 1 1				5FFF
RAM I	0 1 1	0	0 0 0 0	0 0 0 0	0 0 0 0				6000
	0 1 1	0	0 0 0 0	0 0 0 0	0 0 0 1				6001
	0 1 1	0	0 0 0 0	0 0 0 0	0 0 1 0				6002
	: : :	:	: : : :	: : : :	: : : :				:
	0 1 1	1	1 1 1 1	1 1 1 1	1 1 1 1				7FFF
RAM II	1 0 0	0	0 0 0 0	0 0 0 0	0 0 0 0				8000
	1 0 0	0	0 0 0 0	0 0 0 0	0 0 0 1				8001
	1 0 0	0	0 0 0 0	0 0 0 0	0 0 1 0				8002
	: : :	:	: : : :	: : : :	: : : :				:
	1 0 0	1	1 1 1 1	1 1 1 1	1 1 1 1				9FFF
RAM III	1 0 1	0	0 0 0 0	0 0 0 0	0 0 0 0				A000
	1 0 1	0	0 0 0 0	0 0 0 0	0 0 0 1				A001
	1 0 1	0	0 0 0 0	0 0 0 0	0 0 1 0				A002
	: : :	:	: : : :	: : : :	: : : :				:
	1 0 1	1	1 1 1 1	1 1 1 1	1 1 1 1				BFFF
RAM IV	1 1 0	0	0 0 0 0	0 0 0 0	0 0 0 0				C000
	1 1 0	0	0 0 0 0	0 0 0 0	0 0 0 1				C001
	1 0 0	0	0 0 0 0	0 0 0 0	0 1 0 0				C002
	: : :	:	: : : :	: : : :	: : : :				:
	1 1 0	1	1 1 1 1	1 1 1 1	1 1 1 1				DFFF
RAM V	1 1 1	0	0 0 0 0	0 0 0 0	0 0 0 0				E000
	1 1 1	0	0 0 0 0	0 0 0 0	0 0 0 1				E001
	1 1 1	0	0 0 0 0	0 0 0 0	0 0 1 0				E002
	: : :	:	: : : :	: : : :	: : : :				:
	1 1 1	1	1 1 1 1	1 1 1 1	1 1 1 1				FFFF

In this system the full memory capacity of $64\,kB$ is utilized for memory. Hence the peripheral ICs and the IO ports should be IO-mapped in the system. The EPROM is mapped from 0000_H to $5FFF_H$. The RAM is mapped from 6000_H to $FFFF_H$. The EPROM capacity is $24\,kB$. The RAM capacity is $4\,kB$.

DESIGN EXAMPLE - 3

In a microprocessor system using 8085, the memory requirement is 8 kB EPROM and 8 kB RAM. For interfacing IO devices, three numbers of 8255 are required. Select suitable memories and explain how they are interfaced to the system. Interface the 8255 by memory mapping.

Solution

The IC 2764 is selected for EPROM memory and the IC 6264 is selected for RAM memory. Both the memory IC's have time compatibility with 8085 processor.

The 8 kB EPROM, 2764 requires 13 address lines ($2^{13} = 8$ k). The 8 kB RAM, 6264 also requires 13 address lines ($2^{13} = 8$ k). The address lines A_0 to A_{12} are connected to both EPROM and RAM memory ICs. The 8255 requires four internal addresses. Let us connect A_1 of 8085 to A_0 of 8255 and A_2 of 8085 to A_1 of 8255. The 8255 is memory-mapped in the system.

> *Note:*   *The internal devices of 8255 can be selected by connecting any two address lines of the processor to A_0 and A_1 of 8255.*

For the memories and 8255's we require 5 chip select signals. Hence we can use a 3-to-8 decoder 74LS138 for generating eight chip select signals by decoding the unused address lines A_{13}, A_{14} and A_{15}. The decoder enabled pins are permanently tied to appropriate levels. In the eight chip select signals, five are used for selecting memory ICs and 8255 and the remaining three can be used for future expansion. The memory/8255 interface diagram is shown in Fig. DE3.

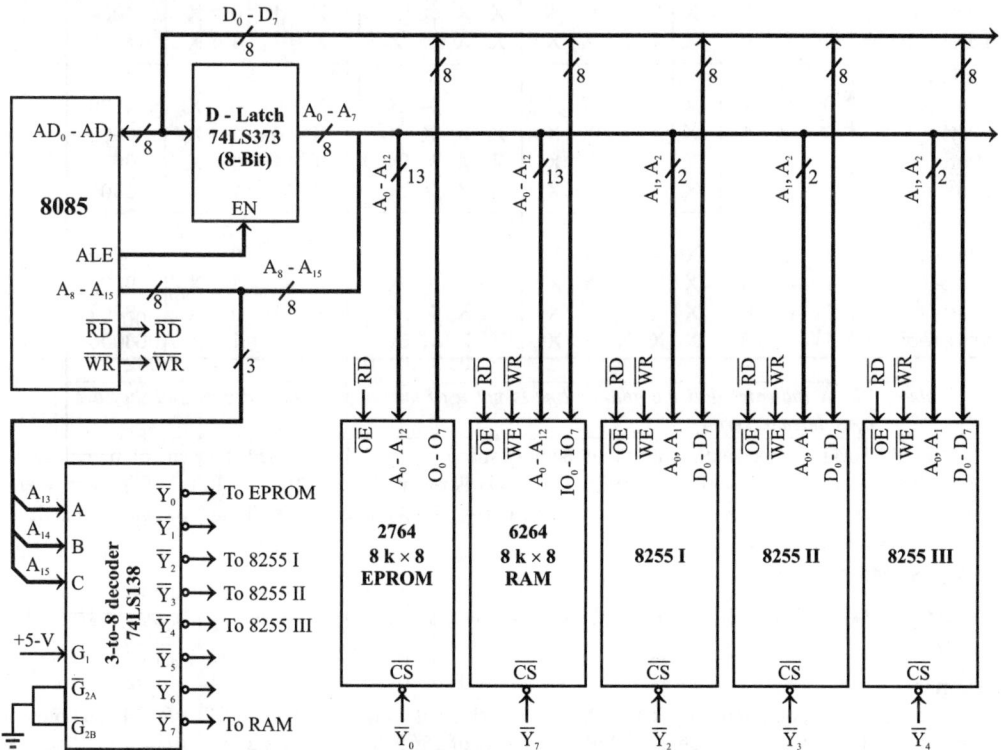

Fig. DE3: Memory interface diagram for Design Example - 3.

Table DE3: Address Allocation Table for Design Example - 3

Device	Decoder input A_{15} A_{14} A_{13}	A_{12}	A_{11} A_{10} A_9 A_8	A_7 A_6 A_5 A_4	A_3 A_2 A_1 A_0	Hexa address
	0 0 0	0	0 0 0 0	0 0 0 0	0 0 0 0	0000
2764	0 0 0	0	0 0 0 0	0 0 0 0	0 0 0 1	0001
EPROM	0 0 0	0	0 0 0 0	0 0 0 0	0 0 1 0	0002
	. . .	.	. . . .	. . . .	. . . .	.
	. . .	.	. . . .	. . . .	. . . .	.
	. . .	.	. . . .	. . . .	. . . .	.
	0 0 0	1	1 1 1 1	1 1 1 1	1 1 1 1	1FFF
	1 1 1	0	0 0 0 0	0 0 0 0	0 0 0 0	E000
	1 1 1	0	0 0 0 0	0 0 0 0	0 0 0 1	E001
6264	1 1 1	0	0 0 0 0	0 0 0 0	0 0 1 0	E002
RAM	. . .	.	. . . .	. . . .	. . . .	.
	. . .	.	. . . .	. . . .	. . . .	.
	. . .	.	. . . .	. . . .	. . . .	.
	1 1 1	1	1 1 1 1	1 1 1 1	1 1 1 1	FFFF
8255 I						
Port-A	0 1 0	X	X X X X	X X X X	X 0 0 X	4000
Port-B	0 1 0	X	X X X X	X X X X	X 0 1 X	4002
Port-C	0 1 0	X	X X X X	X X X X	X 1 0 X	4004
Control register	0 1 0	X	X X X X	X X X X	X 1 1 X	4006
8255 II						
Port-A	0 1 1	X	X X X X	X X X X	X 0 0 X	6000
Port-B	0 1 1	X	X X X X	X X X X	X 0 1 X	6002
Port-C	0 1 1	X	X X X X	X X X X	X 1 0 X	6004
Control register	0 1 1	X	X X X X	X X X X	X 1 1 X	6006
8255 III						
Port-A	1 0 0	X	X X X X	X X X X	X 0 0 X	8000
Port-B	1 0 0	X	X X X X	X X X X	X 0 1 X	8002
Port-C	1 0 0	X	X X X X	X X X X	X 1 0 X	8004
Control register	1 0 0	X	X X X X	X X X X	X 1 1 X	8006

> *Note:* *The X indicates that the address line is not used for the particular device and they are considered as zero.*

The EPROM is mapped at the starting of memory space. The RAM is mapped at the end of memory space. The EPROM is mapped from 0000_H to $1FFF_H$. The RAM is mapped from $E000_H$ to $FFFF_H$. The four internal devices of 8255 are control register, port-A, port-B and port-C. A 16-bit address is allotted to each internal device of 8255 as shown in Table-DE3.

DESIGN EXAMPLE - 4

Interface 2 kB RAM and 256 × 8 ROM with 8085 processor to satisfy the total memory requirement of 8 kB RAM and 1 kB ROM.

Solution

The memory requirement of 8 kB RAM can be achieved with 4 numbers of 2 kB RAM. The memory requirement of 1kB ROM can be achieved with 4 numbers of 256 × 8 ROM. (4 × 256 = 1024 = 1k). The 2 kB RAM requires 11 address lines ($2^{11} = 2$ k). The 256 × 8 ROM requires 8 address lines ($2^8 = 256$).

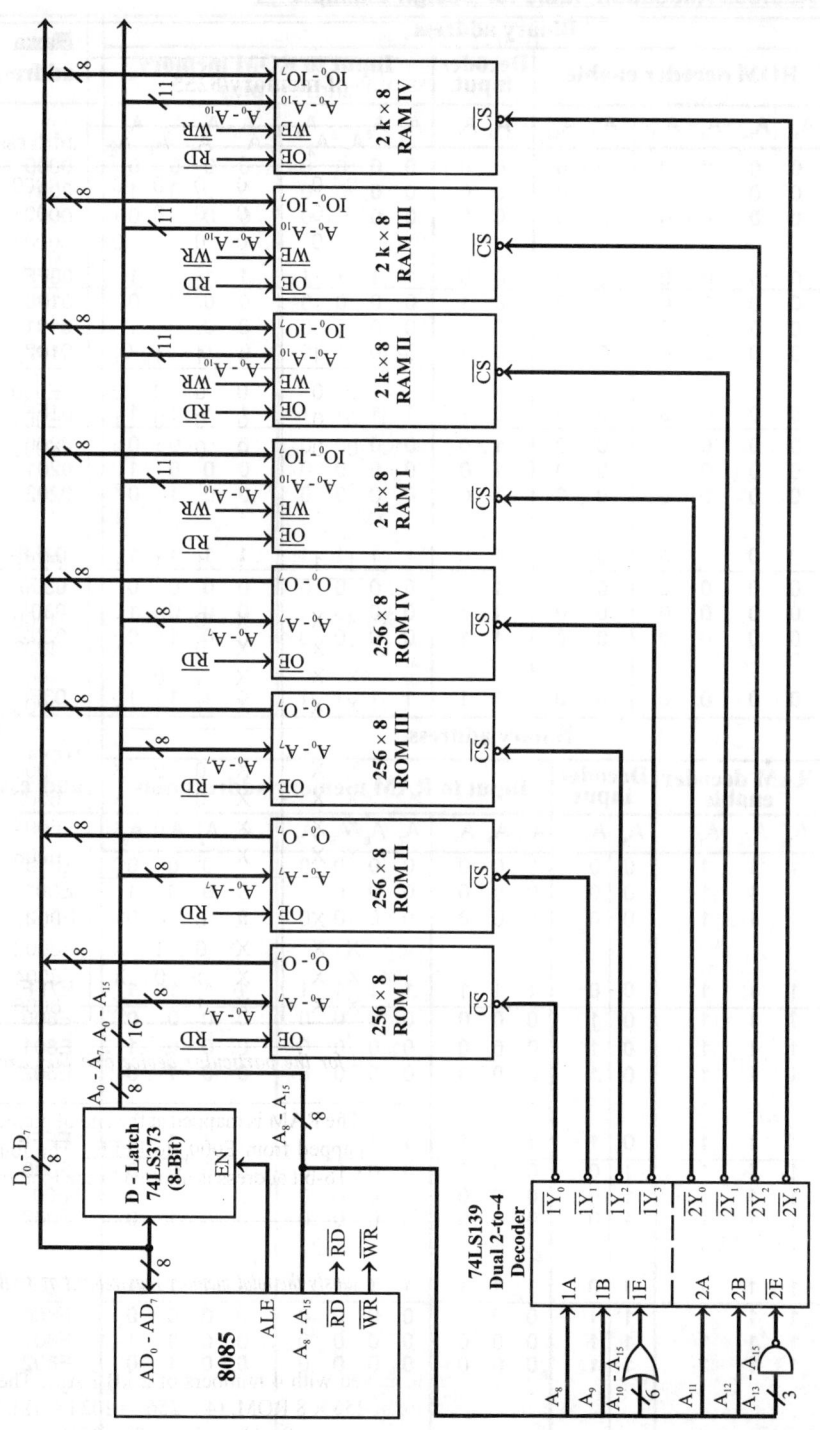

Fig. DE4: Memory interface diagram for Design Example - 4.

Table DE4: Address Allocation Table for Design Example - 4

Device	ROM decoder enable						Decoder input		Input to ROM memory address pins								Hexa address
	A_{15}	A_{14}	A_{13}	A_{12}	A_{11}	A_{10}	A_9	A_8	A_7	A_6	A_5	A_4	A_3	A_2	A_1	A_0	
256 × 8 ROM I	0	0	0	0	0	0	0	0	0	0	0	0	0	0	0	0	0000
	0	0	0	0	0	0	0	0	0	0	0	0	0	0	0	1	0001
	0	0	0	0	0	0	0	0	0	0	0	0	0	0	1	0	0002
	·	·	·	·	·	·	·	·	·	·	·	·	·	·	·	·	·
	0	0	0	0	0	0	0	0	1	1	1	1	1	1	1	1	00FF
256 × 8 ROM II	0	0	0	0	0	0	0	1	0	0	0	0	0	0	0	0	0100
	0	0	0	0	0	0	0	1	0	0	0	0	0	0	0	1	0101
	0	0	0	0	0	0	0	1	0	0	1	0	0	0	1	0	0102
	·	·	·	·	·	·	·	·	·	·	·	·	·	·	·	·	·
	0	0	0	0	0	0	0	1	1	1	1	1	1	1	1	1	01FF
256 × 8 ROM III	0	0	0	0	0	0	1	0	0	0	0	0	0	0	0	0	0200
	0	0	0	0	0	0	1	0	0	0	0	0	0	0	0	1	0201
	0	0	0	0	0	0	1	0	0	0	0	0	0	0	1	0	0202
	·	·	·	·	·	·	·	·	·	·	·	·	·	·	·	·	·
	0	0	0	0	0	0	1	0	1	1	1	1	1	1	1	1	02FF
256 × 8 ROM IV	0	0	0	0	0	0	1	1	0	0	0	0	0	0	0	0	0300
	0	0	0	0	0	0	1	1	0	0	0	0	0	0	0	1	0301
	0	0	0	0	0	0	1	1	0	0	0	0	0	0	1	0	0302
	·	·	·	·	·	·	·	·	·	·	·	·	·	·	·	·	·
	0	0	0	0	0	0	1	1	1	1	1	1	1	1	1	1	03FF

Device	RAM decoder enable			Decoder input		Input to RAM memory address pins											Hexa address
	A_{15}	A_{14}	A_{13}	A_{12}	A_{11}	A_{10}	A_9	A_8	A_7	A_6	A_5	A_4	A_3	A_2	A_1	A_0	
2 k × 8 RAM I	1	1	1	0	0	0	0	0	0	0	0	0	0	0	0	0	E000
	1	1	1	0	0	0	0	0	0	0	0	0	0	0	0	1	E001
	1	1	1	0	0	0	0	0	0	0	0	0	0	0	1	0	E002
	·	·	·	·	·	·	·	·	·	·	·	·	·	·	·	·	·
	1	1	1	0	0	1	1	1	1	1	1	1	1	1	1	1	E7FF
2 k × 8 RAM II	1	1	1	0	1	0	0	0	0	0	0	0	0	0	0	0	E800
	1	1	1	0	1	0	0	0	0	0	0	0	0	0	0	1	E801
	1	1	1	0	1	0	0	0	0	0	0	0	0	0	1	0	E802
	·	·	·	·	·	·	·	·	·	·	·	·	·	·	·	·	·
	1	1	1	0	1	1	1	1	1	1	1	1	1	1	1	1	EFFF
2 k × 8 RAM III	1	1	1	1	0	0	0	0	0	0	0	0	0	0	0	0	F000
	1	1	1	1	0	0	0	0	0	0	0	0	0	0	0	1	F001
	1	1	1	1	0	0	0	0	0	0	0	0	0	0	1	0	F002
	·	·	·	·	·	·	·	·	·	·	·	·	·	·	·	·	·
	1	1	1	1	0	1	1	1	1	1	1	1	1	1	1	1	F7FF
2 k × 8 RAM IV	1	1	1	1	1	0	0	0	0	0	0	0	0	0	0	0	F800
	1	1	1	1	1	0	0	0	0	0	0	0	0	0	0	1	F801
	1	1	1	1	1	0	0	0	0	0	0	0	0	0	1	0	F802
	·	·	·	·	·	·	·	·	·	·	·	·	·	·	·	·	·
	1	1	1	1	1	1	1	1	1	1	1	1	1	1	1	1	FFFF

The address lines A_0 - A_{10} are connected to RAM ICs. Hence, they will select the required memory location in that ICs. The address lines A_0 - A_7 are connected to ROM ICs. Hence, they will select the required memory location in those ICs.

Totally there are 8-memory ICs hence we require 8-chip select signals. The 8-chip select signals can be generated by using a dual 2-to-4 decoder 74LS139. One of the 2-to-4 decoder is used to generate chip select signals for ROM memory ICs and the other decoder is used to generate chip select signals for RAM memory ICs. The address lines A_8 and A_9 are used to generate chip select signals for ROM memory. The address lines A_{10} to A_{15} are logically ORed and used as enable for ROM decoder. The address lines A_{11} and A_{12} are used to generate chip select signals for RAM memory. The address lines A_{13} to A_{15} are logically NANDed and used as enable for RAM decoder.

ROM memories are mapped in the beginning of memory space. The RAM memories are mapped at the end of memory space. The ROM memories are mapped from 0000_H to $03FF_H$. The RAM memories are mapped from $E000_H$ to $FFFF_H$.

DESIGN EXAMPLE - 5

A system requires 16 kB EPROM and 16 kB RAM. Also the system has 2 numbers of 8255, one number of 8279, one number of 8251 and one number of 8254.

(8255 - Programmable peripheral interface, 8279-Keyboard/display controller, 8251 - USART and 8254 - Timer)

Draw the Interface diagram. Allocate addresses to all the devices. The peripheral IC's should be IO-mapped.

Solution

The IO devices in the system should be mapped by standard IO mapping. Hence separate decoders can be used to generate chip select signals for memory IC's and peripheral IC's.

For 16 *kB* EPROM, we can provide 2 numbers of 2764 (8 k × 8) EPROM. For 16 *kB* RAM we can provide 2 numbers of 6264 (8 k × 8) RAM.

The 8 *kB* memory requires 13 address lines ($2^{13} = 8$ k). Hence the address lines A_0 - A_{12} are used for selecting the memory locations. The unused address lines A_{13}, A_{14} and A_{15} are used as input to decoder 74LS138 (3-to-8-decoder) of memory IC. The logic **low** enables of this decoder are tied to IO/M of 8085, so that this decoder is enabled for memory read/write operation. The other enable pins of decoder are tied to appropriate logic levels permanently. The 4 outputs of the decoder are used to select memory IC's and the remaining 4 are kept for future expansion.

The EPROM is mapped in the beginning of memory space from 0000_H to $3FFF_H$. The RAM is mapped at the end of memory space from $C000_H$ to $FFFF_H$.

There are five peripheral IC's to be interfaced to the system. The chip select signals for these IC's are given through another 3-to-8 decoder 74LS138 (IO decoder). The input to this decoder is A_{10}, A_{11} and A_{12}. The address lines A_{13}, A_{14} and A_{15} are logically ORed and applied to **low** enable of IO decoder. The logic **high** enable of IO decoder is tied to IO/$\overline{M}$ signal of 8085, so that this decoder is enabled for IO read/write operation.

Here, the high order address lines can be used for decoding because the processor outputs the 8-bit port address both on AD_0 to AD_7 and A_8 to A_{15}. The address lines A_0 and A_1 are used to select the internal devices of the peripheral ICs. The output of the decoder are used to select the ICs. Three outputs of the decoder will be spare for future expansion.

> Note: *Since the IO devices are IO-mapped in the system, 8-bit addresses have been allotted to them.*

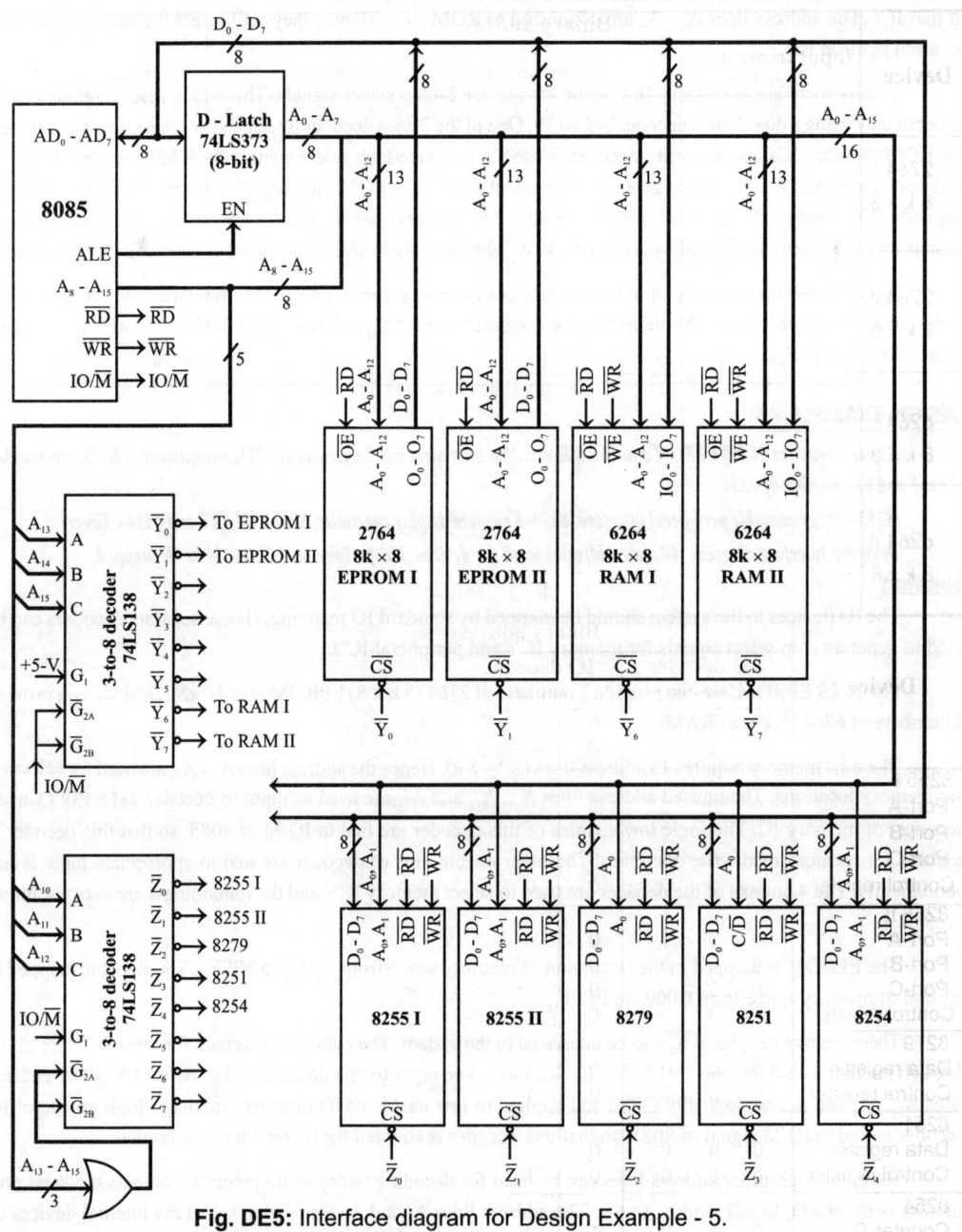

Fig. DE5: Interface diagram for Design Example - 5.

Table DE5: Address Allocation Table for Design Example - 5

Device	Input to memory decoder A_{15} A_{14} A_{13}	A_{12}	A_{11} A_{10} A_9 A_8	A_7 A_6 A_5 A_4	A_3 A_2 A_1 A_0	Hexa address
2764 I 8 k × 8	0 0 0 . . . . . . 0 0 0	0 . . 1	0 0 0 0 . . . . 1 1 1 1	0 0 0 0 . . . . 1 1 1 1	0 0 0 0 . . . . 1 1 1 1	0000 . . 1FFF
2764 II 8 k × 8	0 0 1 . . . . . . 0 0 1	0 . . 0	0 0 0 0 . . . . 1 1 1 1	0 0 0 0 . . . . 1 1 1 1	0 0 0 0 . . . . 1 1 1 1	2000 . . 3FFF
6264 I 8 k × 8	1 1 0 . . . . . . 1 1 0	0 . . 1	0 0 0 0 . . . . 1 1 1 1	0 0 0 0 . . . . 1 1 1 1	0 0 0 0 . . . . 1 1 1 1	C000 . . DFFF
6264 II 8 k × 8	1 1 1 . . . . . . 1 1 1	0 . . 1	0 0 0 0 . . . . 1 1 1 1	0 0 0 0 . . . . 1 1 1 1	0 0 0 0 . . . . 1 1 1 1	E000 . . FFFF

Device	IO decoder enable A_{15} A_{14} A_{13} / A_7 A_6 A_5	IO decoder input A_{12} A_{11} A_{10} / A_4 A_3 A_2	Input to IO device address pins A_9 / A_1	A_8 / A_0	Hexa address
8255 I					
Port-A	0 0 0	0 0 0	0	0	00
Port-B	0 0 0	0 0 0	0	1	01
Port-C	0 0 0	0 0 0	1	0	02
Control register	0 0 0	0 0 0	1	1	03
8255 II					
Port-A	0 0 0	0 0 1	0	0	04
Port-B	0 0 0	0 0 1	0	1	05
Port-C	0 0 0	0 0 1	1	0	06
Control register	0 0 0	0 0 1	1	1	07
8279					
Data register	0 0 0	0 1 0	X	0	08
Control register	0 0 0	0 1 0	X	1	09
8251					
Data register	0 0 0	0 1 1	X	0	0C
Control register	0 0 0	0 1 1	X	1	0D
8254					
Counter-0	0 0 0	1 0 0	0	0	10
Counter-1	0 0 0	1 0 0	0	1	11
Counter-2	0 0 0	1 0 0	1	0	12
Control register	0 0 0	1 0 0	1	1	13

Note: Don't care (X) is considered as zero.

DESIGN EXAMPLE - 6

In a microprocessor-based system 8085, 8 kB EPROM and 8 kB RAM are needed. For interfacing IO devices two numbers of 8155 are required. Select suitable memories and explain how they are interfaced in the system. Interface the 8155 ports by IO mapping.

Solution

The IC 2764 (8 k × 8) is selected for EPROM memory and IC 6264 (8 k × 8) is selected for RAM memory. Both the memory IC's have time compatibility with 8085 processor.

The 8 *kB* memories require 13 address lines ($2^{13} = 8$ k). Hence, the address lines A_0 - A_{12} are used to select memory locations.

In addition to 6264, each one of the 8155 chip provides a static RAM capacity of 256 bytes. The RAM locations of 8155 are selected by address lines A_0-A_6.

A 3-to-8 decoder, 74LS138 is used for generating chip select signals by decoding the address lines A_{13}, A_{14} and A_{15}.

The 8155 has internal address latch and decoder to differentiate memory operation and IO operation. To utilize this facility, the control signals ALE and IO/M are connected to 8155.

The 8155 ports and memory locations can be selected from the decoder used for memory devices. It differentiates the memory and IO operation from IO/M signal. Eight bit addresses are allotted to ports of 8155 and sixteen bit addresses are allotted to RAM memory locations of 8155.

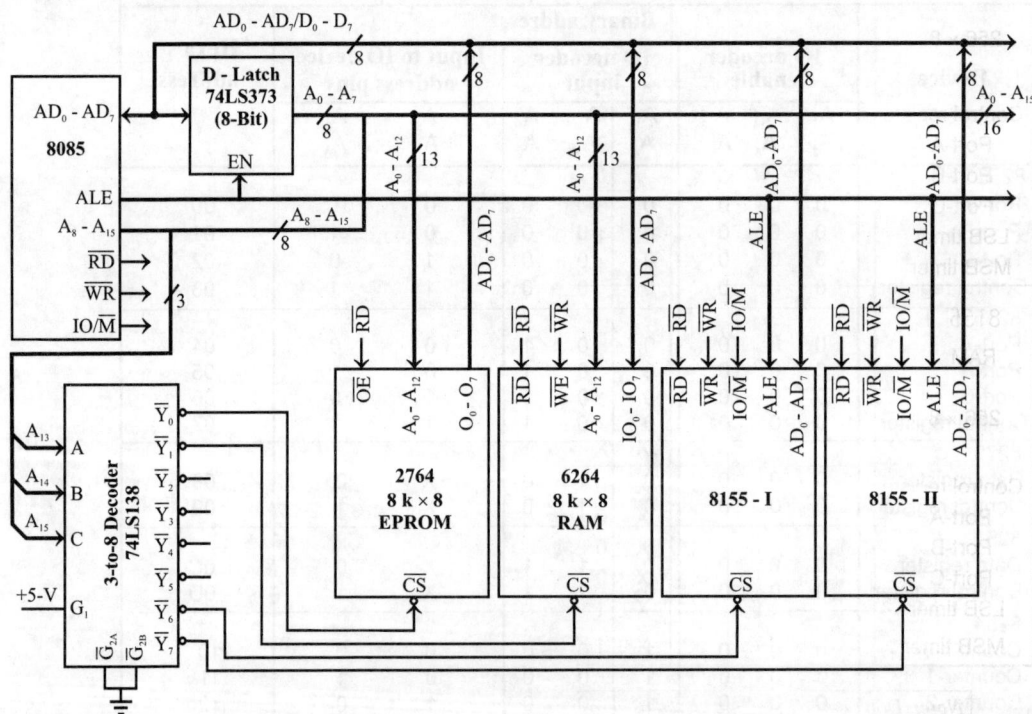

Fig. DE6: Interface diagram for Design Example 6.

Table DE6: Address Allocation Table for Design Example - 6

Device	Binary address						Hexa address
	Decoder input A_{15} A_{14} A_{13}	A_{12}	A_{11} A_{10} A_9 A_8	A_7 A_6 A_5 A_4	A_3 A_2 A_1 A_0		
2764	0 0 0	0	0 0 0 0	0 0 0 0	0 0 0 0		0000
	0 0 0	0	0 0 0 0	0 0 0 0	0 0 0 1		0001
8k × 8	. . .	.	. . . .	. . . .	. . . .		.
EPROM	. . .	.	. . . .	. . . .	. . . .		
	. . .		. . . .	. . . .	. . . .		
	0 0 0	1	1 1 1 1	1 1 1 1	1 1 1 1		1FFF
	0 0 1	0	0 0 0 0	0 0 0 0	0 0 0 0		2000
	0 0 1	0	0 0 0 0	0 0 0 0	0 0 0 1		2001
6264	. . .	.	. . . .	. . . .	. . . .		.
8k × 8	. . .	.	. . . .	. . . .	. . . .		.
RAM	. . .	.	. . . .	. . . .	. . . .		.
	0 0 1	1	1 1 1 1	1 1 1 1	1 1 1 1		3FFF
8155 I RAM	1 1 0	X	X X X X	0 0 0 0	0 0 0 0		C000
	. . .	.	. . . .	. . . .	. . . .		.
256 × 8	. . .	.	. . . .	. . . .	. . . .		.
	1 1 0	X	X X X X	1 1 1 1	1 1 1 1		C0FF
Control register	1 1 0	X	X 0 0 0				C0
Port-A	1 1 0	X	X 0 0 1				C1
Port-B	1 1 0	X	X 0 1 0				C2
Port-C	1 1 0	X	X 0 1 1				C3
LSB timer	1 1 0	X	X 1 0 0				C4
MSB timer	1 1 0	X	X 1 0 1				C5
8155 II RAM	1 1 1	X	X X X X	0 0 0 0	0 0 0 0		E000
	. . .	.	. . . .	. . . .	. . . .		.
256 × 8	. . .	.	. . . .	. . . .	. . . .		.
	1 1 1	X	X X X X	1 1 1 1	1 1 1 1		E0FF
Control register	1 1 1	X	X 0 0 0				E0
Port-A	1 1 1	X	X 0 0 1				E1
Port-B	1 1 1	X	X 0 1 0				E2
Port-C	1 1 1	X	X 0 1 1				E3
LSB timer	1 1 1	X	X 1 0 0				E4
MSB timer	1 1 1	X	X 1 0 1				E5

Note: Don't care (X) is considered as zero.

1.4.2 PARALLEL DATA COMMUNICATION INTERFACE

In microprocessor-based systems, digital information can be transmitted from one system to another system either by parallel or serial data transfer scheme.

In parallel data transfer, a group of bits (for eg., 8 bits) is transmitted from one device to another at any one time. To achieve parallel data transfer scheme, a group of data lines will be connecting the processor and peripheral devices. Normally in microprocessor-based systems the parallel data transfer schemes are adopted to transfer data between various devices inside the system.

Basically the microprocessor-based system has been fabricated on a PCB (**P**rinted **C**ircuit **B**oard) in which a bus is formed with the required number of data lines and the bus connects all the devices in the system. The data transmitted over the bus in a PCB are highly reliable. In a well designed board, there will not be any loss of data and the data will not be corrupted.

When data has to be transmitted over longer distances (i.e., greater than 0.5m), we require high current signals to drive the data for longer distance. In such cases data is transmitted bit by bit through a single data line.

Parallel Data Transfer Schemes

Data transfer schemes refer to the method of data transfer between the processor and peripheral devices. In a typical microcomputer, data transfer takes place between any two devices: microprocessor and memory, microprocessor and IO devices, or memory and IO devices. For effective data transfer between these devices, the timing parameters of the devices should be matched. But most of the devices have incompatible timings. For example, an IO device may be slower than the processor due to which it cannot send data to the processor at the expected time.

Semiconductor memories are available with compatible timings. Moreover, slow memories can be interfaced using additional hardware to introduce wait states in machine cycles. The microprocessor system designer often faces difficulties while interfacing IO devices and magnetic memories (like floppy or hard disk) to achieve efficient data transfer to or from the microprocessor. Several data transfer schemes have been developed to solve the interfacing problems with IO devices.

The data transfer schemes have been broadly classified into the following two categories :

    1. Programmed data transfer.

    2. Direct memory access (DMA) data transfer.

In programmed data transfer, a memory resident routine (subroutine) requests the device for data transfer to or from one of the processor register.

Programmed data transfer scheme is used when relatively small amount of data is to be transferred. In these schemes, usually one byte or word of data is transferred at a time. Examples of devices using programmed data transfer are ADC, DAC, Hex-keyboard, 7-segment LEDs, etc.

Programmed data transfer scheme can be further classified into the following three types:

    a) Synchronous data transfer scheme.

    b) Asynchronous data transfer scheme.

    c) Interrupt driven data transfer scheme.

In DMA data transfer, the processor is forced to **HOLD** state (**high impedance** state) by an IO device until the data transfer between the device and the memory is complete. The processor does not execute any instruction during the **HOLD** period.

The DMA data transfer is used for large block of data transfer between IO device and memory. Typical examples of devices using DMA are CRT controller, floppy disk, hard disk, high speed line printer, etc.

The different types of DMA data transfer schemes are as follows:

    a) Cycle stealing DMA or Single transfer mode DMA.

    b) Block or Burst mode DMA.

    c) Demand transfer mode DMA.

Figure 1.18 shows the various types of data transfer schemes. All the data transfer schemes discussed above requires both software and hardware for their implementation. Within a microcomputer, more than one scheme can be used for interfacing different IO devices. However, some of these schemes require specific hardware features in the microprocessor for implementing the scheme.

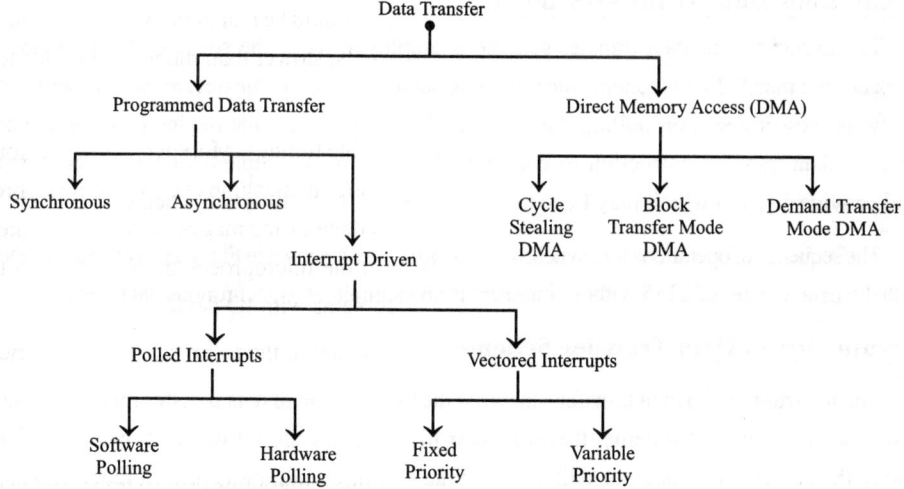

Fig. 1.18: Types of data transfer schemes.

Synchronous Data Transfer Scheme

The synchronous data transfer scheme is the simplest of all data transfer schemes. In this scheme the processor does not check the readiness of the device. The IO device or peripheral should have matched timing parameters. Whenever data is to be obtained from the device or transferred to the device, the user program can issue a suitable instruction for the device. At the end of the execution of this instruction, the transfer would have been completed.

The synchronous data transfer scheme can also be implemented with a small delay (if the delay is tolerable) after the request has been made. The sequence of operations for synchronous data transfer scheme is shown in Fig. 1.19. The mode-0 input/output in 8255 is an example of synchronous data transfer.

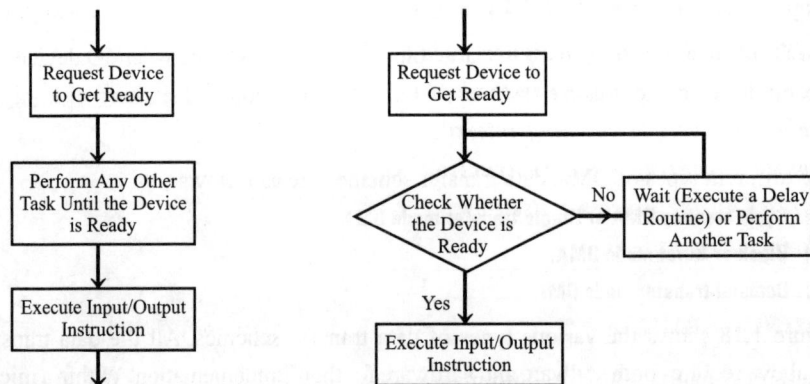

Fig. 1.19: Synchronous data transfer scheme.   **Fig. 1.20:** Asynchronous data transfer scheme.

Asynchronous Data Transfer Scheme

The asynchronous data transfer scheme is employed when the speed of the processor and IO device does not match. In this scheme the processor sends a request to the device for read/write operation. Then the processor keeps on polling the status of the device. Once the device is ready, the processor executes a data transfer instruction to complete the process. To implement this scheme, the device should provide a signal which may be tested by the processor to ascertain whether it is ready or not.

The sequence of operations for asynchronous data transfer is shown in Fig. 1.20. The mode-1 and mode-2 handshake data transfer of 8255 without interrupt is an example of asynchronous data transfer.

Interrupt Driven Data Transfer Scheme

The interrupt driven data transfer scheme is the best method of data transfer for efficient utilization of processor time. In this scheme, the processor first initiates the IO device for data transfer. After initiating the device, the processor will continue the execution of instructions in the program. Also, at the end of every instruction the processor will check for a valid interrupt signal. If there is no interrupt then the processor will continue the execution.

When the IO device is ready, it will interrupt the processor. On receiving an interrupt signal the processor will complete the current instruction execution and save the processor status in stack. Then the processor calls an **I**nterrupt **S**ervice **R**outine (ISR) to service the interrupting device. At the end of ISR, the processor status is retrieved from the stack and the processor starts executing its main program. The sequence of operations for an interrupt driven data transfer scheme is shown in Fig. 1.21.

> *Note: The user/system designer need not write any subroutine/procedure to check for an interrupt. The logic of checking interrupt signals while executing each instruction is incorporated in the processor itself by the manufacturer of the processor.*

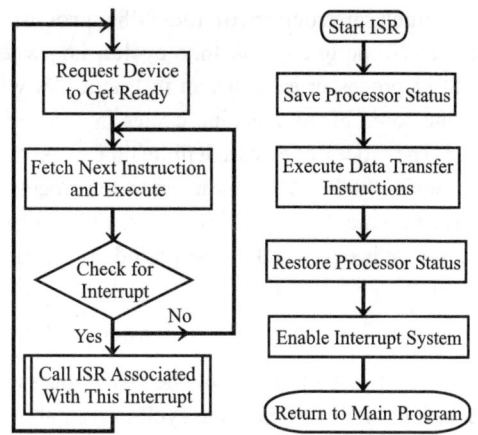

Fig. a: Main program execution sequence.

Fig. b: ISR execution sequence.

Fig. 1.21: Interrupt driven data transfer scheme.

1.5 TIMING DIAGRAM
(AU, Nov/Dec' 19, 13 Marks)

The timing diagram provides information about the various condition (**high** state or **low** state or **high impedance** state) of the signals while a machine cycle is executed. The timing diagrams are supplied by the manufacturer of the microprocessor. The timing diagrams are essential for a system designer. Only from the knowledge of timing diagrams, the matched peripheral devices like memories, ports, etc., can be selected to form a system with microprocessor as CPU.

1.5.1 PROCESSOR CYCLES

The sequence of operations that a processor has to carry out while executing the instruction is called instruction cycle. Each instruction cycle of a processor in turn consists of a number of machine cycles. The machine cycles are the basic operations performed by the processor. To execute an instruction, the processor executes one or more machine cycles in a particular sequence. The machine cycles of a processor are also called processor cycles. The manufacturers of microprocessors define the timings and status of various signals during the processor cycles.

In general, the instruction cycle of an instruction can be divided into two sub cycles: Fetch cycle and Execute cycle. The fetch cycle is executed to fetch the opcode from memory and the execute cycle is executed to decode the instruction and to perform the work specified by the instruction.

1.5.2 MACHINE CYCLES OF 8085

The 8085 microprocessor has seven basic machine cycles. They are as follows:

1. **Opcode fetch cycle (4T or 6T)**
2. **Memory read cycle (3T)**
3. **Memory write cycle (3T)**
4. **IO read cycle (3T)**
5. **IO write cycle (3T)**
6. **Interrupt acknowledge cycle (6T or 12T)**
7. **Bus idle cycle (2T or 3T)**

Each instruction of the 8085 processor consists of one to five machine cycles, i.e., when the 8085 processor executes an instruction, it will execute some of the machine cycles in a specific order. The processor takes a definite time to execute the machine cycles. The time taken by the processor to execute a machine cycle is expressed in T-states.

Fig. 1.22: Clock Signal.

One T-state is equal to the time period of the internal clock signal of the processor. The T-state starts at the falling edge of a clock.

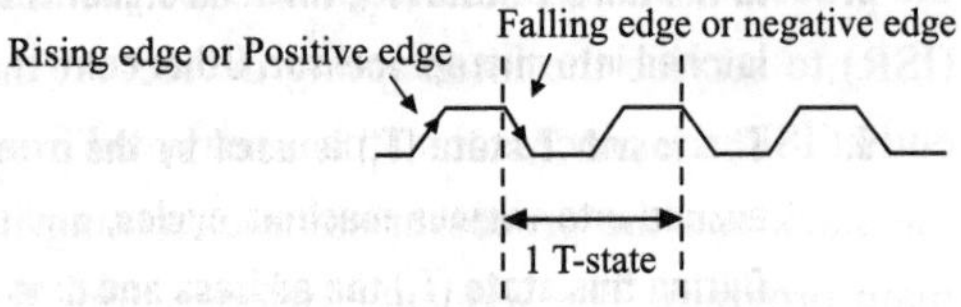

> *Note: Time period, T = 1/f ; where f = Internal clock frequency.*

The T-states required by the 8085 processor to execute each machine cycle are mentioned within brackets in the list of machine cycles given above.

1.5.3 OPCODE FETCH MACHINE CYCLE OF 8085

Each instruction of the processor has one-byte opcode. The opcodes are stored in memory. The opcode fetch machine cycle is executed by the processor to fetch the opcode from memory. Hence, every instruction starts with opcode fetch machine cycle.

The time taken by the processor to execute the opcode fetch cycle is either 4T or 6T. In this time, the first 3T states are used for fetching the opcode from memory and the remaining T-states are used for internal operations by the processor. The timings of various signals during opcode fetch cycle are shown in Fig. 1.23.

1. At the falling edge of first T-state (T_1), the microprocessor outputs the low byte address on AD_0-AD_7 lines and high byte address on A_8 to A_{15} lines. ALE is asserted high to enable the external address latch. The other control signals are asserted as follows:

   $IO/\overline{M}$=0, S_0 = 1, S_1 = 1. ($IO/\overline{M}$ is asserted low to indicate memory access.)

2. At the middle of T_1, the ALE is asserted low and this enables the external address latch to take low byte of the address and keep on its output lines.

3. In the second T-state (T_2), the memory is requested for read by asserting read line low. When read is asserted low, the memory is enabled for placing the opcode on the data bus. The time allowed for memory to output the opcode is the time during which read remains low.

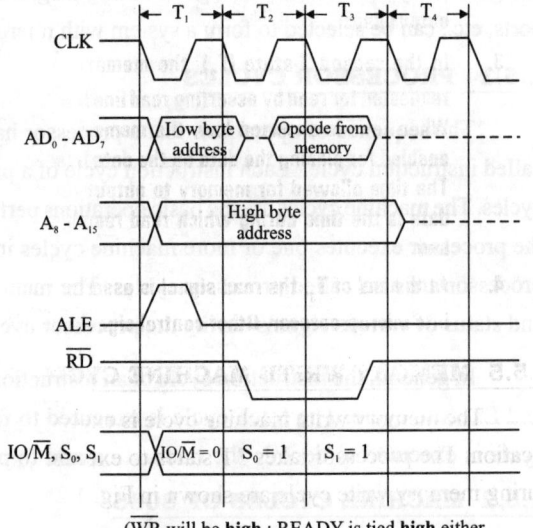

($\overline{WR}$ will be **high** ; READY is tied **high** either permanently or temporarily in the system.)

Fig. 1.23: Opcode fetch machine cycle of 8085.

4.   In the third T-state (T_3), the read signal is asserted high. On the rising edge of read signal, the opcode is latched into microprocessor. Other control signals remain in the same state until the next machine cycle.

5.   The fourth T-state (T_4) is used by the processor for internal operations to decode the instruction and encode into various machine cycles, and also for completing the task specified by 1-byte instruction. During this state (T_4) the address and data bus will be in high impedance state.

1.5.4 MEMORY READ MACHINE CYCLE OF 8085

The memory read machine cycle is executed by the processor to read a data byte from memory. The processor takes 3T states to execute this cycle. The timings of various signals during memory read cycle are shown in Fig. 1.24.

1.   At the falling edge of T_1, the microprocessor outputs the low byte address on AD_0 - AD_7 lines and high byte address on A_8 to A_{15} lines. ALE is asserted high to enable the external address latch. The other control signals are asserted as follows:

      $IO/\overline{M}=0$, $S_0 = 0$, $S_1 = 1$. ($IO/\overline{M}$ is asserted low to indicate memory access.)

2.   At the middle of T_1, the ALE is asserted low and this enables the external address latch to take low byte of address and keep on its output lines.

3.   In the second T-state (T_2), the memory is requested for read by asserting read line low. When read is asserted low, the memory is enabled for placing the data on the data bus. The time allowed for memory to output the data is the time during which read remains low.

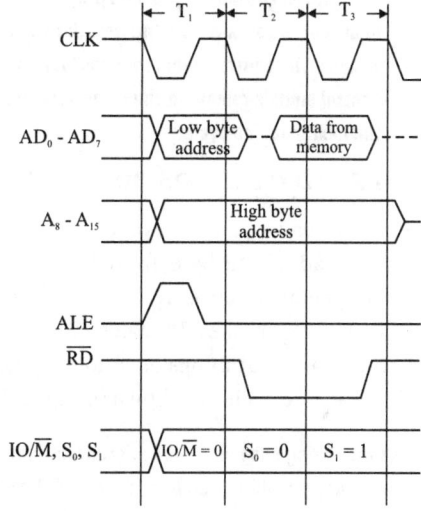

(WR will be **high** ; READY is tied **high** either permanently or temporarily in the system.)

Fig 1.24: Memory read machine cycle of 8085.

4.   At the end of T_3, the read signal is asserted high. On the rising edge of read signal, the data is latched into microprocessor. Other control signals remain in the same state until the next machine cycle.

1.5.5 MEMORY WRITE MACHINE CYCLE OF 8085

The memory write machine cycle is executed by the processor to write a data byte in a memory location. The processor takes 3T states to execute this machine cycle. The timings of various signals during memory write cycle are shown in Fig. 1.25.

1.   At the falling edge of T_1, the microprocessor outputs the low byte address on AD_0 - AD_7 lines and high byte address on A_8 to A_{15} lines. ALE is asserted high to enable the external address latch. The other control signals are asserted as follows:

      $IO/\overline{M}=0$, $S_0 = 1$, $S_1 = 0$. ($IO/\overline{M}$ is asserted low to indicate memory access.)

2. At the middle of T$_1$, the ALE is asserted low and this enables the external address latch for latching the low byte address into its output lines.

3. In the falling edge of T$_2$, the processor output data on AD$_0$ to AD$_7$ lines and then request memory for write operation by asserting the write control signal $\overline{\text{WR}}$ to low.

4. At the end of T$_3$, the processor asserts $\overline{\text{WR}}$ high. This enables the memory to latch the data into it. The memory should prepare itself to accept the data within the time duration in which write control signal remains low. Other control signals remain in the same state until the next machine cycle.

1.5.6 IO READ CYCLE OF 8085

The IO read cycle is executed by the processor to read a data byte from IO port or from the peripheral which is IO-mapped in the system. The processor takes 3T states to execute this machine cycle. The timings of various signals during this machine cycle are shown in Fig. 1.26.

1. At the falling edge of T$_1$, the microprocessor output the 8-bit port address on both the low order address lines (AD$_0$ - AD$_7$) and high order address lines (A$_8$ to A$_{15}$). ALE is asserted high to enable the external address latch. The other control signals are asserted as follows:

    IO/M = 1, S$_0$ = 0 and S$_1$ = 1. (IO/M is asserted high to indicate IO access.)

2. At the middle of T$_1$, the ALE is asserted low and this enables the external address latch to take the port address and keep on its output lines.

3. In the second T-state (T$_2$) the IO device is requested for read by asserting read line low. When read is asserted low, the IO port is enabled for placing the data on the data bus. The time allowed for IO port to output the data is the time during which read remains low.

4. At the end of T$_3$, the read signal is asserted high. On the rising edge of read signal the data is latched into microprocessor. Other control signals remains in the same state until the next machine cycle.

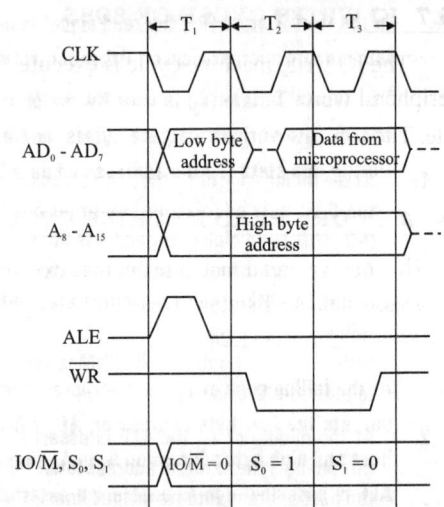

(RD will be **high** ; READY is tied **high** either permanently or temporarily in the system.)

Fig 1.25: Memory write machine cycle of 8085.

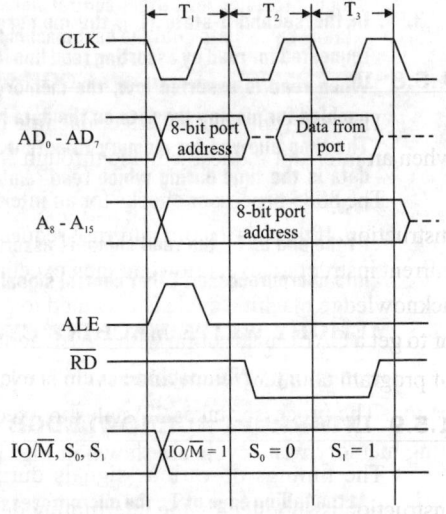

($\overline{\text{WR}}$ will be **high** ; READY is tied **high** either permanently or temporarily in the system.)

Fig 1.26: i/O read machine cycle of 8085.

1.5.7 IO WRITE CYCLE OF 8085

The IO write machine cycle is executed by the processor to write a data byte in an IO port or to a peripheral which is IO-mapped in the system. The processor takes 3T states to execute this machine cycle. The timings of the various signals of IO write cycle are shown in Fig. 1.27.

1. At the falling edge of T_1, the microprocessor outputs the 8-bit port address on low order address line (AD_0 - AD_7) and high order address lines (A_8 to A_{15}). ALE is asserted high to enable the external address latch. The other control signals are asserted as follows:
$I0/\overline{M} = 1$, $S_0 = 1$ and $S_1 = 0$. ($I0/\overline{M}$ is asserted high to indicate IO access.)

2. At the middle of T_1, the ALE is asserted low and this enables the external address latch for latching the port address into its output lines.

3. In the falling edge of T_2, the processor output data on AD_0 - AD_7 lines and then request IO port for write operation by asserting the write control signal $\overline{WR}$ to low.

4. At the end of T_3, the processor asserts $\overline{WR}$ high. This enables the IO port to latch the data into it. The IO port should prepare itself to accept the data within the time duration in which write control signal remains low. Other control signals remains in the same state until the next machine cycle.

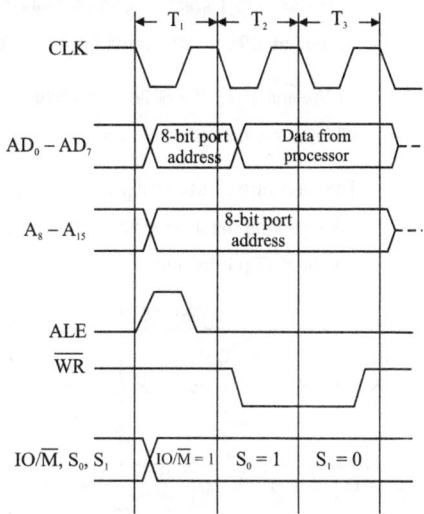

($\overline{RD}$ will be **high** ; READY is tied **high** either permanently or temporarily in the system.)

Fig 1.27: I/O write machine cycle of 8085.

1.5.8 INTERRUPT ACKNOWLEDGE MACHINE CYCLE OF 8085

The interrupt acknowledge machine cycle is executed by the processor to service an interrupt when an interrupt request is made through INTR pin of the processor.

The 8085 processor checks for an interrupt at the second T-state of the last machine cycle of every instruction. If there is a valid interrupt request and if INTR is enabled then the processor completes the current instruction execution and then executes an interrupt acknowledge machine cycle. The interrupt acknowledge machine cycle is executed to get either a **RST n** instruction from the interrupting device or to get a CALL instruction with CALL address from the interrupting device. It also stores the content of program counter (return address) in stack.

1.5.9 INTERRUPT ACKNOWLEDGE CYCLE OF 8085 WITH RST N INSTRUCTION

The timings of various signals during interrupt acknowledge cycle of 8085 when **RST n** instruction is supplied by the interrupting device are shown in Fig. 1.28.

1. In the first T-state of interrupt acknowledge cycle, the address is placed on the AD_0 - AD_7 and A_8-A_{15} lines and ALE is asserted high. But the address is not used to read from memory. The other control signals are asserted as follows:

IO/$\overline{\text{M}}$ = 1, S_0 = 1 and S_1 = 1.

In the middle of T_1, ALE is asserted low. The INTR signal can remain high or it can go low once the interrupt is accepted.

2.   In the second T-state (T_2), $\overline{\text{INTA}}$ is asserted low, and this enables the interrupting device to place the opcode of RST n instruction on the data bus.

3.   At the end of T_3, the $\overline{\text{INTA}}$ is asserted high and the RST n opcode is latched into the processor. The time allowed for the external hardware to place the RST n opcode is the time during which $\overline{\text{INTA}}$ remains low.

4.   The next three T states T_4, T_5 and T_6 are used for internal operations. The internal operations performed are decoding the instruction and encoding into various machine cycles and generation of vector address for the RST n interrupt.

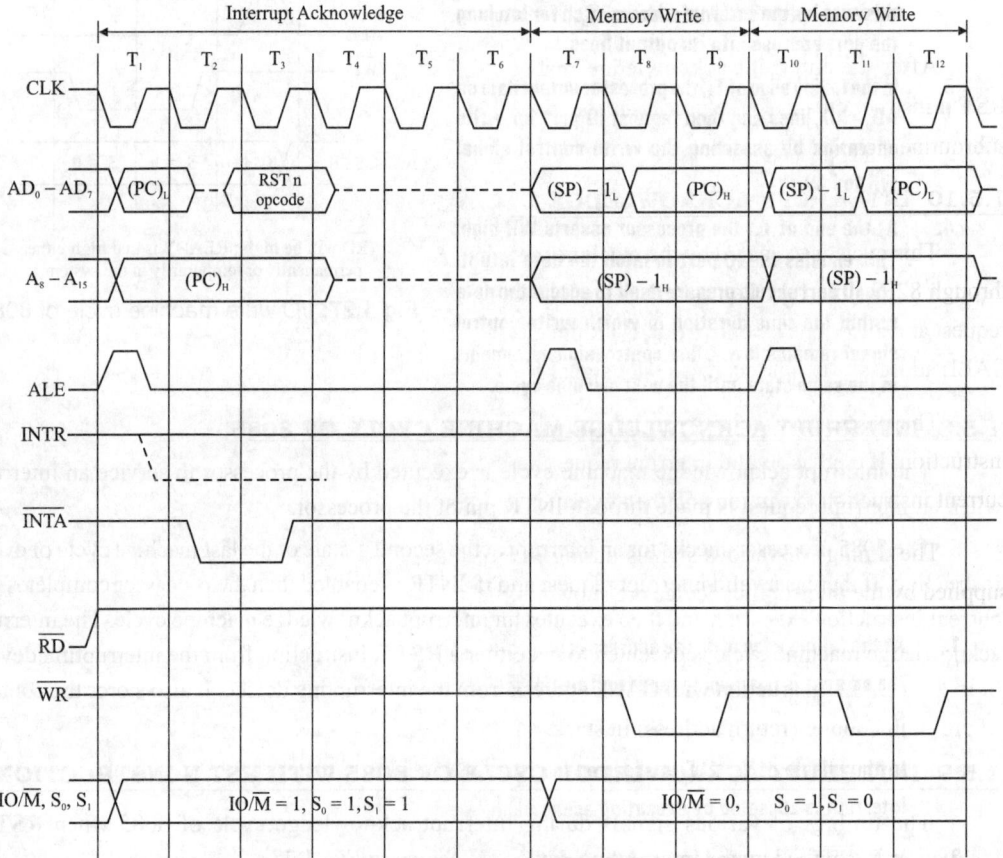

Fig. 1.28: Interrupt acknowledge cycle with **RST n** opcode.

5.   The T states T_7, T_8 and T_9 are used to store the high byte of the Program Counter (PC) in stack (using the content of Stack Pointer (SP) as address).

In T_7, the content of SP is decremented by one and placed on AD_0-AD_7 and A_8-A_{15} lines. ALE is asserted high and then low, to latch the low byte of address into external latch. The status signals are asserted as $IO/\overline{M}=0$, $S_0 = 1$ and $S_1 = 0$.

In T_8, the high byte of PC is placed on AD_0 - AD_7 lines and $\overline{WR}$ is asserted low to enable the stack memory for write operation. At the end of T_9, $\overline{WR}$ is asserted high.

6.   The T states T_{10}, T_{11} and T_{12} are used to store the low byte of the program counter into stack.

In T_{10}, the content of SP is again decremented by one and placed on AD_0-AD_7 and A_8 - A_{15} lines. ALE is asserted high and then low, to latch the low byte of address into external latch. The status signals are asserted as $IO/\overline{M}=0$, $S_0 = 1$ and $S_1 = 0$.

In T_{11}, the low byte of PC is placed on AD_0 - AD_7 lines and $\overline{WR}$ is asserted low to enable the stack memory for write operation. At the end of T_{12} $\overline{WR}$ is asserted high.

After the interrupt acknowledge machine cycle, the PC will have the vector address of **RST n** instruction and so the processor starts servicing the interrupt by executing the interrupt service subroutine stored at this address.

1.5.10 INTERRUPT ACKNOWLEDGE CYCLE OF 8085 WITH CALL INSTRUCTION

This cycle is executed by the machine to service an interrupt, when an interrupt request is made through 8259 (Interrupt Controller) to the INTR pin of 8085. The INTEL 8259 can accept 8 interrupt request and allow one by one to the INTR pin of the 8085 processor. It also supplies CALL opcode and CALL address, when it receives INTA signal from the processor.

The processor checks for an interrupt at the second T-state of the last machine cycle of every instruction. If there is a valid interrupt request and if INTR is enabled then the processor completes the current instruction execution and then executes an interrupt acknowledge machine cycle.

The timings of various signals during interrupt acknowledge cycle when CALL instruction is supplied by the interrupting device are shown in Fig. 1.29.

1.   At the falling edge of T_1 the address is placed on AD_0 - AD_7 and A_8 - A_{15} lines and ALE is asserted high. But the address is not used to read from memory. The other control signals are asserted as $IO/\overline{M}=1$, $S_0 = 1$ and $S_1 = 1$.

In the middle of T_1, ALE is asserted low. The INTR signal can remain high or it can go low once the interrupt is accepted by executing acknowledge cycle.

2.   In T_2, $\overline{INTA}$ is asserted low and this enables the interrupt controller 8259 to place a CALL opcode on the data bus.

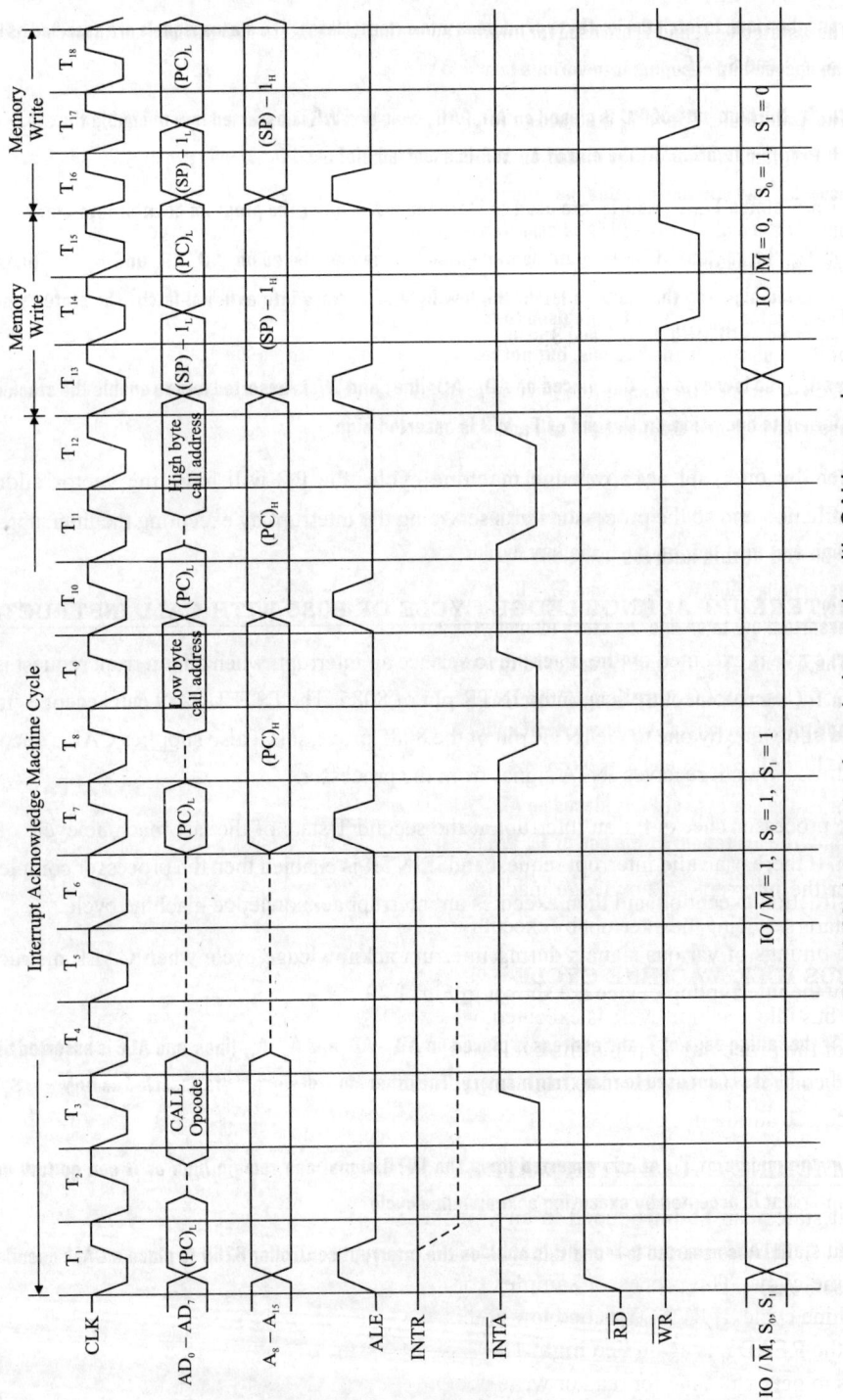

Fig. 1.29: Interrupt acknowledge cycle with CALL opcode.

3. At the end of T_3, the INTA is asserted high and the CALL opcode is latched into the processor.

4. The T states T_4, T_5 and T_6 are used for internal operations. The internal operations performed are decoding the opcode and encoding into various machine cycles.

5. The T states T_7, T_8 and T_9 are used to fetch the low byte of call address from 8259. In T_7, the content of Program Counter (PC) is placed on address bus but not used for memory operation. In T_8 the $\overline{INTA}$ is asserted low and this enables the interrupt controller 8259 to place the low byte of call address on data bus. At the end of T_9 the $\overline{INTA}$ is asserted high and the low byte call address on the data bus is latched into the processor.

6. The T states T_{10}, T_{11} and T_{12} are used to fetch the high byte of call address from 8259. In T_{10} the content of PC is placed on address bus, but not used for memory operation. In T_{11} the INTA is asserted low and 8259 is enabled for placing the high byte of call address on data bus. At the end of T_{12}, the $\overline{INTA}$ is asserted high and the high byte call address on the data bus is latched into the processor.

7. The T states T_{13}, T_{14} and T_{15} are used to store the high byte of the program counter in stack memory. In T_{13}, the content of Stack Pointer (SP) is decremented by one and placed on address bus. ALE is asserted high and then low, to latch the low byte of address into external latch. The other control signals are asserted as $IO/\overline{M}=0$, $S_0 = 1$ and $S_1 = 0$. In T_{14}, the high byte of PC is placed on $AD_0 - AD_7$ lines and $\overline{WR}$ is asserted low to enable the stack memory for write operation. At the end of T_{15}, $\overline{WR}$ is asserted high.

8. The T states T_{16}, T_{17} and T_{18} are used to store the low byte of the program counter in stack memory. In T_{16}, the content of SP is again decremented by one and placed on address bus. ALE is asserted high and then low, to latch the low byte of address into external latch. The other control signals are asserted as $IO/\overline{M}=0$, $S_0 = 1$ and $S_1 = 0$.

    In T_{17} the low byte of PC is placed on $AD_0 - AD_7$ lines and WR is asserted low to enable the stack memory for write operation. At the end of T_{15} $\overline{WR}$ is asserted high.

After the interrupt acknowledge machine cycle, the PC will have the call address and so the processor starts servicing the interrupt by executing the interrupt service subroutine stored at this address.

1.5.11 BUS IDLE MACHINE CYCLE

The bus idle machine cycle is executed, when extra time or more time is needed for an internal operation of the processor. During this cycle, the status signals S_0 and S_1 are asserted **low**. The data, address and control pins are driven to **high impedance** state. The READY signal will not be sampled by the processor during this cycle.

1.5.12 MACHINE CYCLE WITH WAIT STATES

Wait states can be introduced in any machine cycle except bus idle cycle between T_2 and T_3. The wait states are introduced in the machine cycle if READY pin is tied **low** at the second T-state of a machine cycle. The processor samples (or check) the READY signal at the second T-state of every machine cycle. If READY is tied **low** at this time, then the processor keeps on introducing wait state until the READY is again tied **high**. This facility is used by the slow memories, IO devices and peripherals to get extra time for read or write operations.

In the system when the peripheral timings are matched with processor timings, then the READY pin is permanently tied **high**. If the system peripherals require more time for read or write cycles, then using additional hardware the READY pin should be tied **low** for the required number of T-states.

The circuit shown in Fig. 1.30 can be used to introduce one wait state in the machine cycles. The working of the circuit shown in Fig. 1.30 can be explained as follows:

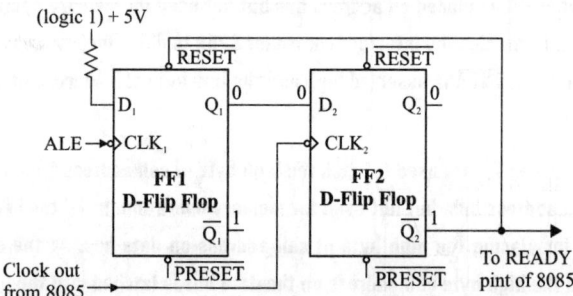

(The values shown at the input and output of the flip-flops are initial conditions)

Fig. 1.30: Circuit to introduce one wait state in 8085 machine cycle.

1. Initially $Q_2 = 0$ and $\overline{Q}_2 = 1$. The input D_1 is permanently tied high. The flip-flops are negative edge sensitive and so they are clocked (recognizes the clock) at the falling edges.

2. In the beginning of every machine cycle (except bus idle), ALE is asserted high and then low. At the falling edge of ALE, FF1 is clocked and its output Q_1 changes to 1. Also the input to FF2, D_2 changes to 1.

3. Now $D_1 = 1$, $Q_1 = 1$, $D_2 = 1$, $Q_2 = 0$, $\overline{Q}_2 = 1$ and $\overline{RESET} = 1$.

4. At the falling edge (beginning) of T_2, FF2 is clocked and so its output Q_2 changes to 1 and changes to 0.

5. Now, $D_1 = 1$, $Q_1 = 1$, $D_2 = 1$, $Q_2 = 1$, $\overline{Q}_2 = 0$ and $\overline{RESET} = 0$.

6. Since $\overline{Q}_2$ is connected to READY pin of 8085, the READY will be tied low. The $\overline{Q}_2$ is also used to reset FF1 and so when $\overline{Q}_2$ goes to 0 the FF1 is resetted or cleared. Now $Q_1 = 0$ and since $Q_1 = D_2$, the D_2 is also equal to 0.

7. Now, $D_1 = 1$, $Q_1 = 0$, $D_2 = 0$, $Q_2 = 1$, $\overline{Q}_2 = 0$ and $\overline{RESET} = 0$.

8. At the falling edge of next T-state (i.e., in wait state) again FF2 is clocked and so the output of FF2 will change.

9. Now, $D_1 = 1$, $Q_1 = 0$, $D_2 = 0$, $Q_2 = 0$, $\overline{Q}_2 = 1$ and $\overline{RESET} = 1$.

10. Since $\overline{Q}_2 = 1$, again READY is tied high. When the processor checks the READY at the falling edge of next cycle (T_3), it will be high and it will continue the machine cycle.

Thus, the hardware shown in Fig. 1.30 introduces one wait state in the machine cycles. A machine cycle with one wait state is shown in Fig. 1.31.

Truth Table of D-flip-flop

Clock	Input	Output	
	D	**Q̄**	
↓	1	1	0
↓	0	0	1

Preset and reset/clear facility in D-flip-flop

PRESET	RESET	Q	Q̄
0	1	1	0
1	0	0	1
1	1	Clock and D input decide the output	
0	0	Should not occur	

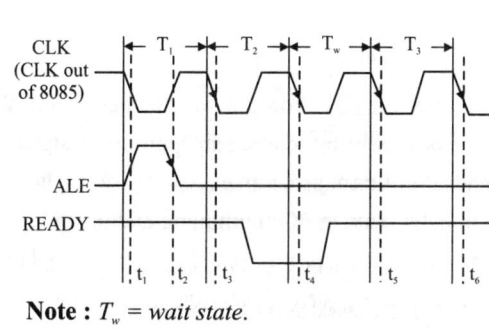

Time	D₁	Q₁	D₂	Q₂	Q̄₂	RESET
t₁	1	0	0	0	1	1
t₂	1	1	1	0	1	1
t₃	1	1	1	1	0	0
	1	0	0	1	0	0
t₄	1	0	0	0	1	1
t₅	1	0	0	0	1	1
t₆	1	0	0	0	1	1

Note : T_w = wait state.

Fig. 1.31: Machine cycle with one wait state.

1.6 INTERRUPTS

Microprocessors allow normal program execution to be interrupted in order to carry out a specific task/work. A processor can be interrupted in the following ways :

(i)   **by an external signal generated by a peripheral,**

(ii) **by an internal signal generated by a special instruction in the program,**

(iii) **by an internal signal generated due to an exceptional condition which occurs while executing an instruction.**
      (For example, in 8086 processor, 'divide by zero' is an exceptional condition which initiates type-0 interrupt and such an interrupt is also called exception.)

In general, the process of interrupting the normal program execution to carry out a specific task/work is referred to as interrupt.

The interrupt is initiated by a signal generated by an external device or by a signal generated internal to the processor. When a microprocessor receives an interrupt signal, it stops executing the current normal program, saves the status (or content) of various registers (PC in case of 8085) in stack and then executes a subroutine/procedure in order to perform the specific task/work requested by the interrupt. The subroutine/procedure that is executed in response to an interrupt is also called Interrupt

Service Routine (ISR). At the end of ISR, the stored status of registers in stack are restored to respective registers and the processor resumes the normal program execution from the point (instruction) where it was interrupted.

The external interrupts are used to implement interrupt driven data transfer scheme. The interrupts generated by special instructions are called software interrupts and they are used to implement system services/calls (or monitor services/calls). The system/monitor services are procedures developed by the system designer for various operations and stored in memory. The user can call these services through software interrupts. The interrupts generated by exceptional conditions are used to implement error conditions in the system.

Interrupt Driven Data Transfer Scheme

Interrupts are useful for efficient data transfer between the processor and the peripheral. When a peripheral is ready for data transfer, it interrupts the processor by sending an appropriate signal. Upon receiving an interrupt signal, the processor suspends the current program execution, saves the status in a stack and executes an ISR to perform the data transfer between the peripheral and the processor. At the end of ISR the processor status is restored from stack and the processor resumes its normal program execution. This type of data transfer scheme is called interrupt driven data transfer scheme.

The data transfer between the processor and peripheral devices can be implemented either by polling technique or by interrupt method. In polling technique, the processor has to periodically poll or check the status/readiness of the device and can perform data transfer only when the device is ready. In polling technique the processor time is wasted, because the processor has to suspend its work and check the status of the device in predefined intervals.

Alternatively, if the device interrupts the processor to initiate a data transfer whenever it is ready then the processor time is effectively utilized because the processor need not suspend its work and check the status of the device in predefined intervals.

For example, consider the data transfer from a keyboard to the processor. Normally a keyboard has to be checked by the processor once in every 10 millisecond for a key press. Therefore, once in every 10 milliseconds the processor has to suspend its work and then check the keyboard for a valid key code. Alternatively, the keyboard can interrupt the processor, whenever a key is pressed and a valid key code is generated. In this way the processor need not waste its time to check the keyboard once in every 10 milliseconds.

1.6.1 CLASSIFICATION OF INTERRUPTS

In general interrupts can be classified in the following three ways:

- **Hardware and software interrupts.**
- **Vectored and non-vectored interrupts.**
- **Maskable and non-maskable interrupts.**

Interrupts initiated by external hardware by sending an appropriate signal to the interrupt pin of the processor is called hardware interrupt. The 8085 processor has five interrupt pins TRAP, RST 7.5, RST 6.5, RST 5.5 and INTR and the interrupts initiated by applying appropriate signal to these pins are called hardware interrupts of 8085.

Software interrupts are program instructions. These instructions are inserted at desired locations in a program. While running a program, if a software interrupt instruction is encountered then the processor initiates an interrupt. The 8085 processor has 8 types of software interrupts. The software interrupt instruction is INT n, where n is the type number in the range 0 to 7.

When an interrupt signal is accepted by the processor, and the program control automatically branches to a specific address (called vector address) then the interrupt is called vectored interrupt. The automatic branching to a vector address is predefined by the manufacturer of the processor. (In these vector addresses the interrupt service subroutines(ISR) are stored.) In non-vectored interrupts the interrupting device should supply the address of the ISR to be executed in response to the interrupt. All the 8085 interrupts excepts INTR are vectored interrupts.

The processors have the facility for accepting or rejecting hardware interrupts. Programming the processor to reject an interrupt is referred to as masking or disabling and programming the processor to accept an interrupt is referred to as unmasking or enabling. In 8085 the hardware interrupts RST 7.5, RST 6.5, and RST 5.5 can be masked/unmasked using SIM instruction. All the hardware interrupts except TRAP are disabled by executing DI instruction and they are enabled by executing EI instruction.

The interrupts whose request can be either accepted or rejected by the processor are called maskable interrupts. The interrupts whose request has to be definitely accepted (i.e., it cannot be rejected) by the processor are called non-maskable interrupts. Whenever a request is made by a non-maskable interrupt, the processor has to definitely accept that request and service that interrupt by suspending its current program and executing an ISR. In 8085 processor all the hardware interrupts except TRAP are maskable. The interrupt initiated through TRAP pin and all software interrupts are non-maskable.

1.6.2   INTERRUPTS OF 8085

The interrupt in 8085 can come from one of the following two sources:

1. **One source is from an external signal applied to TRAP, RST7.5, RST6.5, RST5.5 or INTR pin of the processor. The interrupts initiated by applying appropriate signals to these pins are called hardware interrupts.**

2. **The second source of an interrupt is the execution of the interrupt instruction "RST n" where n can take values from 0 to 7. The interrupts initiated by "RST n" instructions are called software interrupts.**

Software Interrupts Of 8085

Software interrupts are program instructions. When a software interrupt instruction is executed, the processor executes an Interrupt Service Routine(ISR) stored in the vector address of that software interrupt instruction. The software interrupts of 8085 are RST0, RST1, RST2, RST3, RST4, RST5, RST6 and RST7. The software interrupts of 8085 are vectored interrupts. Software interrupts cannot be masked or be disabled. The Vector addresses of software interrupts are given in Table 1.14.

Table 1.14

Interrupt	Vector address
RST 0	0000_H
RST 1	0008_H
RST 2	0010_H
RST 3	0018_H
RST 4	0020_H
RST 5	0028_H
RST 6	0030_H
RST 7	0038_H

Software interrupt instructions are included at the appropriate (or required) place in the main program. When the processor encounters the software instruction, it pushes the content of PC (**P**rogram **C**ounter) to stack. Then, it loads the vector address in to the PC and starts executing an ISR stored in this address. The last instruction of the ISR will be RET instruction. When the RET instruction is executed, the processor POPs the content of top of stack to PC. Hence, the processor control returns to main program after servicing the interrupt. *[Execution of ISR is referred to as servicing of interrupt.]*

Hardware Interrupts of 8085

(AU, Nov/Dec' 19, 2 Marks)

The hardware interrupts of 8085 are initiated by an external device by placing an appropriate signal at the interrupt pin of the processor. The processor keeps on checking the interrupt pins at the second T-state of the last machine cycle of every instruction. If the processor finds a valid interrupt signal and if the interrupt is unmasked and enabled, then the processor accepts the interrupt. The acceptance of the hardware interrupt is acknowledged by sending an $\overline{INTA}$ signal to the interrupting device.

When the interrupt is accepted, the processor saves the content of the PC (**P**rogram **C**ounter) in stack and then loads the vector address of the interrupt to the PC. (If the interrupt is non-vectored, then the interrupting device has to supply the address of ISR when it receives $\overline{INTA}$ signal.) Then the processor starts executing ISR in this address. The last instruction of ISR will be an RET instruction. When the processor executes the RET instruction, it POP the content of top of stack to PC. Thus the processor control returns to the main program after servicing the interrupt.

The hardware interrupts of 8085 are TRAP, RST 7.5, RST6.5, RST5.5 and INTR. TRAP, RST7.5, RST6.5 and RST5.5 are vectored interrupts. In vectored interrupts the address to which the program control is transferred (when the interrupt is accepted) is fixed by the manufacturer. The vector addresses of hardware interrupts are given in Table 1.15. The INTR is a non-vectored interrupt. Hence when a device interrupts through INTR, it has to supply the address of ISR after receiving interrupt acknowledge signal.

Table 1.15

Interrupt	Vector address
RST 7.5	$003C_H$
RST 6.5	0034_H
RST 5.5	$002C_H$
TRAP	0024_H

The type of signal that has to be placed on the interrupt pin of hardware interrupts of 8085 are defined by INTEL. The TRAP interrupt is edge and level sensitive. Hence, to initiate TRAP, the interrupt signal has to make a **low** to **high** transition and then it has to remain **high** until the interrupt is recognized. The RST 7.5 interrupt is edge sensitive (positive edge). In order to initiate the RST7.5, the interrupt signal has to make a **low** to **high** transition and it need not remain **high** until it is recognized. The RST6.5, RST5.5 and INTR are level sensitive interrupts. Hence, for these interrupts the interrupt signal should remain **high**, until it is recognized.

TRAP is a non-maskable interrupt and RST7.5, RST6.5 and RST5.5 are maskable interrupts, which use the SIM (Set Interrupt Mask) instruction. Interrupts can be masked by moving an appropriate data (or code) to the accumulator and then executing the SIM instruction. The status of maskable interrupts can be read into the accumulator by executing the RIM instruction (RIM - **R**ead **I**nterrupt **M**ask).

All the hardware interrupts, except TRAP are disabled when the processor is reset and they can also be disabled by executing the DI instruction. (DI - **D**isable **I**nterrupt). When an interrupt is disabled, it will not be accepted by the processor (i.e., INTR, RST5.5, RST6.5 and RST7.5 are disabled by the DI instruction and upon hardware reset). In order to enable (or to allow) the disabled interrupts, the processor has to execute the EI instruction (EI - **E**nable **I**nterrupt).

Priorities of Interrupts of 8085

When all the interrupts are enabled, the priority sequence of hardware interrupts from highest to lowest is TRAP, RST 7.5, RST 6.5, RST 5.5 and INTR. When the 8085 processor accepts an interrupt it will disable all the hardware interrupts except TRAP. Hence in order to allow the higher priority interrupt while executing Interrupt Service Subroutine (ISR) for lower priority interrupt, enable the interrupt system in the beginning of ISR of lower priority interrupt, by executing EI instruction.

For example, if the processor accepts RST 5.5 interrupt, then it will disable RST 7.5, RST 6.5 and INTR interrupts. In order to allow the higher priority interrupt RST 7.5 and RST 6.5 while executing ISR of RST 5.5, the EI instruction should be executed in the beginning of ISR of RST 5.5.

The execution of software interrupt will not disable any hardware interrupt. Therefore while executing ISR of software interrupts, the processor will recognize or allow the hardware interrupts.

1.6.3 ENABLING, DISABLING AND MASKING OF 8085 INTERRUPTS

TRAP

The interrupt TRAP is non-maskable and it cannot be disabled by DI instruction. Also the TRAP is not disabled by system (processor) reset or after recognition of another interrupt. The only signal which can override TRAP is HOLD signal. (i.e., If the processor receives HOLD and TRAP at the same time then HOLD is recognized first and only then is TRAP recognized.)

INTR

The interrupt INTR is disabled by any one of the following operations:

- **Executing DI instruction.**
- **System or processor reset.**
- **After recognition (acceptance) of an interrupt.**

The interrupt INTR can be enabled by executing EI instruction.

RST 7.5, RST 6.5 and RST 5.5

The interrupt RST 7.5, RST 6.5 and RST 5.5 are disabled by any one of the following operations.

- **Executing DI instruction.**
- **System or processor reset.**
- **After recognition (acceptance) of an interrupt.**

These hardware interrupts can be enabled by executing EI instruction.

The 8085 provides additional masking facility for RST 7.5, RST 6.5 and RST 5.5 using SIM instruction. The status of these interrupts can be read by executing RIM instruction.

The masking or unmasking of RST 7.5, RST 6.5 and RST 5.5 interrupts can be performed by moving an 8-bit data to accumulator and then executing SIM instruction. The format of the 8-bit data is shown in Fig. 1.32.

The status of pending interrupts can be read from accumulator after executing RIM instruction. When RIM instruction is executed, an 8-bit data is loaded to the accumulator, which can be interpreted as shown in Fig. 1.33.

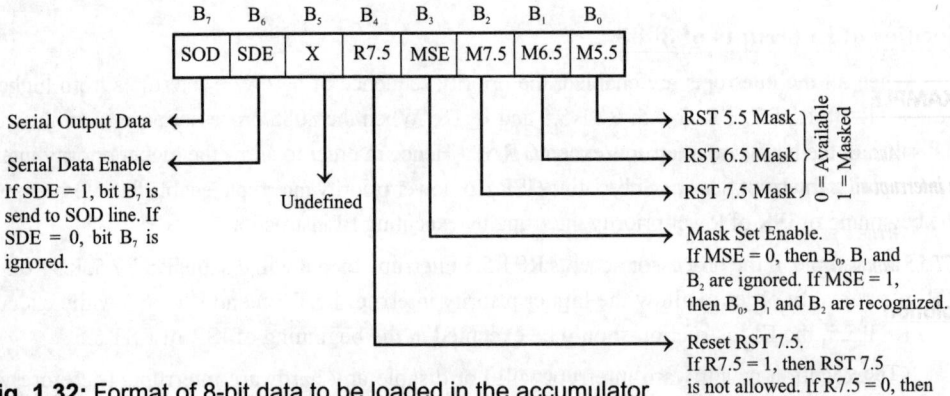

Fig. 1.32: Format of 8-bit data to be loaded in the accumulator before executing a SIM instruction.

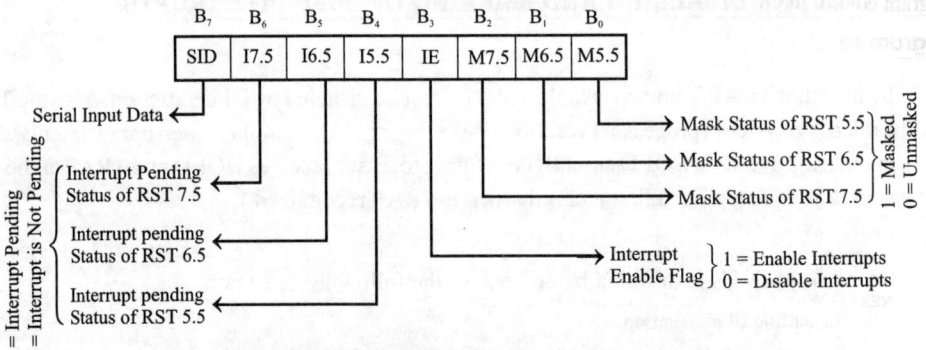

Fig. 1.33: Interpretation of the content of the accumulator after executing a RIM instruction.

EXAMPLE 1

Write a program segment to mask RST 6.5 and RST 5.5 interrupts and enable RST 7.5 interrupt.

Solution

The 8-bit data format to be loaded in the accumulator for enabling RST 7.5 and masking RST 6.5 and RST 5.5 is shown below. The data to be loaded in accumulator is $0B_H$.

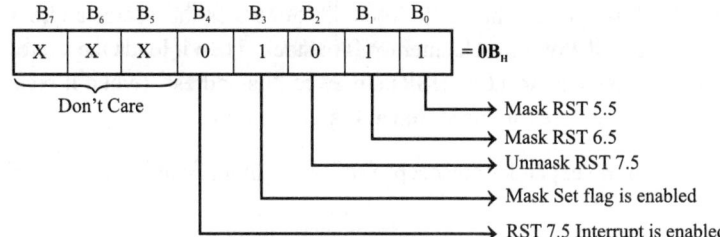

Program segment

```
EI          ; Enable all interrupts of 8085
MVI A,0BH   ; Move 0B_H to A-register
SIM         ; Mask 6.5 and 5.5, Enable 7.5
```

EXAMPLE 2

Assume that the 8085 microprocessor returns to the main program after servicing RST 6.5. (Remember that while servicing an interrupt all other interrupts are disabled.)

Write a program segment to check whether RST 5.5 interrupt is pending. If it is pending then the program has to enable RST 5.5 without affecting any other interrupts. Otherwise the program has to enable all interrupts and return to main program.

Solution

The status of pending interrupts can be read by executing RIM instruction. This will load an 8-bit data in accumulator. If RST 5.5 is pending then the bit D_4 in accumulator will be **1** and if it is not pending then bit D_4 will be **0**. The following program segment have been written to check whether bit D_4 is **1** or **0**. If it is **1** then the program control jumps to another part of program to enable RST 5.5 and mask other interrupts.

Program segment

```
            RIM        ;Read the status of interrupts.
            MOV C,A    ;Save the status in C-register.
            ANI 10H    ;Check whether RST 5.5 is pending.
            JNZ NEXT   ;If  RST 5.5 is pending, go to NEXT.
            EI         ;If RST 5.5 is not pending, enable
            RET        ;all interrupts and return to main program.
NEXT:       MOV A,C    ;Get the interrupt status in A-register.
            ANI FEH    ;Set D_0 = 0, for enabling RST 5.5
            ORI 08H    ;Set D_3 =1, for enabling interrupt enable flag.
            SIM        ;Enable RST 5.5
            JMP ISR55  ;Jump to Interrupt Service Routine of RST 5.5
```

1.6.4 INTR AND ITS EXPANSION

The INTR is general interrupt request. An external device can interrupt the processor by placing a **high** signal on INTR pin of 8085. If the processor accepts the interrupt, then it will send an acknowledge signal $\overline{\text{INTA}}$ to the interrupting device. On receiving the acknowledge signal, the interrupting device has to place either an **RST n** opcode (or CALL opcode followed by 16-bit address) on the data bus.

On receiving the **RST n** opcode, the 8085 processor generates the vector address of **RST n** instruction. It saves the content of **P**rogram **C**ounter (PC) in stack. Then it loads the vector address in PC and executes an **I**nterrupt **S**ervice **R**outine (ISR) stored at this address. (when it receives CALL opcode it executes an interrupt service routine stored at CALL address.)

The INTR interrupt can be expanded to accept 8-interrupt inputs using 8-to-3 priority encoder as shown in Fig. 1.34.

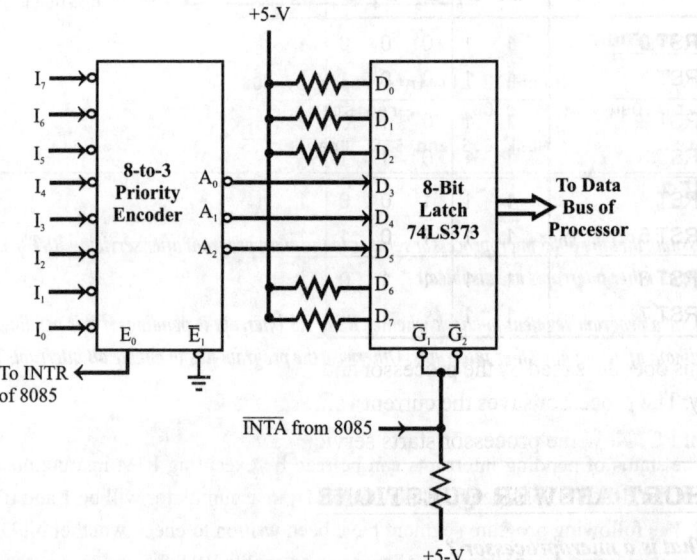

Fig. 1.34: Expanding an INTR of the 8085 using an 8-to-3 priority encoder.

The priority encoder has 8 inputs I_0 to I_7 and three outputs A_0 to A_2. It also has an output control signal, E_0. If the priority encoder receives a logic **low** at one of the inputs, for example I_n, then it asserts E_0 **high** and outputs the binary value of n on the output lines A_0, A_1 and A_2 lines (i.e., if input I_0 is **low** then output is 000; if input I_1 is **low** then output is 001 and so on). In this scheme I_7 has the highest priority and I_0 has the lowest priority.

Eight external devices can interrupt the processor through I_0 to I_7 lines, by placing a logic **low** on these pins. On receiving a valid interrupt signal the priority encoder allows the highest priority interrupt by asserting E_0 **high** and sending the corresponding binary value on A_0, A_1 and A_2 lines. The E_0 is connected to INTR of 8085 and A_0, A_1 and A_2 are connected to the inputs D_3, D_4 and D_5 of an 8-bit latch. All other inputs of the latch are tied to +5 V (logic 1) permanently.

The opcodes and vector addresses of RST n instructions are shown in Table 1.16. If we carefully look at the opcode of RST instruction, the binary bits D_3, D_4, D_5 constitutes the binary value of n in RST n instruction and all other bits are 1's. The priority encoder helps in placing the RST opcodes at the input of latch (74LS373). *[The priority encoder places the **RST n** opcode for the interrupt I_n.]*

When the processor accepts the interrupt, it sends $\overline{INTA}$ signal to the interrupting device. This signal is used to enable the latch. When the latch is enabled, the RST opcode available at the input is latched into output lines. The output of latch is connected to data bus of the processor. Hence, the opcode will be placed on the data bus.

Table 1.16: Opcodes of RST Instructions

RST instruction	Opcode in binary								Opcode in hexa	Vector address
	D_7	D_6	D_5	D_4	D_3	D_2	D_1	D_0		
RST 0	1	1	0	0	0	1	1	1	C7	0000_H
RST 1	1	1	0	0	1	1	1	1	CF	0008_H
RST 2	1	1	0	1	0	1	1	1	D7	0010_H
RST 3	1	1	0	1	1	1	1	1	DF	0018_H
RST 4	1	1	1	0	0	1	1	1	E7	0020_H
RST 5	1	1	1	0	1	1	1	1	EF	0028_H
RST 6	1	1	1	1	0	1	1	1	F7	0030_H
RST 7	1	1	1	1	1	1	1	1	FF	0038_H

This opcode is read by the processor and then it generates the vector address of the RST instruction internally. The processor saves the current value of **P**rogram **C**ounter (PC) in stack and loads the vector address in PC. Now the processor starts servicing the interrupt.

1.7 SHORT-ANSWER QUESTIONS

Q1.1 *What is a microprocessor ?*

A microprocessor is a program controlled semiconductor device (IC), which fetches, decodes and executes instructions.

Q1.2 *What are the basic functional blocks of a microprocessor ?*

The basic functional blocks of a microprocessor are ALU, an array of registers and control unit.

Q1.3 *What is a bus ?*

Bus is a group of conducting lines that carries data, addresses and control signals.

Q1.4 *Define bit, byte and word.*

A digit of the binary number or code is called bit. The bit is also the fundamental storage unit of computer memory.

The 8-bit (8-digit) binary number or code is called byte and 16-bit binary number or code is called word. (Some microprocessor manufacturers refer to the basic data size operated by the processor as word.)

Q1.5   State the relation between the number of address pins and physical memory space?

The size of the binary number used to address the memory decides the physical memory space. If a microprocessor has n-address pins then it can directly address 2^n memory locations. (The memory locations that are directly addressed by the processor are called physical memory space.)

Q1.6   Why is data bus is bidirectional?

The microprocessor has to fetch (read) the data from memory or input device for processing and after processing it has to store (write) the data in memory or output device. Hence, the data bus is bidirectional.

Q1.7   Why is address bus unidirectional?

The address is an identification number used by the microprocessor to identify or access a memory location or IO device. It is an output signal from the processor. Hence, the address bus is unidirectional.

Q1.8   State the difference between CPU and ALU.

The ALU is the unit that performs the arithmetic or logical operations. The CPU is the unit that includes ALU and control unit. Apart from processing the data, the CPU controls the entire system functioning. Usually, a microprocessor will be the CPU of a system and it is called the brain of the computer.

Q1.9   What is a tristate logic? Why it is needed in microprocessor system?

In a tristate logic, three logic levels are used **high, low** and **high impedance** state. The **high** and **low** are normal logic levels and **high impedance** state is electrical open circuit condition.

In a microprocessor system, all the peripheral/slave devices are connected to a common bus. But communication (data transfer) takes place between the master (microprocessor) and one slave (peripheral) at any time instant. During this time instant, all other devices should be isolated from the bus. Therefore, normally all the slaves (peripherals) will remain in **high impedance** state (i.e., in electrical isolation). The master will select a slave by sending address and chip select signal. When the slave is selected, it comes to normal logic and it can communicate with the master.

Q1.10   What is HMOS and HCMOS.

The HMOS is High density n-type Metal Oxide Silicon field effect transistors. The third generation microprocessors are fabricated using HMOS transistors.

The HCMOS is High density n-type Complementary Metal Oxide Silicon field effect transistors. It is the low power version of HMOS and the fourth generation microprocessors are fabricated using HCMOS transistors.

Q1.11   What are the drawbacks of first generation microprocessors.

The first generation processors are fabricated using PMOS technology and it has the drawbacks like slow speed, provides low output currents and was not compatible with TTL logic levels.

Q1.12   What is a microcomputer? Explain the difference between a microprocessor and a microcomputer.

A system designed using a microprocessor as its CPU is called microcomputer. The term microcomputer refers to the whole system, whereas the microprocessor is the CPU of the system.

Q1.13 What is the function of microprocessor in a system?

The microprocessor is the master in the system, which controls all the activity of the system. It issues address and control signals and fetches the instruction and data from memory. Then it executes the instruction to take appropriate action.

Q1.14 List the components of microprocessor-based (single board microcomputer) system.

The microprocessor-based system consist of microprocessor as CPU, semiconductor memories like EPROM and RAM, input device, output device and interfacing devices.

Q1.15 Why interfacing is needed for IO devices?   *(AU, Nov/Dec' 19, 2 Marks)*

Generally IO devices are slow devices. Therefore, the speed of IO devices does not match with the speed of microprocessor. And so an interface is provided between system bus and IO devices.

Q1.16 What is the difference between CPU bus and system bus?

The CPU bus has multiplexed lines but the system bus has separate lines for each signal. (The multiplexed CPU lines are demultiplexed by the CPU interface circuit to form system bus.)

Q1.17 What is multiplexing and what is its advantage?

Multiplexing is transferring different information at different well-defined times through same lines. A group of such lines is called multiplexed bus. The advantage of multiplexing is that fewer pins are required for microprocessors to communicate with the outside world.

Q1.18 How the address and data lines are demultiplexed in 8085?

The low order address and data lines of 8085 are demultiplexed using an external 8-bit D-Latch (74LS373) and the ALE signal of 8085, as shown in Fig. Q1.18.

At the beginning of every machine cycle, ALE is asserted **high** and then **low**. Also, the low byte of address is given out through AD_0 - AD_7 lines. Since, the ALE is connected to enable of latch, whenever ALE is asserted **high** and then **low,** the addresses are latched into the output lines of the latch then the lines AD_0 - AD_7 are free for data transfer.

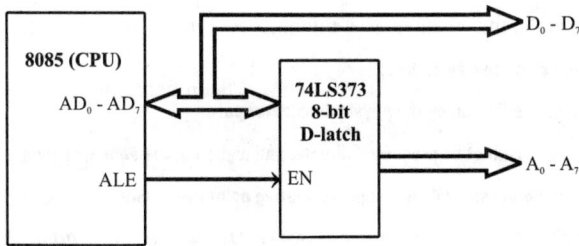

Fig. Q1.18: Demultiplexing of address and data lines in an 8085 processor.

Q1.19 On what basis the computers are classified into micro, mini and large/mainframe?

The classification of computers into micro, mini and mainframes are based on the following factors:

1. Speed of execution
2. Size and type of data
3. Memory capacity
4. IO devices and peripheral support devices
5. Application programs it can run.

Q1.20 What is the difference between micro and minicomputers?

. The microcomputers are systems built using a single microprocessor and has single motherboard as CPU. The minicomputers will have multiple microprocessors connected in a particular configuration. The minicomputer can handle large data words, address more memory space, supports a variety of IO devices and can be used for multiuser applications. In today's technology, the features of microcomputers have exceeded the capability of minicomputers.

Q1.21 What is mainframe?

The largest and most powerful computers are called mainframes.

Q1.22 What is a supercomputer?

The computer built using very high speed devices (or devices with very low switching speeds) and can execute instructions at very high speeds are called supercomputers. The speed of supercomputers are measured in MIPS (**M**illions of **I**nstructions **P**er **S**econd) or Megaflops (Millions of floating point operations per second). A typical supercomputer can execute 3000 MIPS. The speed of Cray X-MP2 supercomputer is 500 Megaflops.

Q1.23 List the applications of microcomputer.

1. Personal computing   4. Control applications
2. Calculators   5. Instrumentation systems
3. Small business system.

Q1.24 What are the advantages of microprocessor-based system?

The advantages of microprocessor-based system are the following:

1. Computational or Processing speed is high.
2. Intelligence has been brought to systems.
3. Automation of industrial processes and office administration.
4. Both operation and maintenance are easier.

Q1.25 What are the disadvantages of microprocessor-based system?

The following are the disadvantages of microprocessor-based system:

1. It has limitations on the size of data.
2. The applications are limited by the physical address space.
3. The analog signals cannot be processed directly and digitizing the analog signals introduces errors.
4. Most of the microprocessors do not support floating point operations.

Q1.26 What do you mean by 16 and 8-bit processors? Mention a few 8-bit and 16-bit processors.

The processors are classified into 8-bit or 16-bit depending on the basic data size handled by the ALU of the processor.

8-bit microprocessors : 8085, Z80, Motorola 6800.
16-bit microprocessors : 8086, Z8000, MC68000.

Q1.27 What is the fabrication technology used for 8085?

The 8085A is fabricated used NMOS technology and 8085AH is fabricated using HMOS technology.

Q1.28 What is the physical memory space in 8085?

The 8085 uses 16-bit address to access memory locations. Hence, it can directly address 64 k memory locations (2^{16} = 65,536 = 64 k). Since 8085 has 8 data lines, it can read or write 8-data bits from a memory address. Therefore, the physical memory space is 64 k × 1byte = 64 kilobytes (64 kB).

Q1.29 What is ALE?

The ALE (**A**ddress **L**atch **E**nable) is a signal used to demultiplex the address and data lines using an external latch. It is used as enable signal for the external latch.

Q1.30 Explain the function of IO/$\overline{M}$ in 8085.

The IO/$\overline{M}$ is used to differentiate memory access and IO access. For IN and OUT instruction it is asserted **high**. For memory reference instructions it is asserted **low**.

Q1.31 How the READY signal is used in microprocessor system?

The READY is an input signal that can be used by slow peripherals to get extra time in order to communicate with 8085. The 8085 will work only when READY is tied to logic **high**. Whenever READY is tied to logic **low**, the 8085 will enter a wait state. When the system has slow peripheral devices, additional hardware is provided in the system to make the READY input **low** during the required extra time while executing a machine cycle, so that the processor will remain in wait state during this extra time.

Q1.32 What is HOLD and HLDA? How is it used?

The HOLD and HLDA signals are used for the **D**irect **M**emory **A**ccess (DMA) type of data transfer. These type of data transfers are achieved by employing a DMA controller in the system. When DMA is required, the DMA controller will place a **high** signal on the HOLD pin of 8085. When HOLD input is asserted **high,** the processor will enter a wait state and drive all its tristate pins to **high impedance** state and send an acknowledge signal to DMA controller through HLDA pin. Upon receiving the acknowledge signal, the DMA controller will take control of the bus and perform DMA transfer and at the end it asserts HOLD signal **low**. When HOLD is asserted **low,** the processor will resume its execution.

Q1.33 How clock signals are generated in 8085 and what is the frequency of the internal clock?

The 8085 has the clock generation circuit on the chip but an external quartz crystal or LC circuit or RC circuit should be connected at the pins X_1 and X_2 in order to generate a clock signal. The 8085 clock generation circuit, generate a clock whose frequency is double as compared to that of internal clock. The generated clock is divided by two and then used as internal clock. The maximum internal clock frequency of 8085A is 3.03 MHz.

Q1.34 What happens to the 8085 processor when it is resetted?

When $\overline{\text{RESET IN}}$ pin is asserted **low**, the program counter, instruction register, interrupt mask bits and all internal registers are cleared or resetted. Also the RESET OUT signal is asserted **high** to clear or reset all the peripheral devices in the system. After a reset, the content of program counter will be 0000_H and so the processor will start executing the program stored at 0000_H.

Q1.35 What are the operations performed by ALU of 8085?

The operations performed by ALU of 8085 are addition, subtraction, logical AND, OR, Exclusive-OR, compare, complement, increment, decrement and left/right shift.

Q1.36 *Mention the names of various registers in 8085 along with its size.*

Register		Size (bits)	Register		Size (bits)
Accumulator (A)	-	8	Stack pointer	-	16
Temporary register	-	8	Program counter	-	16
Instruction register	-	8			
General purpose register	-	8			

(B, C, D, E, H and L)

Q1.37 *What is a flag?*

Flag is a flip-flop used to store the information about the status of the processor and the status of the instruction executed most recently.

Q1.38 *List the flags of 8085.*

There are five flags in 8085. They are sign flag, zero flag, auxiliary carry flag, parity flag and carry flag.

Q1.39 *Show the bit positions of various flags in 8085 flag register.*

The bit positions of various flags in the flag register of 8085 is shown in Fig. Q1.39.

D_7	D_6	D_5	D_4	D_3	D_2	D_1	D_0
SF	ZF		AF		PF		CF

SF - Sign Flag
PF - Parity Flag
ZF - Zero Flag
AF - Auxiliary Carry Flag
CF - Carry Flag

Fig. Q1.39: Bit positions of various flags in the flag register of 8085.

Q1.40 *What are the hardware interrupts of 8085?*

The hardware interrupts in 8085 are TRAP, RST 7.5, RST 6.5 and RST 5.5.

Q1.41 *Which interrupt has highest priority in 8085? What is the priority of other interrupts?*

The TRAP has the highest priority, followed by RST 7.5, RST 6.5, RST 5.5 and INTR.

Q1.42 *Define stack.*

Stack is a sequence of RAM memory locations defined by the programmer.

Q1.43 *What is program counter? How is it useful in program execution?*

The program counter keeps a track of program execution. To execute a program, the starting address of the program is loaded in program counter. The PC sends out an address to fetch a byte of instruction from memory and increment its content automatically.

Q1.44 *How is the microprocessor synchronized with peripherals ?*

The timing and control unit synchronizes all the microprocessor operations with clock and generates control signals necessary for communication between the microprocessor and peripherals.

Q1.45 *What is memory ?*

A memory is a storage device in a microprocessor-based system and its primary function is to store programs and data.

Q1.46 What is physical memory space?

The memory locations that are directly addressed by the microprocessor is called physical memory space.

Q1.47 What is memory word size?

The size of data that can be stored in memory location is called memory word size.

Q1.48 What is meant by memory mapping?

Memory mapping is the process of interfacing memories to microprocessor and allocating addresses to each memory location.

Q1.49 What is memory access time?

Memory access time is the time taken by the processor to read or write a memory location. Read operation is the time between a valid address on the bus and the end of read control signal. Write operation is the time between a valid address on the bus and the end of write control signal.

Q1.50 What is bus contention?

If two devices drive the data bus simultaneously then it is called bus contention. It may lead to following undesirable events:

1. **Damaging one or both the IC chip.**
2. **The high current may cause a voltage spike in the supply system leading to data loss.**

Q1.51 Why is EPROM mapped at the beginning of memory space in 8085 ?

When EPROM is mapped at the beginning of memory space, then 0000_H address will be allotted to EPROM. The monitor program can be stored from 0000_H address. Whenever the processor is reset, the program counter will be cleared (i.e it will have 0000_H address) and the monitor program will be executed automatically.

Q1.52 What is chip select and how it is generated?

Chip select is the control signal that has to be asserted TRUE to bring an IC from **high impedance** state to normal state. Generally the chip select signals are generated in a system by decoding the unused address lines with the help of decoders.

Q1.53 What are the typical control signals involved in EPROM interfacing ?

The control signals needed for EPROM are chip select and output enable.

Q1.54 What are the typical control signals involved in RAM interfacing ?

The control signals needed for RAM interfacing are chip enable, output enable and write enable.

Q1. 55 What is programmed IO ?

If the data transfer between an IO device and the processor is accomplished through an IO port and controlled by a program then the IO device is called programmed IO.

Q1.56 What is interrupt IO?

If the IO device initiates the data transfer through interrupt, then the IO is called interrupt driven IO.

Q1.57 What is DMA?

The direct data transfer between the IO device and the memory is called DMA.

Q1.58 What is the need for Port?

IO devices are generally slow devices and their timing characteristics do not match with processor timings. Hence, the IO devices are connected to a system bus through the ports.

Q1.59 What is a port?

A port is a buffered I/C which is used to hold the data transmitted from the microprocessor to IO device or vice versa.

Q1.60 Give some examples of port devices used in a 8085 microprocessor-based system.

The various INTEL IO port devices used in 8085 microprocessor-based system are 8212, 8155, 8156, 8255, 8355 and 8755.

Q1.61 What are the different methods of interfacing IO devices to 8085-based system.

There are two methods of interfacing IO devices to 8085 system. They are memory mapping of IO device and standard IO mapping.

Q1.62 Compare the memory-mapped IO with the standard IO-mapped IO.

Memory-mapped IO	Standard IO-mapped IO
1. 16-bit address is allotted to an IO device.	1. 8-bit address is allotted to an IO device.
2. The devices are accessed by memory read or memory write cycle.	2. The devices are accessed by IO read or IO write cycle.
3. All instructions related to memory can be used for data.	3. Only IN and OUT instructions can be used for data transfer.
4. A large number of IO ports can be interfaced.	4. Only 256 ports can be interfaced.

Q1.63 What is the drawback in memory-mapped IO?

When IO devices are memory-mapped, some of the addresses are allotted to IO devices. So the full address space cannot be used for addressing memory (i.e., physical memory address space will be reduced). Hence, memory mapping is useful only for small systems, where the memory requirement is less.

Q1.64 What is Software and Hardware?

The software is a set of instructions or commands needed for performing a specific task by a programmable device or a computing machine.

The hardware refers to the components or devices used to form computing machine in which the software can be run and tested. Without software the hardware is an idle machine.

Q1.65 What is machine language?

The language that can be understood by a programmable machine is called machine language. The machine language program are developed using 1s and 0 s.

Q1.66 What is assembly language?

The language in which the mnemonics (short-hand form of instructions) are used to write a program is called assembly language. The mnemonics are given by the manufacturers of microprocessor.

Q1.67 Draw a simple circuit to decode the three control signals $\overline{RD}, \overline{WR}$ and $IO/\overline{M}$ and to produce separate read/write control signals for memory and IO devices.

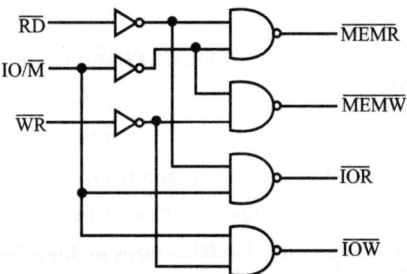

Fig. Q1.67: Circuit to generate separate read and write signals for memory and IO devices in an 8085-based system.

Q1.68 What are machine language and assembly language programs?

The software developed using 1s and 0 s are called machine language programs. The software developed using mnemonics are called assembly language programs.

Q1.69 What is the drawback in machine language and assembly language programs?

The machine language and assembly language programs are machine dependent. The programs developed using these languages for a particular machine cannot be directly run on another machine. (But after conversion using suitable conversion software it can be run on another machine.)

Q1.70 Define mnemonics.

The short-hand form of describing the instructions are called mnemonics. The mnemonics are given by the manufacturers of microprocessors and programmable devices.

Q1.71 What is processor cycle (machine cycle)?

The processor cycle or machine cycle is the basic external operation performed by the processor. To execute an instruction, the processor will run one or more machine cycles in a particular order.

Q1.72 What is instruction cycle?

The sequence of operations that a processor has to carry out while executing an instruction is called instruction cycle. Each instruction cycle of a processor in turn consists of a number of machine cycles.

Q1.73 What is fetch and execute cycle?

In general, the instruction cycle of an instruction can be divided into fetch and execute cycles. The fetch cycle is executed to fetch the opcode from memory. The execute cycle is executed to decode the instruction and to perform the work instructed by the instruction.

Q1.74 *List the various machine cycles of 8085.*

The various machine cycles of 8085 are as follows:

    (i)   **Opcode fetch cycle**

    (ii) **Memory read cycle**

    (iii) **Memory write cycle**

    (iv) **IO read cycle**

    (v) **IO write cycle**

    (vi) **Interrupt acknowledge cycle**

    (vii) **Bus idle cycle.**

Q1.75 *What is the need for timing diagram?*

The timing diagram provides information regarding the status of various signals, when a machine cycle is executed. The knowledge of timing diagram is essential for system designer to select matched peripheral devices like memories, latches, ports, etc., to form a microprocessor system.

Q1.76 *What is T-state?*

The T-state is the time period of the internal clock signal of the processor. The time taken by the processor to execute a machine cycle is expressed in T-state.

Q1.77 *How many machine cycles constitute one instruction cycle in 8085?*

Each instruction of the 8085 processor consist of one to five machine cycles.

Q1.78 *Define opcode and operand.*

Opcode (**Op**eration **Code**) is the part of an instruction/directive that identifies a specific operation.

Operand is a part of an instruction/directive that represents a value on which the instruction acts.

Q1.79 *What is opcode fetch cycle?*

The opcode fetch cycle is a machine cycle executed to fetch the opcode of an instruction stored in memory. The first machine cycle of every instruction is opcode fetch machine cycle.

Q1.80 *What operation is performed during first T-state of every machine cycle in 8085?*

In 8085, during the first T-state of every machine cycle the low byte address is latched into an external latch using ALE signal.

Q1.81 *Why status signals are provided in microprocessor?*

The status signals can be used by the system designer to track the internal operations of the processor. Also, it can be used for memory expansion (by providing separate memory banks for program and data, and selecting the banks using status signals).

Q1.82 *How the 8085 processor differentiates memory access (read/write) and IO access (read/write)?*

The memory access and IO access is differentiated using IO/$\overline{\text{M}}$ signal. The 8085 processor asserts IO/$\overline{\text{M}}$ **low** for memory read/write operation and IO/$\overline{\text{M}}$ is asserted **high** for IO read/write operation.

Q1.83 In which lines the 8085 processor gives the output of IO port address during IO read/write operation?

When the processor executes an IO read or write cycle, 8-bit port address is sent out both on low order address bus and high order address bus. This facility offers a flexibility for system designer to use either low-order address lines or high-order address lines for addressing ports and generating chip select signals for IO devices.

Q1.84 When the 8085 processor checks for an interrupt?

In the second T-state of the last machine cycle of every instruction, the 8085 processor checks whether an interrupt request is made or not.

Q1.85 What is interrupt acknowledge cycle?

The interrupt acknowledge cycle is a machine cycle executed by 8085 processor after acceptance of the interrupt to get the address of the interrupt service routine in-order to service the interrupting device.

Q1.86 What will be the status of the processor during bus idle cycle?

During bus idle cycle, the status signals S_0 and S_1 are both asserted **low** and data, address and control pins are driven to **high impedance** state. Also, the processor will not sample the READY signal.

Q1.87 How the slow peripherals are interfaced with 8085 processor?

The slow peripherals require longer read/write time than allowed by the processor. Hence to interface slow peripherals, an extra hardware should be designed so that it introduces required number of wait states in machine cycles between T_1 and T_2. An alternate solution is to interface the slow peripherals using ports.

Q1.88 When is the READY signal sampled by the processor?

The 8085 processor samples or checks the READY signal at the second T-state of every machine cycle.

Q1.89 What are wait states?

The T states introduced between T_2 and T_3 of a machine cycle by the slow peripherals (to get extra time for read/write operation) are called wait states.

Q1.90 When the 8085 processor will enter wait state?

The 8085 processor will check the READY signal at the second T-state of a machine cycle. If the READY is tied **low** at this time, then it will enter into wait state (i.e., after second T-state). The processor will come out of wait state only when READY is again made **high**.

Q1.91 What is the difference between wait state and bus idle condition?

During bus idle condition, the tristate pins of the processor are driven to **high impedance** state, but during wait state they are in normal states (either **low** or **high**). The READY is not sampled during bus idle condition but it is sampled during wait state.

Q1.92 How many instructions are available in 8085 instruction set?

The 8085 instruction set consists of 74 basic instructions and 246 total instructions.

Q1.93 What is an Interrupt ?

Interrupt is a signal sent by an external device to the processor so as to request the processor to perform a particular task or work.

Q1.94 How are interrupts classified ?

There are three methods of classifying interrupts.

      Method I   :   The interrupts are classified into Hardware and Software interrupts.

      Method II   :   The interrupts are classified into Vectored and Non-vectored interrupt.

      Method III :   The interrupts are classified into Maskable and Non-maskable interrupts.

Q1.95 How does a microprocessor service an interrupt request ?

When the processor recognizes an interrupt, it saves the processor status in stack. Then it calls and executes an Interrupt Service Routine (ISR). At the end of ISR, it restores the processor status and the program control is transferred to the main program.

Q1.96 What is the function of interrupt service routine?

For each interrupt the processor has to perform a specific job. An interrupt service routine has been developed in order to perform the operations required for a device that is interrupting the processor.

Q1.97 How are interrupts affected by system reset?

Whenever the processor or system is reset, all the interrupts except TRAP are disabled. In order to enable the interrupts, EI instruction has to be executed after a reset.

Q1.98 What are Software interrupts?

Software interrupts are program instructions. These instructions are inserted at desired locations in a program. While running a program, if a software interrupt instruction is encountered then the processor executes an interrupt service routine.

Q1.99 What is Hardware interrupt?

If an interrupt is initiated in a processor by applying an appropriate signal to an interrupt pin, then the interrupt is called Hardware interrupt.

Q1.100 What is the difference between Software and hardware interrupts?

Software interrupt is initiated by the main program, but a hardware interrupt is initiated by an external device.

In 8085, the software interrupt cannot be disabled or masked but the hardware interrupt except TRAP can be disabled or masked.

Q1.101 What are vectored and non-vectored interrupt?

When an interrupt is accepted, if the processor control branches to a specific address defined by the manufacturer, then the interrupt is called vectored interrupt.

In non-vectored interrupt, there is no specific address for storing the interrupt service routine. Hence, the interrupting device should give the address of the interrupt service routine.

Q1.102 List the software and hardware interrupts of 8085.

Software interrupts :   RST 0, RST1, RST 2, RST 3, RST 4, RST 5, RST 6 and RST 7.

Hardware interrupts :   TRAP, RST 7.5, RST 6.5, RST 5.5 and INTR.

Q1.103 **What is TRAP?**

TRAP is a non-maskable interrupt of 8085. It is not disabled by processor reset or after recognition of interrupt.

Q1.104 **Does HOLD has higher priority than TRAP or not?**

The interrupts including TRAP are recognized only if the HOLD is not valid, hence TRAP has lower priority than HOLD.

Q1.105 **What is masking and why it is required?**

Masking is preventing the interrupt from disturbing the current program execution. When the processor is performing an important job (process) and if the process should not be interrupted then all the interrupts should be masked or disabled.

In processor with multiple interrupts, the lower priority interrupt can be masked so as to prevent it from interrupting, the execution of interrupt service routine of higher priority interrupt.

Q1.106 **When does the 8085 processor accept a hardware interrupt?**

The processor keeps on checking the interrupt pins at the second T-state of the last machine cycle of every instruction. If the processor finds a valid interrupt signal and if the interrupt is unmasked and enabled then the processor accepts the interrupt. The acceptance of the interrupt is acknowledged by sending an $\overline{INTA}$ signal to the interrupting device.

Q1.107 **List the type of signals that have to be applied to initiate a hardware interrupt in 8085.**

The TRAP is level and edge-sensitive and so the interrupt signal has to take a **low** to **high** transition and then remain **high** until it is recognized. The RST 7.5 is edge-sensitive and so the interrupt signal has to take a **low** to **high** transition and need not remain **high** until it is recognized. The RST 6.5, RST 5.5 and INTR are level-sensitive and so the interrupt signal should be **high** until the interrupt is recognized.

Q1.108 **What are maskable and non-maskable interrupts of 8085?**

The TRAP is non-maskable interrupt. The RST 7.5, RST 6.5 and RST 5.5 are maskable interrupts. The INTR of 8085 can also be disabled by DI instruction.

Q1.109 **When will the 8085 processor disable the interrupt system ?**

The interrupts of 8085 except TRAP are disabled after any one of the following operations.

- **Executing EI instruction.**
- **System or processor reset.**
- **After recognition (acceptance) of an interrupt.**

Q1.110 **What is the function performed by DI instruction?**

The function of DI instruction is to disable the entire interrupt system.

Q1.111 **What is the function performed by EI instruction?**

The EI instruction can be used to enable all the interrupts after disabling.

Q1.112   How can the interrupt INTR of 8085 be expanded?

The interrupt INTR of 8085 can be expanded upto eight interrupts using 8-to-3 priority encoder. It can also be expanded to eight interrupts using one number of 8259 (Programmable interrupt controller) or upto 64 interrupts using 8259's in cascaded mode.

Q1.113   How can we check whether an 8085 interrupt is masked or not?

The masking status of an 8085 interrupt can be obtained by executing RIM instruction. When RIM instruction is executed, a 8-bit data is loaded in the accumulator. The bits B_0, B_1 and B_2 will give the masking status of RST 5.5, RST 6.5 and RST 7.5 respectively. If this bit is 1, then the corresponding interrupt is masked, otherwise it is unmasked.

Q1.114   How can we check the interrupt request pending status of 8085 interrupt?

The pending status of an 8085 interrupt can be obtained by executing RIM instruction. When the RIM instruction is executed an 8-bit data is loaded in accumulator. The bits B_4, B_5 and B_6 will give the pending status of RST 5.5, RST 6.5 and RST 7.5 respectively. If this bit is 1, then the interrupt is pending, otherwise it is not pending.

Q1.115   What is vectoring?

Vectoring is the process of generating the address of interrupt service routine to be loaded in program counter.

Q1.116   How are vector addresses generated for hardware interrupts of 8085?

For the hardware interrupts TRAP, RST 7.5, RST 6.5 and RST 5.5 the vector addresses are generated by the processor itself. These addresses are fixed by the manufacturer.

Q1.117   How are vector addresses generated for software interrupts of 8085?

For the software interrupts RST 0 to RST 7, the vector addresses are generated internal to the processor. These vector addresses are fixed by the manufacturer.

Q1.118   How can the hardware interrupt of 8085 be masked or unmasked?

The masking or unmasking of RST 7.5, RST 6.5 and RST 5.5 interrupts can be performed by moving an 8-bit data to accumulator and then executing SIM instruction. The format of the 8-bit data is shown in Fig. Q1.118.

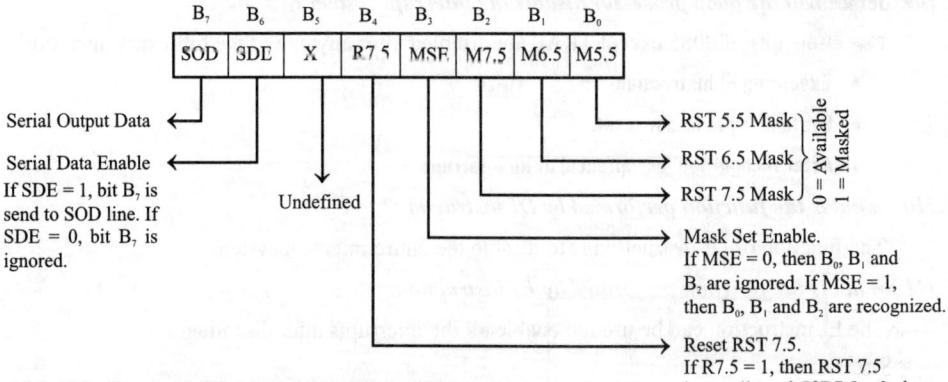

Fig. Q1.118: Format of 8-bit data to be loaded in the accumulator before executing a SIM instruction.

Q1.119 What is synchronous data transfer scheme?

In synchronous data transfer scheme, the processor does not check the readiness of the device after a command has been issued for read/write operation. In this scheme the processor will request the device to get ready and then read/write to the device immediately after the request. In some synchronous schemes a small delay is allowed after the request.

Q1.120 How can the status of maskable interrupts be read in 8085 processor?

The status of hardware interrupts like interrupt request pending or not, interrupts enabled or not, and masked or unmasked can read from accumulator after executing RIM instruction. When RIM instruction is executed an 8-bit data is loaded in accumulator which can be interpreted as shown in Fig. Q1.120.

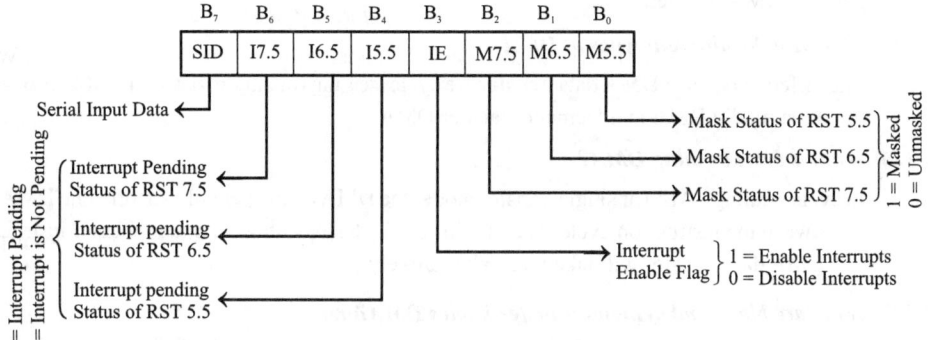

Fig. Q1.120: Interpretation of the content of the accumulator
after executing a RIM instruction.

Q1.121 How is a vector address generated for the INTR interrupt of 8085?

For the INTR interrupt, the interrupting device has to place either RST opcode or CALL opcode followed by a 16-bit address. If RST opcode is placed, then the corresponding vector address is generated by the processor. In case of CALL opcode the given 16-bit address will be the vector address.

Q1.122 What is data transfer scheme and what are its types?

The data transfer scheme refers to the method of data transfer between the processor and peripheral devices.

The different types of data transfer schemes are shown below :

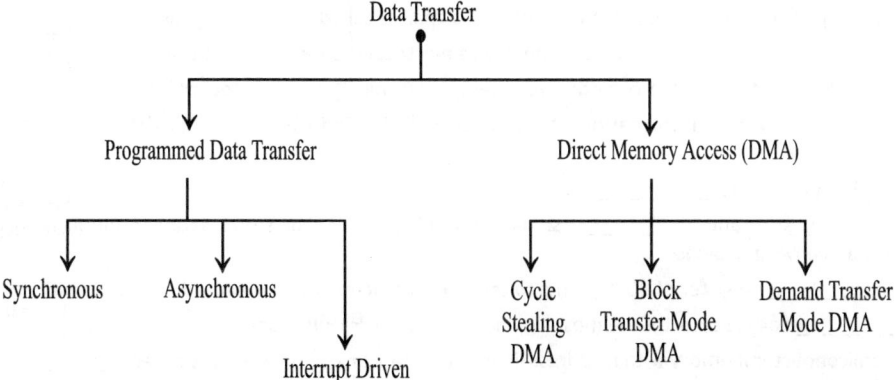

Q1.123 What is asynchronous data transfer scheme?

In asynchronous data transfer scheme, first the processor sends a request to the device for read/write operation. Then the processor keeps on polling the status of the device. Once the device is ready, the processor executes a data transfer instruction to complete the process.

Q1.124 How is DMA initiated?

When the IO device needs a DMA transfer, it will send a DMA request signal to the DMA controller. The DMA controller inturn sends a HOLD request to the processor. When the processor receives a HOLD request, it will drive its tristate pins to **high impedance** state at the end of current instruction execution and sends an acknowledge signal to the DMA controller. Now the DMA controller will perform DMA transfer.

Q1.125 What are the different types of DMA?

The different types of DMA data transfer are cycle stealing (or single transfer) DMA, Block transfer (or Burst mode) DMA and Demand transfer DMA.

Q1.126 What is cycle stealing DMA?

In cycle stealing DMA (or single transfer mode) the DMA controller will perform one DMA transfer in between the instruction cycles (i.e., In this mode the execution of one processor instruction and one DMA data transfer will take place alternatively).

Q1.127 What are block and demand transfer modes DMA?

In block transfer mode, the DMA controller will transfer a block of data and relieve the bus to the processor. After sometime another block of data is transferred by the DMA and so on.

In demand transfer mode the DMA controller will complete the entire data transfer at a stretch and then relieve the bus to the processor.   *(AU, Nov/Dec' 19, 2 Marks)*

Q1.128 List two major difference between INTR and the other hardware interrupts.

1. INTR is non-vectored interrupt but all other hardware interrupts are vectored interrupt.

2. INTR is expandable using 8-to-3 priority encoder or using INTEL 8259 programmable interrupt controller, but other hardware interrupts can not be expanded.

1.8 EXERCISES

I. Fill in the blanks with appropriate words

1. A digit of the binary number or code is callled _____ .

2. The group of conducting lines that carry control signals is called _____ bus.

3. The _____ state is used to keep the device electrically isolated from the system.

4. The third generation microprocessors were designed using _____ technology.

5. Transfering different information at different well defined times through the same lines is called _____ .

6. 1 mil is equivalent to _____ inch.

7. The _____ and _____ signals of 8085 are used for serial data communication between 8085 and any serial device.

8. The _____ register of 8085 points to the next instruction to be executed.

9. _____ flag is set to 1 if the most significant bit of the result is one.

10. A semiconductor memory with n address lines and m data lines has a memory capacity of _____ bits.

11. The memory which retains data even after the power supply to the chip is switched OFF is called _____ memory.

12. RAM is a _____ memory.

13. The ROM which can be erased by passing electrical current or UV light is called _____.

14. The memory which requires frequent refreshing is called _____.

15. The 1024 × 4 bit memory chip has _____ address lines and _____ data lines.

16. The 2^{11} × 8 bit memory chip has _____ number of memory locations and each location can store _____ number of bits.

17. The number of address lines needed to address a location in 8 kB memory is _____.

18. The _____ decoder is used to generate four chip select signals.

19. If the starting address of 4 kB RAM chip is mapped to 2000_H the end address will be _____.

20. The address range of 32 kB EPROM chip interfaced to 8085 system is from _____ to _____ if $\overline{CS} = \overline{A}_{15}$.

21. The number of 256 × 4 bits RAM chips required to make up 1 kB memory is _____.

22. The _____ is the data transfer scheme which enables direct data transfer between memory and IO devices.

23. Only _____ and _____ instructions can be used for data transfer between the IO devices and the processor in IO mapped devices.

24. The process of interrupting the normal program execution to carry out a specific task is referred to as _____.

25. The interrupts initiated by external hardware by sending an appropriate signal to the interrupt pin of the processor are called _____ interrupts.

26. The interrupt for which interrupt service subroutine address is predefined is called _____ interrupt.

27. The interrupt which cannot be rejected by the processor is called _____ interrupt.

28. The vector address of RST 5.5 is _____.

29. The non-maskable interrupt of 8085 is _____.

30. The second highest priority interrupt of 8085 is _____.

31. The _____ interrupt of 8085 is used during power failure.

Answers

1. bit	9. Sign	17. 13	25. hardware
2. control	10. 2^n × m	18. 2 to 4	26. vectored
3. high impedance	11. non-volatile	19. $2FFF_H$	27. non-maskable
4. High Density MOS(HMOS)	12. volatile	20. 0000_H, $7FFF_H$	28. $002C_H$
5. multiplexing	13. EPROM	21. 8	29. TRAP
6. 10^{-3}	14. dynamic RAM	22. direct memory access (DMA)	30. RST 7.5
7. SID, SOD	15. 10, 4	23. IN, OUT	31. TRAP
8. Program Counter(PC)	16. 2048, 8	24. interrupt	

II. State whether the following statements are True/False.

1.   The address bus is bidirectional.

2.   The CPU bus is directly connected to the microprocessor.

3.   The high impedance state is an electrical open-circuit condition .

4.   The NMOS technology offers faster speed and higher density than HMOS technology.

5.   Registers can be read/written faster than memory chips.

6.   The 8085 has an internal clock oscillator.

7.   The 8085 interrupts RST 7.5, RST 6.5 and RST 5.5 are non-maskable interrupts.

8.   To reset the 8085 microprocessor the $\overline{\text{RESET IN}}$ pin should be held low for atleast three clock pulses.

9.   The stack pointer (SP) holds the address of the stack top.

10. The registers of microprocessors can be accessed faster than the main memory.

11. The primary memory is slower than secondary memory.

12. The volatile memory loses its data when the power supply to the chip is switched OFF.

13. The EPROM is a non-volatile memory.

14. All the semiconductor memories are non-destructive readout memory.

15. PROMs are many time programmable by the user.

16. Static RAM can pack more bits than dynamic RAM in a given physical area.

17. The dynamic RAM (DRAM) is faster than static RAM and consumes less power.

18. The flash memory is a type of non-volatile RAM.

19. The unused address lines of microprocessor are generally used for generating chip select signals.

20. All the memory/peripheral ICs always remain in active state.

21. IO devices can be directly connected to microprocessors.

22. In memory mapped I/0 technique, both memory and IO devices are treated in a same way.

23. In I/O mapped ports, the data can be moved from any registers to the ports and vice versa.

24. A separate decoder is required to generate the chip select signals for IO mapped devices.

25. In polling technique, the processor has to check the readiness of the device periodically for data transfer.

26. In interrupt driven data transfer scheme, the external device interrupts the processor only when the data is ready.

27. The Interrupt Service Subroutine (ISR) address is predefined by the processor manufacture for non vectored interrupts.

28. The processor should compulsorily accept all interrupts any time.

29. The software interrupts cannot be masked.

30. The vector address of 8085 software interrupt RSTn is equivalent to n×8.

31. All the hardware interrupts are disabled when the 8085 processor is reset.

32. INTR interrupt of 8085 is the highest priority interrupt.

Answers

1. False	7. False	13. True	19. True	25. True	31. False
2. True	8. True	14. True	20. False	26. True	32. False
3. True	9. True	15. False	21. False	27. False	
4. False	10. True	16. False	22. True	28. False	
5. True	11. False	17. True	23. False	29. True	
6. True	12. True	18. True	24. True	30. True	

III. Choose the right answer for the following questions.

1. *Microprocessors are intended to be a _____ computer.*

    a) general-purpose   b) special purpose   c) hybrid   d) analog

2. *Group of 4-bits is called*

    a) byte   b) nibble   c) word   d) double word

3. *What would be the total memory capacity of a microprocessor with 10 address lines*

    a) 1 MB   b) 1 GB   c) 1 KB   d) 512 MB

4. *The first 8-bit processor introduced by INTEL is*

    a) 8080   b) 8008   c) 8085   d) 8086

5. *Which of the following is used to store temporary programs and data?*

    a) EPROM   b) ROM   c) RAM   d) all the three

6. *The 1-bit register provided to store the results of certain program instructions is*

    a) status register   b) instruction register   c) program counter   d) flag

7. *Which of the following is not a 16-bit processor?*

    a) 8086   b) 80186   c) 8088   d) 8096

8. *Which of the following signals is used to demultiplex the address/data lines in 8085 and 8086 processors?*

    a) DT/$\overline{R}$ b) ALE   c) SOD   d) READY

9. *The maximum internal clock frequency of 8085A microprocessor is*

    a) 5.03 MHz   b) 6 MHz   c) 10.6 MHz   d) 3.03 MHz

10. *The total memory capacity of 8085 processor is*

    a) 64 KB   b) 1 MB   c) 64 MB   d) 10 MB

11. *Which of the following 8085 signals are used for DMA operation?*

    a) $\overline{RD}, \overline{WR}$ b) $\overline{RESET\ IN}, \overline{RESET\ OUT}$ c) HOLD, HLDA   d) SOD, SID

12. *When initiated, certain interrupts can be delayed or rejected but when allowed, the program execution starts from a fixed location. Such an interrupt is known as,*

    a) non maskable and non vectored   b) maskable and non vectored
    c) non maskable and vectored   d) maskable and vectored

13. *The vector address for the 8085 interrupt TRAP is,*

    a) 0028_H b) 0020_H c) 0024_H d) 0000_H

14. *Which of the following is volatile memory?*

    a) RAM   b) ROM   c) PROM   d) EPROM

15. What is the range of address (in hexadecimal) for 16 KB memory?

   a) 0000_H to $FFFF_H$ b) 0000_H to $03FF_H$ c) 0000_H to $3FFF_H$ d) 0000 to $7FFF_H$

16. In 8085 processor, how many 1024×4 bits memory chips are required for a system memory of 4 kB?

   a) 4   b) 8   c) 12   d) 16

17. The starting address of a 32 kB memory chip in 8085 processor is 0000H. What is the end address?

   a) $FFFF_H$ b) $7FFF_H$ c) $3FFF_H$ d) $1FFF_H$

18. A memory chip has 11 address lines and 4 data lines. How many such chips are required to design a memory of 4 kB?

   a) 4   b) 8   c) 16   d) 32

19. In 8085 microprocessor, a 4 kB EPROM and 4 kB RAM are selected when $A_{15} = A_{14} = A_{13} = A_{12} = 0$ and $A_{15} = A_{14} = A_{13} = A_{12} = 1$ respectively. What is the unused address space?

   a) 1000_H to $EFFF_H$ b) 2000_H to $DFFF_H$ c) 4000_H to $FFFF_H$ d) 6000_H to $CFFF_H$

20. An 8 kB EPROM and 8 kB RAM are interfaced to 8085 microprocessor. Which of the following can be possible correct address range for this memory devices,

   i) EPROM = 0000_H to $1FFF_H$ ii) EPROM = 0000_H to $2FFF_H$

       RAM = 2000_H to $3FFF_H$ RAM = 3000_H to $3FFF_H$

   iii) EPROM = 0000_H to $0FFF_H$ iv) EPROM = 0000_H to $1FFF_H$

       RAM = 1000_H to $2FFF_H$ RAM = $D000_H$ to $FFFF_H$

   a) i   b) iii   c) i and iv   d) ii and iv

21. A 16 kB RAM in an 8085 system has address range 4000_H to $7FFF_{H'}$ If the unused address lines are used to generate the chip select signals the $\overline{CS}$ = ?

   a) $\overline{CS} = \overline{A}_{15}\overline{A}_{14}$ b) $\overline{CS} = \overline{A}_{15}A_{14}$ c) $\overline{CS} = A_{15}$ d) $\overline{CS} = A_{14}$

22. The interrupts whose request can be either accepted or rejected by the processor are called as

   a) vectored interrupts   b) Non-vectored interrupts
   c) maskable interrupts   d) non-maskable interrupts

23. Which of the following is not a hardware interrupt of 8085?

   a) TRAP   b) RST 7.5   c) INTR   d) RST n

24. The vector address of the 8085 software interrupt RST 3 is

   a) 0008_H b) $000C_H$ c) 0010_H d) 0018_H

25. Which of the following signal on 8085 TRAP line initiates the interrupt?

   a)   b)   c)   d) none

26. Which of the following 8085 interrupt is not a level sensitive interrupt?

   a) RST 7.5   b) RST 6.5   c) RST 5.5   d) INTR

27. **The 8085 receives both DMA request and an interrupt on TRAP pin. Which of the following statement is true?**

   a) It recognizes TRAP first and then DMA request

   b) It recognizes DMA request first and then TRAP

   c) It recognizes both TRAP and DMA request simultaneously.

   d) It rejects both requests and execute the next instruction in the main program.

28. **In the 8085 microprocessor, the RST 5 instruction transfers the program control to the following location.**

   a) 0010_H b) 0018_H c) 0020_H d) 0028_H

Answers

1. a	8. b	15. c	22. c
2. b	9. d	16. b	23. d
3. c	10. a	17. b	24. d
4. b	11. c	18. a	25. c
5. c	12. d	19. a	26. a
6. d	13. c	20. c	27. b
7. d	14. a	21. c	28. d

IV. Answer the following questions.

E1.1 What is meant by addressability of a microprocessor?

E1.2 State the significance of clock pulse in microprocessor based system.

E1.3 List some of the 4-bit microprocessors with specifications.

E1.4 Define the term speed power product(SPP).

E1.5 Mention the packing density of NMOS and HMOS technology

E1.6 Define the term MIPS.

E1.7 Draw the basic functional blocks of a microprocessor.

E1.8 What is meant by active low signal? Explain with an example.

E1.9 Write down the significance of bus status signals $IO/\overline{M}$, S_0 and S_1.

E1.10 What is the use of signals SOD and SID in 8085?

E1.11 Define maskable and non-maskable interrupts.

E1.12 What is meant by hardware interrupt?

E1.13 What is meant by software interrupt?

E1.14 What is meant by vectored and non-vectored interrupts?

E1.15 Write down the vector address of all the interrupts of 8085.

E1.16 When power on, how does the CPU know the starting address of the first instruction it has to execute? What is that first instruction? why?

E1.17 What is the use of stack pointer in 8085 microprocessor?

E1.18 Why the program counter and stack pointer of 8085 are 16-bit registers?

E1.19 A semiconductor memory has 16 address lines and 8 data lines. What is the capacity of the memory?

E1.20 Show the memory interfacing of 32 kB RAM and 32 kB ROM with 8085 microprocessor.

E1.21 Write an 8085 ALP to enable all the interrupts of 8085.

E1.22 The 8085 processor is executing the following ISR of RST 6.5. Rearrange the program to allow the higher priority interrupt RST 7.5 to interrupt the processor while execution of RST 6.5 ISR.

```
;ISR of RST 6.5
        MVI  A,00H
Back:   INR  A
        CPI  A,0FFH
        JNZ  BACK
        RET
```

CHAPTER 2

PROGRAMMING OF 8085 PROCESSOR

2.1 INSTRUCTION FORMAT OF 8085

The 8085 has 74 basic instructions and 246 total instructions. The instruction set of 8085 is defined by the manufacturer INTEL Corporation. Each instruction of 8085 has one-byte opcode. With 8-bit binary code, we can generate 256 different binary codes. In this, 246 codes have been used for opcodes of 8085 instructions.

The size of 8085 instruction can be one-byte, two-byte or three-byte. The one-byte instruction has an opcode alone and the two-byte instruction has an opcode followed by an eight bit address or data. The 3-byte instruction has an opcode followed by 16-bit address or data. While storing the 3-byte instruction in memory, the sequence of storage is, opcode first followed by low byte of address or data and then high byte of address or data. The data or address specified in the instruction is also known as operand. The format of 8085 instructions are shown in Fig. 2.1.

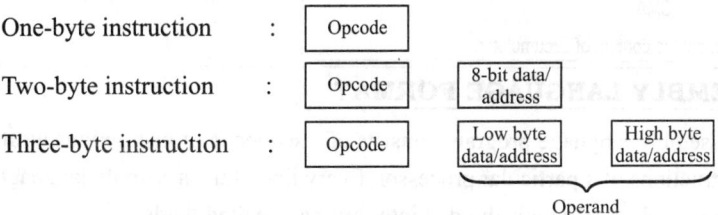

Fig 2.1: Format of 8085 instructions.

2.2 ADDRESSING MODES

(AU, Nov/Dec' 19, 10 Marks)

Every instruction of a program has to operate on a data. The method of specifying the data to be operated by the instruction is called Addressing. The 8085 supports the following five addressing modes:

1. **Immediate Addressing**
2. **Direct Addressing**
3. **Register Addressing**
4. **Register Indirect Addressing**
5. **Implied Addressing**

Immediate Addressing

In immediate addressing mode, the data is specified in the instruction itself. The data will be a part of the program instruction.

Example : MVI B, 3E$_H$

Move the data 3E$_H$ given in the instruction to B-register.

Direct Addressing

In direct addressing mode, the address of the data is specified in the instruction. The data will be in memory. In this addressing mode, the program instructions and data can be stored in different memory blocks.

Example :   LDA 1050$_H$

Load the data available in memory location 1050$_H$ in accumulator.

Register Addressing

In register addressing mode, the instruction specifies the name of the register in which the data is available.

Example :   MOV A, B

Move the content of B-register to A-register.

Register Indirect Addressing

In register indirect addressing mode, the instruction specifies the name of the register in which the address of the data is available. Here the data will be in memory and the address will be in a register pair.

Example :   MOV A, M

The memory data addressed by HL pair is moved to A-register.

Implied Addressing

In implied addressing mode, the instruction itself specifies the data to be operated.

Example :   CMA

Complement the content of accumulator.

2.3 ASSEMBLY LANGUAGE FORMAT

The assembly language program consists of a sequence of instructions written using assembly language instructions of a particular processor. Every line of the assembly language program is called a statement and each statement is divided into four parts called fields.

The four fields of a statement are label, operation code or opcode, operand and comment.

Label field is not compulsory for all the statements and necessary only when the instruction is referred for jump and call operations. Every statement should have an opcode and operand that specify the operation performed by the instruction or statement. The comment in the statement is optional and written for human or user understanding.

The four fields are separated by delimiters which specify the end or boundary of the field. Also, any number of blank spaces can be given in between fields which is also known as free-field format.

The commonly used delimiters are colon, space, comma and semicolon. The usage of these delimiters are as follows:

- Colon is used at the end of label field
- Space is used between opcode and operand
- Comma is used between two operands
- Semicolon is used at the start of a comment.

A simple example program (Example Program -10) with parts of statements placed under various fields is given below.

Label	Opcode	Operand	Comment
START:	LDA	4200H	;Get 1st data in A and save in B.
	MOV	B,A	
	LDA	4201H	;Get 2nd data in A-register.
	MVI	C,00H	;Clear C-register to account for carry.
	ADD	B	;Get the sum in A-register.
	JNC	AHEAD	;If CF=0, go to AHEAD.
	INR	C	;If CF=1, increment C-register.
AHEAD:	STA	4202H	;Store the sum in memory.
	MOV	A,C	
	STA	4203H	;Store the carry in memory.
STOP:	HLT		;Halt program execution.

2.4 INSTRUCTION SET

The 8085 instructions can be broadly classified into the following two functional groups.

1. Data Transfer Instructions

2. Data Manipulation and Control Instructions.

The data manipulation and control instructions can be classified as follows:

(a) Arithmetic Instructions.

(b) Logical Instructions.

(c) Branching Instructions.

(d) Machine Control Instructions.

Data Transfer Instructions : Includes the instructions that moves (copies) data between registers or between memory location and register. In all data transfer operations, the content of source register or memory is not altered. Hence the data transfer is copying operation.

Arithmetic Instructions : Includes the instructions which performs addition, subtraction, increment or decrement operations. The flag conditions are altered after execution of an instruction in this group.

Logical Instructions : The instructions which performs the logical operations like AND, OR, EXCLUSIVE-OR, complement, compare and rotate instructions are grouped under this heading. The flag conditions are altered after execution of an instruction in this group.

Branching Instructions : The instructions that are used to transfer the program control from one memory location to another memory location are grouped under this heading.

Machine Control instructions : Includes the instructions related to interrupts and the instruction used to halt program execution.

The 74 basic instructions of 8085 are listed in Table 2.1. The opcode of each instruction, size, machine cycles, number of T-state and the total number of instructions in each type are also shown in Table 2.1. The instructions affecting the status flag are listed in Table 2.2.

Table 2.1: Summary of 8085 Instruction Set

S.No.	Mnemonic	Opcode	Number of bytes	Machine cycles	Number of T-states	Total number of instructions
Data transfer instructions						
1.	MOV Rd, Rs	0 1 D D D S S S	1	F	4T	49
2.	MOV Rd, M	0 1 D D D 1 1 0	1	F, R	7T	7
3.	MOV M, Rs	0 1 1 1 0 S S S	1	F, W	7T	7
4.	MVI Rd, d8	0 0 D D D 1 1 0	2	F, R	7T	7
5.	MVI M, d8	0 0 1 1 0 1 1 0	2	F, R, W	10T	1
6.	LDA addr16	0 0 1 1 1 0 1 0	3	F, R, R, R	13T	1
7.	LDAX rp	0 0 R P 1 0 1 0	1	F, R	7T	2
8.	LXI rp, d16	0 0 R P 0 0 0 1	3	F, R, R	10T	4
9.	LHLD addr16	0 0 1 0 1 0 1 0	3	F,R,R,R,R	16T	1
10.	STA addr16	0 0 1 1 0 0 1 0	3	F, R, R,W	13T	1
11.	STAX rp	0 0 R P 0 0 1 0	1	F, W	7T	2
12.	SHLD addr16	0 0 1 0 0 0 1 0	3	F,R,R,W,W	16T	1
13.	SPHL	1 1 1 1 1 0 0 1	1	S	6T	1
14.	XCHG	1 1 1 0 1 0 1 1	1	F	4T	1
15.	XTHL	1 1 1 0 0 0 1 1	1	F,R,R,W,W	16 T	1
16.	PUSH rp	1 1 R P 0 1 0 1	1	S, W, W	12T	3
17.	PUSH PSW	1 1 1 1 0 1 0 1	1	S, W, W	12T	1
18.	POP rp	1 1 R P 0 0 0 1	1	F, R, R	10T	3
19.	POP PSW	1 1 1 1 0 0 0 1	1	F, R, R	10T	1
20.	IN addr8	1 1 0 1 1 0 1 1	2	F, R, I	10T	1
21.	OUT addr8	1 1 0 1 0 0 1 1	2	F, R, O	10T	1
Arithmetic instructions						
22.	ADD reg	1 0 0 0 0 S S S	1	F	4T	7
23.	ADD M	1 0 0 0 0 1 1 0	1	F, R	7T	1

Table 2.1 continued...

S.No.	Mnemonic	Opcode	Number of bytes	Machine cycles	Number of T-states	Total number of instructions
24.	ADI d8	1 1 0 0 0 1 1 0	2	F, R	7T	1
25.	ADC reg	1 0 0 0 1 S S S	1	F	4T	7
26.	ADC M	1 0 0 0 1 1 1 0	1	F, R	7T	1
27.	ACI d8	1 1 0 0 1 1 1 0	2	F, R	7T	1
28.	DAA	0 0 1 0 0 1 1 1	1	F	4T	1
29.	DAD rp	0 0 R P 1 0 0 1	1	F, B, B	10T	4
30.	SUB reg	1 0 0 1 0 S S S	1	F	4T	7
31.	SUB M	1 0 0 1 0 1 1 0	1	F, R	7T	1
32.	SUI d8	1 1 0 1 0 1 1 0	2	F, R	7T	1
33.	SBB reg	1 0 0 1 1 S S S	1	F	4T	7
34.	SBB M	1 0 0 1 1 1 1 0	1	F, R	7T	1
35.	SBI d8	1 1 0 1 1 1 1 0	2	F, R	7T	1
36.	INR reg	0 0 S S S 1 0 0	1	F	4T	7
37.	INR M	0 0 1 1 0 1 0 0	1	F, R, W	10T	1
38.	INX rp	0 0 R P 0 0 1 1	1	S	6T	4
39.	DCR reg	0 0 S S S 1 0 1	1	F	4 T	7
40.	DCR M	0 0 1 1 0 1 0 1	1	F, R, W	10T	1
41.	DCX rp	0 0 R P 1 0 1 1	1	S	6T	4
Logical instructions						
42.	ANA reg	1 0 1 0 0 S S S	1	F	4T	7
43.	ANA M	1 0 1 0 0 1 1 0	1	F, R	7T	1
44.	ANI d8	1 1 1 0 0 1 1 0	2	F, R	7T	1
45.	ORA reg	1 0 1 1 0 S S S	1	F	4T	7
46.	ORA M	1 0 1 1 0 1 1 0	1	F, R	7T	1
47.	ORI d8	1 1 1 1 0 1 1 0	2	F, R	7T	1

Table 2.1 continued...

S.No.	Mnemonic	Opcode	Number of bytes	Machine cycles	Number of T-states	Total number of instructions
48.	XRA reg	1 0 1 0 1 S S S	1	F	4T	7
49.	XRA M	1 0 1 0 1 1 1 0	1	F,R	7T	1
50.	XRI d8	1 1 1 0 1 1 1 0	2	F, R	7T	1
51.	CMP reg	1 0 1 1 1 S S S	1	F	4T	7
52.	CMP M	1 0 1 1 1 1 1 0	1	F, R	7T	1
53.	CPI d8	1 1 1 1 1 1 1 0	2	F, R	7T	1
54.	CMA	0 0 1 0 1 1 1 1	1	F	4T	1
55.	CMC	0 0 1 1 1 1 1 1	1	F	4T	1
56.	STC	0 0 1 1 0 1 1 1	1	F	4T	1
57.	RLC	0 0 0 0 0 1 1 1	1	F	4T	1
58.	RAL	0 0 0 1 0 1 1 1	1	F	4T	1
59.	RRC	0 0 0 0 1 1 1 1	1	F	4T	1
60.	RAR	0 0 0 1 1 1 1 1	1	F	4T	1
Branching instructions						
61.	JMP addr16	1 1 0 0 0 0 1 1	3	F,R,R	10T	1
62.	J<condition> addr16	1 1 C C C 0 1 0	3	F,R/F,R,R	7T/10T	8
63.	CALL addr16	1 1 0 0 1 1 0 1	3	S,R,R,W,W	18T	1
64.	C<condition> addr16	1 1 C C C 1 0 0	3	S, R or S,R,R,W,W	9T/18T	8
65.	RET	1 1 0 0 1 0 0 1	1	F,R,R	10T	1
66.	R<condition>	1 1 C C C 0 0 0	1	S/S,R,R	6T/12T	8
67.	RST n	1 1 N N N 1 1 1	1	S,W,W	12T	8
68.	PCHL	1 1 1 0 1 0 0 1	1	S	6T	1

Table 2.1 continued...

S.No.	Mnemonic	Opcode	Number of bytes	Machine cycles	Number of T-states	Total number of instruction

Machine control instructions

S.No.	Mnemonic	Opcode	Number of bytes	Machine cycles	Number of T-states	Total number of instruction
69.	SIM	0 0 1 1 0 0 0 0	1	F	4T	1
70.	RIM	0 0 1 0 0 0 0 0	1	F	4T	1
71.	DI	1 1 1 1 0 0 1 1	1	F	4T	1
72.	EI	1 1 1 1 1 0 1 1	1	F	4T	1
73.	HLT	0 1 1 1 0 1 1 0	1	F,B	5T	1
74.	NOP	0 0 0 0 0 0 0 0	1	F	4T	1
						246

Meanings of various symbols used in Table 2.1

Symbol	Meaning
rp, RP	Register pair
Rs, SSS	Source register
Rd, DDD	Destination register
M	Memory
d8	8-bit data
d16	16-bit data
addr8	8-bit address
addr16	16-bit address
reg	Register
PSW	Program status word
n, NNN	Type number of restart instruction
<condition>, CCC	Flag condition
F	4T-Opcode fetch cycle
S	6T-Opcode fetch cycle
R	Memory read cycle
W	Memory write cycle
I	IO read cycle
O	IO write cycle
B	Bus idle cycle

Flag condition can be any one of the conditions given below

Z	→	Zero flag = 1	M	→	Sign flag = 1
NZ	→	Zero flag = 0	P	→	Sign flag = 0
C	→	Carry flag = 1	PE	→	Parity flag = 1
NC	→	Carry flag = 0	PO	→	Parity flag = 0

The binary codes for the symbols used in opcode of 8085 instructions are given below:

Register	DDD or SSS
B	0 0 0
C	0 0 1
D	0 1 0
E	0 1 1
H	1 0 0
L	1 0 1
A	1 1 1

Register	RP
BC	0 0
DE	0 1
HL	1 0
SP	1 1

Flag condition	C C C
NZ	0 0 0
Z	0 0 1
NC	0 1 0
C	0 1 1
PO	1 0 0
PE	1 0 1
P	1 1 0
M	1 1 1

n	N N N
0	0 0 0
1	0 0 1
2	0 1 0
3	0 1 1
4	1 0 0
5	1 0 1
6	1 1 0
7	1 1 1

Table 2.2: 8085 Instructions Affecting the Status Flags

Instructions	Status flags				
	CF	AF	ZF	SF	PF
ACI d8	+	+	+	+	+
ADC reg	+	+	+	+	+
ADC M	+	+	+	+	+
ADD reg	+	+	+	+	+
ADD M	+	+	+	+	+
ADI d8	+	+	+	+	+
ANA reg	0	1	+	+	+
ANA M	0	1	+	+	+
ANI d8	0	1	+	+	+
CMC	+				
CMP reg	+	+	+	+	+
CMP M	+	+	+	+	+
CPI d8	+	+	+	+	+
DAA	+	+	+	+	+
DAD rp	+				
DCR reg		+	+	+	+
DCR M		+	+	+	+
INR reg		+	+	+	+
INR M		+	+	+	+
ORA reg	0	0	+	+	+
ORA M	0	0	+	+	+

Table - 2.2 continued...

Instructions	Status flags				
	CF	AF	ZF	SF	PF
ORI d8	0	0	+	+	+
RAL	+				
RAR	+				
RLC	+				
RRC	+				
SBB reg	+	+	+	+	+
SBB M	+	+	+	+	+
SBI d8	+	+	+	+	+
STC	+				
SUB reg	+	+	+	+	+
SUB M	+	+	+	+	+
SUI d8	+	+	+	+	+
XRA reg	0	0	+	+	+
XRA M	0	0	+	+	+
XRI d8	0	0	+	+	+

> *Note :*
>
> $+ \rightarrow$ *Indicates that the particular flag is affected.*
> $0 \rightarrow$ *Indicates that the particular flag is always zero.*
> $1 \rightarrow$ *Indicates that the particular flag is always one.*

Table 2.3: Meaning/Expansion of Mnemonics used in an 8085 Instruction Set

S.No.	Mnemonic	Meaning
1.	ACI	Add the immediate data and the carry to the accumulator.
2.	ADC	Add the register/memory and the carry to the accumulator.
3.	ADD	Add the register/memory to the accumulator.
4.	ADI	Add the immediate data to the accumulator.
5.	ANA	AND register/memory with the accumulator.
6.	ANI	AND immediate data with the accumulator.
7.	CALL	Call a subroutine/procedure.
8.	CC	Call on carry.
9.	CM	Call on minus.
10.	CMA	Complement accumulator.
11.	CMC	Complement carry.
12.	CMP	Compare register/memory with accumulator.
13.	CNC	Call on no carry.
14.	CNZ	Call on not zero.

Table 2.3 continued...

S.No.	Mnemonic	Meaning
15.	CP	Call on positive.
16.	CPE	Call on parity even.
17.	CPI	Compare immediate data with the accumulator.
18.	CPO	Call on parity odd.
19.	CZ	Call on zero.
20.	DAA	Decimal adjust accumulator after addition.
21.	DAD	Double addition.
22.	DCR	Decrement the register/memory.
23.	DCX	Decrement the register pair.
24.	DI	Disable interrupt.
25.	EI	Enable interrupt.
26.	HLT	Halt program execution.
27.	IN	Input data from specified port to accumulator.
28.	INR	Increment the register/memory.
29.	INX	Increment the register pair.
30.	JC	Jump on carry.
31.	JM	Jump on minus.
32.	JMP	Jump to specified address to get the next instruction.
33.	JNC	Jump on no carry.
34.	JNZ	Jump on not zero.
35.	JP	Jump on positive.
36.	JPE	Jump on parity even.
37.	JPO	Jump on parity odd.
38.	JZ	Jump on zero.
39.	LDA	Load the accumulator.
40.	LDAX	Load accumulator indirectly using the address in the specified register pair.
41.	LHLD	Load HL direct.
42.	LXI	Load the immediate data in the register pair.
43.	MOV	Move (copy) the content of register/memory to another register/memory.
44.	MVI	Move the immediate data to register/memory.
45.	NOP	No operation.
46.	ORA	OR register/memory with accumulator.
47.	ORI	OR immediate data with accumulator.
48.	OUT	Output the content of accumulator to specified port.
49.	PCHL	Move the content of HL to PC.

Table 2.3 continued...

S.No.	Mnemonic	Meaning
50.	POP	Move the top of stack to the specified register pair.
51.	PUSH	Push the content of the specified register pair to top of stack.
52.	RAL	Rotate the accumulator left along with carry.
53.	RAR	Rotate the accumulator right along with carry.
54.	RC	Return on carry.
55.	RET	Return from subroutine/procedure to calling program.
56.	RIM	Read interrupt mask status.
57.	RLC	Rotate accumulator left to carry.
58.	RM	Return on minus.
59.	RNC	Return on no carry.
60.	RNZ	Return on not zero.
61.	RP	Return on positive.
62.	RPE	Return on parity even.
63.	RPO	Return on parity odd.
64.	RRC	Rotate accumulator right to carry.
65.	RST	Restart the program execution from the specified vector address.
66.	RZ	Return on zero.
67.	SBB	Subtract register/memory and the carry (borrow) from accumulator.
68.	SBI	Subtract the immediate data and the carry (borrow) from accumulator.
69.	SHLD	Store HL direct.
70.	SIM	Set interrupt mask.
71.	SPHL	Move HL to SP.
72.	STA	Store accumulator.
73.	STAX	Store accumulator indirectly by using the address in specified register pair.
74.	STC	Set carry.
75.	SUB	Subtract register/memory from accumulator.
76.	SUI	Subtract the immediate data from accumulator.
77.	XCHG	Exchange DE and HL.
78.	XRA	Exclusive-OR register/memory with accumulator.
79.	XRI	Exclusive-OR the immediate data with accumulator.
80.	XTHL	Exchange the top of stack and HL.

2.5   DATA TRANSFER INSTRUCTIONS

1.   **MOV Rd, Rs**   **(Rd) ← (Rs)**

The content of source register (Rs) is copied to the destination register (Rd). The registers Rd and Rs can be any one of the general purpose registers A, B, C, D, E, H or L. No flags are affected.

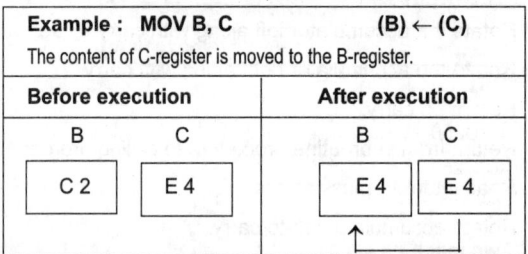

Example : MOV B, C	**(B) ← (C)**
The content of C-register is moved to the B-register.	

Before execution		**After execution**	
B	C	B	C
C 2	E 4	E 4	E 4

One-byte instruction   **One machine cycle :** Opcode fetch - 4T

Register addressing

Total number of instructions = 49

MOV A, A	MOV B, A	MOV D, A	MOV H, A
MOV A, B	MOV B, B	MOV D, B	MOV H, B
MOV A, C	MOV B, C	MOV D, C	MOV H, C
MOV A, D	MOV B, D	MOV D, D	MOV H, D
MOV A, E	MOV B, E	MOV D, E	MOV H, E
MOV A, H	MOV B, H	MOV D, H	MOV H, H
MOV A, L	MOV B, L	MOV D, L	MOV H, L
	MOV C, A	MOV E, A	MOV L, A
	MOV C, B	MOV E, B	MOV L, B
	MOV C, C	MOV E, C	MOV L, C
	MOV C, D	MOV E, D	MOV L, D
	MOV C, E	MOV E, E	MOV L, E
	MOV C, H	MOV E, H	MOV L, H
	MOV C, L	MOV E, L	MOV L, L

2.   **MOV Rd, M**   **(Rd) ← (M)   or   (Rd) ← ((HL))**

The content of memory (M) addressed by the HL pair is moved to the destination register (Rd). The register Rd can be any one of the general purpose registers A, B, C, D, E, H or L. No flags are affected.

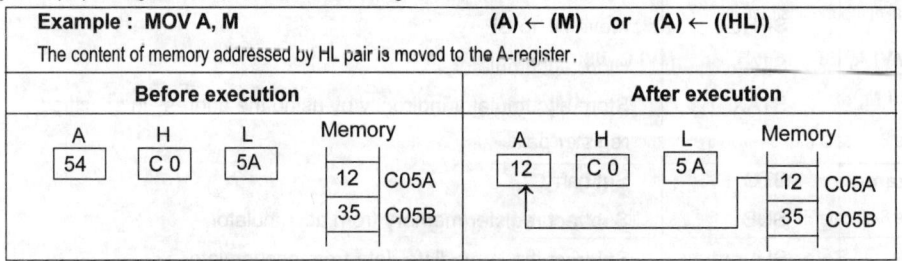

Example : MOV A, M		**(A) ← (M) or (A) ← ((HL))**	
The content of memory addressed by HL pair is movod to the A-register.			

Before execution				**After execution**			
A	H	L	Memory	A	H	L	Memory
54	C 0	5A	12 C05A	12	C 0	5 A	12 C05A
			35 C05B				35 C05B

One-byte instruction   **Two machine cycles:** Opcode fetch - 4T

Register indirect addressing   Memory read   - 3T

                                                                                                              7T

Total number of instructions = 7

MOV A, M	MOV B, M	MOV C, M	MOV D, M	MOV E, M	MOV H, M	MOV L, MOV

3.   **MOV M, Rs**   **(M) ← (Rs)**   or   **((HL)) ← (Rs)**

The content of source register (Rs) is moved to the memory location addressed by HL pair. The register Rs can be any one of the general purpose registers A, B, C, D, E, H or L. No flags are affected.

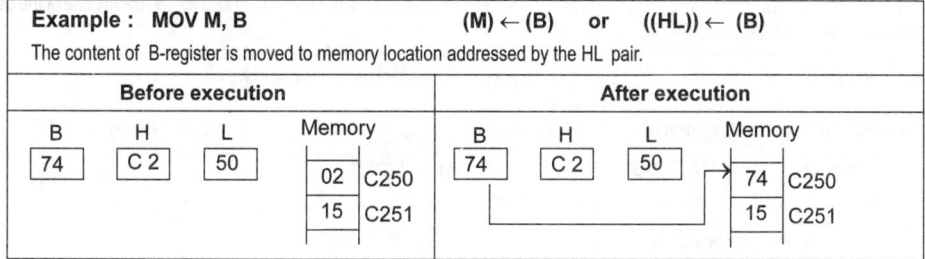

Example : MOV M, B   **(M) ← (B)**   or   **((HL)) ← (B)**

The content of B-register is moved to memory location addressed by the HL pair.

Before execution	After execution

One-byte instruction   **Two machine cycles :**   Opcode fetch - 4T

Register indirect addressing   Memory write - 3T

                                                           7 T

Total number of instructions = 7

     MOV M,A   **MOV M,B**   **MOV M, C**   **MOV M, D**   **MOV M, E**   **MOV M, H**   **MOV M, L**

4.   **MVI Rd, d8**   **(Rd) ← d8**

The 8-bit data (d8) given in the instruction is moved to the destination register (Rd). The register Rd can be any one of the general purpose registers A, B, C, D, E, H or L. No flags are affected.

Example : MVI D,09H   **(D) ← 09$_H$**

The 8-bit data 09$_H$ given in the instruction is moved to the D-register.

Before execution	After execution
D	D
C2	09

Two-byte instruction   **Two machine cycles:**   Opcode fetch - 4T

Immediate addressing   Memory read - 3T

                                                             7 T

Total number of instructions = 7

     MVI A, d8   **MVI B, d8**   **MVI C, d8**   **MVI D, d8**   **MVI E, d8**   **MVI H, d8**   **MVI L, d8**

5.   **MVI M, d8**   **(M) ← d8**   or   **((HL)) ← d8**

The 8-bit data (d8) given in the instruction is moved to the memory location addressed by the HL pair. No flags are affected.

Example : MVI M, E7H   **(M) ← E7$_H$**   or   **((HL)) ← E7$_H$**

The 8-bit data E7$_H$ given in the instruction is moved to the memory location addressed by the HL pair.

Before execution	After execution

Two-byte instruction	**Three machine cycles :**	Opcode fetch	- 4T
Register indirect addressing or		Memory read	- 3T
Immediate addressing		Memory write	- 3T
			10 T

Total number of instructions = 1

---

6.   LDA addr16   **(A) ← (M)   or   (A) ← (addr16)**

The content of the memory location whose address is given in the instruction, is moved to accumulator. No flags are affected.

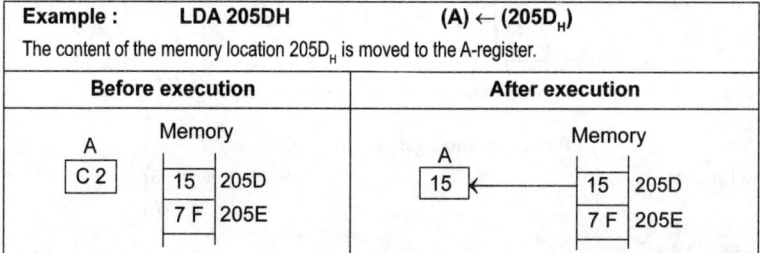

Example :   **LDA 205DH**   **(A) ← (205D_H)**
The content of the memory location 205D_H is moved to the A-register.

Three-byte instruction   **Four machine cycles :**

Three-byte instruction	Opcode fetch	- 4T
Direct addressing	Memory read	- 3T
	Memory read	- 3T
	Memory read	- 3T
		13T

Total number of instructions = 1

---

7.   LHLD addr16   **(L) ← (M)   or   (L) ← (addr16)**

$\qquad\qquad\qquad\qquad\qquad\qquad$ **(H) ← (M)   (H) ← (addr16 + 01)**

The content of the memory location whose address is given in the instruction, is moved to the L-register. The content of the next memory location is moved to the H-register. No flags are affected.

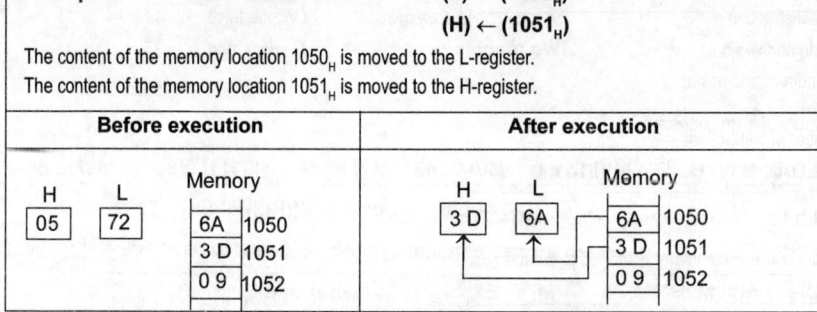

Example :   **LHLD 1050H**   **(L) ← (1050_H)**
$\qquad\qquad\qquad\qquad\qquad\qquad\qquad\qquad$ **(H) ← (1051_H)**
The content of the memory location 1050_H is moved to the L-register.
The content of the memory location 1051_H is moved to the H-register.

Five machine cycles:

Three-byte instruction	Opcode fetch -	4T
Direct addressing	Memory read -	3T
	Memory read -	3T
	Memory read -	3T
	Memory read -	3T
		16T

Total number of instructions = 1

8.   **LXI rp, d16**   **(rp) ← d16**   *(AU, Nov/Dec' 19, 2 Marks)*

The 16-bit data given in the instruction is moved to the register pair (rp). The register pair can be BC, DE, HL or SP.

Example : **LXI H, 1050H**	**(L) ← 50$_H$**
	(H) ← 10$_H$

The 16-bit data 1050$_H$ given in the instruction is moved to the HL register pair.

Before execution	**After execution**
H L	H L
xx yy	10 50
(some arbitrary value)	

Three-byte instruction   **Three machine cycles :** Opcode fetch - 4T

Immediate addressing   Memory read - 3T

                                                            Memory read - 3T

                                                                      10T

Total number of instructions = 4

        LXI B, d16   **LXI D, d16**   **LXI H, d16**   **LXI SP, d16**

9.   **LDAX rp**   **(A) ← (M)**   or   **(A) ← ((rp))**

The content of the memory addressed by the register pair (rp) is moved to the accumulator. (The content of the register pair is the memory address). The register pair can be either BC or DE.

Example : LDAX B   **(A) ← (M)   or   (A) ← ((BC))**

The content of the memory location addressed by the BC pair is moved to the A-register.

Before execution	**After execution**
A B C Memory	A B C Memory
02 20 5A 1 E 205A	1E 20 5A 1 E 205A
3 C 205B	3 C 205B

One-byte instruction   **Two machine cycles:** Opcode fetch - 4T

Register indirect addressing   Memory read - 3T

                                                                    7T

Total number of instructions = 2

        LDAX B   **LDAX D**

10.   **STA addr16**   **(M) ← (A)**   or   **(addr16) ← (A)**

The content of the accumulator is moved to the memory. The address of the memory location is given in the instruction. No flags are affected.

Example : STA 2050H   **(2050$_H$) ← (A)**

The content of the accumulator is moved to memory location 2050$_H$.

Before execution	**After execution**
A Memory	A Memory
F4 0 6 2050	F4 → F4 2050
7A 2051	7A 2051

Three-byte instruction	**Four machine cycles:**	Opcode fetch	-	4T
Direct addressing		Memory read	-	3T
		Memory read	-	3T
		Memory write	-	3T
				13T

Total number of instructions = 1

11.   STAX rp   (M) ← (A)   or   ((rp)) ← (A)

The content of the accumulator is moved to the memory addressed by the register pair (rp). (The content of the register pair is the memory address.) The register pair can be either BC or DE.

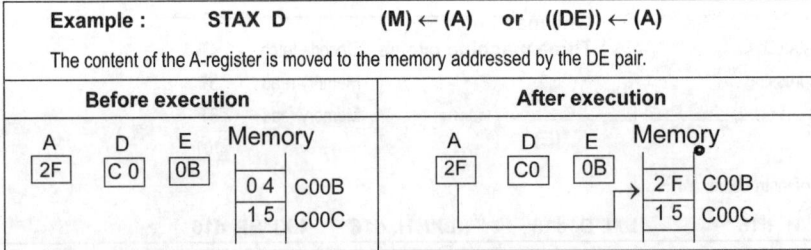

| **Example :** | **STAX D** | (M) ← (A) or ((DE)) ← (A) |

The content of the A-register is moved to the memory addressed by the DE pair.

One-byte instruction	**Two machine cycles:**	Opcode fetch	-	4 T
Register indirect addressing		Memory write	-	3T
				7 T

Total number of instructions = 2

                              STAX B   **STAX D**

12.   SHLD addr16   (M) ← (L)   or   (addr16) ← (L)
                                                       (M) ← (H)   (addr16+1) ← (H)

The content of the L-register is stored in the memory location, whose address is given in the instruction. The content of the H-register is stored in the next memory location. No flags are affected.

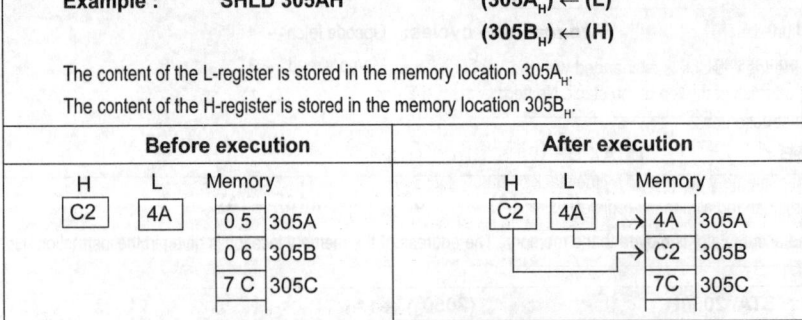

| **Example :** | **SHLD 305AH** | $(305A_H) ← (L)$ |
| | | $(305B_H) ← (H)$ |

The content of the L-register is stored in the memory location $305A_H$.
The content of the H-register is stored in the memory location $305B_H$.

Three-byte instruction	**Five machine cycles:**	Opcode fetch	-	4T
Direct addressing		Memory read	-	3T
		Memory read	-	3T
		Memory write	-	3T
		Memory write	-	3T
				16T

Total number of instructions = 1

13.   SPHL   (SP) ← (HL)

The content of the HL pair is moved to the Stack Pointer (SP). No flags are affected.

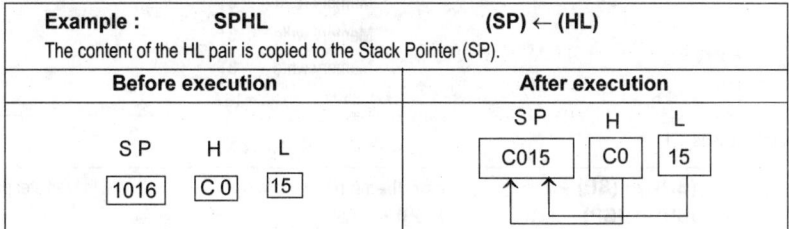

Example :	**SPHL**	(SP) ← (HL)
The content of the HL pair is copied to the Stack Pointer (SP).		

One-byte instruction   **One machine cycle:** Opcode fetch - 6T

Implied addressing

Total number of instructions = 1

---

14.   XCHG   (E) ↔ (L)

                                                                (D) ↔ (H)

The content of the HL pair is exchanged with the DE pair. No flags are affected.

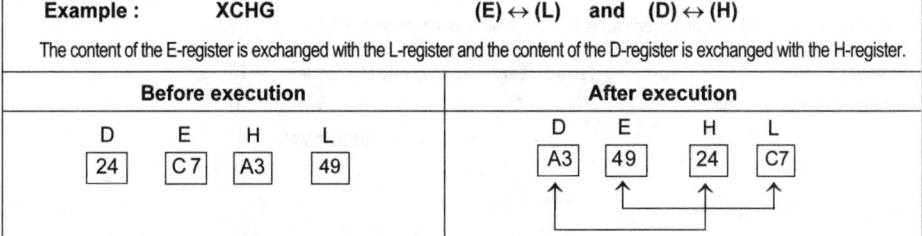

Example :	**XCHG**	(E) ↔ (L) and (D) ↔ (H)
The content of the E-register is exchanged with the L-register and the content of the D-register is exchanged with the H-register.		

One-byte instruction   **One machine cycle:** Opcode fetch - 4T

Implied addressing

Total number of instructions = 1

---

15.   XTHL   (HL) ↔ (M)   or   (HL) ↔ ((SP))

        The content of the top of stack is exchanged with the HL pair. Stack is a portion of memory (RAM memory). The content of the Stack Pointer (SP) is the address of the top of the stack. No flags are affected.

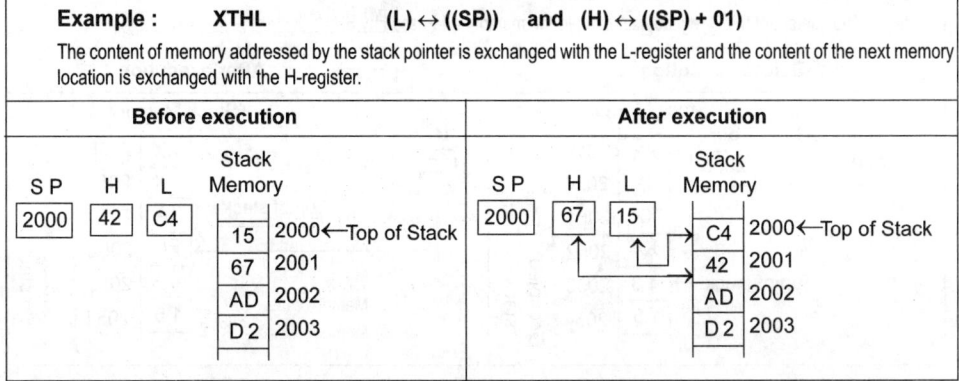

Example :	**XTHL**	(L) ↔ ((SP)) and (H) ↔ ((SP) + 01)
The content of memory addressed by the stack pointer is exchanged with the L-register and the content of the next memory location is exchanged with the H-register.		

One-byte instruction   **Five machine cycles:** Opcode fetch - 4T

Implied addressing   Memory read - 3T

                                             Memory read - 3T

                                            Memory write - 3T

                                            Memory write - 3T

                                                        16T

Total number of instructions = 1

---

16.   **PUSH rp**   **$(SP) \leftarrow (SP) - 1$**   ;   **$((SP)) \leftarrow (rp)_H$**

                              $(SP) \leftarrow (SP) - 1$;   **$((SP)) \leftarrow (rp)_L$**

The content of the register pair (rp) is pushed to the stack. After execution of this instruction, the content of the Stack Pointer (SP) will be 02 less than the earlier value. The register pairs can be BC, DE , HL and PSW. No flags are affected.

[PSW (Program Status Word) : Accumulator and Flag register together called PSW. Accumulator is high order register and Flag register is low order register.]

The instruction is executed as follows:

    (i)   The content of the SP is decremented by one.

    (ii)   The content of the high order register is moved to memory addressed by SP.

    (iii)   The content of the SP is decremented by one.

    (iv)   The content of the low order register is moved to memory addressed by SP.

One-byte instruction   **Three machine cycles:** Opcode fetch - 6T

Register indirect addressing   Memory write - 3T

                                             Memory write - 3T

                                                        12T

Total number of instructions = 4

PUSH PSW   PUSH B   PUSH D   PUSH H

Example : PUSH B   **$(SP) \leftarrow (SP) - 01$**

                                    $((SP)) \leftarrow (B)$

                                    $(SP) \leftarrow (SP) - 01$

                                    $((SP)) \leftarrow (C)$

(i)   The content of the SP is decremented by one.

(ii)   The content of the B-register is moved to the memory addressed by the Stack Pointer (SP).

(iii)   Again the content of SP is decremented by one.

(iv)   The content of the C-register is moved to the memory addressed by SP.

Before execution	After execution

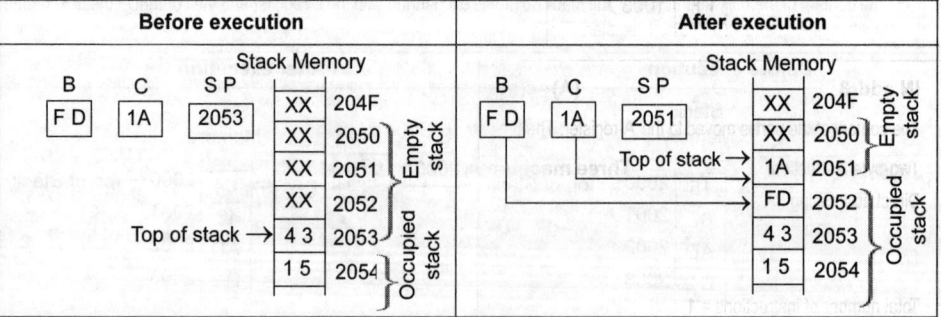

17.   POP rp   (rp)$_L$ ←((SP))   ;   (SP) ← (SP) + 1

                     (rp)$_H$ ← ((SP))   ;   (SP) ← (SP) + 1

The content of top of stack memory is moved to the register pair. After execution of this instruction the content of the Stack Pointer (SP) will be 02 greater than the earlier value. The register pairs can be BC, DE , HL and PSW. No flags are affected. [PSW (Program Status Word) : Accumulator and Flag register are together called PSW. The accumulator is a high order register and the flag register is a low order register.]

The pop instruction is executed as follows:

(i)   The content of the memory addressed by the SP is moved to the low order register.

(ii)   The content of the SP is incremented by one.

(iii)   The content of the memory addressed by the SP is moved to the high order register.

(iv)   The content of the SP is incremented by one.

One-byte instruction	**Three machine cycles**: Opcode fetch	-	4T
Register indirect addressing	Memory read	-	3T
	Memory read	-	3T
			10T

Total number of instructions = 4

POP PSW   POP B   POP D   POP H

Example : POP D   (E) ← ((SP))

                          (SP) ← (SP) + 01

                          (D) ← ((SP))

                          (SP) ← (SP) + 01

(i)   The content of the memory addressed by the SP is moved to the E-register.

(ii)   The content of the SP is incremented by one.

(iii) The content of the memory addressed by the SP is moved to the D-register.

(iv)   The content of the SP is incremented by one.

Before execution	**After execution**

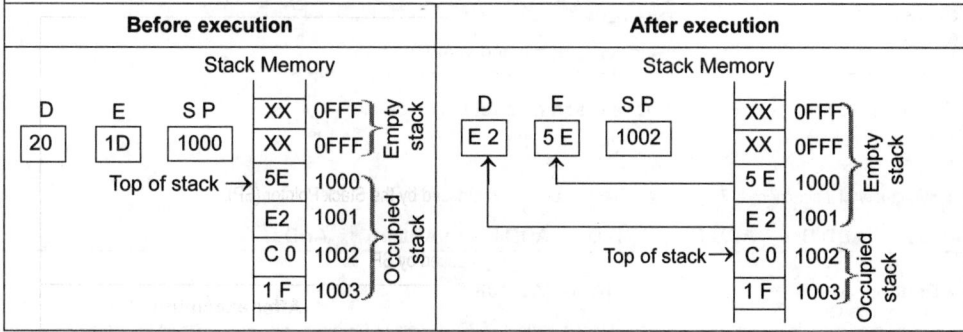

---

18.   IN addr8   (A)← (addr8)

The content of the port is moved to the A-register. The 8-bit port address will be given in the instruction. No flags are affected.

Two-byte instruction	**Three machine cycles:**	Opcode fetch	-	4T
Direct addressing		Memory read	-	3T
		IO read	-	3T
				10T

Total number of instructions = 1

19.   **OUT addr8**   (addr8) ← (A)

The content of the A-register is moved to the port. The 8-bit port address will be given in the instruction. No flags are affected.

Two-byte instruction	**Three machine cycles:**	Opcode fetch	-	4T
Direct addressing		Memory read	-	3T
		IO write	-	3T
				10T

Total number of instructions = 1

Note : In an 8085 processor-based system when the IO devices are mapped by IO mapping then the processor can communicate with these IO devices only by using IN and OUT instructions. The processor uses an 8-bit address to select IO-mapped IO devices. With 8-bit address the processor can generate $2^8 = 256_{10}$ IO addresses.

2.6 DATA MANIPULATION AND CONTROL INSTRUCTIONS

2.6.1 ARITHMETIC INSTRUCTIONS

1.   **ADD reg**   (A) ← (A) + (reg)

The content of the register is added to the content of the accumulator (A-register). After addition the result is stored in the accumulator. All flags are affected. The register can be any one of the general purpose register A, B, C, D, E, H or L.

Example : ADD E	**(A) ← (A) + (E)**	
The content of the E-register is added to the content of the A-register.		
The result will be in the A-register. All flags are affected.		
Before execution	**Addition**	**After execution**
A E C 2 B8 C F = 0 P F = 0 A F = 0 Z F = 0 S F = 0	$C2_H$ = 1100 0010 $B8_H$ = 1011 1000 ────────── 1 0111 1010 Sum = 0111 1010 =7A_H Carry = 1 (Addition is performed in ALU)	A E 7A B8 C F = 1 P F = 0 A F = 0 Z F = 0 S F = 0

| One-byte instruction | **One machine cycle:** Opcode fetch - 4T |
| Register addressing | |

Total number of instructions = 7

| ADD A   ADD B   ADD C   ADD D   ADD E   ADD H   ADD L |

2.   **ADI d8**   (A) ← (A) + d8

The 8-bit data given in the instruction is added to the content of the A-register (Accumulator). After addition, the result is stored in the accumulator. All flags are affected.

Two-byte instruction	Two machine cycles:	Opcode fetch	-	4T
Immediate addressing		Memory read	-	3T
				7T

Total number of instructions = 1

3.   **ADD M**   (A) ← (A) + (M)   or   (A) ← (A) + ((HL))

The content of memory addressed by HL pair is added to the content of the A-register. After addition, the result is stored in the A-register. All flags are affected.

Example : ADD M (A) ← (A) + (M)   or   (A) ← (A) + ((HL))

Let the content of A be 44$_H$.
Let the content of memory location C00A$_H$ be 73$_H$.
The content of the memory location C00A$_H$ is added to the content of the A-register. The result is put back in the A-register.

Before execution	Addition	After execution
A H L Memory	44$_H$ = 0100 0100	A H L Memory
44 C00A 73 C00A	73$_H$ = 0111 0011	B7 C00A 73 C00A
C F = 0 14 C00B	————	C F = 0 14 C00B
P F = 0 27 C00C	1011 0111	P F = 1 27 C00C
A F = 0	————	A F = 0
Z F = 0	Sum = B7	Z F = 0
S F = 0	Carry = 0	S F = 1
	(Addition is performed in ALU)	

One-byte instruction   **Two machine cycles:** Opcode fetch - 4T
Register indirect addressing   Memory read - 3T
                                                                                    ———
                                                                                     7T

Total number of instructions = 1

---

4.   **ACI d8**   **(A) ← (A) + d8 + CF**

The 8-bit data given in the instruction and the carry flag (the value of carry flag before executing this instruction) are added to the content of the A-register (Accumulator). After addition, the result is stored in the accumulator. All flags are affected.

Two-byte instruction   **Two machine cycles :** Opcode fetch - 4T
Immediate addressing   Memory read - 3T
                                                                                    ———
                                                                                     7T

Total number of instructions = 1

---

5.   **ADC reg**   **(A) ← (A) + (reg) + CF**

The content of the register and the carry flag are added to the content of the A-register. After addition, the result is stored in the A-register. All flags are affected. The register can be any one of the general purpose register A, B, C, D, E, H or L.

Example : ADC H   **(A) ← (A) + (H) + CF**

The content of the H-register and the value of the carry flag (before executing this instruction) are added to the content of the A-register. After addition, the result will be in the A-register.

Before execution	Addition	After execution
A H	43$_H$ = 0100 0011	A H
43 7A	7A$_H$ = 0111 1010	BE 7A
	CF = 1	
C F = 1	————	C F = 0
P F = 0	1011 1110	P F = 1
A F = 0	————	A F = 0
Z F = 0	Sum = BE$_H$	Z F = 0
S F = 1	Carry = 0	S F = 1
	(Addition is performed in the ALU)	

One-byte instruction   **One machine cycle :** Opcode fetch - 4T

Register addressing

Total number of instructions = 7

| ADC A | ADC B | ADC C | ADC D | ADC E | ADC H | ADC L |

---

6.   **ADC M** $(A) \leftarrow (A) + (M) + CF$ or $(A) \leftarrow (A) + ((HL)) + CF$

The content of the memory addressed by the HL pair and the value of the carry flag (before executing this instruction) are added to the content of A-register. After addition, the result is stored in the A-register. All flags are affected.

One-byte instruction   **Two machine cycles:**   Opcode fetch -   4T

Register indirect addressing   Memory read -   3T
                                                                          ───
                                                                          7 T

Total number of instructions = 1

---

7.   **SUB reg** $(A) \leftarrow (A) - (reg)$

The content of the register is subtracted from the content of the accumulator (A-register). After subtraction the result is stored in the A-register. All flags are affected. The register can be any one of the general purpose register A, B, C, D, E, H or L.

Example: SUB C $(A) \leftarrow (A) - (C)$

The content of the C-register is subtracted from A-register. The result will be in the A-register.

Case i

Before execution	Subtraction
A C C 4 89 C F = 0 P F = 0 A F = 0 Z F = 0 S F = 1	$C4_H = 1100\ 0100$ $89_H = 1000\ 1001$ 1's complement of $89_H = 0111\ 0110$ 2's complement of $89_H = 0111\ 0110 + 1$ $= 0111\ 0111 = 77_H$
A C 3B 89 C F = 0 P F = 0 A F = 0 Z F = 0 S F = 0	$C4_H = 1100\ 0100$ $+77_H = 0111\ 0111$ ─────── $\boxed{1}0011\ 1011$ Complement ↓ 3 B Carry $\boxed{0}$ Result = $3B_H$ CF = 0

Case ii

Before execution	Subtraction
A C 89 C 4 C F = 0 P F = 0 A F = 0 Z F = 1 S F = 1	$89_H = 1000\ 1001$ $C4_H = 1100\ 0100$ 1's complement of $C4_H = 0011\ 1011$ 2's complement of $C4_H = 0011\ 1011 + 1$ $= 0011\ 1100 = 3C_H$

Case ii continued ...

After execution	Subtraction
A C $\boxed{C\,5}$ $\boxed{C\,4}$ CF = 1 PF = 1 AF = 1 ZF = 0 SF = 1	89_H = 1000 1001 $+3C_H$ = 0011 1100 $\boxed{0}$ 1100 0101 Complement ↓ C 5 Carry $\boxed{1}$ Result = $C5_H$ CF = 1 *Note : 2's complement of $C5_H$ = $3B_H$*

Note : *The 8085 microprocessor performs 2's complement subtraction. But after subtraction, it will complement the carry alone. In 2's complement subtraction, if CF =1, then the result is positive and if CF =0, then the result is negative. Since, the 8085 processor complements the carry after subtraction, here if CF = 0, then the result is positive and if CF = 1, then the result is negative. If the result is negative, then it will be in 2's complement form.*

One-byte instruction   **One machine cycle:** Opcode fetch - 4T

Register addressing

Total number of instructions = 7

    SUB A   SUB B   SUB C   SUB D   SUB E   SUB H   SUB L

8.   **SUI d8**   **(A) ← (A) – d8**

     The 8-bit data given in the instruction is subtracted from the A-register (accumulator). After subtraction, the result is stored in the A-register. All flags are affected.

Two-byte instruction   **Two machine cycles :**   Opcode fetch - 4T

Immediate addressing   Memory read - 3T

                                                          7T

Total number of instructions = 1

9.   **SUB M**   **(A) ← (A) – (M)**   or   **(A) ← (A) – ((HL))**

     The content of the memory addressed by the HL pair is subtracted from the A-register. After subtraction, the result is stored in the A-register. All flags are affected.

One-byte instruction   **Two machine cycles :**   Opcode fetch - 4T

Register indirect addressing   Memory read - 3T

                                                          7 T

Total number of instructions = 1

10.   **SBB reg**   **(A) ← (A) – (reg) – CF**

     The content of the register and the value of carry (before executing this instruction) are subtracted from the accumulator (A-register). After subtraction, the result is stored in the accumulator. All flags are affected. The register can be any one of the general purpose register A, B, C, D, E, H or L.

One-byte instruction   One machine cycle : Opcode fetch - 4T

Register addressing

Total number of instructions = 7

    SBB A   SBB B   SBB C   SBB D   SBB E   SBB H   SBB L

11.   SBI d8   **(A) ← (A) − d8 − CF**

The 8-bit data given in the instruction and the value of carry (before executing this instruction) are subtracted from accumulator. After subtraction, the result is stored in the accumulator. All flags are affected.

Two-byte instruction	**Two machine cycles :**	Opcode fetch -	4T
Immediate addressing		Memory read -	3T
			7T

Total number of instructions = 1

---

12.   SBB M   **(A) ← (A) − (M) − CF**   or   **(A) ← (A) − ((HL)) − CF**

The content of the memory addressed by HL and the value of carry (before executing this instruction) are subtracted from accumulator (A-register). After subtraction, the result is stored in the A-register. All flags are affected.

One-byte instruction	**Two machine cycles:**	Opcode fetch -	4T
Register indirect addressing		Memory read -	3T
			7T

Total number of instructions = 1

---

13.   DAA

(DAA - Decimal Adjust Accumulator)

After BCD addition, the DAA instruction is executed to get the result in BCD. When DAA instruction is executed, the content of the accumulator is altered or adjusted as explained below :

i)   If the sum of the lower nibbles exceeds 09_H or auxiliary carry is set, then a correction 06_H (0110) is added to sum of lower nibbles.

ii)   If the sum of the upper nibbles exceeds 09_H or carry is set, then a correction 06_H (0110) is added to sum of upper nibble.

After executing this instruction all flags are modified to indicate the status of the result.

One-byte instruction   **One machine cycle:** Opcode fetch - 4T

Implied addressing

Total number of instructions = 1

---

14.   DAD rp   **(HL) ← (HL) + (rp)**

(DAD - Double Addition)

The content of the register pair is added to the content of the HL pair. After addition, the result is stored in the HL pair. Only the carry flag is affected. The register pair can be BC, DE, HL or SP.

One-byte instruction	**Three machine cycles:**	Opcode fetch -	4T
Register addressing		Bus idle -	3T
		Bus idle -	3T
			10T

Total number of instructions = 4

            DAD B   **DAD D**   **DAD H**   **DAD SP**

15.   **INR reg**   **(reg) ← (reg) + 01**

The content of the register is incremented by one. Except carry flag, all other flags are affected. The register can be any one of the general purpose register A, B, C, D, E, H or L.

Example :   **INR B**   **(B) ← (B) + 01**

The content of the B-register is incremented by one. The increment opertation is performed by adding 01_H to the content of B-register.

Before execution	Increment Operation	After execution
B [4 A] CF = 0 PF = 0 AF = 0 ZF = 0 SF = 0	$4A_H$ = 0100 1010 $+ 01_H$ = 0000 0001 ——————— 0100 1011 4 B	B [4B] CF = 0 PF = 1 AF = 0 ZF = 0 SF = 0

One-byte instruction   **One machine cycle:** Opcode fetch - 4T

Register addressing

Total number of instructions = 7

     INR A   INR B   INR C   INR D   INR E   INR H   INR L

16.   **INR M**   **(M) ← (M) + 01**   or   **((HL)) ← ((HL)) + 01**

The content of the memory addressed by the HL pair is incremented by one. Except carry, all other flags are affected.

Example :   **INR M**   **(M) ← (M) + 01**

Let the content of the HL pair be $C00A_H$. Let the content of memory location $C00A_H$ be $C5_H$. The content of the memory location $C00A_H$ is incremented by one. The increment operation is performed by adding 01_H to the content of the memory.

Before execution	Increment Operation	After execution
H L Memory [C00A] [C 5] C00A CF = 0 [A2] C00B PF = 0 [0 7] C00C AF = 0 ZF = 0 SF = 0	$C5_H$ = 1100 0101 $+ 01_H$ = 0000 0001 ——————— 1100 0110 C 6	H L Memory [C00A] [C 6] C00A CF = 0 [A2] C00B PF = 1 [0 7] C00C AF = 0 ZF = 0 SF = 1

One-byte instruction   **Three machine cycles :**   Opcode fetch - 4T

Register indirect addressing   Memory read - 3T

                                                     Memory write - 3T
                                                              —————
                                                                10T

Total number of instructions = 1

17.   **DCR reg**   **(reg) ← (reg) – 01**

The content of the register is decremented by one. Except carry, all other flags are affected. The register can be A, B, C, D, E, H or L.

Example : DCR D   **(D) ← (D) – 01**

The content of the D-register is decremented by one. The decrement operation is performed by subtracting 01_H from the content of the D-register.

Before execution	Decrement operation
D [60] CF = 0 PF = 0 AF = 0 ZF = 0 SF = 0	01_H = 0000 0001 1's complement of 01_H = 1111 1110 2's complement of 01_H = 1111 1110 + 1 = 1111 1111 = FF_H
After execution	60_H = 0110 0000

D [5F]   CF = 0
 PF = 1
 AF = 0
 ZF = 0
 SF = 0
	+ FF_H = 1111 1111
	         [1] 0101 1111
	           5 F
	 Carry is discarded

One-byte instruction   **One machine cycle :** Opcode fetch - 4T

Register addressing

Total number of instructions = 7

DCR A	**DCR B**	**DCR C**	**DCR D**	**DCR E**	**DCR H**	**DCR L**

18.   **DCR M**   **(M) ← (M) – 01**   or   **((HL)) ← ((HL)) – 01**

The content of memory addressed by the HL pair is decremented by one. Except carry, all other flags are affected.

Example: DCR M   **(M) ← (M) – 01**

Let the content of the HL pair be 2010_H. Let the content of memory location 2010_H be FA_H. The content of memory location 2010_H is decremented by one.

Before execution	Decrement operation

H L [2010]   Memory
              FA 2010
 CF = 0   02 2011
 PF = 0
 AF = 0
 ZF = 0
 SF = 0
	01_H = 0000 0001
	 1's complement of 01_H = 1111 1110
	 2's complement of 01_H = 1111 1110 + 1
	                     = 1111 1111 = FF_H

After execution	FA_H = 1111 1010

H L [2010]   Memory
              F9 2010
 CF = 0   02 2011
 PF = 1
 AF = 1
 ZF = 0
 SF = 1
	+ FF_H = 1111 1111
	        [1] 1111 1001
	           F 9
	 Carry is discarded

One-byte instruction   **Three machine cycles :** Opcode fetch - 4T

Register indirect addressing   Memory read - 3T

                                                    Memory write - 3T

                                                               10T

Total number of instructions = 1

19.   INX rp   **(rp) ← (rp) + 01**

The content of the register pair is incremented by one. The register pair can be BC, DE, HL or SP. No flags are affected.

Example : INX H (HL) ← (HL) + 01		
The content of the HL pair is incremented by one.		
Before execution	**After execution**	
H L 00FF	H L 0100	

One-byte instruction   **One machine cycle :** Opcode fetch - 6T

Register addressing

Total number of instructions = 4

INX B   INX D   INX H   INX SP

---

20.   DCX rp   **(rp) ← (rp) – 01**

The content of the register pair is decremented by one. The register pair can be BC, DE, HL or SP. No flags are affected.

Example : DCX SP (SP) ← (SP) – 01		
The content of the stack pointer is decremented by one.		
Before execution	**After execution**	
S P 1000	S P 0FFF	

One-byte instruction   **One machine cycle :** Opcode fetch - 6T

Register addressing

Total number of instructions = 4

DCX B   DCX D   DCX H   DCX SP

2.6.2 LOGICAL INSTRUCTIONS

1.   ANA reg   **(A) ← (A) & (reg)**

(& is the symbol used for logical AND operation)

The content of the register is logically ANDed bit by bit with the content of the accumulator. In bit by bit AND operation, the bit D_0 of register is ANDed with the bit D_0 of A-register, the bit D_1 of register is ANDed with bit D_1 of A-register, and so on. The register can be any one of the general purpose register A, B, C, D, E, H or L. After execution of the instruction, carry flag is always reset and auxiliary carry flag is always set. Other flags are altered (according to the results). After AND operation, result is stored in accumulator.

Example : ANA E	**(A) ← (A) & (E)**	
The content of E-register is logically ANDed bit by bit with the content of accumulator.		
Before execution	**AND operation**	**After execution**

A   E   C F = 0
15   E 2   P F = 0
              A F = 0
              Z F = 0
              S F = 0
	15_H = 0001 0101
	
	$E2_H$ = 1110 0010
	————————
	        0000 0000
	          0 0
		A E C F = 0
		00 E 2 P F = 1
		              A F = 1
		              Z F = 1
		              S F = 0

One-byte instruction   **One machine cycle:** Opcode fetch - 4T

Register addressing

Total number of instructions = 7

| ANA A | ANA B | ANA C | ANA D | ANA E | ANA H | ANA L |

2. **ANI d8**   **(A) ← (A) & d8**

The 8-bit data given in the instruction is logically ANDed bit by bit with the content of the accumulator. The result is stored in the accumulator. After execution of this instruction, CF = 0 and AF = 1. Other flags are affected.

Two-byte instruction   **Two machine cycles :**   Opcode fetch - 4T

Immediate addressing   Memory read - 3T

                                                                    7T

Total number of instructions = 1

3. **ANA M**   **(A) ← (A) & (M)   or   (A) ← (A) & ((HL))**

The content of the memory addressed by the HL pair is logically ANDed bit by bit with the content of the accumulator. The result is stored in the accumulator. After execution, CF = 0 and AF = 1. Other flags are affected.

Example : ANA M   **(A) ← (A) & (M)**

Let the content of HL be $105A_H$. Let the content of the memory location $105A_H$ be $4C_H$. The content of the memory location $105A_H$ is logically ANDed bit by bit with the content of the accumulator. The result is stored in the accumulator.

Before execution	AND operation	After execution
A HL Memory	27_H = 0010 0111	A HL Memory
[27] [105A] [14] 1059	$4C_H$ = 0100 1100	[04] [105A] [14] 1059
CF = 0 [4C] 105A	————————	CF = 0 [4C] 105A
PF = 0	0000 0100	PF = 0
AF = 0	0 4	AF = 1
ZF = 0		ZF = 0
SF = 0		SF = 0

One-byte instruction   **Two machine cycles:**   Opcode fetch - 4T

Register indirect addressing   Memory read - 3T

                                                                    7T

Total number of instructions = 1

4. **ORA reg**   **(A) ← (A) | (reg)**

(| is the symbol used for logical OR operation)

The content of the register is logically ORed bit by bit with the content of the accumulator. In bit by bit OR operation, the bit D_0 of the register is ORed with bit D_0 of the A-register, the bit D_1 of the register is ORed with bit D_1 of the A-register, and so on. The register can be any one of the general purpose register A, B, C, D, E, H or L. After execution of the instruction, both the carry and auxiliary flags are always reset (AF = 0, CF = 0). Other flags are modified (according to the result). After OR operation, the result is stored in the accumulator.

One-byte instruction   **One machine cycle:** Opcode fetch - 4T

Register addressing

Example : ORA B		$(A) \leftarrow (A) \mid (B)$				
The content of the B-register is logically ORed bit by bit with the content of the accumulator.						

Before execution	OR operation	After execution
A B C F = 0 04 7A P F = 0 AF = 0 ZF = 0 S F = 0	04_H = 0000 0100 $7A_H$ = 0111 1010 ─────────── 0111 1110 ─────────── 7 E	A B C F = 0 7E 7A P F = 1 AF = 0 ZF = 0 S F = 0

Total number of instructions = 7

ORA A   ORA B   ORA C   ORA D   ORA E   ORA H   ORA L

5.   **ORA M** $(A) \leftarrow (A) \mid (M)$ or $(A) \leftarrow (A) \mid ((HL))$

The content of the memory addressed by the HL pair is logically ORed bit by bit with the content of the accumulator. The result is stored in the accumulator. After execution, CF = AF = 0. Other flags are affected.

Example : ORA M		$(A) \leftarrow (A) \mid (M)$		
Let the content of the HL pair be 2050_H. Let the content of memory location 2050_H be $1B_H$. The content of the memory location 2050_H is logically ORed bit by bit with the content of the accumulator. The result is stored in the accumulator.				

Before execution	OR operation	After execution

A   HL   Memory
45   2050
          1B   2050
          07   2051
CF   = 0
PF   = 0
AF   = 0
ZF   = 0
SF   = 0
	45_H = 0100 0101
	$1B_H$ = 0001 1011
	───────────
	0101 1111
	───────────
	5 F
		A HL Memory
		5F 2050
		          1B 2050
		          07 2051
		C F = 0
		P F = 1
		AF = 0
		ZF = 0
		SF = 0

One-byte instruction	**Two machine cycles:**	Opcode fetch - 4 T
Register indirect addressing		Memory read - 3 T
		───── 7 T

Total number of instructions = 1

6.   **ORI d8** $(A) \leftarrow (A) \mid d8$

The 8-bit data given in the instruction is logically ORed bit by bit with the content of the accumulator. The result is stored in the accumulator. After execution of this instruction, CF = AF = 0. Other flags are affected.

Two-byte instruction	**Two machine cycles :**	Opcode fetch - 4 T
Immediate addressing		Memory read - 3 T
		───── 7 T

Total number of instructions = 1

7.   **XRA reg** $(A) \leftarrow (A) \wedge (reg)$

(^ is the symbol used for logical EXCLUSIVE-OR operation).

The content of the register is logically EXCLUSIVE-ORed bit by bit with the content of the accumulator. In bit by bit EXCLUSIVE-OR operation, the bit D_0 of register is EXCLUSIVE-ORed with bit D_0 of A-register, the bit D_1 of register is EXCLUSIVE-ORed with bit D_1 of A-register, and so on. The result is stored in the accumulator. The register can be any one of the general purpose register A, B, C, D, E, H or L. After execution AF = CF = 0. Other flags are modified (according to the result).

Example :   XRA A   (A) ← (A) ∧ (A)

The content of the A-register is EXCLUSIVE-ORed bit by bit with the content of the A-register itself.

Before execution	EXCLUSIVE-OR operation	After execution

A   CF = 1
 74 PF = 0
      AF = 1
      ZF = 0
      SF = 1
	74$_H$ = 0111 0100
	
	 74$_H$ = 0111 0100
	      0000 0000
		A CF = 0
		 00 PF = 1
		      AF = 0
		      ZF = 1
		      SF = 0

One-byte instruction   **One machine cycle**: Opcode fetch - 4T

Register addressing

Total number of instructions = 7

XRA A	XRA B	XRA C	XRA D	XRA E	XRA H	XRA L

8.   **XRI d8**   **(A) ← (A) ∧ d8   or   (A) ← (A) ∧ d8**

The 8-bit data given in the instruction is logically EXCLUSIVE-ORed bit by bit with the content of the accumulator. The result is stored in the accumulator. After execution of this instruction, CF = AF = 0. Other flags are affected.

Two-byte instruction   **Two machine cycles :**   Opcode fetch - 4T

Immediate addressing   Memory read - 3T

                                                                    7T

Total number of instructions = 1

9.   **XRA M**   **(A) ← (A) ∧ (M)   or   (A) ← (A) ∧ ((HL))**

The content of the memory addressed by the HL pair is logically EXCLUSIVE-ORed bit by bit with the content of accumulator. The result is stored in accumulator. After execution, CF = AF = 0. Other flags are affected.

Example :   XRA M   (A) ← (A) ∧ (M)

Let the content of the HL pair be 805A$_H$. Let the content of memory location 805A$_H$ be C4$_H$. The content of the memory location 805A$_H$ is logically EXCLUSIVE-ORed bit by bit with the content of the accumulator. The result will be in the accumulator.

Before execution	Exclusive-OR operation	After execution

A   H L   Memory
 B7   805A   1 C 8059
 CF = 1   C 4 805A
 PF = 1   2 0 805B
 AF = 1   51 805C
 ZF = 0
 SF = 1
	B7$_H$ = 1011 0111
	 C4$_H$ = 1100 0100
	      0111 0011
	       7 3
		A H L Memory
		 73 805A 1 C 8059
		 CF = 0 C 4 805A
		 PF = 0 2 0 805B
		 AF = 0 51 805C
		 ZF = 0
		 SF = 0

One-byte instruction   **Two machine cycles :**Opcode fetch - 4T

Register indirect addressing   Memory read - 3T

                                                                    7T

Total number of instructions = 1

10. **CMP reg** (A) – (reg) $\Rightarrow$ **Modify flags**

The content of the register is compared with the accumulator. The comparison is performed by subtracting the content of register from the A-register. The subtraction is performed in the ALU, and the result is used to modify flags and then the result is discarded (i.e., it is not stored in any register). After execution of this instruction, the content of accumulator and the register are not altered. All flags are affected by this instruction. The register can be any one of the general purpose register A, B, C, D, E, H or L.

The status of carry and zero flag after comparison are given below :

i) If (A) < (reg) then the carry flag is set (i.e., CF = 1)

ii) If (A) > (reg) then the carry flag is reset or cleared (i.e., CF = 0)

iii) If (A) = (reg) then the zero flag is set (i.e., ZF = 1).

Example : CMP B (A) – (B) $\Rightarrow$ **Modify flags.**
The content of the B-register is compared with the accumulator. The comparison is performed by subtracting the content of the B-register from the content of the accumulator. The subtraction is performed in the ALU and the result is used to modify the flags and then discarded. The content of the accumulator and the B-register are not altered.

Before execution	Comparison	After execution
A B 15 C2 C F = 0 P F = 0 A F = 0 Z F = 0 S F = 0	$C2_H = 1100\ 0010$ 1's complement of $C2_H = 0011\ 1101$ 2's complement of $C2_H = 0011\ 1101+1$ $= 0011\ 1110 = 3E_H$ $15_H = 0001\ 0101$ $+3E_H = 0011\ 1110$ ──────── Complement $\overline{0}$0101 0011 Carry $\downarrow$ 5 3 1	A B 15 C2 C F = 1 P F = 1 A F = 1 Z F = 0 S F = 0

One-byte instruction   **One machine cycle:** Opcode fetch - 4T

Register addressing

Total number of instructions = 7

CMP A   **CMP B**   **CMP C**   **CMP D**   **CMP E**   **CMP H**   **CMP L**

11. **CPI d8** (A) – d8 $\Rightarrow$ **Modify flags.**

The 8-bit data given in the instruction is compared with the accumulator. The comparison is performed by subtracting the 8-bit data from the A-register. The subtraction is performed in ALU and the result is used to modify flags and then discarded. After execution of the instruction, the content of the accumulator is not altered. All flags are affected.

The status of carry and zero flag after comparision are given below :

i) If (A) < d8 then the carry flag is set (i.e., CF = 1)

ii) If (A) > d8 then the carry flag is reset or cleared (i.e., CF = 0)

iii) If (A) = d8 then the zero flag is set (i.e., ZF = 1).

Two-byte instruction	**Two machine cycles :**	Opcode fetch	-	4T
Immediate addressing		Memory read	-	3T
				7T

Total number of instructions = 1

12. **CMP M** **(A) – (M) ⇒ Modify flags or (A) – ((HL)) ⇒Modify flags.**

The content of the memory addressed by HL pair is compared with the accumulator. The comparison is performed by subtracting the content of memory from the A-register. The subtraction is performed in the ALU and the result is used to modify flags and then discarded. After execution of the instruction, the content of the accumulator and the memory are not altered. All flags are affected by this instruction.

The status of carry and zero flag after comparison are given below:

    i) If (A) < (M) then the carry flag is set (i.e., CF = 1).

    ii) If (A) > (M) then the carry flag is reset or cleared (i.e., CF = 0).

    iii) If (A) = (M) then the zero flag is set (i.e., ZF = 1).

Example : CMP M

Let the content of the HL pair be $C050_H$. Let the content of the memory location $C050_H$ be $7A_H$. The content of the memory location $C050_H$ is compared with the content of the accumulator. Only flags are altered. The content of the accumulator and the memory remains the same.

Before execution	Comparison	After execution

A   HL
$\boxed{25}$ $\boxed{C050}$
 	
Memory
$\boxed{7A}$ C050
$\boxed{10}$ C051
 	
C F = 0
P F = 0
A F = 0
Z F = 0
S F = 0
	25_H = 0010 0101
	$7A_H$ = 0111 1010
	
	1'complement of $7A_H$= 1000 0101
	2'complement of $7A_H$ = 1000 0101 +1
	                = 1000 0110 =86_H
	
	25_H = 0010 0101
	$+86_H$ = 1000 0110
	—————————
	$\boxed{0}$ 1010 1011
	Complement
	Carry ↓ A B
	   $\boxed{1}$
		A HL
		$\boxed{25}$ $\boxed{C050}$
		
		Memory
		$\boxed{7A}$ C050
		$\boxed{10}$ C051
		
		C F = 1
		P F = 0
		A F = 0
		Z F = 0
		S F = 1

One-byte instruction   **Two machine cycles:** Opcode fetch - 4T

Register indirect addressing   Memory read - 3T

                                                                             7T

Total number of instructions = 1

---

13. **CMA** **(A) ← $(\overline{A})$**

(CMA - Complement Accumulator)

The content of the accumulator is complemented. No flags are affected.

One-byte Instruction   **One machine cycle:** Opcode fetch - 4T

Implied addressing

---

14. **STC** **(CF) ←1**

(STC - Set Carry)

The carry flag is set to 1. Only carry flag affected by this instruction.

One-byte instruction   **One machine cycle :** Opcode fetch - 4T

Implied addressing

15. CMC $(CF) \leftarrow (\overline{CF})$

(CMC - Complement Carry)

The carry flag is complemented. Only the carry flag is affected by this instruction.

One-byte instruction   **One machine cycle:** Opcode fetch - 4T

Implied addressing

---

16. RLC $D_{n+1} \leftarrow D_n$; $D_0 \leftarrow D_7$ and $(CF) \leftarrow D_7$

(RLC - Rotate Accumulator Left to carry)

The content of the A-register is rotated left by one bit and the left most bit of A-register is rotated to the carry. [The left most bit is most significant bit.] Only the carry flag is affected.

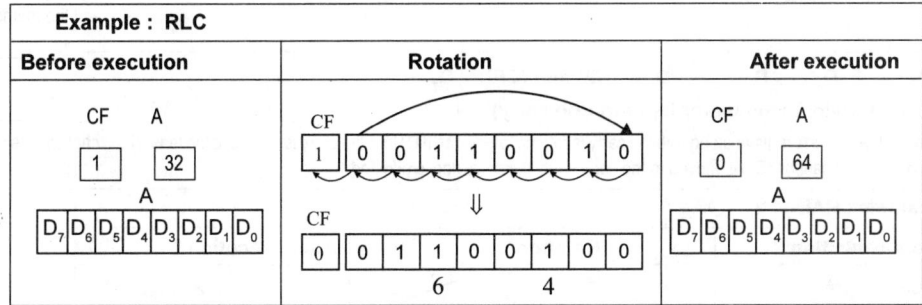

One-byte instruction   **One machine cycle:** Opcode fetch - 4T

Implied addressing

---

17. RRC $D_n \leftarrow D_{n+1}$; $D_7 \leftarrow D_0$ and $(CF) \leftarrow D_0$

(RRC - Rotate Accumulator Right to Carry)

The content of A-register is rotated right by one bit and the right most bit of A-register is rotated to carry. [The right most bit is least significant bit.] Only carry flag is affected.

Example : RRC		
Before execution	**Rotation**	**After execution**

Before execution: CF [1]   A [32]

A: $D_7 D_6 D_5 D_4 D_3 D_2 D_1 D_0$

Rotation: CF [1] 0 0 1 1 0 0 1 0

CF [0] 0 1 1 0 0 1 0 0

6   4

After execution: CF [0]   A [19]

A: $D_7 D_6 D_5 D_4 D_3 D_2 D_1 D_0$

One-byte instruction   **One machine cycle:** Opcode fetch - 4T

Implied addressing

---

18. RAR $D_n \leftarrow D_{n+1}$; $D_7 \leftarrow (CF)$ and $(CF) \leftarrow D_0$

(RAR - Rotate Accumulator Right through carry)

The content of the A-register along with the carry is rotated right by one bit. Here the carry is moved to the most significant bit position (D_7) and the least significant bit (D_0) is moved to the carry. Only the carry flag is affected.

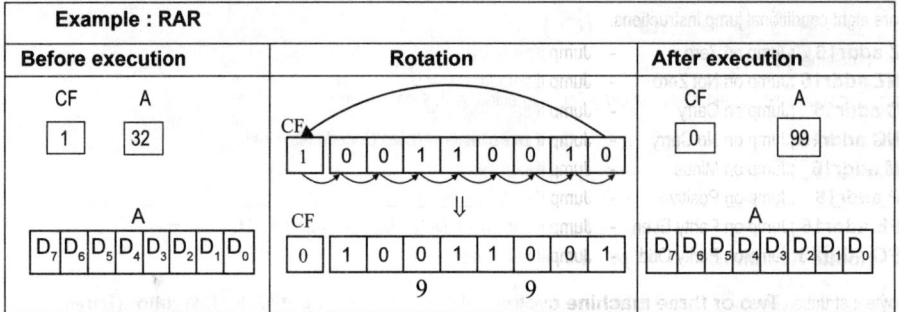

Example : RAR

One-byte instruction

Implied addressing

One machine cycle: Opcode fetch - 4T

---

19. **RAL** $D_{n+1} \leftarrow D_n$; $D_0 \leftarrow (CF)$ and $(CF) \leftarrow D_7$

(RAL - Rotate Accumulator Left through carry)

The content of the A-register along with the carry is rotated left by one bit. Here the carry is moved to the least significant bit position (D_0) and the most significant bit (D_7) is moved to the carry. Only the carry flag is affected.

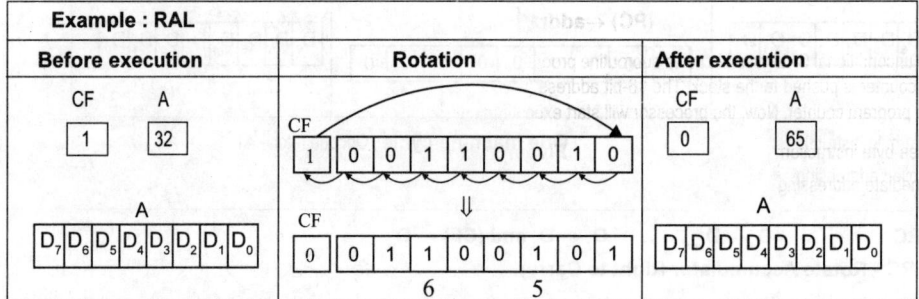

Example : RAL

One-byte instruction

Implied addressing

One machine cycle: Opcode fetch - 4T

2.6.3 BRANCHING INSTRUCTIONS

1. **JMP addr16** **(PC) ← addr16**

It is unconditional jump instruction. When this instruction is executed, the address given in the instruction is moved to the program counter. Now, the processor starts executing the instructions stored in this address.

Three-byte instruction

Immediate addressing

Three machine cycles: Opcode fetch - 4 T

Memory read - 3 T

Memory read - 3 T

10T

---

2. **J <condition> addr16**

If <condition> is TRUE then,

(PC) ← addr16

It is conditional jump instruction. The conditional jump instruction will check a flag condition. If the flag condition is true, then the address given in the instruction is moved to the program counter. Thus the program control is branched to the jump address. If the flag condition is false, then the next instruction is executed.

There are eight conditional jump instructions.

i) **JZ addr16** ;Jump on Zero   -   Jump if zero flag   = 1.
ii) **JNZ addr16** ;Jump on Not Zero   -   Jump if zero flag   = 0.
iii) **JC addr16**   ;Jump on Carry   -   Jump if carry flag   = 1.
iv) **JNC addr16** ;Jump on No Carry   -   Jump if carry flag   = 0.
v) **JM addr16**   ;Jump on Minus   -   Jump if sign flag   = 1.
vi) **JP addr16**   ;Jump on Positive   -   Jump if sign flag   = 0.
vii) **JPE addr16** ;Jump on Parity Even   -   Jump if parity flag   = 1.
viii) **JPO addr16** ;Jump on Parity Odd   -   Jump if parity flag   = 0.

Three-byte instruction **Two or three machine cycles:**

Immediate addressing

	Condition False	Condition True
	Opcode fetch - 4T	Opcode fetch - 4T
	Memory read - 3T	Memory read - 3T
		Memory read - 3T
	7T	10T

3. **CALL addr16** $(SP) \leftarrow (SP) - 1$; $((SP)) \leftarrow (PC)_H$
                                    $(SP) \leftarrow (SP) - 1$; $((SP)) \leftarrow (PC)_L$
                                    $(PC) \leftarrow addr16$

It is unconditional CALL used to call a subroutine program. When this instruction is executed, the address of the next instruction in the program counter is pushed to the stack. The 16-bit address (which is the address of the subroutine program) given in the instruction is loaded in the program counter. Now, the processor will start executing the instructions stored in this call address.

Three-byte instruction	**Five machine cycles:**	Opcode fetch	-	6T
Immediate addressing		Memory read	-	3T
		Memory read	-	3T
		Memory write	-	3T
		Memory write	-	3T
				18T

4.   **C<condition> addr16**

If <condition> is TRUE then,

         $(SP) \leftarrow (SP) - 1$; $((SP)) \leftarrow (PC)_H$
         $(SP) \leftarrow (SP) - 1$; $((SP)) \leftarrow (PC)_L$
         $(PC) \leftarrow addr16$

It is conditional subroutine call instruction. The conditional CALL instruction will check for a flag condition. If the flag condition is true, then the address of the next instruction is pushed to the stack and the call address (address given in the instruction) is loaded in the program counter. Now, the processor will start executing the instructions stored in this address. If the flag condition is false, then the next instruction is executed.

There are eight conditional CALL instructions. These are:

i) **CZ addr16**   ;Call on Zero   -   Call if zero flag   = 1.
ii) **CNZ addr16** ;Call on Not Zero   -   Call if zero flag   = 0.
iii) **CC addr16**   ;Call on Carry   -   Call if carry flag   = 1.
iv) **CNC addr16** ;Call on No Carry   -   Call if carry flag   = 0.
v) **CM addr16**   ;Call on Minus   -   Call if sign flag   = 1.
vi) **CP addr16**   ;Call on Positive   -   Call if sign flag   = 0.
vii) **CPE addr16** ;Call on Parity Even   -   Call if parity flag   = 1.
viii) **CPO addr16** ;Call on Parity Odd   -   Call if parity flag   = 0.

Three-byte instruction **Two or five machine cycles:**	**Condition False**		**Condition True**	
Immediate addressing	Opcode fetch	- 6T	Opcode fetch	- 6T
	Memory read	- 3T	Memory read	- 3T
		9T	Memory read	- 3T
			Memory write	- 3T
			Memory write	- 3T
				18T

5.   RET $(PC)_L \leftarrow ((SP))$; $(SP) \leftarrow (SP) + 1$

                         $(PC)_H \leftarrow ((SP))$; $(SP) \leftarrow (SP) + 1$

(RET - Return to the main program)

It is an unconditional return instruction. This instruction is placed at the end of the subroutine program, in order to return to the main program. When this instruction is executed, the top of the stack is poped to (loaded in) the program counter .

> *Note :   While calling the subroutine using CALL instruction, the return address of the main program is pushed to the stack. The return instruction, (RET) pops that to the program counter. Thus the processor resumes the execution of main program.*

One-byte instruction	**Three machine cycles:** Opcode fetch	- 4 T
Register indirect addressing	Memory read	- 3 T
	Memory read	- 3 T
		10 T

6.   R<condition>

If <condition> is TRUE then,

$(PC)_L \leftarrow ((SP))$; $(SP) \leftarrow (SP) + 1$

$(PC)_H \leftarrow ((SP))$; $(SP) \leftarrow (SP) + 1$

It is conditional return instruction.

In a conditional return instruction a flag condition is tested. If the flag condition is true, then the program control return to main program by poping the top of the stack to the program counter . If the flag condition is false, then the next instruction is executed.

There are eight conditional return instructions:

i)	**RZ**	;Return on Zero	-	Return if zero flag	= 1.
ii)	**RNZ**	;Return on Not Zero	-	Return if zero flag	= 0.
iii)	**RC**	;Return on Carry	-	Return if carry flag	= 1.
iv)	**RNC**	;Return on No Carry	-	Return if carry flag	= 0.
v)	**RM**	;Return on Minus	-	Return if sign flag	= 1.
vi)	**RP**	;Return on Positive	-	Return if sign flag	= 0.
vii)	**RPE**	;Return on Parity Even	-	Return if parity flag	= 1.
viii)	**RPO**	;Return on Parity Odd	-	Return if parity flag	= 0.

One-byte instruction **One or three machine cycles:**	**Condition False**	**Condition True**	
Register indirect addressing	Opcode fetch - 6T	Opcode fetch	- 6T
		Memory read	- 3T
		Memory read	- 3T
			12T

7. **RST n**

It is a restart instruction. The restart instructions are also called software interrupts. Each restart instruction has a vector address. The vector address is fixed by the manufacturer (INTEL).

When a restart instruction is executed, the content of the program counter is pushed to the stack and the vector address is loaded in the program counter. The vector address is internally generated (computed) by the processor. The vector address for RST n is obtained by multiplying n by 8. Thus the program control is branched to a subroutine program stored in this vector address.

One-byte instruction

Register indirect addressing

Three machine cycles: Opcode fetch - 6 T

Memory write - 3 T

Memory write - 3 T

12T

There are eight restart instructions.

RST 0   RST 1   RST 2   RST 3   RST 4   RST 5   RST 6   RST 7

The vector addresses for the restart instructions are listed in the table given below:

Restart instruction	Vector address	Computation of vector address
RST 0	0000_H	$0 \times 8 = 0_{10} = 0_H$
RST 1	0008_H	$1 \times 8 = 8_{10} = 8_H$
RST 2	0010_H	$2 \times 8 = 16_{10} = 10_H$
RST 3	0018_H	$3 \times 8 = 24_{10} = 18_H$
RST 4	0020_H	$4 \times 8 = 32_{10} = 20_H$
RST 5	0028_H	$5 \times 8 = 40_{10} = 28_H$
RST 6	0030_H	$6 \times 8 = 48_{10} = 30_H$
RST 7	0038_H	$7 \times 8 = 56_{10} = 38_H$

8. **PCHL**   **(PC) ← (HL)**

The content of the HL register pair is moved to the program counter. Since this instruction alters the content of the program counter, the program control is transferred to a new address. This instruction is used by the system designer to implement the system subroutine to execute a program.

One-byte instruction

Implied addressing

One machine cycle: Opcode fetch - 6T

2.6.4 MACHINE CONTROL INSTRUCTIONS

1. **DI**

(DI - Disable Interrupts)

When this instruction is executed, all the interrupts except TRAP are disabled. [When the interrupts are disabled the processor will not accept or recognize the interrupt request made by the external devices through the interrrupt pins. When the processor is doing an emergency work, it can execute DI instruction to prevent the interrupts from interrupting the processor.]

One-byte instruction

One machine cycle: Opcode fetch - 4T

2.   EI

(EI - Enable Interrupts)

This instruction is used (or executed) to allow the interrupts after disabling. (The interrupts except TRAP are disabled after processor reset or after execution of DI instruction. When we want to allow the interrupts, we have to execute EI instructions.)

One-byte instruction   **One machine cycle:** Opcode fetch - 4T

3.   SIM

(SIM - Set Interrupt Mask)

The SIM instruction is used to mask the hardware interrupts RST 7.5, RST 6.5 and RST 5.5. It is also used to send data through the SOD line. (SOD: Serial Output Data pin of the 8085 processor.) The execution of SIM instruction uses the content of the accumulator to perform the following functions:

i)   Program the interrupt mask for the hardware interrupts RST 5.5, RST 6.5 and RST 7.5.

ii)   Reset the edge-triggered RST 7.5 input latch.

iii)   Load the SOD output latch.

The bits in the accumulator before execution of the SIM instruction are defined as shown in the Fig. 2.2.

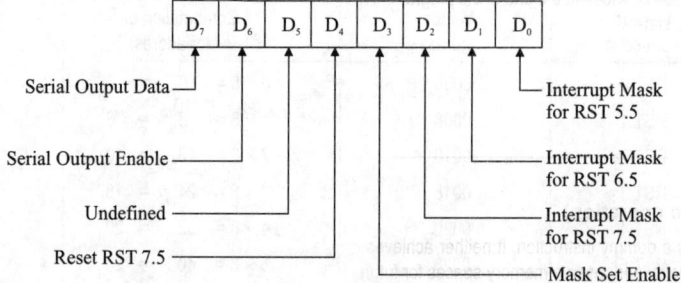

Fig. 2.2: Accumulator content before execution of SIM instruction.

If the mask set enable bit is set to "1" then the interrupt mask bits for RST 7.5, RST 6.5 and RST 5.5 (D_0, D_1 and D_2) are recognized and if it is "0" then these bits are not recognized by the processor. The interrupt mask bits D_0, D_1 and D_2 can be independently set to "1" to mask the particular interrupt and reset to "0" to unmask the particular interrupt.

If the bit D_4 is set to "1", then an internal flip-flop is reset to "0" in order to disable the RST 7.5 interrupt. If the serial output enable is "1", the serial output data is sent to the SOD pin.

One-byte Instruction   **One machine cycle:** Opcode fetch - 4T

4.   RIM

(RIM - Read Interrupt Mask)

The RIM instruction is used to check whether an interrupt is masked or not. It is also used to read data from the SID line. (SID: Serial Input Data pin of 8085 processor).

When a RIM instruction is executed, the accumulator is loaded with 8-bit data. The 8-bit data in the accumulator (content of accumulator) can be interpretted as shown in Fig. 2.3.

Bits D_0, D_1 and D_2 provide the mask status of the RST 5.5, RST 6.5 and RST 7.5 interrupts respectively. If the mask bit corresponding to a particular RST is "1", then the interrupt is masked and if the mask bit is "0" then the interrupt is unmasked.

If the interrupt enable bit (D_3) is "0", the 8085's maskable interrupts are disabled. The interrupts are enabled if this bit is "1".

A "1" in a particular interrupt pending bit indicates that an interrupt is being requested on the identified RST line. When this bit is "0", no interrupt is waiting to be serviced. The serial input data (bit D_7) indicate the value of the signal at the SID pin.

One-byte instruction   **One machine cycle:** Opcode fetch - 4T

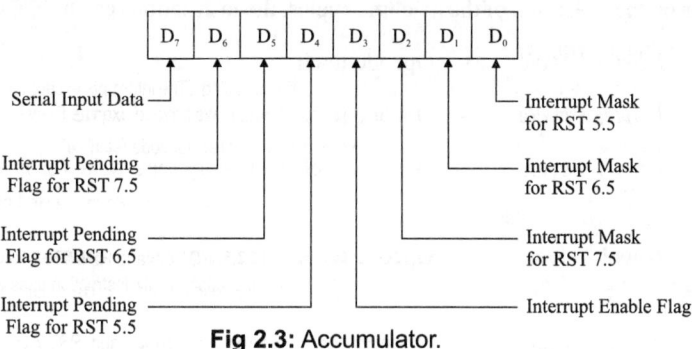

Fig 2.3: Accumulator.

5. **HLT**
   (HLT - Halt program Execution)

   This instruction is placed at the end of the program. When this instruction is executed, the processor suspends program execution and bus will be in idle state.

   One-byte instruction   **Two machine cycle:**Opcode fetch - 3T

   Bus idle   -   2T

   5T

6. **NOP**
   (NOP - No operation)

   The NOP is a dummy instruction, it neither achieves any result nor affects any CPU registers. This is an useful instruction for producing software delay and reserve memory spaces for future software modifications.

   One-byte instruction   **One machine cycle :** Opcode fetch - 4T

2.7 TIMING DIAGRAM OF 8085 INSTRUCTIONS

The 8085 instructions is one to five machine cycles. (Refer Table 2.1 for the machine cycles of instructions.) Actually, the execution of an instruction is the execution of the machine cycles of that instruction in a predefined order. Therefore, from the knowledge of the timing diagrams of machine cycles, the timing diagram of an instruction can be obtained.

The machine cycles of an 8085 instuction can be divided into two parts as shown below:

Machine cycles of an instruction

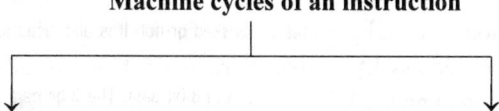

Machine cycles to fetch instruction
        bytes from memory.

One-byte instruction   : **Opcode fetch**

Two-byte instruction   : **Opcode fetch**
                              + memory read

Three-byte instruction   : **Opcode fetch**
                              + memory read
                              + memory read

Additional machine cycles for external read/ write with memory/IO in order to complete instruction execution. These machine cycles depend on instruction execution logic.

Based on the execution of the machine cycles, the instructions can be classified as shown below:

Case(i) : *1-byte, 1-cycle*   - Opcode fetch.

Case(ii) : *1-byte, 2-cycle*   - Opcode fetch + memory read/write.

Case(iii) : *1-byte, 3-cycle*   - Opcode fetch + memory read/write (or Bus idle)
                                                          + memory read/write (or Bus idle).

Case(iv) : *1-byte, 5-cycle*   - Opcode fetch + memory read + memory read
                                                          + memory write + memory write.

Case(v) : *2-byte, 2-cycle*   - Opcode fetch + memory read (read second byte of instruction.)

Case(vi) : *2-byte, 3-cycle*   - Opcode fetch + memory read (read second byte of instruction)
                                                          + memory read/write/or IO read/write.

Case(vii) : *3-byte, 3-cycle*   - Opcode fetch + memory read (read second byte of instruction)
                                                          + memory read (read third byte of instruction.)

Case(viii) : *3-byte, 4-cycle*   - Opcode fetch + memory read (read second byte of instruction)
                                                          + memory read (read third byte of instruction)
                                                          + memory read/write.

Case(ix) : *3-byte, 5-cycle*   - Opcode fetch + memory read (read second byte of instruction)
                                                          + memory read (read third byte of instruction)
                                                          + memory read/write + memory read/write.

The timing diagram of an instruction is obtained by drawing the timing diagrams of the machine cycles of that instruction one by one in the order of execution. The timing diagrams of few instructions are presented from Figs. 2.4 to 2.10.

2.7.1 TIMING DIAGRAM OF STA INSTRUCTION

The **"STA addr16"** instruction is used to store the content of the accumulator to a memory location. This instruction employs direct addressing. Let the content of the accumulator be $C7_H$ and it is desired to store the content of the accumulator to a memory location $526A_H$.

The STA addr16 instruction is a three byte instruction. The first byte is the opcode of the instruction 32_H. The second byte is low byte address $6A_H$ and the third byte is high byte address 52_H. Let the three bytes of the instructions be stored in memory locations $41FF_H$, 4200_H and 4201_H.

In order to execute this instruction, the 8085 microprocessor will first execute opcode fetch machine cycle to get the opcode, followed by two memory read cycles to read the address of data (i.e., to read second and third byte of instruction). Then, the processor executes the memory write cycle to store the content of the accumulator in the memory. The status of various signals during execution of this instruction are shown in Fig. 2.4.

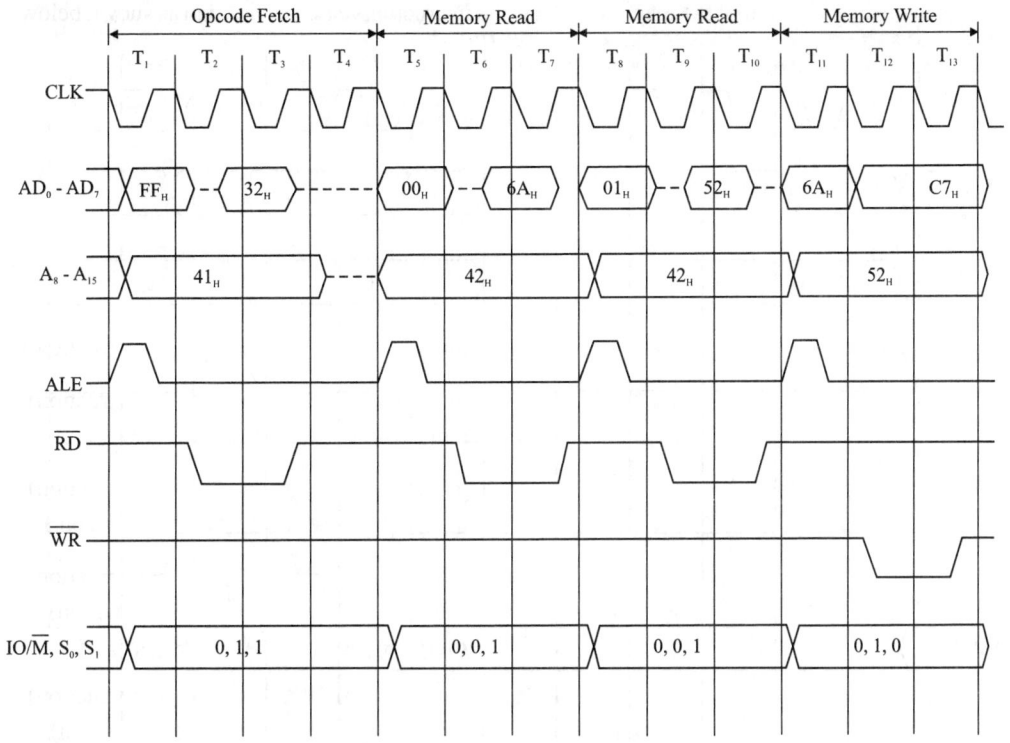

Fig. 2.4: Timing diagram of STA 526AH instruction.

2.7.2 TIMING DIAGRAM OF PUSH INSTRUCTION

The **"PUSH rp"** instruction is used to store the content of a register pair in the stack memory. This instruction employs register indirect addressing using Stack Pointer (SP). Let us consider PUSH B instruction. On execution of this instruction, the content of the BC pair is pushed to the stack. Let the content of the BC pair be $E25D_H$ and the content of SP be $A100_H$.

The PUSH rp is one-byte instruction and it is the opcode of the instruction. The opcode of PUSH B instruction is $C5_H$ and let it be stored in memory location $C010_H$. In order to execute this instruction, the processor will first execute the opcode fetch cycle to get the opcode $C5_H$. Then the processor executes two memory write cycles to store the content of the BC pair in the stack memory. The status of various signals during execution of this instruction are shown in Fig. 2.5.

During the memory write cycles in PUSH rp instruction, the content of the SP is used as the memory address. In the first write cycle, the content of the SP is decremented by one ($A100_H - 1 = A0FF_H$) and output on the address lines and in this address, the content of B-register ($E2_H$) is stored. In the second write cycle, the content of the SP is again decremented by one ($A0FF_H - 1 = A0FE_H$) and output on the address lines and in this address, the content of C-register ($5D_H$) is stored.

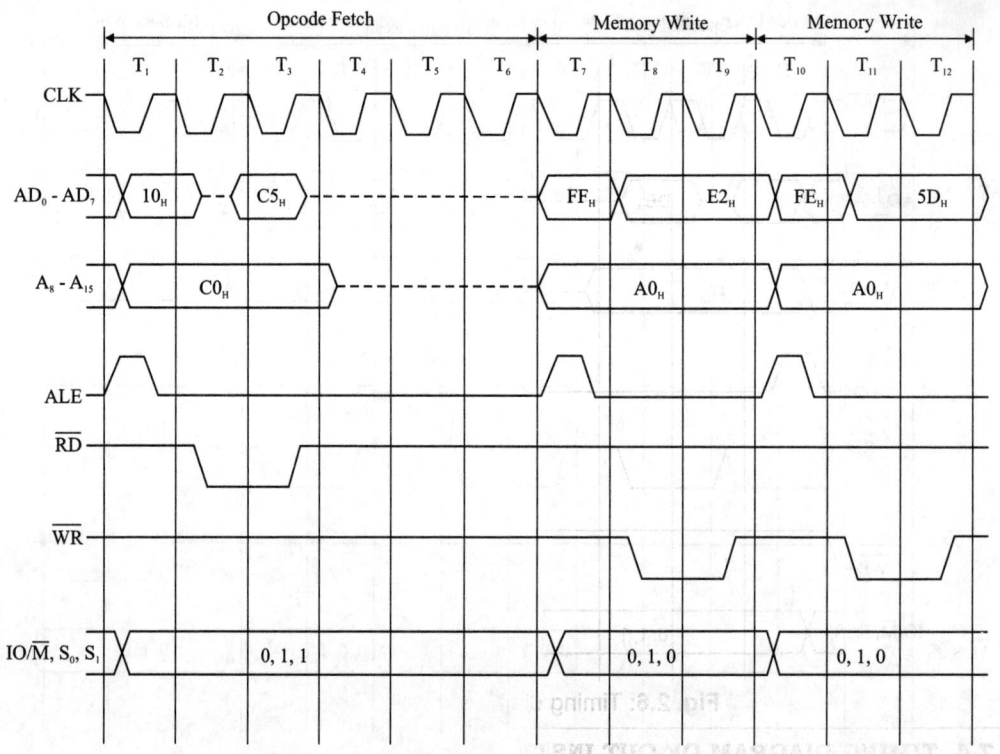

Fig. 2.5: Timing diagram of PUSH B instruction.

2.7.3 TIMING DIAGRAM OF IN INSTRUCTION

The **"IN addr8"** instruction is used to read the content of an IO-mapped device/port and store in the accumulator. For addressing IO-mapped devices the 8085 microprocessor employs 8-bit address. Let the 8-bit address of the IO port be CO_H and the content of IO port be $5E_H$.

The IN addr8 instruction, is a two-byte instruction. The first byte is the opcode of the instruction DB_H and the second byte is the IO port address CO_H. Let the two bytes of the instruction be stored in memory locations 4125_H and 4126_H.

In order to execute this instruction, the 8085 microprocessor will first execute the opcode fetch machine cycle to get the opcode, followed by the memory read cycle to read the IO port address (i.e., to read the second byte of the instruction.) Then, the processor executes IO read cycle to read the content of the IO port. The status of various signals during execution of this instruction are shown in Fig. 2.6.

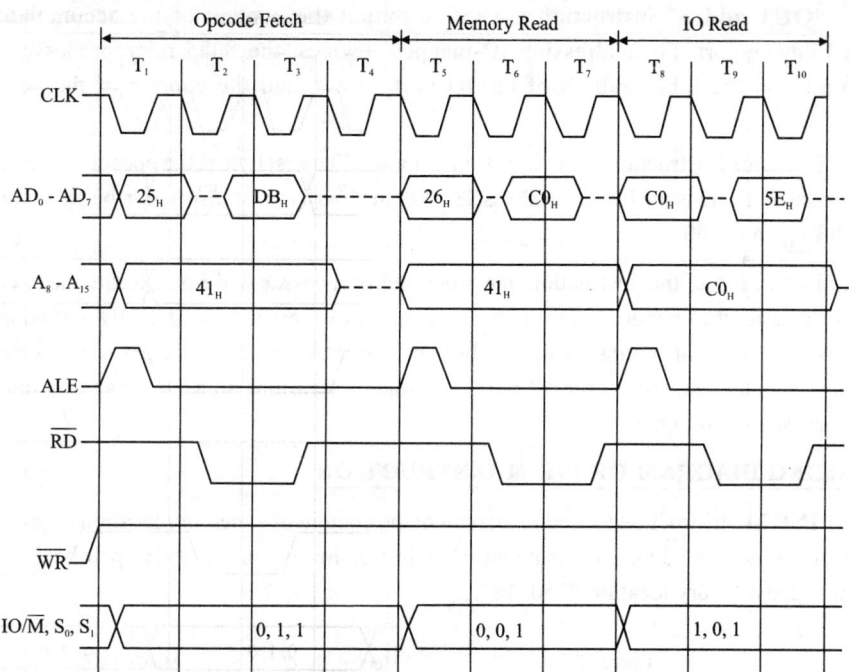

Fig. 2.6: Timing diagram of IN C0H instruction.

2.7.4 TIMING DIAGRAM OF OUT INSTRUCTION

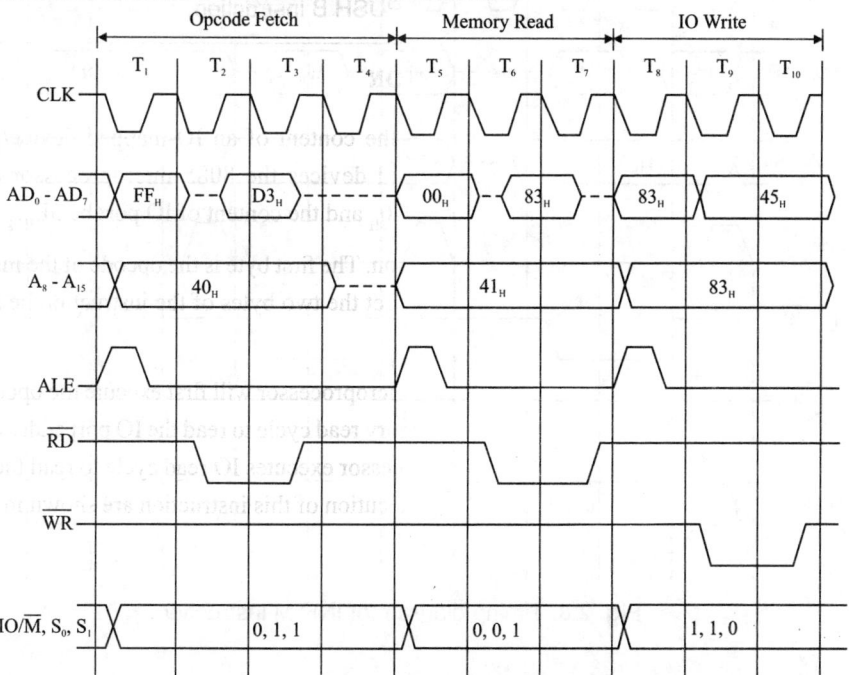

Fig. 2.7: Timing diagram of OUT 83H instruction.

The **"OUT addr8"** instruction is used to output the content of the accumulator to the IO-mapped device/port. For addressing IO-mapped devices, the 8085 microprocessor employs 8-bit address. Let the 8-bit address of the IO port be 83_H and the content of the accumulator be 45_H.

The OUT addr8 instruction is a two-byte instruction. The first byte is the opcode of the instruction $D3_H$, and the second byte is the IO port address 83_H. Let the two bytes of instruction be stored in memory locations $40FF_H$ and 4100_H.

In order to execute this instruction, the 8085 microprocessor will first execute the opcode fetch machine cycle to get the opcode, followed by the memory read cycle to read the IO port address. (i.e., to read the second byte of the instruction.) Then the processor executes the IO write cycle to write the content of the accumulator to the IO port. The status of various signals during execution of this instruction are shown in Fig. 2.7.

2.7.5 TIMING DIAGRAM OF INR M INSTRUCTION

The **"INR M"** instruction is used to increment the content of a memory location. This instruction employs register indirect addressing using an HL pair. Let the content of the HL pair be 4250_H and let the content of the memory location 4250_H be 12_H.

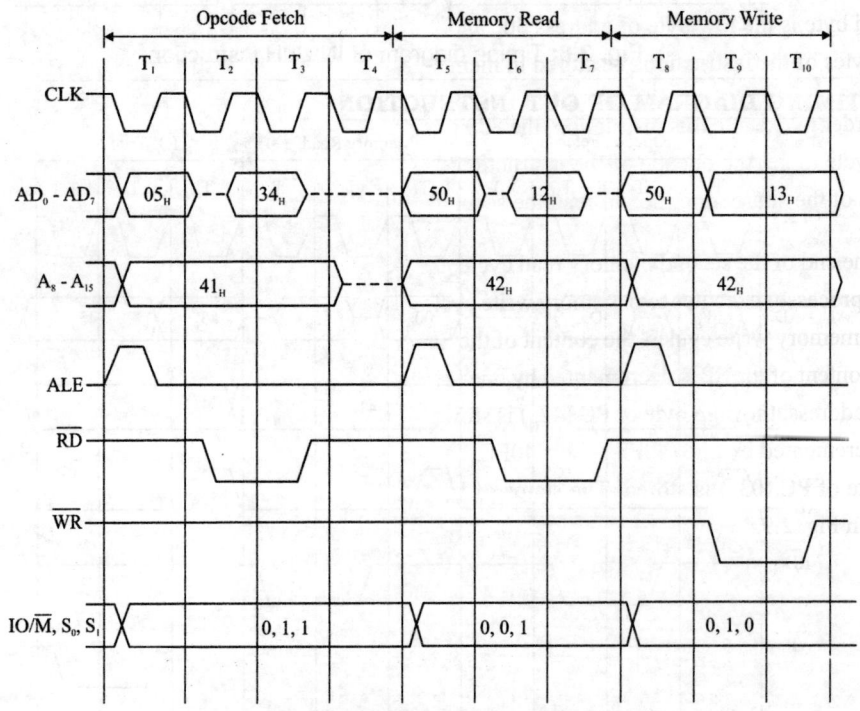

Fig. 2.8: Timing diagram of INR M instruction.

The INR M is one-byte instruction and it is the opcode of instruction 34_H. Let this instruction be stored in the memory location 4105_H. In order to execute the instruction, the 8085 microprocessor will first execute the opcode fetch cycle to get the opcode 34_H. Then, it executes the memory read cycle to read the content of the memory location 4250_H.

The content (12_H) of the memory location is incremented by one in the ALU and then, the processor executes the memory write cycle to store the result (13_H) of the ALU operation in the same memory location 4250_H. The status of various signals during execution of this instruction are shown in Fig. 2.8.

2.7.6 TIMING DIAGRAM OF CALL INSTRUCTION

The "CALL addr16" instruction is used to execute a subroutine/procedure stored at addr16, after saving the address of the next instruction in the stack memory. On execution of this instruction, the addr16 is loaded in the **Program Counter (PC)** and the previous value of the PC is stored in the stack memory pointed by the **Stack Pointer (SP)**. Let the address of the subroutine be $4F50_H$ and the content of SP be 4100_H.

The CALL addr16 is a three-byte instruction. The first byte is the opcode of the instruction CD_H. The second byte is the low byte of address 50_H and the third byte is the high byte of address $4F_H$. Let the three bytes of the instructions be stored in memory locations 4200_H, 4201_H and 4202_H.

In order to execute this instruction, the 8085 microprocessor will first execute the opcode fetch machine cycle to get the opcode of the instruction CD_H, followed by two memory read cycles to get the address of the subroutine (i.e., to read the second and third byte of instruction).

At the end of the second memory read cycle, the content of the PC will be 4203_H. After the read cycles, the processor executes two memory write cycles to store this content (4203_H) of PC in the stack. During the memory write cycles, the content of the SP is used as the memory address. In the first write cycle, the content of the SP is decremented by one ($4100_H - 1 = 40FF_H$) and output on the address lines and in this address, the high byte of PC (42_H) is stored. In the second write cycle, the content of the SP is again decremented by one ($40FF_H - 1 = 40FE_H$) and output on the address lines and in this address the low byte of PC (03_H) is stored. The status of various signals during execution of this instruction are shown in Fig. 2.9.

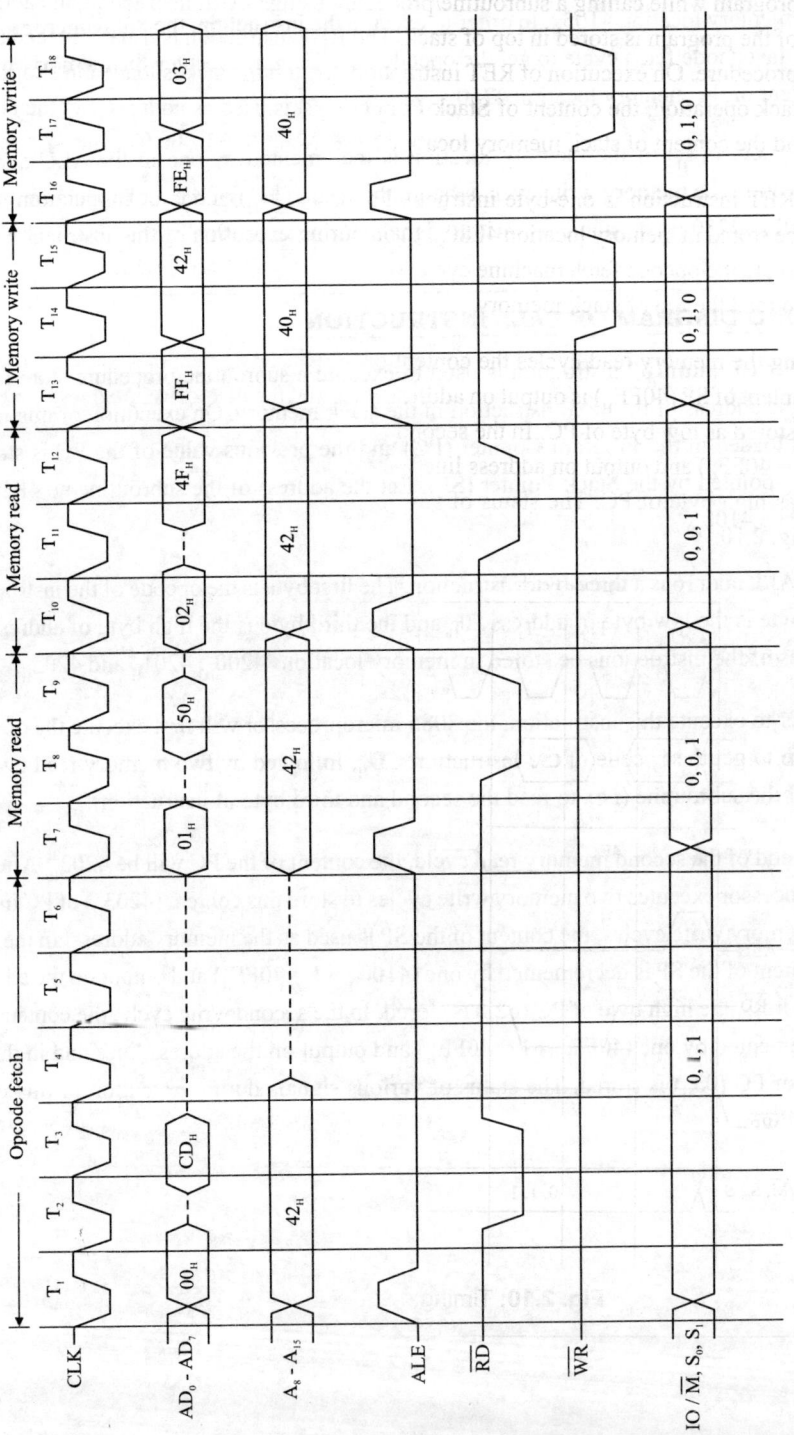

Fig 2.9: Timing diagram of CALL 4F50H instruction.

2.7.7 TIMING DIAGRAM OF RET INSTRUCTION

In a program while calling a subroutine/procedure using CALL instruction, the address of next instruction of the program is stored in top of stack. The RET instruction is usually placed at the end of subroutine/procedure. On execution of RET instruction, the top of stack is loaded in **Program Counter** (PC). For stack operation, the content of **Stack Pointer** (SP) is used as address. Let the content of SP be $40FE_H$ and the content of stack memory locations $40FE_H$ and $40FF_H$ be 03_H and 42_H, respectively.

The RET instruction is one-byte instruction and it is the opcode of instruction $C9_H$. Let this instruction be stored in memory location $4F80_H$. In order to execute this instruction, the 8085 processor will first execute the opcode fetch machine cycle to get the opcode $C9_H$. Then it executes two memory read cycle to read the top of stack memory.

During the memory read cycles the content of SP is used as memory address. In the first read cycle the content of SP ($40FE_H$) is output on address lines and the memory content (03_H) in this address is read and stored as low byte of PC. In the second read cycle the content of SP is incremented by one ($40FE_H + 1 = 40FF_H$) and output on address lines and the memory content (42_H) in this address is read and stored as high byte of PC. The status of various signals during execution of this instruction are shown in Fig. 2.10.

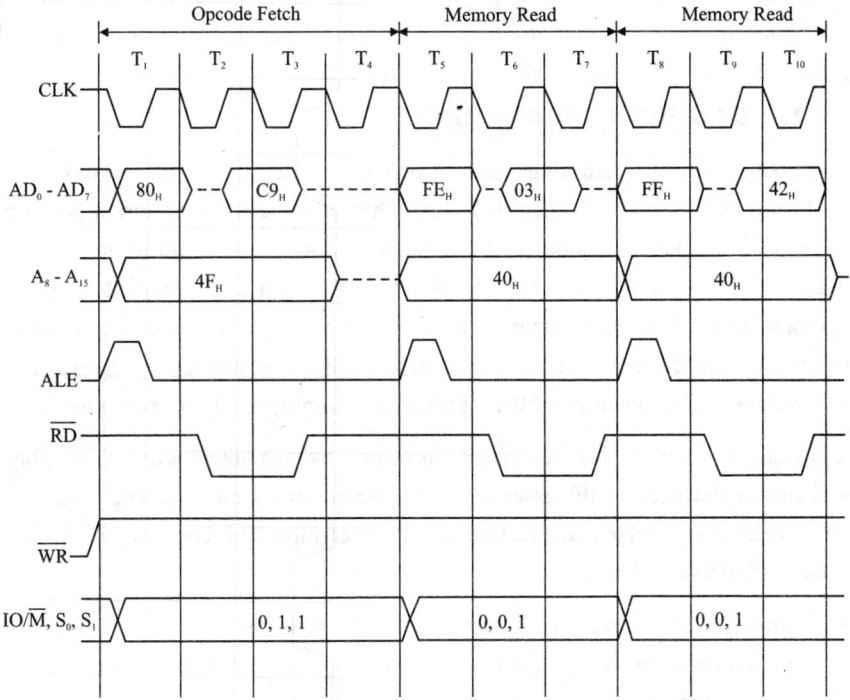

Fig. 2.10: Timing diagram of RET instruction.

2.8   LEVELS OF PROGRAMMING

Programs are a set of instructions or commands needed to perform a specific task by a programmable device such as a microprocessor. The programs needed for a programmable device can be developed at three different levels and they are as follows:

1. **Machine level programming**
2. **Assembly level programming**
3. **High level programming**

2.8.1   MACHINE LEVEL PROGRAMMING

In machine level programming, instructions are written using binary codes which uses only two symbols '0' and '1'. The manufacturer of microprocessors will give a set of instructions for each microprocessor in binary codes, i.e., one binary code will represent one operation performed by the microprocessor. The language in which the instructions are represented by binary codes is called machine language. A microprocessor can understand and execute the machine language programs directly.

The binary instructions of one microprocessor will not be same as that of another microprocessor. Therefore, the machine language programs developed for one microprocessor cannot be used for another microprocessor i.e., the machine level programs are machine dependent. Moreover, it is highly tedious for a programmer to write programs in the machine language.

2.8.2   ASSEMBLY LEVEL PROGRAMMING

In assembly level programming, instructions are written using mnemonics. A mnemonic comprises of a few letters of the English language which represent the operation performed by the instruction. For example, the mnemonic for the instruction which performs **addition** operation is **ADD**. The manufacturer of the microprocessors will provide a set of instructions in the form of a mnemonic for each microprocessor. Also, for each mnemonic a binary code will be specified by the manufacturer. If the program is developed using binary codes then it is called machine level programming and if the program is developed using mnemonics then it is called assembly level programming.

The language in which the instructions are represented by mnemonics is called assembly language. Microprocessors cannot execute the assembly language programs directly. The assembly language programs have to be converted to machine language for execution. This conversion is performed using a software tool called assembler.

The mnemonics of one microprocessor will not be same as that of another microprocessor. Therefore, the assembly language programs developed for one microprocessor cannot be used for another microprocessor directly i.e. the assembly language programs are machine dependent. But certain manufacturers provide upward compatability for the same family of microprocessors. (i.e., the program developed for a the lower version of a microprocessor of a family can be run on the higher version without modifications.) For example, consider the INTEL 80x86 family of microprocessors. The program developed for 8086 microprocessor can be run on 80186, 80286, 80386 or 80486 microprocessor-based system without any modifications.

2.8.3 HIGH LEVEL PROGRAMMING

In high level programming the instructions will be in the form of statements written using symbols, English words and phrases. Each high level language will have its own vocabulary of words, symbols, phrases and sentences. Examples of high level languages are BASIC, C, C++, etc. The programs written in high level languages are easy to understand and machine independent. So they are known as portable programs. A high level language program has to be converted into machine language programs in order to be executed by the microprocessor. This conversion is performed by a software tool called compiler.

2.9 FLOWCHART

Flowchart is a graphical representation of the operation flow of a program. It is also the graphical form of an algorithm. Flowcharts can be a valuable aid in visualizing programs. The various symbols used for drawing flowcharts are shown in Fig. 2.11. The operations represented by various symbols of flowchart are explained in Table 2.4. A sample flowchart is shown in Fig. 2.12.

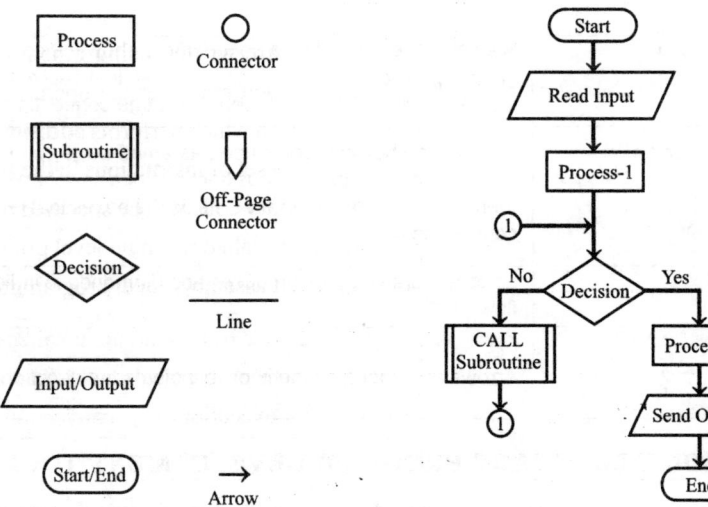

Fig. 2.11: Symbols used in a flowchart.   **Fig. 2.12:** A sample flowchart.

Table 2.4: Operations Represented by the Symbols used in Flowchart

Symbol	Operation
Racetrack shape box	A racetrack shaped symbol is used to indicate the beginning (start) or end of a program.
Parallelogram	A parallelogram is used to represent input or output operation.
Rectangular box	A rectangular box is used to represent simple operations other than input and output operations.
A rectangular box with double lines on vertical sides	A rectangular box with double lines on vertical sides is used to represent a subroutine or procedure.
Diamond shaped box	A diamond shaped box is used to represent a decision point or cross road in the programs
Small circle	A small circle is used as a connector to show the connections between various parts of a flowchart within a page. Identical numbers are entered inside the circles that represent the same connecting points.
Five-sided box	A five-sided box symbol is used as an off-page connector to show the connections between various sections of a flowchart in different pages. Identical numbers are entered inside the boxes that represent the same connecting point.
Line ———	Lines are drawn between boxes and diamonds to indicate the program flow.
Arrow →	Arrows are placed on the lines to indicate the direction of program flow.

2.10 ASSEMBLY LANGUAGE PROGRAM DEVELOPMENT TOOLS

Development system is used by system designers to design and test the software and hardware of a microprocessor-based system before going for practical implementation (or fabrication). The microprocessor development system consists of a set of hardware and software tools. The hardware of a development system usually contain a standard PC (**P**ersonal **C**omputer), printer and an emulator. The software tools are also called program development tools and they are editor, assembler, library builder, linker, debugger and simulator. These software tools can run on a PC in order to write, assemble, debug, modify and test the assembly language programs.

2.10.1 EDITOR (TEXT EDITOR)

Editor is a software tool which, when run on a PC, allows the user to type/enter and modify the assembly language program. The editor provides a set of commands for insertion, deletion and modification of letters, characters, statements, etc. The main function of an editor is to help the user to construct the assembly language program in the right format. The program created using editor is known as source program and it is usually saved with the file extension .ASM For example, if a program for addition is developed using editor then it can be saved as ADDITION.ASM. Some examples of editors are NE (Norton Editor), EDIT (DOS Editor), etc.

2.10.2 ASSEMBLER

The assembler is a software tool which, when run on a PC, converts the assembly language program to a machine language program. Several types of assemblers are available and they are one-pass assembler, two-pass assembler, macro assembler, cross assembler, resident assembler and meta assembler.

In one-pass assembler the source code is processed only once and we can use only backward reference. In a one-pass assembler as the source code is processed, any labels encountered are given an address and stored in a table. Whenever a label in encountered, the assembler may look backward to find the address of the label. If the label is not yet defined then it issues an error message (because the assembler will not look forward). Since only one pass is used to translate the source code, a one-pass assembler is very fast, but because of the forward reference problem, the one-pass assembler is not used often.

Most of the popularly used assemblers are the two-pass assemblers. In a two-pass assembler, the first pass is made through source code for the purpose of assigning an address to all the labels and to store this information in a symbol table. The second pass is made to actually translate the source code into machine code.

The input for the assembler is the source program which is saved with file extension .ASM. The assembler usually generates two output files called object file and list file. The object file consist of relocatable machine codes of the program and it is saved with file extension .OBJ. The list file contains the assembly language statements, the binary codes for each instruction and address of each instruction. The list file is saved with file extension .LST.

The list file also indicates any syntax errors in the source program. The assembler will not identify the logical errors in the source program. In order to correct the errors indicated on the list file, the user have to use the editor again. The corrected source program is saved again and then reassembled. Usually, it may take several times through edit-assemble loop to eliminate the syntax errors from the source program.

Some examples of assemblers are TASM (Borland's Turbo Assembler), MASM (Microsoft's Macro Assembler), ASM86 (INTEL'S 8086 Assembler), ASM85 (INTEL'S 8085 Assembler), etc.

Advantages of the Assembler

1. The assembler translates mnemonics into binary code with speed and accuracy, thus eliminating human errors in looking up the codes.
2. The assembler assigns appropriate values to the variables used in a program. This feature offers flexibility in specifying jump locations.
3. It is easy to insert or delete instructions in a program and reassemble the entire program quickly with new memory locations and modified addresses for jump locations. This avoids rewriting the program manually.
4. The assembler checks syntax errors, such as wrong labels, opcodes, expressions, etc., and provides error messages. However, it cannot check logic errors in a program.
5. The assembler can reserve memory locations for data or results.
6. The assembler provides list file for documentation.

2.10.3 LIBRARY BUILDER

The library builder is used to create library files which are a collection of procedures of frequently used functions. Actually a library file is a collection of assembled object files. While developing a software for a particular application, the programmers can link the library files in their programs. When the library file is linked with a program, only the procedure required by the program are copied from library file and added to the program.

The input to library builder is a set of assembled object files of program modules/procedures. The library builder combines the program modules/procedures into a single file known as library file and it is saved with file extension ".LIB". Some examples of library builder are microsoft's LIB, Borlands TLIB, etc.

2.10.4 DEBUGGER

Debugger is a software tool that allows the execution of a program in a single step or break-point mode under the control of user. The process of locating and correcting the errors in a program using a debugger is known as debugging.

The debugger allows the designer to load the object code program into the memory of the PC, execute the program and troubleshoot or debug it. The debugger allows the designer to look at the contents of registers and memory locations after running the program. It allows the system designer to change the contents of registers and memory locations and return the program.

Some debuggers allow the user to stop execution after each instruction so that the memory/register content can be checked or altered. A debugger also allows the user to set a breakpoint at any point in user program. When the user runs the program, the PC will execute instructions up to this breakpoint and stop. The user can then examine register and memory contents to see whether the results are correct upto that point. If the results are correct, the user can move the breakpoint to a later point in the program. If the results are not correct, the user can check the program up to that point to find out why they are not correct.

Debugger tools can help the user to isolate a problem in the program. Once the problem/errors are identified, the algorithm can be modified. Then the user can use the editor to correct the source program, reassemble the corrected source program, relink and run the program again.

2.10.5 SIMULATOR

The simulator is a program which can run on the development system (Personal computer) to simulate the operations of the newly designed system. Some of the operations that can be simulated are as follows:

- **Execute a program and display result.**
- **Single step execution of a program.**
- **Break-point execution of a program.**
- **Display the contents of register/memory.**

Simulator usually shows the content of registers and memory locations on the screen of the computer and allows the system designer to perform all the operations listed above, with the added advantage of watching the data change as the program operates. This feature saves considerable time because the register/memory contents do not have to be displayed using separate commands. The visual representation also gives the programmer a better feel for what is taking place in the program.

The simulators do not have the ability to perform actual IO or internal hardware operations such as timing or data transmission and reception.

2.10.6 EMULATOR

An emulator is a mixture of hardware and software. It is usually used to test and debug the hardware and software of a newly designed microprocessor-based system. The emulator has a multicore cable which connects the PC of the development system and the newly designed hardware of the microprocessor system. A connector/plug at one end of the cable is plugged into new hardware in place of its microprocessor. The other end of cable is connected to parallel port of PC. Through this connection the software of the emulator allows the designer to download the object code program into RAM in the system being tested and run it.

Like a debugger, an emulator allows the system designer to load and run programs, examine and change the contents of registers, examine and change the contents of memory locations and insert breakpoints in the program.

The emulator also takes a snapshot of the content of registers, activity on the address and data bus and the state of the flags as each instruction executes. Also, the emulator stores this trace data. The user can have a printout of the trace data to see the results that the program produced on a step-by-step basis. Another powerful feature of an emulator is the ability to use either development system memory or the memory on the hardware under test for the program that is being debugged.

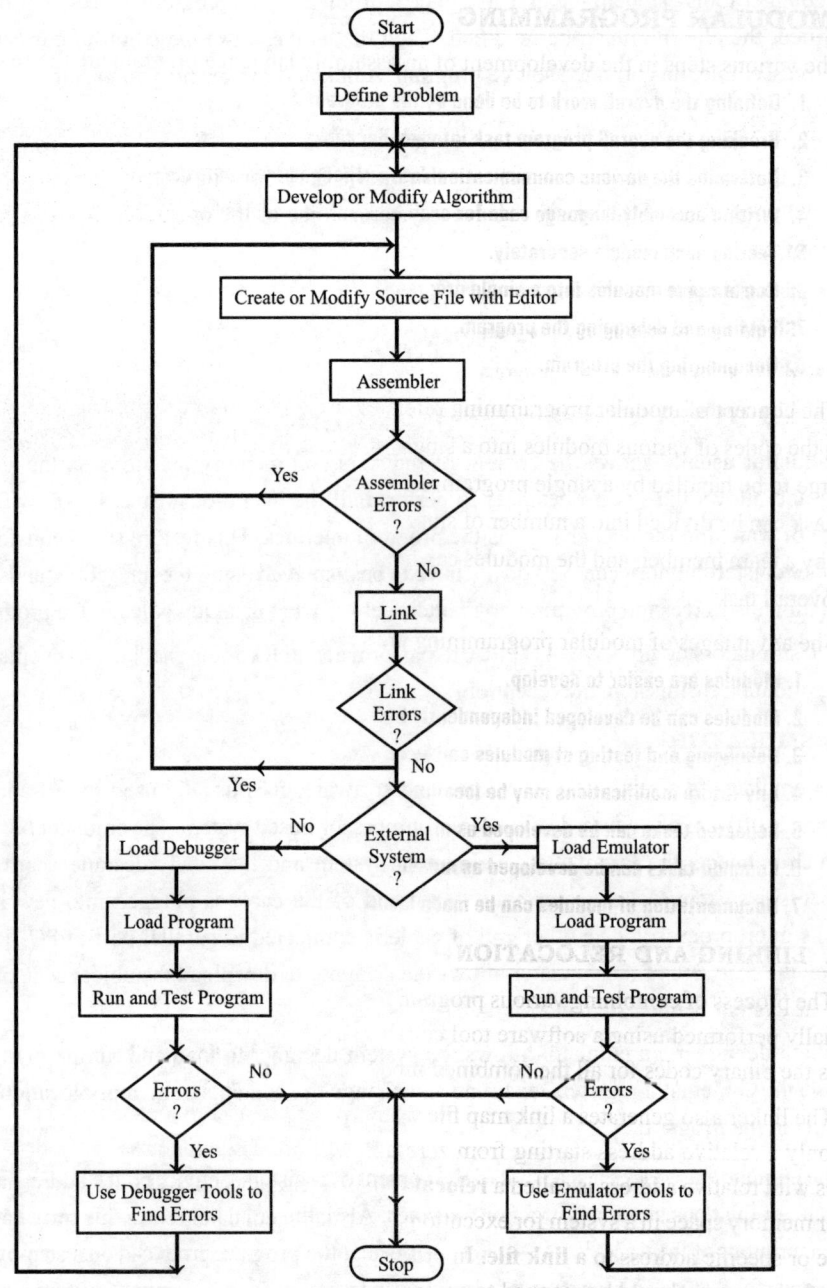

Fig. 2.13: Development process of an assembly language program.

2.11 MODULAR PROGRAMMING

The various steps in the development of an assembly language program are the following:

1. **Defining the overall work to be done by the program.**
2. **Breaking the overall program task into smaller tasks.**
3. **Determine the various communication/data exchange between tasks.**
4. **Writing assembly-language code for each task called modules.**
5. **Testing each module separately.**
6. **Combine the modules into a single program.**
7. **Testing and debugging the program.**
8. **Documenting the program.**

The concept of modular programming refers to development of program codes in modules and merging the codes of various modules into a single program code. When the program to be developed is too large to be handled by a single programmer, a team can be formed to develop the program. The overall task can be divided into a number of smaller tasks and each smaller task can be developed as a module by a team member, and the modules can be integrated by the team leader to obtain the program for the overall task.

The advantages of modular programming are given below:

1. **Modules are easier to develop.**
2. **Modules can be developed independently by different programmers.**
3. **Debugging and testing of modules can be carried out independently.**
4. **Any future modifications may be localized.**
5. **Repeated tasks can be developed as modules and stored as subroutine/macro.**
6. **Common tasks can be developed as modules and stored as library.**
7. **Documentation of modules can be made independently.**

2.11.1 LINKING AND RELOCATION

The process of combining various program modules into a single program is called **linking** and it is usually performed using a software tool called a linker. The linker will generate a link file which contains the binary codes for all the combined modules.

The linker also generates a link map file which contain the address information. The linker will assign only a relative address starting from zero and will not assign an absolute address. The linked modules with relative address is called a **relocatable program,** because this program can be loaded in any user memory space in a system for execution. A software tool called a **locator** can be used to assign absolute or specific address to a **link file**. In order to run a program in an 8085 microprocessor-based system, the program should be mapped to specific user memory address.

In a typical assembly language development process, the source code for various modules can be developed using any text editor like EDIT, NOTEPAD, WORDPAD, etc., and have to be saved as .ASM files. Then .ASM modules can be individually assembled using the software tool MASM assembler to generate .OBJ files then using the software tool LINK, the .OBJ files of all modules can be combined to generate a single relocatable .EXE file, and using a locator software tool called EXE2BIN, the .EXE file can be converted to .BIN file, and this .BIN file can be downloaded to the user memory space of the 8086 microprocessor-based system for execution.

2.11.2 SUBROUTINE

When a group of instructions are to be used several times to perform a same function in a program, then we can write them as a separate subprogram called procedure or subroutine. Whenever required the procedures can be called in a program using CALL instructions.

Procedures are written and assembled as separate program modules and stored in memory. When a procedure is called in the main program, the program control is transferred to the procedure and after executing the procedure the program control is transferred back to the main program. In 8085 processor, the instruction CALL is used to call a procedure in the main program and the instruction RET is used to return the control to the main program.

The main advantage of using a procedure is that the machine codes for the group of instructions in the procedure has to be put in memory only once. The disadvantages of using the procedure are the need for a stack and the overhead time required to call the procedure and return to the calling program.

2.11.3 HANDLING SUBROUTINE

While executing a program, if the 8085 processor encounters a CALL the instruction, then it saves the content of the program counter in a stack and loads the subroutine address in the program counter. (The content of program counter that is saved in stack is the address of the instruction next to CALL in the main program. The subroutine address is the address given in the CALL instruction.)

When the subroutine address is loaded in the program counter, the processor starts executing the subroutine. The last instruction of the subroutine will be RET instruction and when it is executed, the processor moves the top of the stack memory to the program counter. (The top of stack memory is the address which is saved in stack before executing subroutine.) Now the program control (execution) is returned to main program.

The subroutine program may use the registers that are used by the main program. If in the main program the content of these registers are to be preserved then they have to be saved (PUSHed) in stack before calling the subroutine. After returning from subroutine, they can be retrieved (POPed) from the stack back to the respective register. In 8085 the type of stack is LIFO(Last-In-First-Out). Hence, the order of retrieving (POPing) should be opposite to that of storing (PUSHing). For example, if the content of register pair HL is stored first followed by DE then while retrieving the DE pair should be poped first followed by HL pair.

2.11.4 DELAY ROUTINE

Delay routines are the subroutines used for maintaining the timings of various operations in a microprocessor. In control applications, certain equipment need to be ON/OFF after a specified time delay. In some applications, a certain operation has to be repeated after a specified time interval. In such cases simple time delay routines can be used to maintain the timings of the operations.

A delay routine is generally written as a subroutine (It need not be a subroutine always. It can even be a part of the main program.) In a delay routine a count (number) is loaded in a register of microprocessor. Then it is decremented by one and the zero flag is checked to verify whether the content of register is zero or not. This process is continued until the content of the register is zero. When it is zero the time delay is over and the control is transferred to the main program to carry out the desired operation.

The delay time is given by the total time taken to execute the delay routine. It can be computed by multiplying the total number of T-states required to execute the subroutine and the time for one T-state of the processor. The total of number of T-states can be computed from the knowledge of T-states required for each instruction. The time for one T-state of the processor is given by the inverse of the internal clock frequency of the processor. For example, if the 8085 microprocessor has 5 *MHz* quartz crystal then,

The internal clock frequency = $\dfrac{5}{2}$ = 2.5 *MHz*

Time for one T-state = $\dfrac{1}{2.5*10^6}$ = 0.4 *ms*

Two example delay routines that can be used in 8085 assembly language programs are presented in this section with details of timing calculations. For small time delays (< 0.5 millisecond) an 8-bit register can be used as counter, but for large time delays (< 0.5 second) 16-bit register should be used as counter. For very large time delays (>0.5 second), a delay routine can be repeatedly called in the main program. The disadvantage in delay routines is that the processor time is wasted. An alternate solution is to use a dedicated timer like 8253/8254 to produce time delays or to maintain timings of various operations.

2.11.5 STACK   *(AU, Nov/Dec' 19, 3 Marks)*

The stack is a portion of RAM memory defined by the user for temporary storage and retrieval of data while executing a program. The microprocessor will have a dedicated internal register called Stack Pointer (SP) to hold the address of the stack. Also, the processor will have a facility to automatically decrement/increment the content of SP after every write/read operation into stack.

The user can initialize or create a stack by loading a RAM address in the Stack Pointer (SP). Once an address is loaded in SP, the RAM memory locations below the address pointed by SP are reserved for stack. Typically 25 to 100 RAM memory locations are sufficient for stack. The user should take care that the reserved RAM memory locations for stack are not used for any other purpose.

The user has to create/implement a stack whenever the program consists of PUSH, POP, RST n, CALL and RET instructions. Also, the stack is needed whenever the system uses interrupt facility.

In a program, when the number of available registers are not sufficient for storing intermediate result and data, then some of intermediate result and data can be stored in a stack using PUSH instruction and retrieved whenever required using POP instruction.

The CALL instruction and the interrupts store the return address (content of program counter) in stack before executing the subroutine. Usually the subroutines are terminated with RET instruction. When RET instruction is executed, the top of stack is poped to program counter and the program control returns to the main program after the execution of subroutine.

Stack in 8085 Microprocessor

In an 8085 processor, the stack is created by loading a 16-bit address in the stack pointer. Upon reset, the stack pointer is cleared to zero.

In an 8085 processor, for every write operation into stack, the SP is automatically decremented by two and for every read operation from stack, the SP is automatically incremented by two. Hence, data can be stored only in lower addresses from the address pointed by SP. Therefore, we can say that the SP holds the address of the top of stack. All the RAM addresses higher than that pointed by the SP can be considered as occupied stack and all the RAM addresses lower than that pointed by the SP can be considered as empty stack as shown in Fig. 2.14. However, in practice only few memory locations are needed for stack.

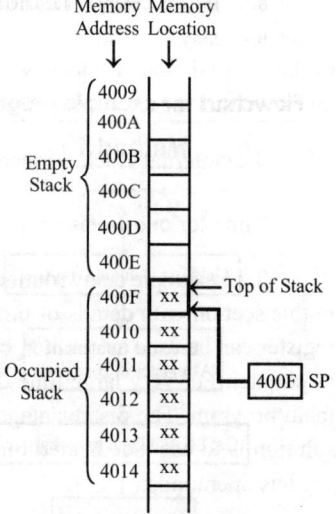

In 8085 processor the content of register pairs can be stored in stack using PUSH instruction and the stored information can be retrieved back to register pair using POP instruction. When a number of register pairs have to be stored and retrieved in the stack, the order of retrieval should be reverse of that of the order of storage. For example, let BC pair be pushed to stack first and DE pair next. When the stored information has to be retrieved to appropriate registers, the top of stack should be poped to the DE pair first and then to the BC pair next. The storage and retrieval in stack are in reverse order, because the SP is decremented for every write operation into stack and SP is incremented for every read operation from stack. Therefore, the stack in 8085 is called Last-In-First-Out (LIFO) stack, i.e., the last stored information can be read first.

Fig. 2.14: Example of stack in an 8085.

2.12 ASSEMBLY LANGUAGE PROGRAMMING

2.12.1 SIMPLE PROGRAM

EXAMPLE PROGRAM - 1: 8-Bit Addition

Write an assembly language program to add two numbers of 8-bit data stored in memory locations 4200_H and 4201_H and store the result in 4202_H and 4203_H

Problem Analysis

In order to perform addition in 8085, one of the data should be in accumulator and another data can be in any one of the general purpose register or in the memory. After addition, the sum is stored in the accumulator. The sum of two 8-bit data can be either 8-bits (sum only) or 9-bits (sum and carry). The accumulator can accommodate only the sum and if there is a carry, the 8085 will indicate by setting the carry flag. Hence, one of the registers is used to account for carry.

In method-1, direct addressing is used to address the data. But in method-2, register indirect addressing is used to the address data. Here, HL-register is used to hold the address of the data and it is called pointer.

Algorithm (Method-1)

1. Load the first data from memory to accumulator and move it to B-register.
2. Load the second data from memory to accumulator.
3. Clear C-register.
4. Add the content of B-register to accumulator.
5. Check for carry. If carry = 1, go to step 6 or if carry = 0, go to step 7.
6. Increment the C-register.
7. Store the sum in memory.
8. Move the carry to accumulator and store in memory.
9. Stop.

Flowchart for example program 1

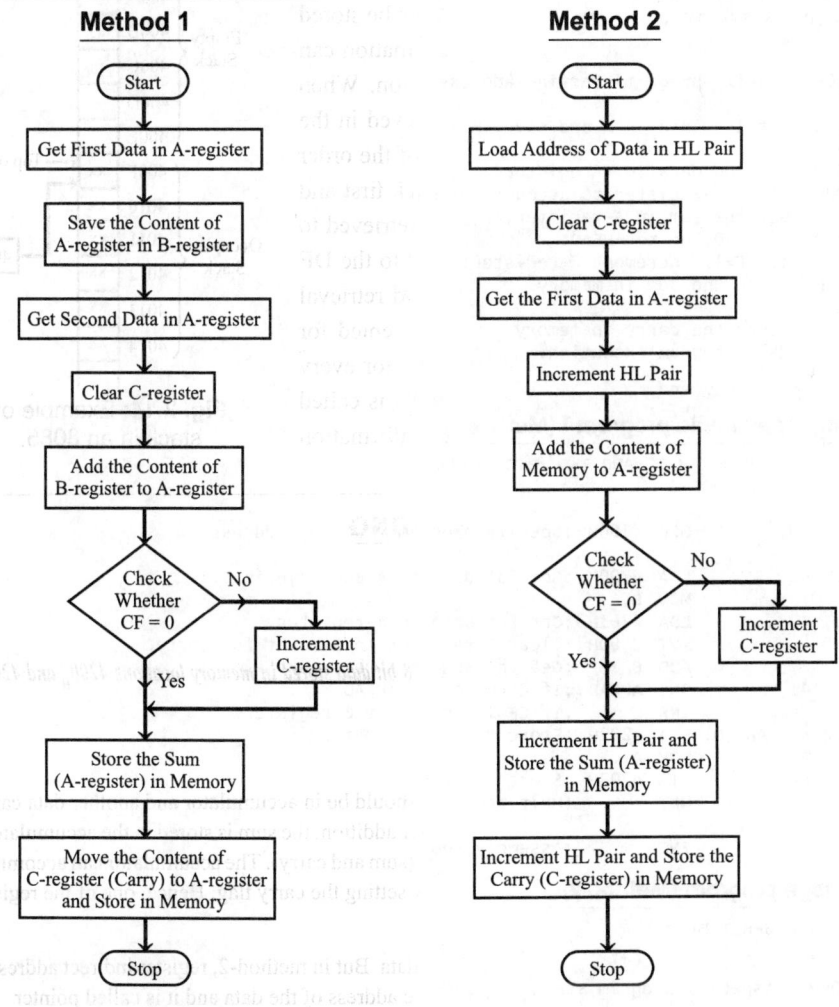

Algorithm (Method-2)

1. Load the address of the data memory in HL pair (i.e., set HL pair as pointer for data).
2. Clear C-register.
3. Move the first data from memory to accumulator.
4. Increment the pointer (HL pair).
5. Add the content of memory addressed by HL with accumulator.
6. Check for carry. If carry = 1, go to step 7 or if carry = 0, go to step 8.
7. Increment the C-register.
8. Increment the pointer and store the sum.
9. Increment the pointer and store the carry.
10. Stop.

Assembly language program (Method 1)

```
;PROGRAM TO ADD TWO 8-BIT DATA
;METHOD-1

        ORG  4100H  ;specify program starting address.

        LDA  4200H  ;Get 1st data in A and save in B.
        MOV  B,A
        LDA  4201H  ;Get 2nd data in A-register.
        MVI  C,00H  ;Clear C-register to account for carry.
        ADD  B      ;Get the sum in A-register.
        JNC  AHEAD  ;If CF=0, go to AHEAD.
        INR  C      ;If CF=1, increment C-register.
AHEAD:  STA  4202H  ;Store the sum in memory.
        MOV  A,C
        STA  4203H  ;Store the carry in memory.
        HLT         ;Halt program execution.

        END         ;Assembly end.
```

Assembler listing for example program 1 (Method 1)

```
 1                          ;PROGRAM TO ADD TWO 8-BIT DATA
 2                          ;METHOD-1
 3  0000
 4  4100                            ORG  4100H  ;specify program starting address.
 5
 6  4100  3A 00 42                  LDA  4200H  ;Get 1st data in A and save in B.
 7  4103  47                        MOV  B,A
 8  4104  3A 01 42                  LDA  4201H  ;Get 2nd data in A-register.
 9  4107  0E 00                     MVI  C,00H  ;Clear C-register to account for carry.
10  4109  80                        ADD  B      ;Get the sum in A-register.
11  410A  D2 0E 41                  JNC  AHEAD  ;If CF=0, go to AHEAD.
12  410D  0C                        INR  C      ;If CF=1, increment C-register.
13  410E  32 02 42    AHEAD:        STA  4202H  ;Store the sum in memory.
14  4111  79                        MOV  A,C
15  4112  32 03 42                  STA  4203H  ;Store the carry in memory.
16  4115  76                        HLT         ;Halt program execution.
17
18  4116                           END         ;Assembly end.
```

Assembly language program (Method 2)

```
;PROGRAM TO ADD TWO 8-BIT DATA
;METHOD-2

        ORG  4100H   ;specify program starting address.

        LXI  H,4200H ;Set pointer for data.
```

```
        MVI  C,00H    ;Clear C-register to account for carry.
        MOV  A,M      ;Get 1st data in A-register.
        INX  H        ;Add 2nd data which is available
        ADD  M        ;in memory to A. Sum in A-register.
        JNC  AHEAD    ;If CF=0, go to AHEAD.
        INR  C        ;If CF=1, increment C-register.
AHEAD:  INX  H
        MOV  M,A      ;Save the sum in memory.
        INX  H
        MOV  M,C      ;Save the carry in memory.
        HLT           ;Halt program execution.

        END           ;Assembly end.
```

Assembler listing for example program 1 (Method 2)

```
1                        ;PROGRAM TO ADD TWO 8-BIT DATA
2                        ;METHOD-2
3
4    4100                ORG  4100H     ;specify program starting address.
5
6    4100  21  00  42    LXI  H,4200H   ;Set pointer for data.
7    4103  0E  00        MVI  C,00H     ;Clear C-register to account for carry.
8    4105  7E            MOV  A,M       ;Get 1st data in A-register.
9    4106  23            INX  H         ;Add 2nd data which is available
10   4107  86            ADD  M         ;in memory to A. Sum in A-register.
11   4108  D2  0C  41    JNC  AHEAD     ;If CF=0, go to AHEAD.
12   410B  0C            INR  C         ;If CF=1, increment C-register.
13   410C  23     AHEAD: INX  H
14   410D  77            MOV  M, A      ;Save the sum in memory.
15   410E  23            INX  H
16   410F  71            MOV  M,C       ;Save the carry in memory.
17   4110  76            HLT            ;Halt program execution.
18
19   4111                END            ;Assembly end.
```

Sample data

		Memory address	content
Input Data :	Data-1 = E2$_H$	4200	E2
	Data-2 = 45$_H$	4201	45
Output data :	Sum = 27$_H$	4202	27
	Carry = 01$_H$	4203	01

EXAMPLE PROGRAM - 2 :   16-Bit Addition

Write an assembly language program to add two numbers of 16-bit data stored in memory locations from 4200$_H$ to 4203$_H$. The data are stored such that the low byte first and then the high byte is stored. Store the result from 4204$_H$ to 4206$_H$.

Problem Analysis

The 16-bit addition can be performed in 8085 either in terms of 8-bit addition or by using DAD instruction. In addition using DAD instruction, one of the data should be in HL pair and another data can be in another register pair. After addition the sum is stored in HL pair. If there is a carry in addition then that is indicated by setting a carry flag. Hence, one of the registers is used to account for carry.

Algorithm

1.   Load the first data in HL-register pair.
2.   Move the first data to DE- register pair.
3.   Load the second data in HL-register pair.
4.   Clear A-register for carry.

5.   Add the content of DE pair to HL pair.
6.   Check for carry. If carry = 1, go to step 7 or If carry = 0, go to step 8.
7.   Increment A-register to account for carry.
8.   Store the sum and carry in memory.
9.   Stop.

Flowchart for example program 2

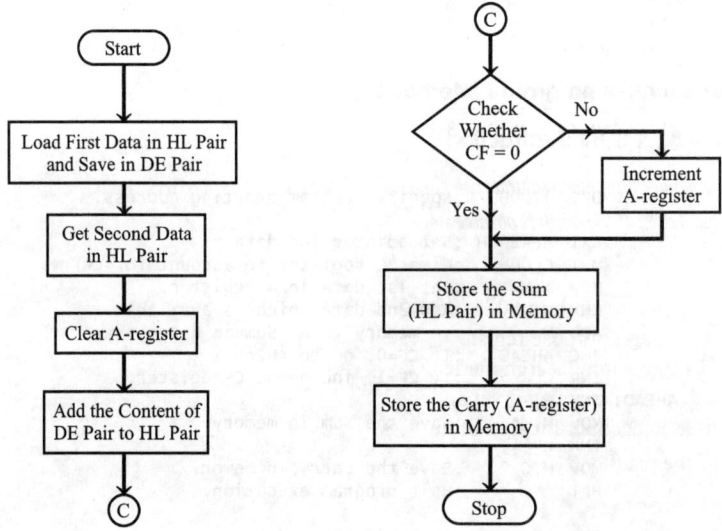

Assembly language program

```
;PROGRAM TO ADD TWO 16-BIT DATA

        ORG   4100H  ;specify program starting address.

        LHLD  4200H  ;Get 1st data in HL pair.
        XCHG         ;Save 1st data in DE pair.
        LHLD  4202H  ;Get 2nd data in HL pair.
        XRA   A      ;Clear A-register for carry.
        DAD   D      ;Get the sum in HL pair.
        JNC   AHEAD  ;If CF=0, go to AHEAD.
        INR   A      ;If CF=1, increment A-register.
AHEAD:  SHLD  4204H  ;Store the sum in memory.
        STA   4206H  ;Store the carry in memory.
        HLT          ;Halt program execution.

        END          ;Assembly end.
```

Assembler listing for example program 2

```
1                              ;PROGRAM TO ADD TWO 16-BIT DATA
2
3  4100                        ORG   4100H  ;specify program starting address.
4
5  4100   2A 00 42             LHLD  4200H  ;Get 1st data in HL pair.
6  4103   EB                   XCHG         ;Save 1st data in DE pair.
7  4104   2A 02 42             LHLD  4202H  ;Get 2nd data in HL pair.
8  4107   AF                   XRA   A      ;Clear A-register for carry.
9  4108   19                   DAD   D      ;Get the sum in HL pair.
```

```
10  4109   D2 0D 41           JNC   AHEAD   ;If CF=0, go to AHEAD.
11  410C   3C                 INR   A       ;If CF=1, increment A-register.
12  410D   22 04 42   AHEAD:  SHLD  4204H   ;Store the sum in memory.
13  4110   32 06 42           STA   4206H   ;Store the carry in memory.
14  4113   76                 HLT           ;Halt program execution.
15
16  4114                      END           ;Assembly end.
```

Sample data

Input Data :	Data-1 = C254$_H$
Output Data :	Data-2 = 8A92$_H$
	Sum = 4CE6$_H$
	Carry = 01

Memory address	Content
4200	54
4201	C2
4202	92
4203	8A

Memory address	Content
4204	E6
4205	4C
4206	01

EXAMPLE PROGRAM - 3 :   8-Bit Subtraction

Write an assembly language program to subtract two numbers of 8-bit data stored in memory locations 4200$_H$ and 4201$_H$. Store the magnitude of the result in 4202$_H$. If the result is positive store 00 in 4203$_H$ or if the result is negative store 01 in 4203$_H$.

Problem Analysis

In order to perform subtraction in 8085, one of the data should be in accumulator and another data can be in any one of the general purpose register or in the memory. After subtraction the result is stored in the accumulator. The 8085 perform 2's complement subtraction and then complement the carry. Therefore, if the result is negative then the carry flag is set and the accumulator will have 2's complement of the result. One of the register is used to account for sign of the result. In order to get the magnitude of the result again take 2's complement of the result.

Flowchart for example program 3

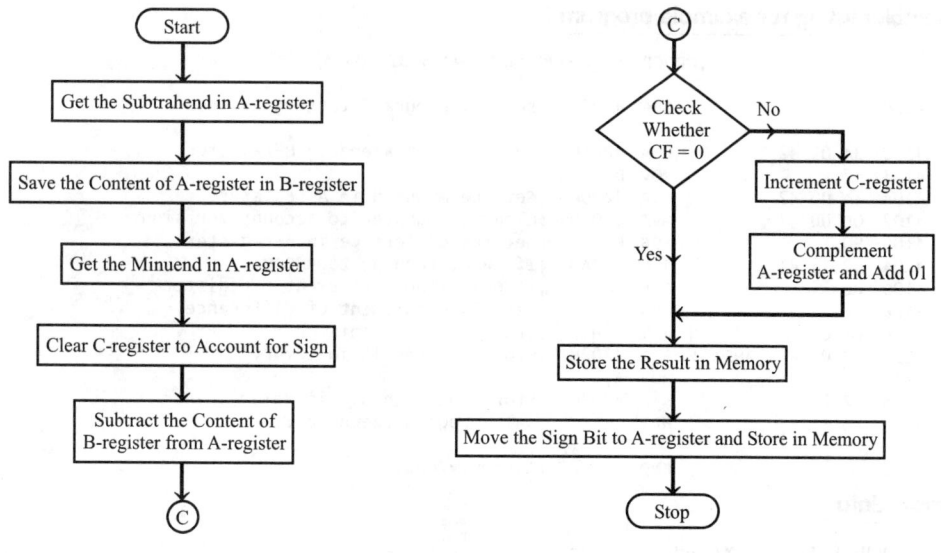

Algorithm

1. Load the subtrahend (the data to be subtracted) from memory to accumulator and move it to B-register.
2. Load the minuend from memory to accumulator.
3. Clear C-register to account for sign of the result.
4. Subtract the content of B-register (subtrahend) from the content of accumulator (minuend).
5. Check for carry. If carry = 1, go to step 6 or if carry = 0, go to step 7.
6. Increment C-register. Complement the accumulator and add 01$_H$.
7. Store the difference (accumulator) in memory.
8. Move the content of C-register (sign bit) to accumulator and store in memory.
9. Stop.

Assembly language program

```
;PROGRAM TO SUBTRACT TWO 8-BIT DATA

        ORG  4100H  ;specify program starting address.

        LDA  4201H  ;Get the subtrahend in B-register.
        MOV  B,A
        LDA  4200H  ;Get the minuend in A-register.
        MVI  C,00H  ;Clear C-register to account for sign.
        SUB  B      ;Get the difference in A-register.
        JNC  AHEAD  ;If CF=0, then go to AHEAD.
        INR  C      ;If CF=1, then increment C-register.
        CMA         ;Get 2's complement of difference
        ADI  01H    ;(result) in A-register.
AHEAD:  STA  4202H  ;Store the result in memory.
        MOV  A,C
        STA  4203H  ;Store the sign bit in memory.
        HLT         ;Halt program execution.

        END         ;Assembly end.
```

Assembler listing for example program 3

```
1                         ;PROGRAM TO SUBTRACT TWO 8-BIT DATA
2
3   4100                      ORG  4100H  ;specify program starting address.
4
5   4100  3A 01 42            LDA  4201H  ;Get the subtrahend in B-register.
6   4103  47                  MOV  B,A
7   4104  3A 00 42            LDA  4200H  ;Get the minuend in A-register.
8   4107  0E 00               MVI  C,00H  ;Clear C-register to account for sign.
9   4109  90                  SUB  B      ;Get the difference in A-register.
10  410A  D2 11 41            JNC  AHEAD  ;If CF=0, then go to AHEAD.
11  410D  0C                  INR  C      ;If CF=1, then increment C-register.
12  410E  2F                  CMA         ;Get 2's complement of difference
13  410F  C6 01               ADI  01H    ;(result) in A-register.
14  4111  32 02 42    AHEAD:  STA  4202H  ;Store the result in memory.
15  4114  79                  MOV  A,C
16  4115  32 03 42            STA  4203H  ;Store the sign bit in memory.
17  4118  76                  HLT         ;Halt program execution.
18
19  4119                      END         ;Assembly end.
```

Sample data

```
    Input Data  :   Minuend     = 5E_H
                    Subtrahend  = 34_H

    Output Data :   Difference  = 2A_H
                    Sign bit    = 00_H
```

EXAMPLE PROGRAM - 4: 16-Bit Subtraction

Write an assembly language program to subtract two numbers of 16-bit data stored in memory locations from 4200$_H$ to 4203$_H$. The data are stored such that the low byte is stored first and then the high byte is stored. Store the result in 4204$_H$ and 4205$_H$.

Problem Analysis

The 16-bit subtraction is performed in terms of 8-bit subtraction. First low bytes of the data are subtracted and the result is stored in memory. Then high bytes of the data are subtracted along with borrow (carry) in the previous subtraction and the result is stored in memory.

Algorithm

1. Load the low byte of subtrahend (the data to be subtracted) in accumulator from memory and move it to B-register.
2. Load the low byte of minuend in accumulator from memory.
3. Subtract the content of B-register (subtrahend) from the content of accumulator (minuend).
4. Store the low byte of result in memory.
5. Load the high byte of subtrahend in accumulator from memory and move it to B-register.
6. Load the high byte of minuend in accumulator from memory.
7. Subtract the content of B-register and the carry (borrow) from the content of accumulator.
8. Store the high byte of the result in memory.
9. Stop.

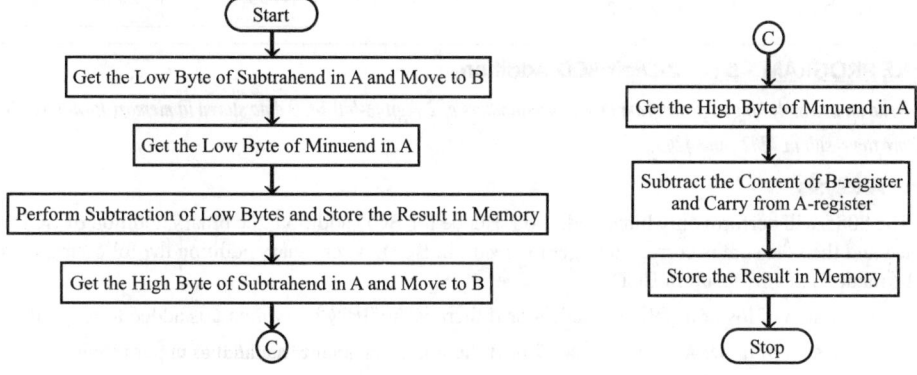

Assembly language program

```
;PROGRAM TO SUBTRACT TWO 16-BIT DATA

        ORG   4100H   ;specify program starting address.

        LDA   4202H
        MOV   B,A     ;Get low byte of subtrahend in B-register.
        LDA   4200H   ;Get low byte of minuend in A-register.
        SUB   B       ;Get difference of low bytes in A-register.
        STA   4204H   ;Store the result in memory.
        LDA   4203H
        MOV   B,A     ;Get high byte of subtrahend in B-register.
        LDA   4201H   ;Get high byte of minuend in A-register.
        SBB   B       ;Get difference of high bytes in A-register.
        STA   4205H   ;Store the result.
        HLT           ;Halt program execution.

        END           ;Assembly end.
```

Assembler listing for example program 4

```
 1                               ;PROGRAM TO SUBTRACT TWO 16-BIT DATA
 2
 3      4100                     ORG 4100H  ;specify program starting address.
 4
 5      4100   3A 02 42   LDA 4202H
 6      4103   47         MOV B,A   ;Get low byte of subtrahend in B-register.
 7      4104   3A 00 42   LDA 4200H ;Get low byte of minuend in A-register.
 8      4107   90         SUB B     ;Get difference of low bytes in A-register.
 9      4108   32 04 42   STA 4204H ;Store the result in memory.
10      410B   3A 03 42   LDA 4203H
11      410E   47         MOV B,A   ;Get high byte of subtrahend in B-register.
12      410F   3A 01 42   LDA 4201H ;Get high byte of minuend in A-register.
13      4112   98         SBB B     ;Get difference of high bytes in A-register.
14      4113   32 05 42   STA 4205H ;Store the result.
15      4116   76         HLT       ;Halt program execution.
16
17      4117             END        ;Assembly end.
```

Memory address	Content
4200	AB
4201	B2
4202	2C
4203	92
4204	7F
4205	20

Sample data

```
Input Data  : Minuend    = B2AB_H
              Subtrahend = 922C_H
Output Data : Difference = 207F_H
```

EXAMPLE PROGRAM - 5 :   2-Digit BCD Addition

Write an assembly language program to add two numbers of 2-digit (8-bit) BCD data stored in memory locations 4200_H and 4201_H. Store the result in 4202_H and 4203_H.

Problem Analysis

The 8085 will perform only binary addition. Hence for BCD addition, the binary addition of BCD data is performed and then the sum is corrected to get the result in BCD. After binary addition the following correction should be made to get the result in BCD.

1. If the sum of lower nibbles exceeds **9** or if there is auxiliary carry then **6** is added to lower nibble.

2. If the sum of upper nibbles exceeds **9** or if there is carry then **6** is added to upper nibble.

The above correction is taken care by DAA (**D**ecimal **A**djust **A**ccumulator) instruction. Therefore after binary addition, execute DAA instruction to do the above correction in the sum.

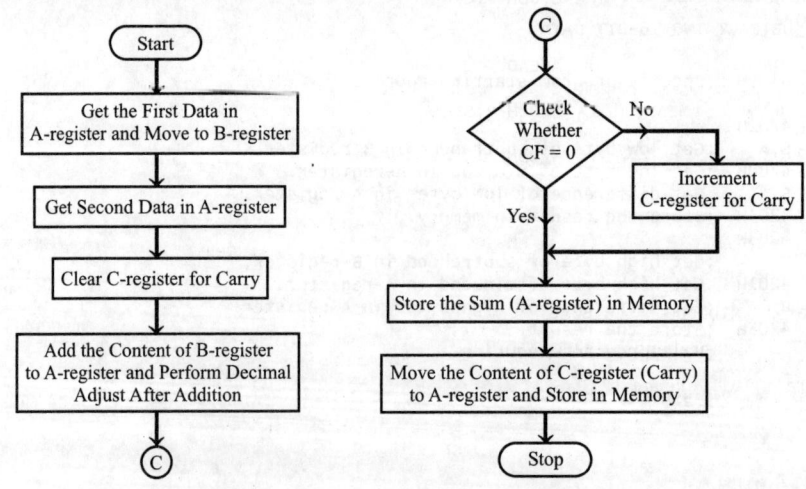

Algorithm

1. Load the first data in accumulator and move it to B-register.
2. Load the second data in accumulator.
3. Clear C-register for storing carry.
4. Add the content of B-register to accumulator.
5. Execute DAA instruction.
6. Check for carry. If carry = 1, go to step 7 or if carry = 0, go to step 8.
7. Increment C-register to account for carry.
8. Store the sum (content of accumulator) in memory.
9. Move the carry (content of C-register) to accumulator and store in memory.
10. Stop.

Assembly language program

```
;PROGRAM TO ADD TWO 2-DIGIT BCD DATA

        ORG  4100H   ;specify program starting address.

        LDA  4200H
        MOV  B,A     ;Get 1st data in B-register.
        LDA  4201H   ;Get 2nd data in A-register.
        MVI  C,00H   ;Clear C-register for accounting carry.
        ADD  B
        DAA          ;Get the sum of BCD data in A-register.
        JNC  AHEAD   ;If CF=0, go to AHEAD.
        INR  C       ;If CF=1, increment C-register.
AHEAD:  STA  4202H   ;Store the sum in memory.
        MOV  A,C
        STA  4203H   ;Store the carry in memory.
        HLT          ;Halt program execution.

        END          ;Assembly end.
```

Assembler listing for example program 5

```
1                              ;PROGRAM TO ADD TWO 2-DIGIT BCD DATA
2
3    4100                      ORG  4100H   ;specify program starting address.
4
5    4100   3A 00 42           LDA  4200H
6    4103   47                 MOV  B,A     ;Get 1st data in B-register.
7    4104   3A 01 42           LDA  4201H   ;Get 2nd data in A-register.
8    4107   0E 00              MVI  C,00H   ;Clear C-register for accounting carry.
9    4109   80                 ADD  B
10   410A   27                 DAA          ;Get the sum of BCD data in A-register.
11   410B   D2 0F 41           JNC  AHEAD   ;If CF=0, go to AHEAD.
12   410E   0C                 INR  C       ;If CF=1, increment C-register.
13   410F   32 02 42    AHEAD: STA  4202H   ;Store the sum in memory.
14   4112   79                 MOV  A,C
15   4113   32 03 42           STA  4203H   ;Store the carry in memory.
16   4116   76 .               HLT          ;Halt program execution.
17
18   4117                      END          ;Assembly end.
```

Sample data

Memory address	content
4200	72
4201	99
4202	71
4203	01

Input Data   :   Minuend   = $B2AB_H$

Subtrahend = $922C_H$

Output Data :   Difference = $20 F_H$

EXAMPLE PROGRAM - 6 :   4-Digit BCD Addition

Write an assembly language program to add two numbers of 4-digit BCD data stored in memory locations from 4200$_H$ to 4203$_H$ and store the result from 4204$_H$ to 4206$_H$

Problem Analysis

The 4-digit BCD addition is performed in terms of 2-digit BCD addition. First lower order two digits are added and the sum is stored in the memory. Then the higher order two digits are added along with previous carry and the sum and final carry are stored in the memory.

Algorithm

1.   Load the low order two digits of first data in accumulator and move it to B-register.
2.   Load the low order two digits of second data in accumulator.
3.   Clear C-register for storing carry.
4.   Add the content of B-register to accumulator.
5.   Execute DAA instruction.
6.   Store the low order two digits of the result in memory.
7.   Load the high order two digits of first data in accumulator and move it to B-register.
8.   Load the high order two digits of second data in accumulator.
9.   Add the content of B-register and carry (from previous addition) to accumulator.
10. Execute DAA instruction.
11. Check for carry. If carry = 1, go to step 12 or if carry = 0, go to step 13.
12. Increment C-register to account for final carry.
13. Store the high order two digits of the result in memory.
14. Move the carry (content of C-register) to accumulator and store in memory.
15. Stop.

Flowchart for example program 6

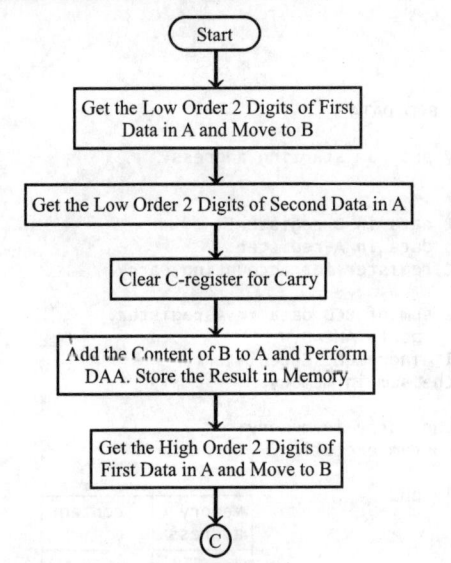
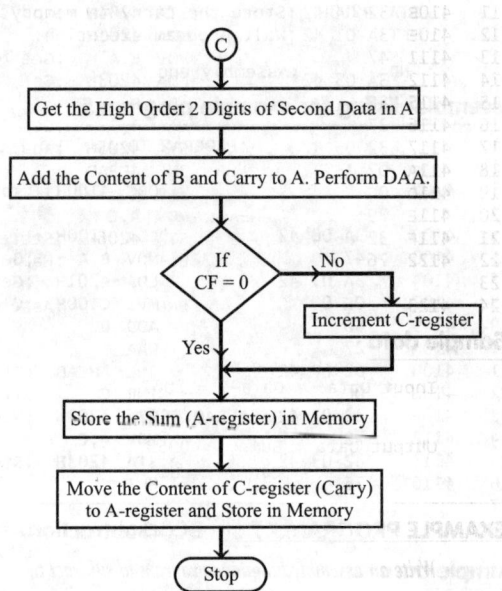

Assembly language program

```
;PROGRAM TO ADD TWO 4-DIGIT BCD DATA

        ORG  4100H  ;specify program starting address.

        LDA  4200H
        MOV  B,A    ;Get low order 2 digits of 1st data in B.
        LDA  4202H  ;Get low order 2 digits of 2nd data in A.
```

```
        MVI  C,00H  ;Clear C-register to account for carry.
        ADD  B
        DAA         ;Get the sum of low order two digits in A
        STA  4204H  ;and store it in memory.
        LDA  4201H
        MOV  B,A    ;Get high order 2 digits of 1st data in B.
        LDA  4203H  ;Get high order 2 digits of 2nd data in A.
        ADC  B
        DAA         ;Get the sum of high order two digits in A
        STA  4205H  ;and store the same in memory.
        JNC  AHEAD
        INR  C      ;If CF=1, increment C-register.
AHEAD:  MOV  A,C
        STA  4206H  ;Store the carry in memory.
        HLT         ;Halt program execution.

        END         ;Assembly end.
```

Assembler listing for example program 6

```
1                        ;PROGRAM TO ADD TWO 4-DIGIT BCD DATA
2
3    4100                ORG 4100H   ;specify program starting address.
4
5    4100  3A 00 42      LDA 4200H
6    4103  47            MOV B,A     ;Get low order 2 digits of 1st data in B.
7    4104  3A 02 42      LDA 4202H   ;Get low order 2 digits of 2nd data in A.
8    4107  0E 00         MVI C,00H   ;Clear C-register to account for carry.
9    4109  80            ADD B
10   410A  27            DAA         ;Get the sum of low order 2 digits in A
11   410B  32 04 42      STA 4204H   ;and store it in memory.
12   410E  3A 01 42      LDA 4201H
13   4111  47            MOV B,A     ;Get high order 2 digits of 1st data in B.
14   4112  3A 03 42      LDA 4203H   ;Get high order 2 digits of 2nd data in A.
15   4115  88            ADC B
16   4116  27            DAA         ;Get the sum of high order 2 digits in A
17   4117  32 05 42      STA 4205H   ;and store the same in memory.
18   411A  D2 1E 41      JNC AHEAD
19   411D  0C            INR C       ;If CF=1, increment C-register.
20   411E  79     AHEAD: MOV A,C
21   411F  32 06 42      STA 4206H   ;Store the carry in memory.
22   4122  76            HLT         ;Halt program execution.
23
24   4123                END         ;Assembly end.
```

Sample data

Input Data : Data-1 = 8067_{10}

Data-2 = 2892_{10}

Output Data : Sum   = 0959_{10}

Carry = 01_{10}

Memory address	Content		Memory address	Content
4200	67		4204	59
4201	80		4205	09
4202	92		4206	01
4203	28			

EXAMPLE PROGRAM - 7 :   BCD Subtraction

Write an assembly language program to subtract two numbers of 2-digit BCD data stored in memory locations 4200_H and 4201_H and store the result in 4202_H.

Problem Analysis

The 8085 will perform only binary subtraction. Hence for BCD subtraction, 10's complement subtraction is performed. First the 10's complement of the subtrahend is obtained and then added to the minuend. The DAA instruction is executed to get the result in BCD.

Algorithm

1. Load the subtrahend in accumulator and move it to B-register.
2. Move 99 to accumulator and subtract the content of B-register from accumulator.
3. Increment the accumulator.
4. Move the content of accumulator to B-register.
5. Load the minuend in accumulator.
6. Add the content of B-register to accumulator.
7. Execute DAA instruction.
8. Store the result in memory.
9. Stop.

Flowchart for example program 7

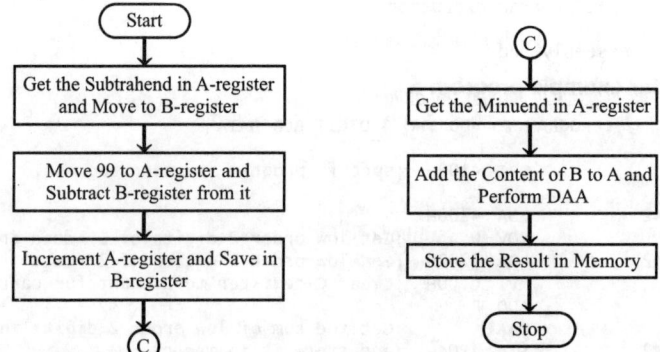

Assembly language program

```
;PROGRAM TO SUBTRACT TWO BCD (2-DIGIT) DATA
        ORG  4100H   ;specify program starting address.

        LDA  4201H
        MOV  B,A     ;Get the subtrahend in B-register.
        MVI  A,99H   ;Get 10's complement of
        SUB  B       ;subtrahend in A.
        INR  A
        MOV  B,A     ;Save 10's complement in B.
        LDA  4200H   ;Get the minuend in A-register.
        ADD  B       ;Get BCD sum of minuend and 10's
        DAA          ;complement of subtrahend. This sum
                     ;is the difference between BCD data.
        STA  4202H   ;Store the result in memory.
        HLT          ;Halt program execution.

        END          ;Assembly end.
```

Assembler listing for example program 7

```
 1                     ;PROGRAM TO SUBTRACT TWO BCD (2-DIGIT) DATA
 2
 3  4100              ORG  4100H   ;specify program starting address.
 4
 5  4100  3A 01 42    LDA  4201H
 6  4103  47          MOV  B,A     ;Get the subtrahend in B-register.
 7  4104  3E 99       MVI  A, 99H  ;Get 10's complement of
 8  4106  90          SUB  B       ;subtrahend in A.
 9  4107  3C          INR  A
10  4108  47          MOV  B,A     ;Save 10's complement in B.
11  4109  3A 00 42    LDA  4200H   ;Get the minuend in A-register.
12  410C  80          ADD  B       ;Get BCD sum of minuend and 10's
13  410D  27          DAA          ;complement of subtrahend. This sum
```

```
14                              ;is the difference between BCD data.
15   410E  32 02 42   STA  4202H  ;Store the result in memory.
16   4111  76         HLT         ;Halt program execution.
17
18   4112             END         ;Assembly end.
```

Sample data

Input Data : Minuend   = 95_{10}

Subtrahend = 32_{10}

Output Data : Difference = 63_{10}

Memory address	Content
4200	95
4201	32
4202	63

2.12.2 PROGRAMS WITH LOOP STRUCTURE AND COUNTER

EXAMPLE PROGRAM - 8 :   8-Bit Multiplication

Write an assembly language program to multiply two numbers of 8-bit data stored in memory locations 4200$_H$ and 4201$_H$ and store the product in 4202$_H$ and 4203$_H$

Problem Analysis

In 8085, the multiplication is performed as repeated additions. The initial value of sum is assumed as zero. One of the data is used as count (N) for the number of additions to be performed. Another data is added to the sum N times, where N is the count. The result of the product of two 8-bit data will be 16 bits. Hence, another register is used to account for the overflow.

Flowchart for example program 8

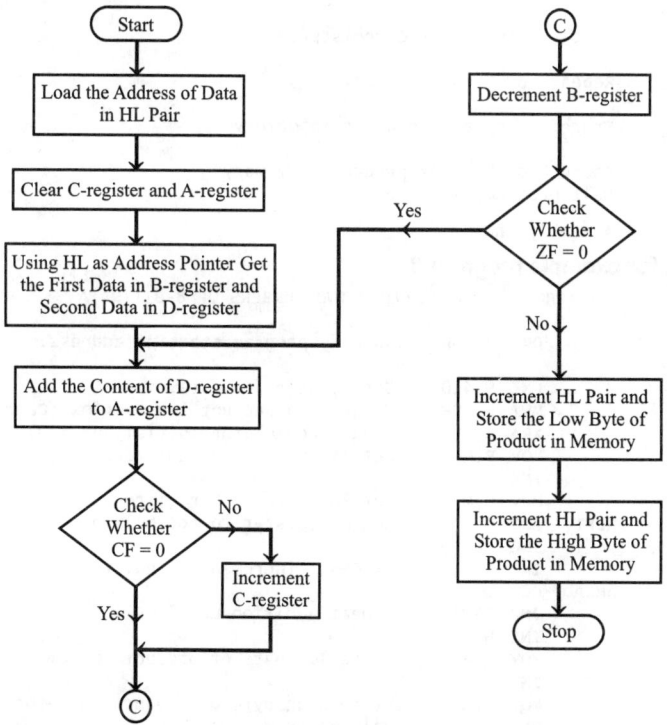

Algorithm

1. Load the address of the first data in HL pair (pointer).
2. Clear C-register for overflow (carry).
3. Clear the accumulator.
4. Move the first data to B-register (count).
5. Increment the pointer.
6. Move the second data to D-register (multiplicand).
7. Add the content of D-register to accumulator.
8. Check for carry. If carry = 1, go to step 9 or If carry = 0, go to step 10.
9. Increment C-register.
10. Decrement B-register (count).
11. Check whether count has reached zero. If ZF = 0 repeat steps 7 through 11, or if ZF = 1 go to next step.
12. Increment the pointer and store low byte of the product in memory.
13. Increment the pointer and store high byte of the product in memory.
14. Stop.

Assembly language program

```
;PROGRAM TO MULTIPLY TWO NUMBERS OF 8-BIT DATA

        ORG  4100H    ;specify program starting address.

        LXI  H,4200H  ;Set pointer for data.
        MVI  C,00H    ;Clear C to account for overflow (Carry).
        XRA  A        ;Clear accumulator(Initial sum = 0).
        MOV  B,M      ;Get 1st data in B-register.
        INX  H
        MOV  D,M      ;Get 2nd data in D-register.
REPT:   ADD  D        ;Add D-register to accumulator.
        JNC  AHEAD
        INR  C        ;If CF=1, increment C-register.
AHEAD:  DCR  B
        JNZ  REPT     ;Repeat addition until ZF=1.
        INX  H
        MOV  M,A      ;Store low byte of product in memory.
        INX  H
        MOV  M,C      ;Store high byte of product in memory.
        HLT           ;Halt program execution.

        END           ;Assembly end.
```

Assembler listing for example program 8

```
1                               ;PROGRAM TO MULTIPLY TWO NUMBERS OF 8-BIT DATA
2
3   4100                        ORG  4100H    ;specify program starting address.
4
5   4100  21 00 42              LXI  H,4200H  ;Set pointer for data.
6   4103  0E 00                 MVI  C,00H    ;Clear C to account for overflow (Carry).
7   4105  AF                    XRA  A        ;Clear accumulator(Initial sum = 0).
8   4106  46                    MOV  B,M      ;Get 1st data in B-register.
9   4107  23                    INX  H
10  4108  56                    MOV  D,M      ;Get 2nd data in D-register.
11  4109  82            REPT:   ADD  D        ;Add D-register to accumulator.
12  410A  D2 0E 41              JNC  AHEAD
13  410D  0C                    INR  C        ;If CF=1, increment C-register.
14  410E  05            AHEAD:  DCR  B
15  410F  C2 09 41              JNZ  REPT     ;Repeat addition until ZF=1.
16  4112  23                    INX  H
17  4113  77                    MOV  M,A      ;Store low byte of product in memory.
18  4114  23                    INX  H
19  4115  71                    MOV  M,C      ;Store high byte of product in memory.
20  4116  76                    HLT           ;Halt program execution.
21
22  4117                        END           ;Assembly end.
```

Sample data

				Memory address	Content
Input Data	: Data-1	= C7$_H$		4200	C7
	Data-2	= 4A$_H$		4201	4A
				4202	86
Output Data	: Product	= 3986$_H$		4203	39

EXAMPLE PROGRAM - 9 :   16-Bit Multiplication

Write an assembly language program to multiply two numbers of 16-bit data stored in memory locations from 4200$_H$ to 4203$_H$. Store the product in memory locations from 4204$_H$ to 4207$_H$.

Problem Analysis

The 16-bit multiplication is performed as repeated 16-bit additions. The initial sum is assumed as zero. One of the data is stored in SP (Stack Pointer) and another data is stored in DE pair. The content of DE pair is used as count for number of additions. The content of SP is added to the sum N times, where N is the count. The maximum size of product will be 32-bit. Hence BC pair is used to account for overflow. In 16-bit decrement no flags are affected. Hence, to check zero of the count (DE pair), move E-register to A-register and logically ORed with D-register.

Flowchart for example program 9

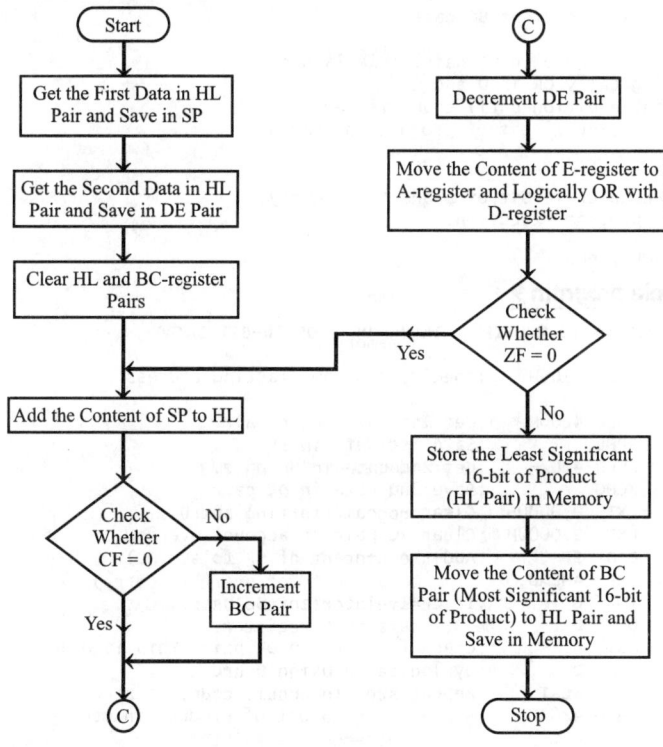

Algorithm

1. Load the first data in HL pair and move to SP.
2. Load the second data in HL and move to DE (count).
3. Clear HL pair (Initial sum).
4. Clear BC pair for overflow (carry).
5. Add the content of SP to HL.
6. Check for carry. If carry=1, go to step 7 or If carry=0, go to step 8.

7. Increment BC pair.
8. Decrement the count.
9. Check whether count has reached zero.
10. To check for zero of the count, move the content of E-register to A-register and logically OR with D-register.
11. Check the zero flag. If ZF=0, repeat steps 5 through 11 or If ZF=1, go to next step.
12. Store the content of HL in memory. (Least significant 16 bits of the product).
13. Move the content of C to L and B to H and store HL in memory. (Most significant 16 bits of the product).
14. Stop.

Assembly language program

```
;PROGRAM TO MULTIPLY TWO NUMBERS OF 16-BIT DATA

          ORG    4100H       ;specify program starting address.

          LHLD   4200H       ;Get 1st data in HL pair.
          SPHL               ;Save 1st data in SP.
          LHLD   4202H       ;Get 2nd data in HL pair.
          XCHG               ;Save 2nd data in DE pair.
          LXI    H,0000H     ;Clear HL pair(initial sum=0).
          LXI    B,0000H     ;Clear BC pair to account overflow.
NEXT:     DAD    SP          ;Add the content of SP to sum(HL).
          JNC    AHEAD
          INX    B           ;If CF=1, increment BC pair.
AHEAD:    DCX    D
          MOV    A,E         ;Check for zero in DE pair. This is done
          ORA    D           ;by logically OR in D and E.
          JNZ    NEXT        ;Repeat addition until count is zero.
          SHLD   4204H       ;Store lower 16-bit of product in memory.
          MOV    L,C
          MOV    H,B
          SHLD   4206H       ;Store upper 16-bit of product in memory.
          HLT                ;Halt program execution.

          END                ;Assembly end.
```

Assembler listing for example program 9

```
1                           ;PROGRAM TO MULTIPLY TWO NUMBERS OF 16-BIT DATA
2
3    4100                    ORG    4100H       ;specify program starting address.
4
5    4100  2A 00 42          LHLD   4200H       ;Get 1st data in HL pair.
6    4103  F9                SPHL               ;Save 1st data in SP.
7    4104  2A 02 42          LHLD   4202H       ;Get 2nd data in HL pair.
8    4107  EB                XCHG               ;Save 2nd data in DE pair.
9    4108  21 00 00          LXI    H,0000H     ;Clear HL pair(initial sum=0).
10   410B  01 00 00          LXI    B,0000H     ;Clear BC pair to account overflow.
11   410E  39         NEXT:  DAD    SP          ;Add the content of SP to sum(HL).
12   410F  D2 13 41         JNC    AHEAD
13   4112  03                INX    B           ;If CF=1, increment BC pair.
14   4113  1B         AHEAD: DCX    D
15   4114  7B                MOV    A,E         ;Check for zero in DE pair. This is done
16   4115  B2                ORA    D           ;by logically ORing D and E.
17   4116  C2 0E 41          JNZ    NEXT        ;Repeat addition until count is zero.
18   4119  22 04 42          SHLD   4204H       ;Store lower 16-bit of product in memory.
19   411C  69                MOV    L,C
20   411D  60                MOV    H,B
21   411E  22 06 42          SHLD   4206H       ;Store upper 16-bit of product in memory.
22   4121  76                HLT                ;Halt program execution.
23
24   4122                    END                ;Assembly end.
```

Sample data

```
Input Data  : Data-1 = 5A 24_H
              Data-2 = 47C2_H
Output Data : Product = 19444B48_H
```

Memory address	Content	Memory address	Content
4200	24	4204	48
4201	5A	4205	4B
4202	C2	4206	44
4203	47	4207	19

EXAMPLE PROGRAM - 10 :   8-Bit Division

Write an assembly language program to divide two numbers of 8-bit data stored in memory locations 4200_H and 4201_H. Store the quotient in 4202_H and the remainder in 4203_H.

Problem Analysis

The division in 8085 is performed as repeated subtraction. The dividend is stored in A-register and divisor in B-register. The initial value of quotient is assumed as zero. Subtraction should be performed only when dividend is greater than divisor. So repeated subtraction is performed until dividend is lesser than the divisor. For each subtraction, the quotient is incremented by one. Then store the quotient and remainder in the memory.

Algorithm

1. Load the divisor in accumulator and move it to B-register.
2. Load the dividend in accumulator.
3. Clear C-register to account for quotient.
4. Check whether divisor is less than dividend. If divisor is less than dividend, go to step 8, otherwise go to next step.
5. Subtract the content of B-register from accumulator.
6. Increment the content of C-register (quotient).
7. Go to step 4.
8. Store the content of accumulator (remainder) in memory.
9. Move the content of C-register (quotient) to accumulator and store in memory.
10. Stop.

Flowchart for example program 10

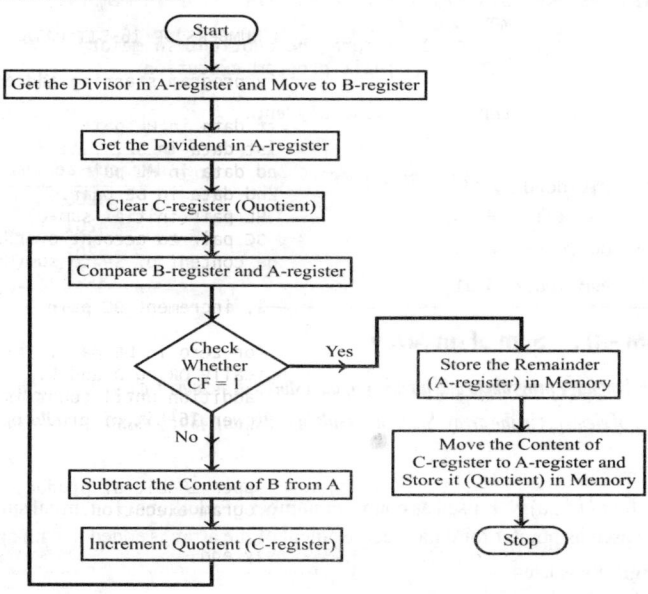

Assembly language program

```
;PROGRAM TO DIVIDE TWO NUMBERS OF 8-BIT DATA

        ORG  4100H   ;specify program starting address.

        LDA  4201H
        MOV  B,A     ;Get the divisor in B-register.
        LDA  4200H   ;Get the dividend in A-register.
        MVI  C,00H   ;Clear C-register for quotient.
AGAIN:  CMP  B
        JC   STORE   ;If divisor is less than dividend go to store.
        SUB  B       ;Subtract divisor from dividend.
        INR  C       ;Increment quotient by one for each subtraction.
        JMP  AGAIN
STORE:  STA  4203H   ;Store the remainder in memory.
        MOV  A,C
        STA  4202H   ;Store the quotient in memory.
        HLT          ;Halt program execution.

        END          ;Assembly end.
```

Assembler listing for example program 10

```
 1                          ;PROGRAM TO DIVIDE TWO NUMBERS OF 8-BIT DATA
 2
 3   4100                   ORG  4100H ;specify program starting address.
 4
 5   4100  3A 01 42         LDA  4201H
 6   4103  47               MOV  B,A     ;Get the divisor in B-register.
 7   4104  3A 00 42         LDA  4200H   ;Get the dividend in A-register.
 8   4107  0E 00            MVI  C,00H   ;Clear C-register for quotient.
 9   4109  B8        AGAIN: CMP  B
10   410A  DA 12 41         JC   STORE   ;If divisor is less than dividend go to store.
11   410D  90               SUB  B       ;Subtract divisor from dividend.
12   410E  0C               INR  C       ;Increment quotient by one for each subtraction.
13   410F  C3 09 41         JMP  AGAIN
14   4112  32 03 42  STORE: STA  4203H   ;Store the remainder in memory.
15   4115  79               MOV  A,C
16   4116  32 02 42         STA  4202H   ;Store the quotient in memory.
17   4119  76               HLT          ;Halt program execution.
18
19   411A                   END          ;Assembly end.
```

Sample data

			Memory address	Content
Input Data	:	Dividend = C9$_H$	4200	C9
		Divisor = 0A$_H$	4201	0A
Output Data	:	Quotient = 14$_H$	4202	14
		Remainder = 01$_H$	4203	01

EXAMPLE PROGRAM - 11 :   Sum of an Array

Write an assembly language program to add an array of data stored in memory from 4200_H to $4200_H + N$. The first element of the array, gives the number of elements in the array. Store the result in 4300_H and 4301_H. Assume that the sum does not exceed 16-bit.

Problem Analysis

The number of bytes (data) N, is used as count for number of additions. The initial sum is assumed as zero. The HL register pair is used as pointer for data. Each element of the array is added to sum and for accounting the overflow one of the registers is used.

Flowchart for example program 11

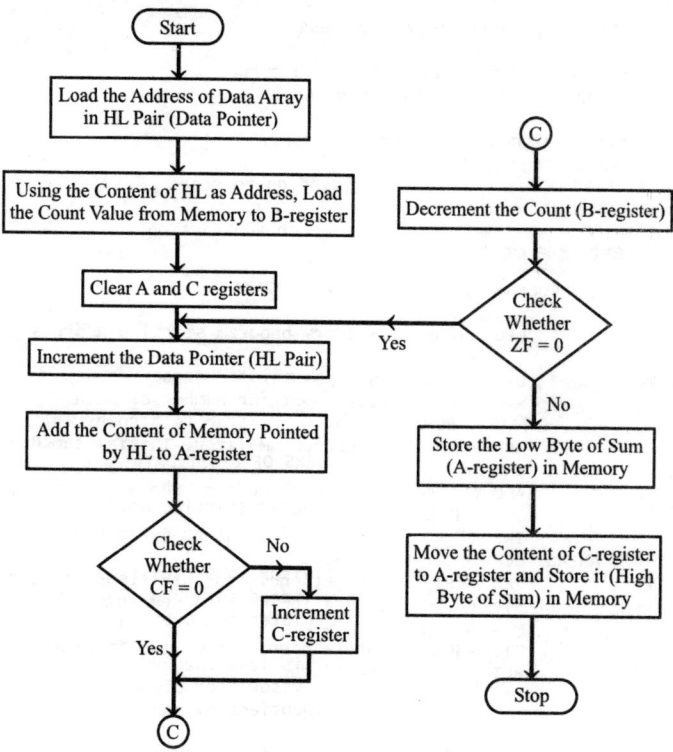

Algorithm

1.   Load the address of the first element of the array in HL pair (pointer).
2.   Move the count to B-register.
3.   Clear C-register for carry.
4.   Clear accumulator for sum.
5.   Increment the pointer (HL pair).
6.   Add the content of memory addressed by HL to accumulator.
7.   Check for carry. If carry = 1, go to step 8, or If carry = 0, go to step 8.
8.   Increment C-register.
9.   Decrement the count.
10. Check for zero of the count. If ZF = 0 go to step 5 or If ZF = 1, go to next step.
11. Store the content of accumulator (low byte of sum).
12. Move the content of C-register (high byte of sum) to accumulator. Store the content of accumulator in memory.
13. Stop.

Assembly language program

```
; PROGRAM TO ADD AN ARRAY OF DATA

        ORG  4100H    ;specify program starting address.

        LXI  H,4200H ;Set pointer for data.
        MOV  B,M     ;Set count for number of data.
        MVI  C,00H    ;Clear C-register to account for carry.
        XRA  A        ;Clear accumulator. Initial sum=0.
```

```
REPT:     INX   H
          ADD   M          ;Add an element of the array to sum.
          JNC   AHEAD
          INR   C          ;If CF=1, increment C-register.
AHEAD:    DCR   B
          JNZ   REPT       ;Repeat addition until count is zero.
          STA   4300H      ;Store low byte of sum in memory.
          MOV   A,C
          STA   4301H      ;Store high byte of sum in memory.
          HLT              ;Halt program execution.

          END              ;Assembly end.
```

Assembler listing for example program 11

```
1                              ;PROGRAM TO ADD AN ARRAY OF DATA
2
3    4100                      ORG   4100H      ;specify program starting address.
4
5    4100   21 00 42           LXI   H,4200H    ;Set pointer for data.
6    4103   46                 MOV   B,M        ;Set count for number of data.
7    4104   0E 00              MVI   C,00H      ;Clear C-register to account for carry.
8    4106   AF                 XRA   A          ;Clear accumulator. Initial sum=0.
9    4107   23          REPT:  INX   H
10   4108   86                 ADD   M          ;Add an element of the array to sum.
11   4109   D2 0D 41           JNC   AHEAD
12   410C   0C                 INR   C          ;If CF=1, increment C-register.
13   410D   05          AHEAD: DCR   B
14   410E   C2 07 41           JNZ   REPT       ;Repeat addition until count is zero.
15   4111   32 00 43           STA   4300H      ;Store low byte of sum in memory.
16   4114   79                 MOV   A,C
17   4115   32 01 43           STA   4301H      ;Store high byte of sum in memory.
18   4118   76                 HLT              ;Halt program execution.
19
20   4119   END                                 ;Assembly end.
```

Sample data

Input Data : Count = 07$_H$
               Array = C2$_H$
                       45$_H$
                       B3$_H$
                       F4$_H$
                       7C$_H$
                       ED$_H$
                       16$_H$

Output Data : Sum = 042D$_H$

Memory address	Content	
4200	07	Count
4201	C2	
4202	45	
4203	B3	
4204	F4	Array
4205	7C	
4206	ED	
4207	16	
4300	2D	Sum
4301	04	

EXAMPLE PROGRAM - 12 :   Search for Smallest Data in an Array

Write an assembly language program to search the smallest data in an array of N data stored in memory locations from 4200$_H$ to (4200$_H$ + N). The first element of the array gives the number of data in the array. Store the smallest data in 4300$_H$

Problem Analysis

The HL register pair is used as pointer for the array. One of the general purpose register is used as count. A data in the array is moved to A-register and compared with next data. After each comparison, the smallest data is brought to accumulator. The comparisons are carried N–1 times. After N–1 comparisons, the smallest data will be in A-register and store it in memory.

Algorithm

1. Load the address of the first element of the array in HL register pair (pointer).
2. Move the count to B-register.
3. Increment the pointer.
4. Get the first data in accumulator.
5. Decrement the count.
6. Increment the pointer.
7. Compare the content of memory addressed by HL pair with that of accumulator.
8. If carry = 1, go to step 10 or If carry = 0, go to step 8.
9. Move the content of memory addressed HL to accumulator.
10. Decrement the count.
11. Check for zero of the count. If ZF = 0, go to step 6, or If ZF = 1 go to next step.
12. Store the smallest data in memory.
13. Stop.

Flowchart for example program 12

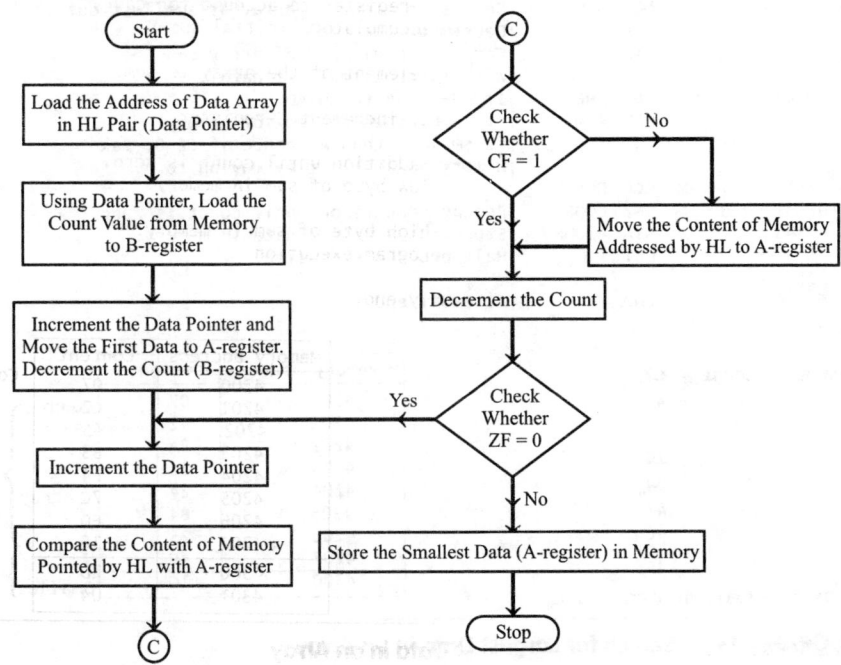

Assembly language program

```
;PROGRAM TO SEARCH SMALLEST DATA IN AN ARRAY

        ORG  4100H    ;specify program starting address.

        LXI  H,4200H  ;Set pointer for array.
        MOV  B,M      ;Set count for number of elements in array.
        INX  H
        MOV  A,M      ;Set 1st element of array as smallest data.
        DCR  B        ;Decrement the count.
```

```
LOOP:    INX  H          ;Compare an element of array
         CMP  M          ;with current smallest data.
         JC   AHEAD      ;If CF=1, go to AHEAD.
         MOV  A,M        ;If CF=0, then content of memory
                         ;is smaller than A, hence if CF=0, make
                         ;memory as smallest by moving to A.
AHEAD:   DCR  B
         JNZ  LOOP       ;Repeat comparison until count is zero.
         STA  4300H      ;Store the smallest data in memory.
         HLT             ;Halt program execution.

         END             ;Assembly end.
```

Assembler listing for example program 12

```
1                              ;PROGRAM TO SEARCH SMALLEST DATA IN AN ARRAY
2
3    4100                      ORG  4100H   ;specify program starting address.
4
5    4100  21 00 42            LXI  H,4200H ;Set pointer for array.
6    4103  46                  MOV  B,M     ;Set count for number of elements in array.
7    4104  23                  INX  H
8    4105  7E                  MOV  A,M     ;Set 1st element of array as smallest data.
9    4106  05                  DCR  B       ;Decrement the count.
10   4107  23           LOOP:  INX  H       ;Compare an element of array
11   4108  BE                  CMP  M       ;with current smallest data..
12   4109  DA 0D 41            JC   AHEAD   ;If CF=1, go to AHEAD.
13   410C  7E                  MOV  A,M     ;If CF=0, then content of memory
14                                          ;is smaller than A, hence if CF=0, make
15                                          ;memory as smallest by moving to A.
16   410D  05           AHEAD: DCR  B
17   410E  C2 07 41            JNZ  LOOP    ;Repeat comparison until count is zero.
18   4111  32 00 43            STA  4300H   ;Store the smallest data in memory.
19   4114  76                  HLT          ;Halt program execution.
20
21   4115                      END          ;Assembly end.
```

Sample data

Input Data : Count = 07$_H$

Array = 42$_H$

3A$_H$

1C$_H$

24$_H$

B4$_H$

25$_H$

4F$_H$

Output Data : Smallest data = 1 C$_H$

Memory address	Content	
4200	07	Count
4201	42	
4202	3A	
4203	1C	
4204	24	Array
4205	B4	
4206	25	
4207	4F	
4300	1C	Smallest data

EXAMPLE PROGRAM - 13 :   Search for Largest Data in an Array

Write an assembly language program to search the largest data in an array of N data stored in memory locations from 4200$_H$ to 4200$_H$ + N. The first element of the array is the number of data (N) in the array. Store the largest data in 4300$_H$

Problem Analysis

The HL register pair is used as pointer for the array. One of the general purpose register is used as count. A data in the array is moved to A-register and compared with next data. After each comparison, the largest data is brought to A-register. The comparisons are performed N−1 times. After N−1 comparisons the largest data will be in A-register and store it in memory.

Algorithm

1. Load the address of the first element of the array in HL register pair (pointer).
2. Move the count to B-register.
3. Increment the pointer.
4. Get the first data in accumulator.
5. Decrement the count.
6. Increment the pointer.
7. Compare the content of memory addressed by HL pair with that of accumulator.
8. If carry = 0, go to step 10 or If carry = 1, go to step 8.
9. Move the content of memory addressed HL to accumulator.
10. Decrement the count.
11. Check for zero of the count. If ZF = 0, go to step 6, or If ZF = 1 go to next step.
12. Store the largest data in memory.
13. Stop.

Flowchart for example program 13

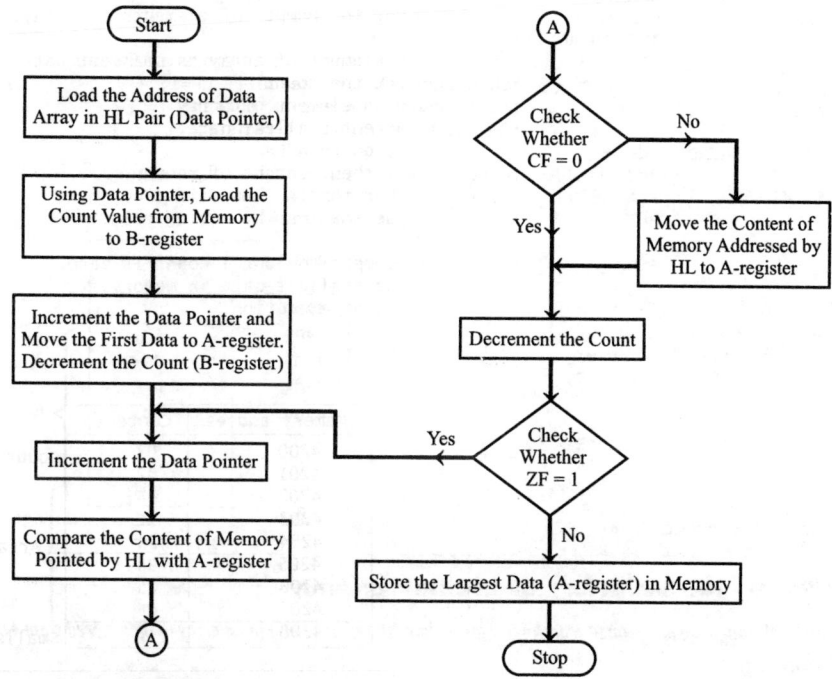

Assembly language program

```
;PROGRAM TO SEARCH LARGEST DATA IN AN ARRAY

        ORG   4100H   ;specify program starting address.

        LXI   H,4200H ;Set pointer for array.
        MOV   B,M     ;Set count for number of elements in array.
        INX   H
        MOV   A,M     ;Set 1st element of array as largest data.
        DCR   B       ;Decrement the count.
LOOP:   INX   H       ;Compare an element of array with
        CMP   M       ;current largest data.
        JNC   AHEAD   ;If CF=0, go to AHEAD.
        MOV   A,M     ;If CF=1,then content of memory is larger
                      ;than accumulator.Hence if CF=1,
```

```
                              ;make memory content as current
                              ;largest by moving it to A-register.
AHEAD:  DCR  B
        JNZ  LOOP             ;Repeat comparison until count is zero.
        STA  4300H            ;Store the largest data in memory.
        HLT                   ;Halt program execution.

        END                   ;Assembly end.
```

Assembler listing for example program 13

```
1                              ;PROGRAM TO SEARCH LARGEST DATA IN AN ARRAY
2
3   4100                       ORG  4100H     ;specify program starting address.
4
5   4100  21 00 42             LXI  H,4200H   ;Set pointer for array.
6   4103  46                   MOV  B,M       ;Set count for number of elements in array.
7   4104  23                   INX  H
8   4105  7E                   MOV  A,M       ;Set 1st element of array as largest data.
9   4106  05                   DCR  B         ;Decrement the count.
10  4107  23          LOOP:    INX  H         ;Compare an element of array with
11  4108  BE                   CMP  M         ;current largest data.
12  4109  D2 0D 41             JNC  AHEAD     ;If CF=0, go to AHEAD.
13  410C  7E                   MOV  A,M       ;If CF=1,then content of memory is larger
14                                            ;than accumulator. Hence if CF=1,
15                                            ;make memory content as current
16                                            ;largest by moving it toA-register.
17  410D  05          AHEAD:   DCR  B
18  410E  C2 07 41             JNZ  LOOP      ;Repeat comparison until count is zero.
19  4111  32 00 43             STA  4300H     ;Store the largest data in memory.
20  4114  76                   HLT            ;Halt program execution.
21
22  4115                       END            ;Assembly end.
```

Sample data

Input Data	:	Count	=	07_H
		Array	=	62_H
				$7D_H$
				FC_H
				24_H
				$C2_H$
				$0F_H$
				92_H

Output Data : Largest data = FC_H

Memory address	Content	
4200	07	Count
4201	62	
4202	7D	
4203	FC	
4204	24	Array
4205	C2	
4206	0F	
4207	92	
4300	FC	Laragest data

EXAMPLE PROGRAM - 14 : Search for a Given Data in an Array

Write an assembly language program to search for a given data (stored in 4250_H) in an array of data stored from 4200_H. The end of the array is marked by 20_H.

If the data is available store FF_H in 4251_H. Store the position of the data and its address in 4252_H, 4253_H and 4254_H respectively. If the data is not available store 00_H in memory locations from 4251_H to 4254_H.

Problem Analysis

The HL pair is used as pointer for given data. B-register is used as pointer for position of the data. C-register is used to record the availability of given data. The given data is moved to A-register and compared with each element of the array one-by-one. If the data is available, terminate the comparison and store the position, address and FF_H (for availability) in memory.

Algorithm

1. Load the address of the data array in HL register pair.
2. Load the given data in accumulator.
3. Clear B-register.
4. Increment B-register.
5. Compare the content of memory addressed by HL pair with that of accumulator.
6. If ZF = 0, go to next step or if ZF = 1, go to step 8.
7. Check for end of array by comparing the data with 20_H. If ZF = 0, go to step 4, or If ZF = 1, go to next step.
8. Clear B, C, H, L registers and jump to step 10.
9. Move FF_H to C-register.
10. Store the content of H and L registers in memory.
11. Move C to L and B to H and store HL in memory.
12. Stop.

Flowchart for example program 14

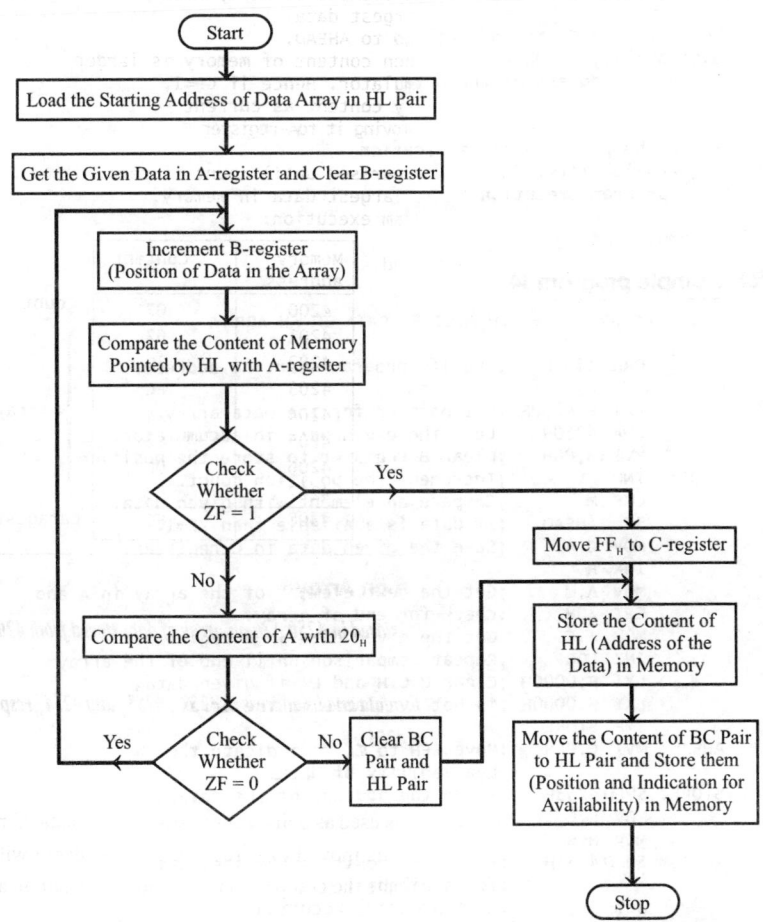

Assembly language program

```
;PROGRAM TO SEARCH A GIVEN DATA IN AN ARRAY

        ORG   4100H    ;specify program starting address.

        LXI   H,4200H  ;Set pointer for the data array.
        LDA   4250H    ;Load the given data in accumulator.
        MVI   B,00H    ;Clear B-register to store the position.
LOOP:   INR   B        ;Increment the position count.
        CMP   M        ;Compare an element with given data.
        JZ    AHEAD    ;If data is available,then ZF=1.
        MOV   C,A      ;Save the given data in C-register.
        INX   H
        MOV   A,M      ;Get the next element of the array in A and
        CPI   20H      ;check for end of array.
        MOV   A,C      ;Get the given data in A-register.
        JNZ   LOOP     ;Repeat comparison until end of the array.
        LXI   B,0000H  ;Clear B,C,H and L if given data
        LXI   H,0000H  ;is not available in the array.
        JMP   STORE
AHEAD:  MVI   C,FFH    ;Move FFH  to C, to indicate the
                       ;availability of data.
STORE:  SHLD  4253H    ;Store the address of the data.
        MOV   L,C
        MOV   H,B
        SHLD  4251H    ;Store the position and indication
                       ;for availability.
        HLT            ;Halt program execution.

        END            ;Assembly end.
```

Assembler listing for example program 14

```
 1                          ;PROGRAM TO SEARCH A GIVEN DATA IN AN ARRAY
 2
 3   4100                   ORG 4100H    ;specify program starting address.
 4
 5   4100  21 00 42         LXI  H,4200H ;Set pointer for the data array.
 6   4103  3A 50 42         LDA  4250H   ;Load the given data in accumulator.
 7   4106  06 00            MVI  B,00H   ;Clear B-register to store the position.
 8   4108  04        LOOP:  INR  B       ;Increment the position count.
 9   4109  BE               CMP  M       ;Compare an element with given data.
10   410A  CA 1F 41         JZ   AHEAD   ;If data is available,then ZF=1.
11   410D  4F               MOV  C,A     ;Save the given data in C-register.
12   410E  23               INX  H
13   410F  7E               MOV  A,M     ;Get the next element of the array in A and
14   4110  FE 20            CPI  20H     ;check for end of array.
15   4112  79               MOV  A,C     ;Get the given data in A-register.
16   4113  C2 08 41         JNZ  LOOP    ;Repeat comparison until end of the array.
17   4116  01 00 00         LXI  B,0000H ;Clear B,C,H and L  if given data
18   4119  21 00 00         LXI  H,0000H ;is not available in the array.
19   411C  C3 21 41         JMP  STORE
20   411F  0E FF     AHEAD: MVI  C,FFH   ;Move FFH to C, to indicate the
21                          ;availability of data.
22   4121  22 53 42  STORE: SHLD 4253H   ;Store the address of the data.
23   4124  69               MOV  L,C
24   4125  60               MOV  H,B
25   4126  22 51 42         SHLD 4251H   ;Store the position and indication
26                          ;for availability.
27   4129  76               HLT          ;Halt program execution.
28
29   412A                   END          ;Assembly end.
```

Sample data

Input Data :

Array = 45$_H$
72$_H$
CA$_H$
2F$_H$
C2$_H$
D1$_H$
4F$_H$
20$_H$

Given Data = 2F$_H$

Memory address	Content
4200	45
4201	72
4202	CA
4203	2F
4204	C2
4205	D1
4206	4F
4207	20

Output Data :

Availability = FF$_H$
Position = 04$_H$
Address = 4203$_H$

Memory address	Content
4250	2F
4251	FF
4252	04
4253	03
4254	42

2.12.3 PROGRAM WITH INDEXING

EXAMPLE PROGRAM - 15 : Sorting an Array in Ascending Order

Write an assembly language program to sort an array of data in ascending order. The array is stored in memory starting from 4200$_H$. The first element of the array gives the count value for the number of elements in the array.

Problem Analysis

The algorithm for bubble sorting is given below. In bubble sorting of N-data, N−1 comparisons are carried by taking two consecutive data at a time. After each comparison, the data is rearranged such that smallest among the two is in the first memory location and the largest in the next memory location. (Here the data is rearranged within the two memory locations whose contents are compared). When we perform N−1 comparisons as mentioned above for N−1, times then the array consisting of N-data will be sorted in the ascending order.

Flowchart for example program 15

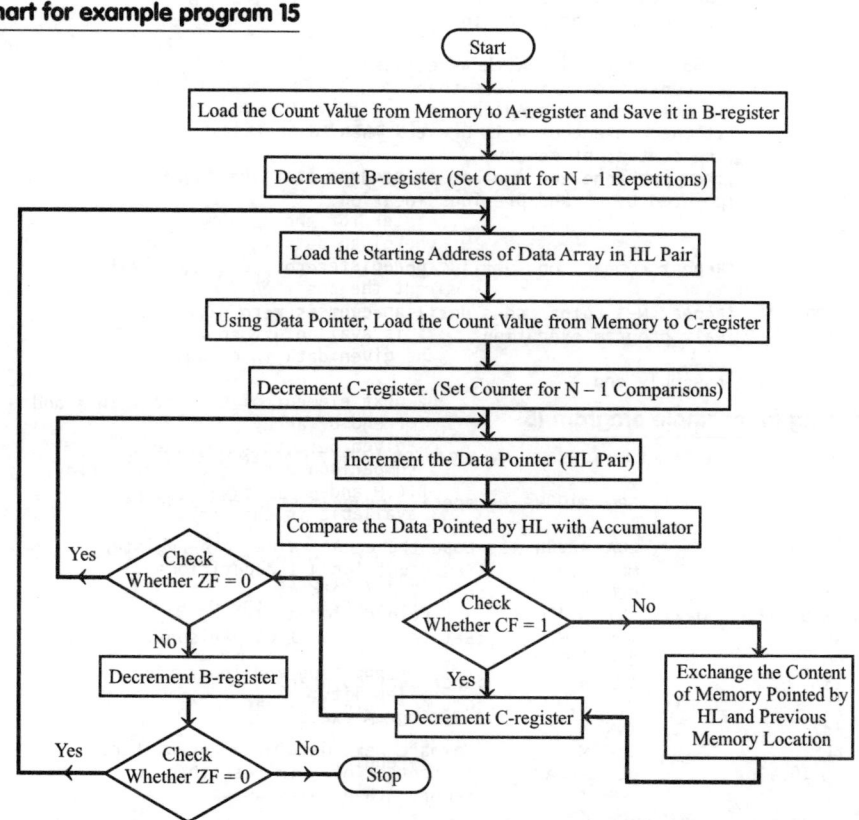

Algorithm

1. Load the count value from memory to A-register and save it in B-register.
2. Decrement B-register (B is counter for N–1 repetitions).
3. Set HL pair as data array address pointer.
4. Set C-register as counter for N–1 comparisons.
5. Load a data of the array in accumulator using the data address pointer.
6. Increment the HL pair (data address pointer).
7. Compare the data pointed by HL with accumulator.
8. If carry flag is set (If the content of accumulator is smaller than memory) then go to step 10, otherwise go to next step.
9. Exchange the content of memory pointed by HL and the accumulator.
10. Decrement C-register. If zero flag is reset go to step 6 otherwise go to next step.
11. Decrement B-register. If zero flag is reset go the step 3 otherwise go to next step.
12. Stop.

Assembly language program

```
;PROGRAM TO SORT AN ARRAY OF DATA IN ASCENDING ORDER

        ORG   4100H     ;specify program starting address.

        LDA   4200H     ;Load the count value in A-register.
        MOV   B,A       ;Set count for N-1 repetitions
        DCR   B         ;of N-1 comparisons.
LOOP2:  LXI   H,4200H   ;Set pointer for array.
        MOV   C,M       ;Set count for N-1 comparisons.
        DCR   C
        INX   H         ;Increment pointer.
LOOP1:  MOV   A,M       ;Get one data of array in A.
        INX   H
        CMP   M         ;Compare next data with A-register.
        JC    AHEAD     ;If content of A is less than
                        ;memory then go to AHEAD.
        MOV   D,M       ;If the content of A is greater than
        MOV   M,A       ;the content of memory,
        DCX   H         ;then exchange the content of memory
        MOV   M,D       ;pointed by HL and previous location.
        INX   H
AHEAD:  DCR   C
        JNZ   LOOP1     ;Repeat comparisons until C count is zero.
        DCR   B
        JNZ   LOOP2     ;Repeat N-1 comparisons until B count is zero.
        HLT             ;Halt program execution.

        END             ;Assembly end.
```

Assembler listing for example program 15

```
 1                              ;PROGRAM TO SORT AN ARRAY OF DATA IN ASCENDING ORDER
 2
 3    4100                      ORG   4100H     ;specify program starting address.
 4
 5    4100   3A 00 42           LDA   4200H     ;Load the count value in A-register.
 6    4103   47                 MOV   B,A       ;Set count for N-1 repetitions
 7    4104   05                 DCR   B         ;of N-1 comparisons.
 8    4105   21 00 42   LOOP2:  LXI   H,4200H   ;Set pointer for array.
 9    4108   4E                 MOV   C,M       ;Set count for N-1 comparisons.
10    4109   0D                 DCR   C
11    410A   23                 INX   H         ;Increment pointer.
12    410B   7E         LOOP1:  MOV   A,M       ;Get one data of array in A.
13    410C   23                 INX   H
14    410D   BE                 CMP   M         ;Compare next data with A-register.
15    410E   DA 16 41           JC    AHEAD     ;If content of A is less than
16                                              ;memory  then go to AHEAD.
```

```
17  4111  56              MOV  D,M    ;If the content of A  is greater than
18  4112  77              MOV  M,A    ;the content of memory,
19  4113  2B              DCX  H      ;then exchange the content of memory
20  4114  72              MOV  M,D    ;pointed by HL and previous location.
21  4115  23              INX  H
22  4116  0D      AHEAD:  DCR  C
23  4117  C2 0B 41        JNZ  LOOP1  ;Repeat comparisons until C count is zero.
24  411A  05              DCR  B
25  411B  C2 05 41        JNZ  LOOP2  ;Repeat N-1 comparisons until B count is zero.
26  411E  76              HLT         ;Halt program execution.
27
28  411F                  END         ;Assembly end.
```

Sample data

Input Data:	07
	AB
	92
	84
	4F
	69
	F2
	34

Memory address	Content
4200	07
4201	AB
4202	92
4203	84
4204	4F
4205	69
4206	F2
4207	34

(Before sorting)

Output data:	07
	34
	4F
	69
	84
	92
	AB
	F2

Memory address	Content
4200	07
4201	34
4202	4F
4203	69
4204	84
4205	92
4206	AB
4207	F2

(After sorting)

EXAMPLE PROGRAM - 16 :   Sorting an Array in Descending Order

Write an assembly language program to sort an array of data in descending order. The array is stored in the memory location starting from 4200$_H$. The first element of the array gives the count value for the number of elements in the array.

Problem Analysis

The algorithm for bubble sorting is given below. In bubble sorting of N-data, N–1 comparisons are carried by taking two consecutive data at a time. After each comparison, the data is rearranged such that largest among the two is in first memory location and the smallest in the next memory location. (Here the data is rearranged within the two memory locations whose contents are compared). When we perform N –1 comparisons as mentioned above for N–1 times, then the array consisting of N-data will be sorted in descending order.

Algorithm

1. Load the count value from memory to A-register and save it in B-register.
2. Decrement B-register (B is counter for N–1 repetitions).
3. Set HL pair as data array address pointer.
4. Set C-register as counter for N–1 comparisons.
5. Load a data of the array in accumulator using the data address pointer.
6. Increment the HL pair (data address pointer).
7. Compare the data pointed by HL with accumulator.
8. If carry flag is reset (If the content of accumulator is larger than memory) then go to step 10, otherwise go to next step.
9. Exchange the content of memory pointed by HL and the accumulator.
10. Decrement C-register. If zero flag is reset go to step 6 otherwise go to next step.
11. Decrement B-register. If zero flag is reset go the step 3 otherwise go to next step.
12. Stop.

Flowchart for example program 16

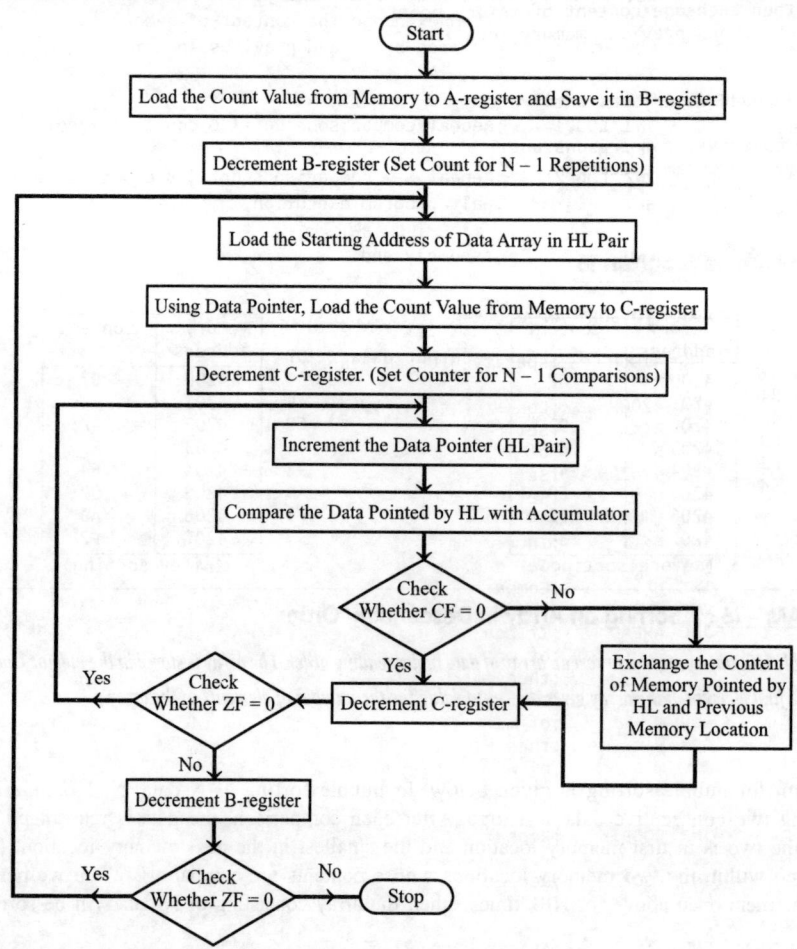

Assembly language program

```
;PROGRAM TO SORT AN ARRAY OF DATA IN DESCENDING ORDER

       ORG  4100H     ;specify program starting address.

       LDA  4200H     ;Load the count value in A-register.
       MOV  B,A       ;Set count for N-1 repetitions
       DCR  B         ;of N-1 comparisons.
LOOP2: LXI  H,4200H   ;Set pointer for array.
       MOV  C,M       ;Set count for N-1 comparisons.
       DCR  C
       INX  H         ;Increment the pointer.
LOOP1: MOV  A,M       ;Get one data of array in A.
       INX  H         ;Compare the next data of array with
       CMP  M         ;the content of A-register.
       JNC  AHEAD     ;If content of A is greater than content
                      ;of memory addressed by HL pair,
                      ;then go to AHEAD.
```

```
        MOV  D,M    ;If the content of A is less than content
        MOV  M,A    ;of memory addressed by HL pair,
        DCX  H      ;then exchange content of memory pointed
        MOV  M,D    ;by HL and previous memory location.
        INX  H
AHEAD:  DCR  C
        JNZ  LOOP1  ;Repeat comparisons until C count is zero.
        DCR  B
        JNZ  LOOP2  ;Repeat N-1 comparisons until B count is zero.
        HLT         ;Halt program execution.

        END         ;Assembly end.
```

Assembler listing for example program 16

```
1                              ;PROGRAM TO SORT AN ARRAY OF DATA IN DESCENDING ORDER
2
3   4100                       ORG  4100H    ;specify program starting address.
4
5   4100  3A 00 42             LDA  4200H    ;Load the count value in A-register.
6   4103  47                   MOV  B,A      ;Set counter for N-1 repetitions
7   4104  05                   DCR  B        ;of N-1 comparisons.
8   4105  21 00 42    LOOP2:   LXI  H,4200H  ;Set pointer for array.
9   4108  4E                   MOV  C,M      ;Set count for N-1 comparisons.
10  4109  0D                   DCR  C
11  410A  23                   INX  H        ;Increment the pointer.
12  410B  7E          LOOP1:   MOV  A,M      ;Get one data of array in A.
13  410C  23                   INX  H        ;Compare the next data of array with
14  410D  BE                   CMP  M        ;the content of A-register.
15  410E  D2 16 41             JNC  AHEAD    ;If content of A is greater than content
16                                           ;of memory addressed by HL pair,
17                                           ;then go to AHEAD.
18  4111  56                   MOV  D,M      ;If the content of A is less than content
19  4112  77                   MOV  M,A      ;of memory addressed by HL pair,
20  4113  2B                   DCX  H        ;then exchange content of memory pointed
21  4114  72                   MOV  M,D      ;by HL and previous memory location.
22  4115  23                   INX  H
23  4116  0D          AHEAD:   DCR  C
24  4117  C2 0B 41             JNZ  LOOP1    ;Repeat comparisons until C count is zero.
25  411A  05                   DCR  B
26  411B  C2 05 41             JNZ  LOOP2    ;Repeat N-1 comparison until B count is zero.
27  411E  76                   HLT           ;Halt program execution.
28
29  411F                       END           ;Assembly end.
```

Sample data

Input Data :

07
C4
84
9A
7B
E2
F4
B2

Memory address	Content
4200	07
4201	C4
4202	84
4203	9A
4204	7B
4205	E2
4206	F4
4207	B2

(Before sorting)

Output Data:

07
F4
E2
C4
B2
9A
84
7B

Memory address	Content
4200	07
4201	F4
4202	E2
4203	C4
4204	B2
4205	9A
4206	84
4207	7B

(After sorting)

2.12.4 PROGRAM WITH LOOKUP TABLE

EXAMPLE PROGRAM - 17 :   BCD to 7-Segment LED Code

Write an assembly language program to find the 7-segment LED code for a 2-digit BCD data, by using the look up table. The BCD data is stored in 4200$_H$. Store the 7-segment code in 4201$_H$ and 4202$_H$.

Problem Analysis

The 7-segment LED codes for decimal digit 0 to 9 are determined and stored in memory locations from 5000$_H$ to 5009$_H$ respectively. The look-up table is created such that the low order address is same as that of decimal digit. Hence, by this method the high order address is fixed (50) and the low order address is the decimal digit itself.

In order to find the 7-segment code, the BCD data is split into lower nibble and upper nibble. The code is determined by taking each nibble as low order address of the **look up table.**

Flowchart for example program 17

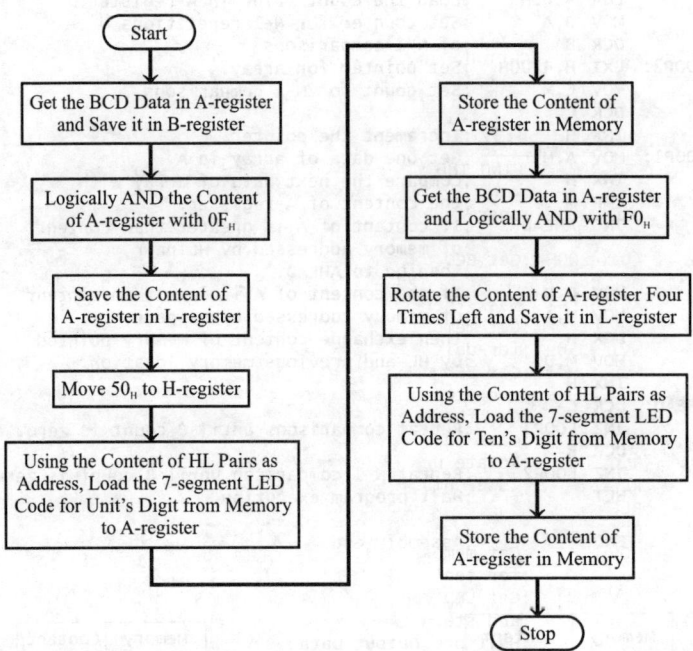

Algorithm

1.   Load the BCD data in A-register and save in B-register.
2.   Logically AND A-register with 0F$_H$ to mask upper nibble (ten's digit).
3.   Move A-register to L-register and move 50$_H$ to H-register.
4.   Get the LED code for lower nibble (unit's digit) in A-register and store in memory.
5.   Move the BCD data from B-register to A-register and mask the lower nibble (unit's digit).
6.   Rotate the upper nibble to lower nibble position.
7.   Move A-register to L-register.
8.   Get the LED code for ten's digit in A-register and store in memory.
9.   Stop.

Assembly language program

```
;PROGRAM TO FIND THE 7-SEGMENT LED CODE FOR A BCD DATA

        ORG 4100H ;specify program starting address.

        LDA 4200H ;Get BCD data in A and save in B.
        MOV B,A
        ANI 0FH   ;Mask the upper nibble (ten's digit).
        MOV L,A   ;Get memory address of LED code
        MVI H,50H ;for unit's digit, in HL pair.
        MOV A,M   ;Get LED code for unit's digit in A
        STA 4201H ;and store in memory.
        MOV A,B   ;Get the BCD data in A-register and
        ANI F0H   ;mask the lower nibble (unit's digit).
        RLC       ;Rotate upper nibble to
        RLC       ;lower nibble position.
        RLC
        RLC
        MOV L,A   ;Get memory address of LED code
                  ;for ten's digit in HL pair.
        MOV A,M   ;Get LED code for ten's digit in A
        STA 4202H ;and store in memory.
        HLT       ;Halt program execution.

        END       ;Assembly end.
```

Assembler listing for example program 17

```
1                           ;PROGRAM TO FIND THE 7-SEGMENT LED CODE FOR A BCD DATA
2
3    4100                   ORG 4100H ;specify program starting address.
4
5    4100   3A 00 42        LDA 4200H ;Get BCD data in A and save in B.
6    4103   47              MOV B,A
7    4104   E6 0F           ANI 0FH   ;Mask the upper nibble (ten's digit).
8    4106   6F              MOV L,A   ;Get memory address of LED code
9    4107   26 50           MVI H,50H ;for unit's digit, in HL pair.
10   4109   7E              MOV A,M   ;Get LED code for unit's digit in A
11   410A   32 01 42        STA 4201H ;and store in memory.
12   410D   78              MOV A,B   ;Get the BCD data in A-register and
13   410E   E6 F0           ANI F0H   ;mask the lower nibble (unit's digit).
14   4110   07              RLC       ;Rotate upper nibble to
15   4111   07              RLC       ;lower nibble position.
16   4112   07              RLC
17   4113   07              RLC
18   4114   6F              MOV L,A   ;Get memory address of LED code
19                                    ;for ten's digit in HL pair.
20   4115   7E              MOV A,M   ;Get LED code for ten's digit in A
21   4116   32 02 42        STA 4202H ;and store in memory.
22   4119   76              HLT       ;Halt program execution.
23
24   411A                   END       ;Assembly end.
```

Sample data 1 :

Loop-up table for Common cathode 7-segment LED

Input Data : 45_{10}
Output Data : $6D_H$
              66_H

Memory address	Content	Memory address	Content
5000	3F	5005	6D
5001	06	5006	7D
5002	5B	5007	07
5003	4F	5008	7F
5004	66	5009	6F

Memory address	Content
4200	45
4201	6D
4202	66

Sample data 2 :

Loop-up table for Common anode 7-segment LED

Memory address	Content
5000	C0
5001	F9
5002	A4
5003	B0
5004	99

Memory address	Content
5005	92
5006	82
5007	F8
5008	80
5009	90

Input Data : 45_{10}
Output Data : 92_H
               99_H

Memory address	Content
4200	45
4201	92
4202	99

2.12.5 PROGRAM WITH SUBROUTINE

EXAMPLE PROGRAM - 18 : Square Root of 8-Bit Binary Number

Write an assembly language program to find the square root of an 8-bit binary number. The binary number is stored in memory location 4200_H and store the square root in 4201_H.

Problem Analysis

Square root can be computed by an iterative technique. First an initial value is assumed. Here the initial value of square root is taken as half the value of given number. The new value of square root is computed by using an expression, XNEW = (X + Y/X)/2 where X is the initial value of square root and Y is the given number. Then XNEW is compared with initial value. If they are not equal then the above process is repeated until X is equal to XNEW after taking XNEW as initial value, (i.e., X ← XNEW).

Algorithm

1.   Load the given data (Y) in A-register.
2.   Save the content of A-register in B-register.
3.   Move 02_H (divisor) to C-register.
4.   Call DIV subroutine to get initial value of square root (X) in D-register.
5.   Save the content of D-register (initial value X) in E-register.
6.   Move the given data (Y) from B-register to A-register.
7.   Move the initial value (X) from D-register to C-register.
8.   Call DIV subroutine to get Y/X in D-register.
9.   Move the Y/X available in D-register to A-register.
10.   Add the value of X in E-register to A-register to get X+Y/X in A-register.
11.   Move 02_H to C-register.
12.   Call DIV subroutine to get now value of square root (XNEW) in D-register.
13.   Compare X and XNEW.
14.   If ZF = 1, go to next step. If ZF = 0, go to step 5.
15.   Store the value of square root (A-register) in memory.
16.   Stop.

Algorithm for subroutine DIV

1.   Clear D-register.
2.   Subtract the content of C-register (divisor) from the content of A-register (dividend).
3.   Increment quotient (D-register).
4.   Compare A-register and C-register.
5.   If CF = 1, go to next step. If CF = 0 go to step 2.
6.   Return to main program.

Flowchart for example program 18

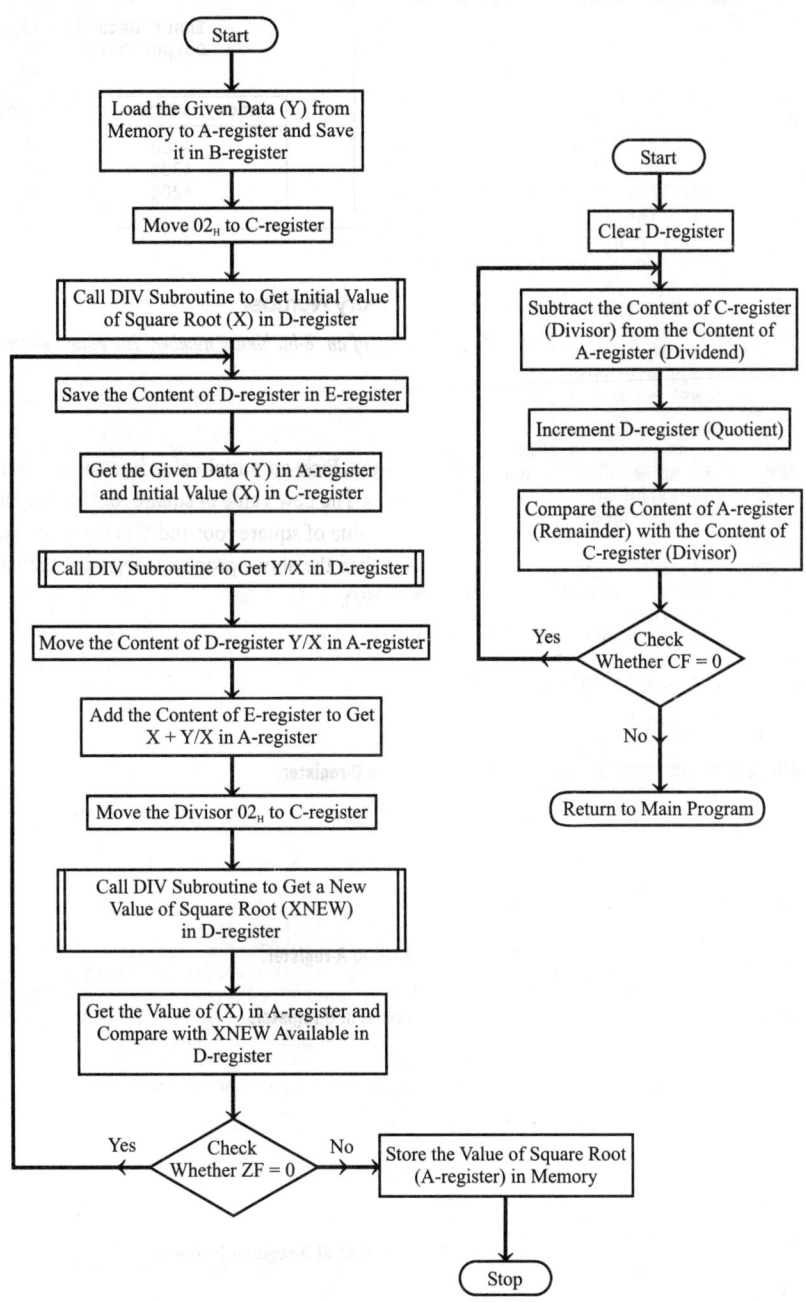

Assembly language program

```
;PROGRAM TO FIND THE SQUARE ROOT OF 8-BIT BINARY NUMBER

        ORG    4100H   ;specify program starting address.


        LDA    4200H   ;Get the given data(Y) in A-register.
        MOV    B,A     ;Save the data in B-register.
        MVI    C,02H   ;Get the divisor(02H) in C-register.
        CALL   DIV     ;Call division subroutine to get initial
                       ;value(X) in the D-register.
REP:    MOV    E,D     ;Save the initial value in E-register.
        MOV    A,B     ;Get the dividend(Y) in A-register.
        MOV    C,D     ;Get the divisor(X) in C-register.
        CALL   DIV     ;Call division subroutine to get Y/X in D.
        MOV    A,D     ;Move Y/X in A-register.
        ADD    E       ;Get((Y/X)+X) in A-register.
        MVI    C,02H   ;Get the divisor (02H) in C-register.
        CALL   DIV     ;Call division subroutine to get
                       ;XNEW in D-register.

        MOV    A,E     ;Get X in A-register.
        CMP    D       ;Compare X and XNEW.
        JNZ    REP     ;If XNEW is not equal to X, then repeat.
        STA    4201H   ;Save the square root in memory.
        HLT            ;Halt program execution.
;DIVISION SUBROUTINE
DIV:    MVI    D,00H   ;Clear D-register for quotient.
NEXT:   SUB    C       ;Subtract the divisor from dividend.
        INR    D       ;Increment the quotient.
        CMP    C       ;Repeat subtraction until the divisor
        JNC    NEXT    ;is less than dividend.
        RET            ;Return to main program.

        END            ;Assembly end.
```

Assembler listing for example program 18

```
1                           ;PROGRAM TO FIND THE SQUARE ROOT OF 8-BIT BINARY NUMBER
2
3    4100                    ORG 4100H  ;specify program starting address.
4
5    4100   3A 00 42         LDA 4200H  ;Get the given data(Y) in A-register.
6    4103   47               MOV B,A    ;Save the data in B-register.
7    4104   0E 02            MVI C,02H  ;Get the divisor(02H) in C-register.
8    4106   CD 1F 41         CALL DIV   ;Call division subroutine to get initial
9                                       ;value(X) in the D-register.
10   4109   5A        REP:   MOV E,D    ;Save the initial value in E-register.
11   410A   78               MOV A,B    ;Get the dividend(Y) in A-register.
12   410B   4A               MOV C,D    ;Get the divisor(X) in C-register.
13   410C   CD 1F 41         CALL DIV   ;Call division subroutine to get Y/X in D.
14   410F   7A               MOV A,D    ;Move Y/X in A-register.
15   4110   83               ADD E      ;Get((Y/X)+X) in A-register.
16   4111   0E 02            MVI C,02H  ;Get the divisor (02H) in C-register.
17   4113   CD 1F 41         CALL DIV   ;Call division subroutine to get
18   4116                               ;XNEW in D-register
19   4116   7B               MOV A,E    ;Get X in A-register.
20   4117   BA               CMP D      ;Compare X and XNEW.
21   4118   C2 09 41         JNZ REP    ;If XNEW is not equal to X, then repeat.
22   411B   32 01 42         STA 4201H  ;Save the square root in memory.
23   411E   76               HLT        ;Halt program execution.
24
25
26                           ;DIVISION SUBROUTINE
27
```

```
28  411F  16 00    DIV:  MVI  D,00H   ;Clear D-register for quotient.
29  4121  91        NEXT: SUB  C       ;Subtract the divisor from dividend.
30  4122  14              INR  D       ;Increment the quotient.
31  4123  B9              CMP  C       ;Repeat subtraction until the divisor
32  4124  D2 21 41        JNC  NEXT    ;is less than dividend.
33  4127  C9              RET          ;Return to main program.
34
35  4128                  END          ;Assembly end.
```

Sample data

Input Data : 64_H

Output Data : $0A_H$

Memory address	Content
4200	64
4201	0A

EXAMPLE PROGRAM - 19 : Binary to ASCII Conversion

Write an assembly language program to convert an 8-bit binary (2-digit hexa) to ASCII code. The binary data is stored in 4200$_H$ and store the ASCII code in 4201$_H$ and 4202$_H$.

Problem Analysis

Each Hexa digit (4-bit binary) is represented by an 8-bit ASCII. The Hexa digit 0 through 9 are represented by 30_H to 39_H in ASCII. Hence, for Hexa 0 to 9, if we add 30_H, we will get the corresponding ASCII. The Hexa digit A through F are represented by 41_H to 46_H in ASCII. Hence, for Hexa digit A to F if we add 37_H we will get the corresponding ASCII.

In the following algorithm the given 8-bit data is split into two nibbles. The ASCII code for each nibble is found by calling a subroutine, which takes care of adding 30_H to the nibble if it is less than $0A_H$, or adding 37_H if the nibble is greater than 09_H.

Algorithm

1. Load the given data in A-register and move to B-register.
2. Mask the upper nibble of the binary (hexa) data in A-register.
3. Call subroutine ACODE to get ASCII code of the lower nibble and store in memory.
4. Move B-register to A-register and mask the lower nibble.
5. Rotate the upper nibble to lower nibble position.
6. Call subroutine ACODE to get the ASCII code of upper nibble and store in memory.
7. Stop.

Algorithm for subroutine code

1. Compare the content of A-register with $0A_H$.
2. If CF = 1, go to step 4. If CF = 0, go to next step.
3. Add 07_H to A-register.
4. Add 30_H to A-register.
5. Return to main program.

Flowchart for example program 19

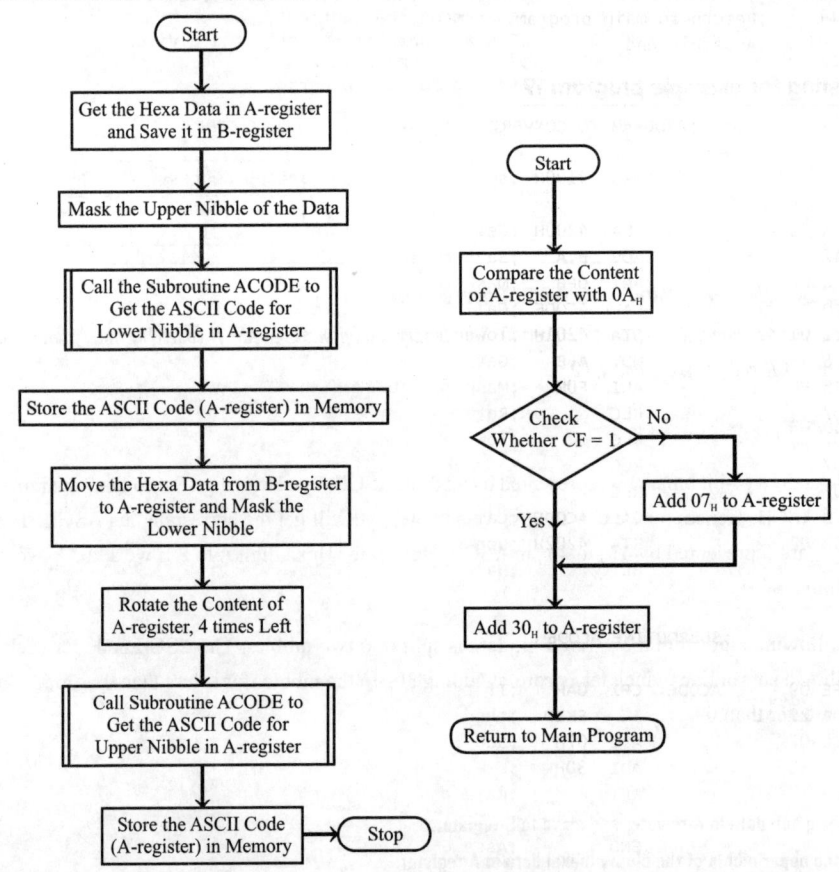

Assembly language program

```
;PROGRAM TO CONVERT 8-BIT BINARY TO ASCII CODE

        ORG 4100H  ;specify program starting address.

        LDA 4200H  ;Get binary data in A.
        MOV B,A    ;Save the binary data in B register.
        ANI 0FH    ;Mask the upper nibble.
        CALL ACODE ;Call subroutine to get ASCII code for
        STA 4201H  ;lower nibble in A and store in memory.
        MOV A,B    ;Get data in A-register.
        ANI F0H    ;Mask the lower nibble.
        RLC        ;Rotate upper nibble to
        RLC        ;lower nibble position.
        RLC
        RLC
        CALL ACODE ;Call subroutine to get ASCII code for
        STA 4202H  ;upper nibble in A and store in memory.
        HLT        ;Halt program execution.
;SUBROUTINE ACODE
ACODE:  CPI 0AH    ;If the content of A is less than 0AH,
        JC  SKIP   ;then add 30H to A  otherwise
```

```
        ADI  07H    ;add 37H to A-register.
SKIP:   ADI  30H
        RET         ;Return to main program.
        END         ;Assembly end.
```

Assembler listing for example program 19

```
1                        ;PROGRAM TO CONVERT 8-BIT BINARY TO ASCII CODE
2
3   4100                 ORG   4100H ;specify program starting address.
4
5   4100  3A 00 42       LDA   4200H ;Get binary data in A.
6   4103  47             MOV   B,A   ;Save the binary data in B-register.
7   4104  E6 0F          ANI   0FH   ;Mask the upper nibble.
8   4106  CD 1A 41       CALL  ACODE ;Call subroutine to get ASCII code for
9   4109  32 01 42       STA   4201H ;lower nibble in A and store in memory.
10  410C  78             MOV   A,B   ;Get data in A-register.
11  410D  E6 F0          ANI   F0H   ;Mask the lower nibble.
12  410F  07             RLC         ;Rotate upper nibble to
13  4110  07             RLC         ;lower nibble position.
14  4111  07             RLC
15  4112  07             RLC
16  4113  CD 1A 41       CALL  ACODE ;Call subroutine  to get ASCII code for
17  4116  32 02 42       STA   4202H ;upper nibble in A and store in memory.
18  4119  76             HLT         ;Halt program execution.
19
20
21                   ;SUBROUTINE ACODE
22  411A
23  411A  FE 09   ACODE: CPI   0AH   ;If the content of A is less than 0AH,
24  411C  DA 21 41       JC    SKIP  ;then add 30H to A otherwise
25  411F  C6 07          ADI   07H   ;add 37H to A-register.
26  4121  C6 30   SKIP:  ADI   30H
27  4123  C9             RET         ;Return to main program.
28
29  4124                 END         ;Assembly end.
```

Sample data

```
    Input Data   : E4_H
    Output Data  : 34 (ASCII code for 4)
                   45 (ASCII code for E)
```

Memory address	Content
4200	E4
4201	34
4202	45

EXAMPLE PROGRAM - 20 : ASCII to Binary Conversion

Write an assembly language program to convert an array of ASCII codes to corresponding binary (hexa) value. The ASCII array is stored starting from 4200_H. The first element of the array gives the number of elements in the array.

Problem Analysis

The hexa digit 0 through 9 are represented by 30_H to 39_H in ASCII. Hence, for ASCII code 30_H to 39_H if we subtract 30_H then we will get the corresponding binary (hexa) value. The hexa digit A through F are represented by 41_H to 46_H in ASCII. Hence for ASCII code 41_H to 46_H we have to subtract 37_H to get corresponding binary (hexa) value. In the following algorithm, a subroutine has been written to subtract either 30_H or 37_H from the given data.

Flowchart for example program 20

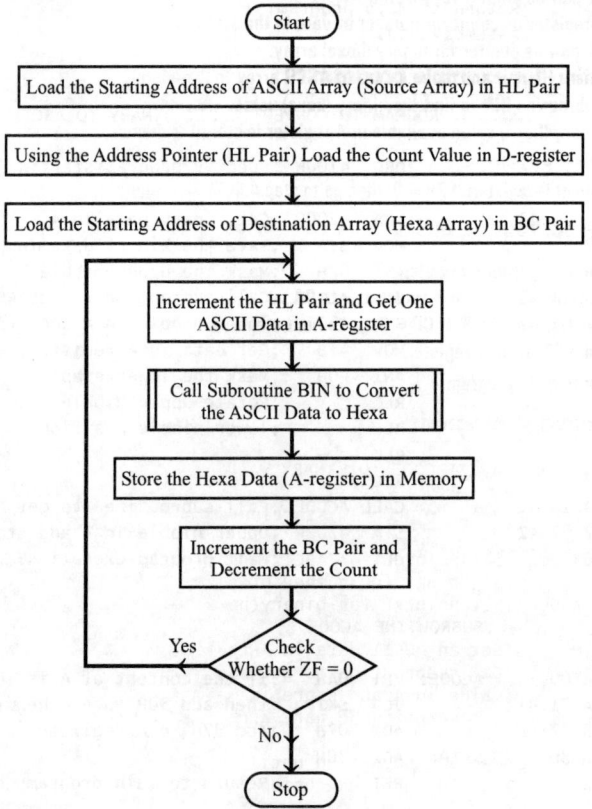

```
                        Start

    Load the Starting Address of ASCII Array (Source Array) in HL Pair

    Using the Address Pointer (HL Pair) Load the Count Value in D-register

    Load the Starting Address of Destination Array (Hexa Array) in BC Pair

            Increment the HL Pair and Get One
                  ASCII Data in A-register

            Call Subroutine BIN to Convert
               the ASCII Data to Hexa

            Store the Hexa Data (A-register) in Memory

              Increment the BC Pair and
                  Decrement the Count

      Yes            Check
      ←            Whether ZF = 0

                     No
                    Stop
```

Flowchart for subroutine BIN

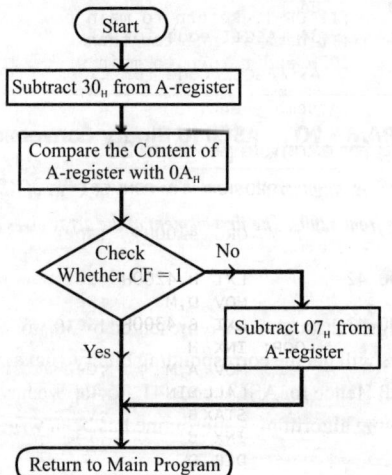

```
                    Start

          Subtract 30_H from A-register

          Compare the Content of
            A-register with 0A_H

              Check              No
           Whether CF = 1    →

                              Subtract 07_H from
              Yes                A-register

          Return to Main Program
```

Algorithm

1. Set HL pair as pointer for ASCII array.
2. Set D-register as count for number of data in the array.
3. Set BC pair as pointer for binary (hexa) array.
4. Increment HL pair and move a data of ASCII array to A-register.
5. Call subroutine BIN to find the binary (hexa) value.
6. The binary (hexa) value available in A-register is stored in memory.
7. Increment BC pair.
8. Decrement D-register. If ZF = 0, then go to step 4. If ZF = 1, then stop.

Algorithm for subroutine BIN

1. Subtract 30_H from A-register.
2. Compare the content of A-register with $0A_H$.
3. If CF = 1, go to step 5. If CF = 0, go to next step.
4. Subtract 07_H from A-register.
5. Return to main program.

Assembly language program

```
;PROGRAM TO CONVERT ASCII CODE TO BINARY VALUE

        ORG  4100H    ;specify program starting address.

        LXI  H,4200H  ;Set pointer for ASCII array.
        MOV  D,M      ;Set count for number of data.
        LXI  B,4300H  ;Set pointer for binary(hexa) array.
LOOP:   INX  H
        MOV  A,M      ;Get an ASCII data in A-register.
        CALL BIN      ;Call subroutine to get binary
        STAX B        ;value in A and store in memory.
        INX  B        ;Increment the binary array pointer.
        DCR  D
        JNZ  LOOP     ;Repeat conversion until count is zero.
        HLT           ;Halt program execution.


;SUBROUTINE BIN
BIN:    SUI  30H      ;Subtract 30H from the data.
        CPI  0AH
        RC            ;If CF=1, Return to main program.
        SUI  07H      ;If data is greater than 0AH, then subtract
        RET           ;07H and return to main program.

        END           ;Assembly end.
```

Assembler listing for example program 20

```
 1                         ;PROGRAM TO CONVERT ASCII CODE TO BINARY VALUE
 2
 3    4100               ORG  4100H    ;specify program starting address.
 4
 5    4100  21 00 42      LXI  H,4200H  ;Set pointer for ASCII array.
 6    4103  56            MOV  D,M      ;Set count for number of data.
 7    4104  01 00 43      LXI  B,4300H  ;Set pointer for binary(hexa) array.
 8    4107  23      LOOP: INX  H
 9    4108  7E            MOV  A,M      ;Get an ASCII data in A-register.
10    4109  CD 13 41      CALL BIN      ;Call subroutine to get binary
11    410C  02            STAX B        ;value in A and store in memory.
12    410D  03            INX  B        ;Increment the binary array pointer.
13    410E  15            DCR  D
14    410F  C2 07 41      JNZ  LOOP     ;Repeat conversion until count is zero.
15    4112  76            HLT           ;Halt program execution.
```

16	4113				
17					
18				;SUBROUTINE BIN	
19					
20	4113	D6 30	BIN:	SUI 30H	;Subtract 30H from the data.
21	4115	FE 0A		CPI 0AH	
22	4117	D8		RC	;If CF = 1, Return to main program.
23	4118	D6 07		SUI 07H	;If data is greater than 0AH then subtract
24	411A	C9		RET	;07H and return to main program.
25					
26	411B			END	;Assembly end.

Sample data

Input Data :

Count : 07

Memory address	Content
4200	07
4201	31
4202	42
4203	35
4204	46
4205	43
4206	39
4207	38

ASCII Array: 31
          42
          35
          46
          43
          39
          38

Output Data :

Binary array= 01
            0B
            05
            0F
            0C
            09
            08

Memory address	Content
4300	01
4301	0B
4302	05
4303	0F
4304	0C
4305	09
4306	08

2.12.6 PROGRAM INVOLVING STACK

EXAMPLE PROGRAM – 21 :   BCD to Binary Conversion

Write an assembly language program to convert a two-digit BCD (8-bit) data to binary data. The BCD data is stored in 4200$_H$ and store the binary data in 4201$_H$

Problem Analysis

The 2-digit BCD data will have units digit and tens digit. When the tens digit (upper nibble) is multiplied by 0A$_H$ and the product is added to units digit (lower nibble), the result will be in binary, because the microprocessor performs binary arithmetic.

Algorithm

1.   Get the BCD data in A-register and save in stack.
2.   Mask the lower nibble (units) of the BCD data in A-register.
3.   Rotate the upper nibble to lower nibble position and save in B-register.
4.   Clear the accumulator.
5.   Move 0A$_H$ to C-register.
6.   Add B-register to A-register.
7.   Decrement C-register. If ZF = 0 go to step 6. If ZF = 1, go to next step.
8.   Save the product in B-register.
9.   Get the BCD data from stack in A-register and mask the upper nibble (tens).
10. Add the units (A-register) to product (B-register).
11. Store the binary value (A-register).
12. Stop.

Flowchart for example program 21

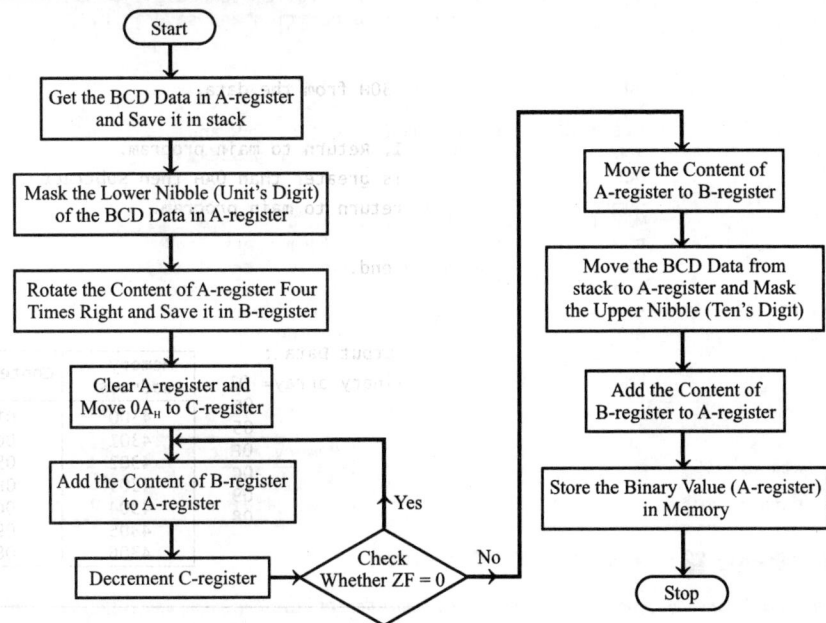

Assembly language program

```
;PROGRAM TO CONVERT 2-DIGIT BCD TO BINARY NUMBER

      ORG  4100H   ;specify program starting address.

      LDA  4200H   ;Get the data in A-register,
      PUSH PSW     ;save data in Stack.
      ANI  F0H     ;Mask the lower nibble (units digit).
      RLC          ;Rotate the upper nibble (tens digit)
      RLC          ;to lower nibble position and save in B.
      RLC
      RLC
      MOV  B,A
      XRA  A       ;Clear accumulator.
      MVI  C,0AH   ;Multiply tens digit by 0AH and
REP:  ADD  B       ;get the product in A-register.
      DCR  C
      JNZ  REP
      MOV  B,A     ;Save the product in B-register.
      POP  PSW     ;Get the BCD data from stack.
      ANI  0FH     ;Mask the upper nibble (tens digit).
      ADD  B       ;Get the binary data in A-register.
      STA  4201H   ;Save the binary data in memory.
      HLT          ;Halt program execution.

      END          ;Assembly end.
```

Assembler listing for example 21

```
1                           ;PROGRAM TO CONVERT 2-DIGIT BCD TO BINARY NUMBER
2
3   4100                    ORG 4100H ;specify program starting address.
4
5   4100  3A 00 42          LDA 4200H ;Get the data in A-register,
6   4103  F5                PUSH PSW  ;save data in stack.
```

```
 7   4104   E6 F0              ANI  F0H    ;Mask the lower nibble (units digit).
 8   4106   07                 RLC         ;Rotate the upper nibble (tens digit)
 9   4107   07                 RLC         ;to lower nibble position and save in B.
10   4108   07                 RLC
11   4109   07                 RLC
12   410A   47                 MOV  B,A
13   410B   AF                 XRA  A      ;Clear accumulator.
14   410C   0E 0A              MVI  C,0AH  ;Multiply tens digit by 0AH and
15   410E   80          REP:   ADD  B      ;get the product in A-register.
16   410F   0D                 DCR  C
17   4110   C2 0E 41           JNZ  REP
18   4113   47                 MOV  B,A    ;Save the product in B-register.
19   4114   F1                 POP  PSW    ;Get the BCD data from stack.
20   4115   E6 0F              ANI  0FH    ;Mask the upper nibble (tens digit).
21   4117   80                 ADD  B      ;Get the binary data in A-register.
22   4118   32 01 42           STA  4201H  ;Save the binary data in memory.
23   411B   76                 HLT         ;Halt program execution.
24
25   411C                      END         ;Assembly end.
```

Sample data

	Memory address	Content
Input Data : 45_{10}	4200	45
Output Data : $2D_H$	4201	2D

EXAMPLE PROGRAM - 22 :   Binary to BCD Conversion

Write an assembly language program to convert an 8-bit binary data to BCD. The binary data is stored in 4200_H. Store the hundred's digit in 4251_H. Store the ten's and unit's digits in 4250_H.

Problem Analysis

The maximum value of 8-bit binary is $FF_H = 256_{10}$. Hence the maximum size of the data will have hundreds, tens and units. The algorithm given below uses two counters to count hundreds and tens. Initially counters are cleared. First let us subtract all hundreds from the binary data. For each subtraction, hundred's register is incremented by one. Then, let us subtract all tens. For each subtraction, ten's register is incremented by one. The remaining will be units. The tens and units are combined to form 2-digit BCD (8-bit binary).

Algorithm

1. **Clear D and E registers to account for hundreds and tens.**

2. **Load the binary data in A-register.**

3. **Compare A-register with 64_H. If carry flag is set, go to step 7 otherwise go to next step.**

4. **Subtract 64_H from A-register.**

5. **Increment E-register (Hundred's register).**

6. **Go to step 3.**

7. **Compare the A-register with $0A_H$. If carry flag is set, go to step 11, otherwise go to next step.**

8. **Subtract $0A_H$ from A-register.**

9. **Increment D-register (ten's register).**

10. **Go to step 7.**

11. **Combine the units and tens to form 8-bit result.**

12. **Save the units, tens and hundreds in memory.**

Flowchart for example program 22

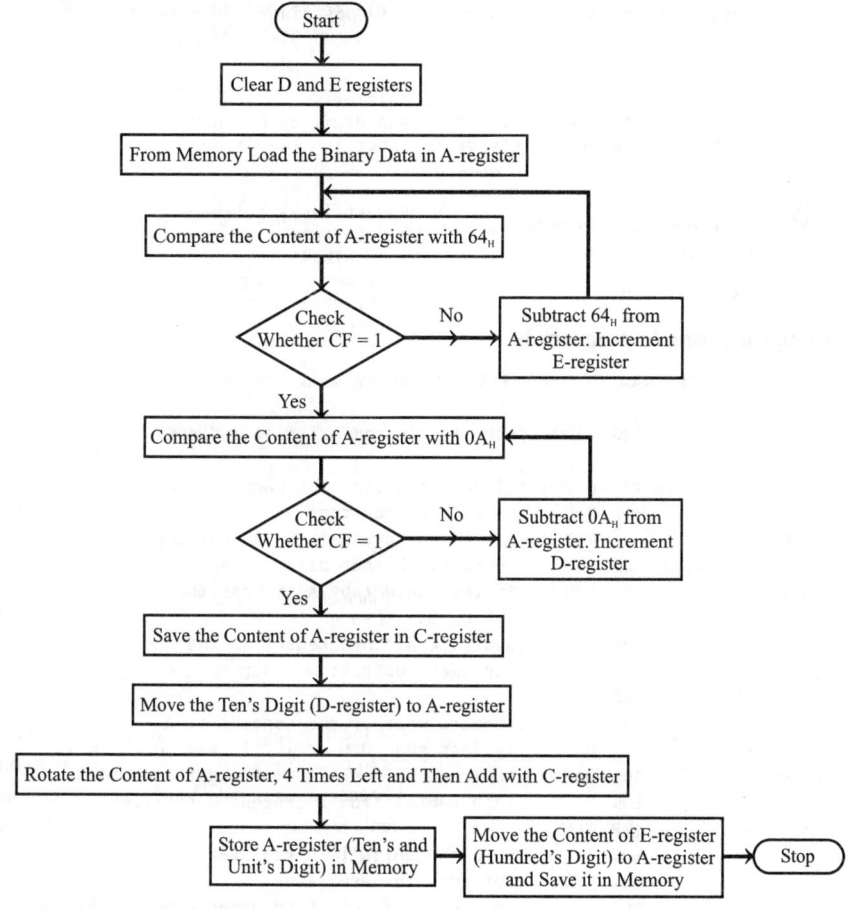

Assembly language program

```
;PROGRAM TO CONVERT 8-BIT BINARY NUMBER TO BCD

        ORG  4100H  ;specify program starting address.

        MVI  E,00H  ;Clear E-register for hundreds and
        MOV  D,E    ;D-register for tens.
        LDA  4200H  ;Get the binary data in A-register.
HUND:   CPI  64H    ;Compare, whether data is less than 64H(100).
        JC   TEN     ;If the content of A is less than
                    ;100 or 64H then go to TEN.
        SUI  64H    ;Subtract all hundreds from the data and
        INR  E      ;for each subtraction increment E-register.
        JMP  HUND
TEN:    CPI  0AH    ;Compare whether the content of A
        JC   UNIT   ;is less than 0AH or 10.If CF=1 go to UNIT.
        SUI  0AH    ;Subtract all tens from the data and for
        INR  D      ;each subtraction increment D-register.
        JMP  TEN
```

```
UNIT:   PUSH PSW    ;Save the units in stack.
        MOV A,D     ;Get tens in A-register.
        RLC         ;Rotate ten's  digit to upper nibble position.
        RLC
        RLC
        RLC
        POP  B      ;get the units from stack in B.
        ADD  B      ;Combine ten's and unit's digits.
        STA 4250H   ;Save tens and units in memory.
        MOV A,E
        STA 4251H   ;Save hundreds in memory.
        HLT         ;Halt program execution.

        END         ;Assembly end.
```

Assembler listing for example program 19

```
1                       ;PROGRAM TO CONVERT 8-BIT BINARY NUMBER TO BCD
2
3    4100               ORG 4100H  ;specify program starting address.
4
5    4100  1E 00        MVI E,00H  ;Clear E-register for hundreds and
6    4102  53           MOV D,E    ;D-register for tens.
7    4103  3A 00 42     LDA 4200H  ;Get the binary data in A-register.
8    4106  FE 64  HUND: CPI 64H    ;Compare, whether data is less than 64H(100).
9    4108  DA 11 41     JC  TEN    ;If the content of A is less than
10                                 ;100 or 64H then go to TEN.
11   410B  D6 64        SUI 64H    ;Subtract all hundreds from the data and
12   410D  1C           INR E      ;for each subtraction increment E-register.
13   410E  C3 06 41     JMP HUND
14   4111  FE 0A  TEN:  CPI 0AH    ;Compare whether the content of A
15   4113  DA 1C 41     JC  UNIT   ;is less than 0AH or 10.If CF=1 go to UNIT.
16   4116  D6 0A        SUI 0AH    ;Subtract all tens from the data and for
17   4118  14           INR D      ;each subtraction increment D-register.
18   4119  C3 11 41     JMP TEN
19   411C  4F     UNIT: MOV C,A    ;Save the units in C-register.
20   411D  7A           MOV A,D    ;Get tens in A-register.
21   411E  07           RLC        ;Rotate ten's  digit to upper nibble position.
22   411F  07           RLC
23   4120  07           RLC
24   4121  07           RLC
25   4122  C1           POP B      ; get the units from stack in B.
26   4123  80           ADD B      ;Combine ten's and unit's digits.
27   4124  32 50 42     STA 4250H  ;Save tens and units in memory.
28   4127  7B           MOV A,E
29   4128  32 51 42     STA 4251H  ;Save hundreds in memory.
30   412B  76           HLT        ;Halt program execution.
31
32   412C               END        ;Assembly end.
```

Sample Data

Input Data : B9$_H$

Output Data : 0185$_{10}$

Memory address	Content	
4200	B9	Binary data
4250	85	BCD data
4251	01	

2.13 SHORT-ANSWER QUESTIONS

Q2.1 How many instructions are available in 8085 instruction set?

The 8085 instruction set consists of 74 basic instructions and 246 total instructions.

Q2.2 What is the instruction format of 8085?

The size of 8085 instruction is 1 to 3 bytes. Each instruction has one-byte opcode. The remaining bytes are either data or address. The format of 8085 instructions are shown below :

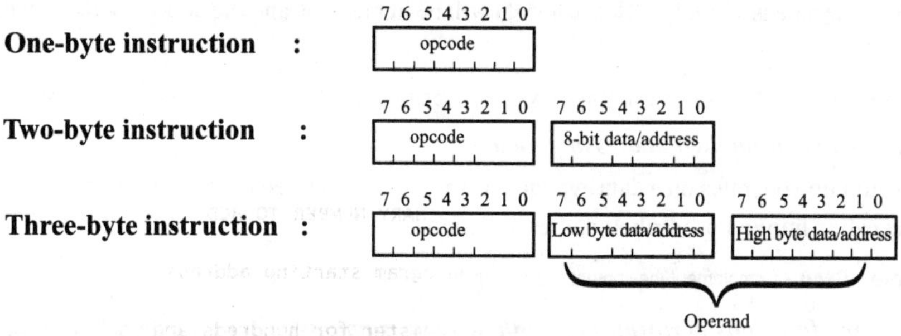

Fig. Q2.2: Format of 8085 instruction.

Q2.3 What is addressing?

The method of specifying the data to be operated (operand) by the instruction is called addressing.

Q2.4 What are the addressing modes available in 8085?

The 8085 has the following five different modes of addressing.

    i)   **Immediate addressing**

    ii)   **Direct addressing**

    iii) **Register addressing**

    iv) **Register indirect addressing**

    v)   **Implied addressing.**

Q2.5 Explain the immediate addressing with an example.

In immediate addressing mode, the data is specified in the instruction itself. The data will be a part of the program instruction.

Example : MVI B, 3EH - Move the data $3E_H$ given in the instruction to B-register.

Q2.6 What is direct addressing? Give an example.

If the address of the data is directly specified in the instruction then the addressing mode is called direct addressing.

Example : LDA 1050H - Load the data available in memory location 1050_H in accumulator.

Q2.7 Explain register addressing with an example.

In register addressing mode, the instruction specifies the name of the register in which the data is available.

Example : MOV A, B - Move the content of B-register to A-register.

Q2.8   What are the functions performed by data transfer instruction? Give an example and explain.

The data transfer instructions can copy the content of one register to another and copy the content of register to memory or vice versa.

Example : MOV B, C - The content of C-register is moved (copied) to B-register.

Q2.9   Explain register indirect addressing with an example.

In register indirect addressing mode, the instruction specifies the name of the register in which the address of the data is available. Here the data will be in memory and the address will be in the register pair.

Example : MOV A, M - The content of memory whose address is available in HL pair is moved to A-register.

Q2.10 What is implied or implicit addressing mode?

If the instruction operates on a data available in the register defined by the opcode then the addressing mode is called implied or implicit addressing mode.

Example : CMA - Complement the content of accumulator.

Q2.11 What are the functions performed by arithmetic instructions? Give an example and explain.

The functions performed by arithmetic instructions are Addition, Subtraction, Increment and Decrement.

Example : ADD E   - The content of E-register is added to accumulator.

Q2.12 What are the operations performed by logical instructions? Give an example and explain.

The operations performed by logical instructions are AND, OR, EXCLUSIVE-OR, Complement, Compare and Shift (Rotate).

Example : ANA D   - The content of D-register is logically ANDed with accumulator.

Q2.13 In which unit the arithmetic and logical operations are performed. Which unit is the destination of result.

The arithmetic and logical operations are performed in ALU. After the operation, the result will be stored in accumulator.

Q2.14 Which group of instruction affects the flags?

The flags are altered after execution of arithmetic and logical instructions.

Q2.15 What are the arithmetic instructions that do not affect the flag?

The 16-bit increment and decrement instructions (INX rp and DCX rp) will not affect any flags.

Q2.16 What are the flags affected by 8-bit increment and decrement instructions?

Except carry, all other flags are affected by 8-bit increment and decrement instructions.

Q2.17 What will be condition of flags after logical AND and OR operations?

After logical AND operation the carry flag is RESET (0), auxiliary carry flag is SET (1) and depending on the result of AND operation other flags are altered.

After logical OR operation the carry flag and auxiliary carry flag are RESET (0). Depending on the result of OR operation other flags are altered.

Q2.18 List the instructions that affect only carry flag.

The instructions that affect only carry flag are the following:

CMC	RAR	STC
DAD rp	RLC	
RAL	RRC	

Q2.19 What is DAA ?

DAA - **D**ecimal **A**djust **A**ccumulator.

After BCD addition, this instruction is executed to get the result in BCD. When DAA instruction is executed, the content of the accumulator is altered or adjusted as explained below:

(i)   If the sum of lower nibbles exceeds 09_H or auxiliary carry is set, a correction 06_H (0110) is added to lower nibble.

(ii)   If the sum of upper nibbles exceeds 09_H or carry is set, a correction 06_H (0110) is added to upper nibble.

Q2.20 What is DAD and what are the flags affected by this instruction?

DAD refers to **D**ouble **Ad**dition. This instruction is used to perform addition of two 16-bit data.

> **Syntax :** DAD rp

The content of register pair (rp) is added to the content of HL pair. After addition, the result will be in HL pair. The register pair can be BC, DE, HL, or SP. On execution of this instruction, only carry flag is affected.

Q2.21 List the various instructions that can be used to clear accumulator ?

The accumulator can be cleared by the following instructions:

1. MVI A,00_H
2. SUB A
3. ANI 00_H
4. XRA A.

Q2.22 What is the similarity and difference between subtract and compare instruction?

Similarity : Both the subtraction and comparison are performed by subtracting two data in ALU and flags are altered depending upon the result.

Difference : After subtract instruction is executed, the result is stored in accumulator, but after the execution of compare instruction the result is discarded (i.e., the subtract instruction alters the content of destination register (accumulator), but the compare instruction will not alter the content of any register or memory).

Q2.23 List the IO instruction in 8085.

- The IO instruction of 8085 are IN addr8 and OUT addr8.
- The IN instruction is used to input a data byte from the IO-mapped device or port. The OUT instruction used to output data byte to IO-mapped device or port.

Q2.24 Explain DI and EI.

DI - **D**isable **I**nterrupt. When this instruction is executed all the interrupts except TRAP are disabled. When the interrupts are disabled the processor will not accept or recognize the interrupt.

EI - **E**nable **I**nterrupt. This instruction is used or executed to allow the interrupts after disabling.

Q2.25 State the difference between LDA and LDAX.

The LDA instruction uses direct addressing mode to load a data byte from memory to accumulator, but LDAX instruction uses register indirect addressing for the same operation.

In LDA instruction, the content of memory location whose address is given in the instruction is moved to accumulator.

In LDAX instruction, a register pair contains the address of memory location. The content of memory location whose address is available in register pair is moved to accumulator.

Q2.26 What is the function performed by SIM instruction?

SIM - **S**et **I**nterrupt **M**ask. The SIM instruction is used to mask the hardware interrupts RST 7.5, RST 6.5 and RST 5.5. The execution of SIM instruction output and the content of the accumulator to program interrupt mask bits are also used to output serial data on the SOD line.

Q2.27 What is the function performed by RIM instruction?

RIM - **R**ead **I**nterrupt **M**ask. The RIM instruction is used to check whether an interrupt is masked or not. It is also used to read data from SID line.

Q2.28 What will be the state of the processor after executing HLT instruction?

When the HLT instruction is executed, the processor suspends program execution and the bus will be in idle state (i.e., the processor keeps on executing bus idle cycle until a reset or interrupt).

Q2.29 What is NOP? State its importance.

The NOP is a dummy instruction, it neither achieves any result nor affects any CPU register. This is used for producing software delay and reserve memory spaces for future software modifications.

Q2.30 What is PSW?

PSW - **P**rogram **S**tatus **W**ord. The flag register and accumulator together is called PSW. Flag register is low order register, Accumulator is high order register.

Q2.31 Explain RET instruction.

RET - Return to main program. This instruction is placed at the end of subroutine program in order to return to the main program. When this instruction is executed, the top of stack is poped to program counter.

Q2.32 Explain the difference between the conditional and unconditional return instructions.

In a conditional return instruction a flag condition is tested. If the flag condition is true, then the program control returns to main program. If the flag condition is false, then the next instruction is executed. In unconditional return instruction, the program control returns to the main program irrespective of the condition of the flag.

Q2.33 State the difference between STA and STAX instructions.

The STA instruction uses direct addressing mode to store the content of accumulator to a memory location, but the STAX instruction uses indirect addressing mode for the same operation.

Q2.34 What will be the content of SP (Stack Pointer) after execution of PUSH and POP instructions?

● After execution of PUSH instruction, the content of Stack Pointer (SP) will be 02 less than the earlier value.

● After execution of POP instruction, the content of Stack Pointer (SP) will be 02 greater than the earlier value.

Q2.35 What is the difference between ADD and ADC instruction?

The ADD instruction will not consider the value of carry flag for addition, but the ADC instruction will consider the value of carry flag (before executing this instruction) for addition. In ADC instruction the content of register or memory and the carry flag are added to the content of accumulator.

Q2.36 How is the subtraction performed in 8085?

The 8085 processor performs 2's complement subtraction and after subtraction, it complements the carry flag.

Q2.37 How the result of subtract operation can be interpreted?

● After subtract operation, if the carry flag is SET (1), then the result is negative will be in 2's complement form.

● After subtract operation, if the carry flag is RESET (0), then the result is positive.

Q2.38 What is the difference in 2's complement subtraction and 8085 subtraction?

In 2's complement subtraction, the result is positive if carry is equal to one (1) and negative if carry is equal to zero (0). But in 8085, the result is negative if carry is equal to one (1) and positive if carry is equal to (0).

Q2.39 What is the difference between CALL and JUMP instruction?

In CALL instruction, the address of next instruction is pushed to stack (i.e., stored in stack memory) before transferring the program control to call address. But in JUMP instruction, the address of next instruction is not saved.

Q2.40 What is the difference between conditional and unconditional branch instructions?

In unconditional branch instructions, the program control is transferred to branch address without checking any flag condition. But in conditional branch instructions, a flag condition is checked and only if the flag condition is true, program control is transferred to branch address, otherwise the next instruction is executed.

Q2.41 What is meant by a program?

A program is a set of instructions written to perform a certain task.

Q2.42 What is assembler, interpreter and compiler?

(a) **Assembler:** It is a software that converts assembly language program codes to machine language codes.

(b) **Compiler:** It is a software that converts the programs written in high level language to machine language.

(c) **Interpreter:** It is similar to a compiler but it converts the instructions one by one.

Q2.43 *What is the need for a assembler?*

An assembler is used to translate assembly language programs to machine language programs (i.e., in the executable format). Without the assembler it is very difficult to convert very large assembly language programs to machine codes.

Q2.44 *What are the advantages of an assembler?*

The advantages of an assembler are:

1. The assembler translates mnemonics into binary code with speed and accuracy.
2. It allows the programmer to use variables in the program.
3. It is easy to alter the program and reassemble.
4. The assembler identifies the syntax errors.
5. The assembler can reserve memory locations for data or result.
6. The assembler provides list file for documentation.

Q2.45 *What is a macro and when is it used?*

A macro is a group of instructions written within brackets and identified by a name. A macro is written when a repeated group of instructions is too short or not appropriate to be written as a subroutine.

Q2.46 *What is the meaning of expanding the macro?*

While assembling a program, the assembler replaces the instructions represented by a macro in the place where macro is called. This is called expanding the macro.

Q2.47 *What is the disadvantage in a macro?*

The disadvantage in a macro is that, if it is expanded or used a number of times in a program then the program may occupy more memory.

Q2.48 *What is a subroutine (or procedure)?*

A subroutine (or procedure) is a group of instructions written separately from the main program to perform a function that occurs repeatedly in the main program.

Q2.49 *What are the advantages of a subroutine?*   *(AU, Nov/Dec' 19, 2 Marks)*

1. Modular programming : The various tasks in a program can be developed as separate modules and called in the main program.
2. Reduction in the amount of work and program development time.
3. Reduces memory requirement for program storage.

Q2.50 *How is a subroutine implemented in 8085?*

A subroutine is written as a separate program (Procedure) and stored in a separate memory location. The subroutine program should be terminated by an RET instruction. It is called in the main program using CALL addr16 instruction. The addr16 should be the starting address of the subroutine program.

When the CALL instruction is executed, the processor saves the return address (address of next instruction) in stack, load the subroutine address in Program Counter(PC) and starts executing the subroutine. At the end of subroutine when the RET instruction is executed, the return address is retrieved from the stack and loaded to the PC.

Q2.51 What is a Delay routine?

A delay routine is a subroutine used for maintaining the timings of various operations in microprocesor.

Q2.52 Write a simple delay subroutine involving a single 8-bit register of 8085.

Delay subroutine

```
        MVI  C,d8 ;Load the count value (d8) in C-register.
LOOP:   DCR  C    ;Decrement the count
        NOP
        NOP
        JNZ  LOOP ;If ZF = 0, go to LOOP.
        RET       ;Return to main program.
```

Q2.53 Write a simple delay subroutine involving a register pair of 8085.

Delay subroutine

```
        LXI  D,d16 ;Load the count value (d16) in DE register pair.
LOOP:   DCX  D     ;Decrement the count.
        MOV  A,E   ;Logically OR the content of
        ORA  D     ;E-register with D-register.
        JNZ  LOOP  ;If ZF = 0, go to LOOP.
        RET        ;Return to main program.
```

Q2.54 What is a flowchart?

A flowchart is graphical representation of the operation flow of a program. It is the graphical (pictorial) form of an algorithm.

Q2.55 List the symbols used for drawing a flowchart.

The following are the symbols used for drawing flowchart:

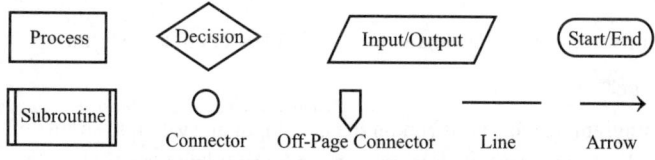

Fig. Q2.55: Symbols used in flowcharts.

Q2.56 What is a development system? What are its components?

A development system is a system used by the microprocessor-based system designer to design and test the software and hardware aspects of a new system under development.

The components of a development system are a microcomputer with standard accessories, emulator and program development tools like editor, assembler, linker, locator, debugger, simulator, etc.

Q2.57 Write a short note on assembly language program development tools.

The program development tools for an assembly language program are editor, assembler, linker, locator, debugger and simulator. These tools are softwares that can be run on the development system in order to write, assemble, debug, modify and test the assembly language programs.

Q2.58 What is an Editor?

An editor is a program which when run on a microcomputer system, allows the user to type and modify the assembly language program statements. The main function of an editor is to help the user to construct the assembly language program in the right format and save as a file.

Q2.59 What is a one-pass assembler?

A one-pass assembler is an assembler in which the source codes are processed only once. A one-pass assembler is very fast and in one-pass assembler only backward reference may be used.

Q2.60 What is a two-pass assembler?

The two-pass assembler is an assembler in which the source codes are processed two times. In the first pass, the assembler assigns addresses to all the labels and attach values to all the variables used in the program. In the second pass it converts the source code into machine code.

Q2.61 What is the drawback of a one-pass assembler?

The drawback of a one-pass assembler is that the program cannot have forward reference, because, a one-pass assembler issues an error message if it encounters a label or variable that is defined at a later part of a program.

Q2.62 What is linker and locator?

(a) A linker is a program used to join together several object files into one large object file.

(b) A locator is a program used to assign specific addresses to the object codes to be loaded into memory.

Q2.63 What is debugging?

The process of locating and correcting an error using a debugger is known as debugging.

Q2.64 What is a debugger?

A debugger is a software used to locate and troubleshoot errors in a program.

Q2.65 What is a simulator?

A simulator is a program which can be run on the development system to simulate the operations of the newly designed system. Some operations that can be simulated are given below:

- **Execute a program and display the result.**
- **Single step execution of a program.**
- **Break-point execution of a program.**
- **Display the content of register/memory.**

Q2.66 What is an emulator?

An emulator is a system that can be used to test the hardware and software of a newly developed microprocessor-based system.

Q2.67 What is the difference between an emulator and a simulator?

A simulator can be used to run and check the software of a newly developed microprocessor-based system but an emulator can be used to run and check both the hardware and software of a newly developed microprocessor-based system.

Q2.68 *Write a subroutine to output the content of flag register to LED's connected to the port of a 8085 microprocessor-based system.*

Subroutine to display the content of flag register

```
PUSH    PSW    ;Push the A-register and flag register to stack.
POP     B      ;POP the top of stack to BC pair.
MOV     A,C    ;Get the content of flag register in A-register.
OUT     PORT   ;Output the content of A-register to PORT.
RET            ;Return to main program.
```

> *Note : The A-register and Flag register are together called PSW (Program Status Word). In this, the A-register is high order register and Flag register is low order register.*

Q2.69 *Write a simple program to find the smallest among the two data stored in memory.*
(Assume that data are stored in 4200_H and 4201_H. Store the result in 5000_H)

```
       LDA  4200H ;Get first data in A-register.
       MOV  B,A   ;Save first data in B-register.
       LDA  4201H ;Get second data in A-register.
       CMP  B     ;Compare the two data.
       JC   AHEAD ;If CF = 1, go to AHEAD.
       MOV  A,B   ;If CF = 0, move B-register to A-register.
AHEAD: STA  5000H ;Store the smallest data in memory
       HLT        ;Stop
```

Q2.70 *Write a simple program to multiply an 8-bit data stored at 4200_H by 02_H and store the result at 4300_H and $4301_{H'}$*

```
       MVI B,00H ;Clear B-register.
       XRA A     ;Clear A-register and carry
       LDA 4200H ;Get the data in A-register.
       RAL       ;Multiply the content of A by 02.
       JNC AHEAD ;If CF = 0, go to AHEAD.
       INR B     ;If CF = 1, increment B-register.
AHEAD: STA 4300H ;Store the product in memory.
       MOV A,B
       STA 4301H ;Store the carry in memory.
       HLT       ;Halt program execution.
```

Q2.71 *Write a simple program to divide an 8-bit data stored at 4200_H by 02_H and store the result at 4300_H and $4301_{H'}$*

```
        MVI B,00H  ;Clear B-register.
        XRA A      ;Clear A-register and carry
        LDA 4200H  ;Get the data in A-register.
        RAR        ;Divide the content of A by 02.
        JNC AHEAD  ;If CF = 0, go to AHEAD.
        INR B      ;If CF = 1, increment B-register.
AHEAD:  STA 4300H  ;Store the quotient in memory.
        MOV A,B
        STA 4301H  ;Store the remainder in memory.
        HLT        ;Halt program execution.
```

Q2.72 *Write a simple program to split a hexa data into two nibbles and store in memory.*

```
LXI  H,4200H  ;Set pointer for data array.
MOV  B,M      ;Get the data in B-register.
MOV  A,B      ;Copy the data to A-register.
ANI  0FH      ;Mask the upper nibble.
INX  H
MOV  M,A      ;Store the lower nibble in memory.
MOV  A,B      ;Get the data in A-register
ANI  F0H      ;Mask the lower nibble.
RRC           ;Bring the upper nibble to lower nibble position.
RRC
RRC
RRC
INX  H
MOV  M,A      ;Store the upper nibble in memory.
HLT           ;Halt program execution.
```

Q2.73 *Explain the mathematical functions performed by the following instructions.*

```
MVI  A,07H
RLC
MOV  B,A
RLC
RLC
ADD  B
```

The operations performed by each of the above given mathematical instruction are as follows:

1. An 8-bit data 07_H is moved to A-register.
2. The content of A-register is multiplied by 02.
3. The content of A-register ($07 \times 02 = 0E_H$) is copied to B-register.
4. The content of A-register is multiplied by 02.
5. The content of A-register is multiplied by 02.
6. The content of B-register is added to B-register.

The result of the above operations is that the content of A-register is multiplied by $0A_H$. Therefore, after executing the above instructions, the content of A-register will be 46_H.

Q2.74 *Write a subroutine to clear a flag register and an accumulator.*

Subroutine to clear flag register

```
LXI  SP,4200H  ;Initialize stack.
LXI  B,0000H   ;Clear BC register pair.
PUSH B         ;Push the content of BC pair to stack.
POP  PSW       ;Pop the top of stack to A-register and flag register (PSW).
RET            ;Return to main program.
```

> *Note :* *The A-register and Flag register are together called PSW (Program Status Word). The A-register is a high order register and the flag register is a low order register.*

Q2.75 *Write a subroutine program to exchange the content of BC pair and DE pair?*

Subroutine to exchange BC pair and DE pair

```
LXI    SP,4200H      ;Initialize stack.
PUSH   B             ;Store the content of BC pair in stack.
PUSH   D             ;Store the content of DE pair in stack.
POP    B             ;Move the content of DE pair stored in stack to BC pair.
POP    D             ;Move the content of BC pair stored in stack to DE pair.
RET                  ;Return to main program.
```

Q2.76 *Write a program to load the accumulator with the value 82_H and complement the accumulator 700 times.*   *(AU, Nov/Dec' 19, 2 Marks)*

```
          MVI  A,82H      ;Load 82H is accumulator.
          LXI  B,02BCH    ;Load count value in BC pair (700₁₀ = 02BCₕ)
AGAIN:    CMA             ;Complement accumulator
          DCX  B          ;Decrement count
          MOV  D,A        ;Save A in D
          Mov  A,C        ;Move C to A
          CMP  B          ;Compare B and C
          MOV  A,D        ;Restore A
          JNZ  AGAIN      ;If count zero, repeat complement
          HLT             ;Stop
```

Q2.77 *Differentiate RAL and RLC instruction.*   *(AU, Nov/Dec' 19, 3 Marks)*

1. The RAL instruction rotates accumulator and carry together so that most significant bit of accumulator is moved to carry and carry is moved to least significant bit of accumulator.

2. But in RLC instruction the content of accumulator alone is rotated such that most significant bit of accumulator is moved to least significant bit and also to carry.

Q2.78 *Write an assembly language program for 8085 microprocessor to count even numbers in series of 10 numbers.*   *(AU, Nov/Dec' 19, 10 Marks)*

Example:

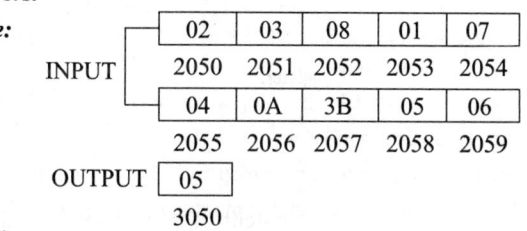

INPUT	02	03	08	01	07
	2050	2051	2052	2053	2054
	04	0A	3B	05	06
	2055	2056	2057	2058	2059

OUTPUT | 05 |
        3050

Solution:

```
          LXI  H,2050H    ;Set HL pair as data pointer
          MVI  C,0AH      ;Set C-register as counter
          MVI  B,00H      ;Set initial count as zero
NEXT:     MOV  A,M        ;Get one data in A
          RAR             ;Check least significant bit
          JC   AHEAD      ;If least significant bit is 1, odd
          INR  B          ;If least significant bit is 0, even
          INX  H          ;Increment even count and data pointer
AHEAD:    DCR  C          ;Decrement count
          JNZ  NEXT       ;Repeat until count is zero
          MOV  A,B        ;Move even number count to A
          STA  3050H      ;Store A in memory
          HLT
```

2.14 EXERCISES

I.   Fill in the blanks with appropriate words

1.   The software which is used to convert assembly language programs into machine language program is called _____ .

2.   The _____ software tool is used to convert assembly language into machine language.

3.   The _____ is used to create library files which are collections of procedures of frequently used functions.

4.   An _____ is a mixture of hardware and software.

5.   The _____ assembler directive is used to define byte type variable.

6.   The number of T-states required to execute the 8085 memory read cycle is _____.

7.   The various ways of specifying data are called _____.

8.   The addressing mode used in the 8085 instruction MOV A,M is _____ addressing.

9.   The 8085 instruction _____ is used to load the accumulator with content of memory.

10. The 8085 assembly language instruction that stores the content of H and L registers into the memory locations 2050_H and 2051_H, respectively is _____.

11. A single byte 8085 instruction which is used to push the memory content into stack memory is _____.

12. The _____ and _____ instructions of 8085 are used to communicate with peripheral devices.

13. The _____ instruction of 8085 is used for BCD addition.

14. The content of A register is equal to 23_H. The _____ instruction of 8085 is used to mask the lower nibble of A register.

15. The status of carry flag, CF = _____ and zero flag, ZF = _____ after executing the 8085 instruction CMP A,B where A = 25_H and B = 52_H.

16. The 8085 instruction XCHG exchanges the content of _____ and _____ register pairs.

17. The single byte 8085 instruction which is used to find the one's complement of the accumulator content is _____.

18. The content of Accumulator, A = 12_H. The execution of 8085 instruction RLC and RRC change the A register as _____ and _____ respectively.

19. The _____ instruction of 8085 is used to transfer the program control back to the main program.

20. The vector address for the 8085 restart instruction RST 5 is _____.

21. When the 8085 instruction DI is executed, all the interrupts except _____ are disabled.

22. The _____ instruction of 8085 is used to read the status of interrupt masks.

23. The _____ instruction performs nothing but simply waste clock cycles in 8085 processor.

24. The language in which instructions are represented by binary codes is called _____ language.

25. The _____ register of 8085 has to be preserved in stack whenever a subroutine is called.

26. The _____ instruction of 8085 is equivalent to the following set of instructions,

```
MOV  A,L
STA  3000H
MOV  A,H
STA  3001H
```

27. The 8085 instruction _____ is used to add two 16-bit numbers which are available in HL and BC register pairs.

Answers

1. assembler	8. register indirect	15. 1, 0	22. RIM
2. assembler	9. LDA	16. HL, DE	23. NOP
3. library builder	10. SHLD 2050H	17. CMA	24. machine
4. emulator	11. SPHL	18. 24_H, 09_H	25. program counter (PC)
5. DB	12. IN, OUT	19. RET	26. SHLD 3000H
6. three	13. DAA	20. 0028_H	27. DAD B
7. addressing mode	14. ANI 0FH	21. TRAP	

II. State whether the following statements are True/False.

1. The assembly language programs are machine dependent.
2. Any programming language has to be converted into machine language finally.
3. The compiler software converts high level language into assembly level language.
4. The simulators do not have the ability to perform timing and data transmission and reception.
5. Assembler directives are pseudo-instructions.
6. No machine codes are generated for assembler directives.
7. The assembler directive DT is used to define ten words of a variable.
8. One T-state is equal to twice the time period of internal clock signal of the 8085 processor.
9. The T-state of 8085 always starts at falling edge of the clock.
10. The READY signal of 8085 will not be sampled by the processor during bus idle cycle.
11. In implied addressing mode, the operand is always in accumulator.
12. The source remains unchanged after executing data transfer instructions.
13. The SP register of 8085 processor always points to top of the stack.
14. The register pair DE of 8085 processor is used to point to the memory.
15. The content of SP register of 8085 processor is incremented by 02 after the execution of PUSH instruction.
16. The DAD instruction of 8085 modifies only the carry flag.
17. The 8085 processor performs subtraction by using 1's complement method.
18. There are no direct instructions to perform multiplication and division in 8085.
19. The source and destination contents remain unchanged after executing CMP instruction of 8085.
20. The CALL instruction of 8085 is a conditional branching instruction.
21. The logical operation AND is generally used to mask the bits.
22. The 8085 instruction INX will not alter any flags.
23. The STC instruction of 8085 will always clear the carry flag.
24. There are no near and far procedures in 8085.
25. Subroutines are stored in memory only once.
26. The stack pointer(SP) is automatically incremented or decremented by the processor after every stack read/ write operation.
27. The SP register of 8085 is incremented by one after read operation from stack.

Answers

1. True	6. True	11. False	16. True	20. False	24. True
2. True	7. False	12. True	17. False	21. True	25. True
3. False	8. False	13. True	18. True	22. True	26. True
4. True	9. True	14. False	19. True	23. False	27. False
5. True	10. True	15. False			

III. Choose the right answer for the following questions.

1. The assembly language programs are written using _____.

    a) binary code   b) Mnemonics   c) english words   d) all the three

2. Which of the following is a high level language?

    a) c   b) c++   c) JAVA   d) all the three

3. Which of the following is used to combine relocatable object files and library functions into a single executable file?

    a) library builder   b) linker   c) debugger   d) simulator

4. Which of the following tools is used to test and run the programs of the newly designed system?

    a) linker   b) debugger   c) simulator   d) builder

5. Identify the valid constants of assembler

    a) A2H   b) F2   c) 112B   d) 8BC2H

6. During which T-state of opcode fetch cycle the higher order address/data lines are demultiplexed in 8085 processor?

    a) T_1 b) T_2 c) T_3 d) T_4

7. The data, address and control pins are driven to high impedance state during _____ machine cycle of 8085.

    a) interrupt acknowledge   b) bus idle   c) wait state   d) all the three

8. The 8085 instruction used for providing direct value to a register through the instruction itself is

    a) MOV   b) MVI   c) ADD   d) JNZ

9. Logical operation in an 8085 processor include _____ operation?

    a) AND,OR,EXOR   b) rotate   c) compare & complement   d) all of the above

10. The binary code for the registers A and B are 111 and 000 respectively. what is the 8-bit opcode of the 8085 instruction MOV A,B?

    a) 47_H b) 78_H c) 38_H d) 37_H

11. The _____ instruction is used to store the accumulator content into memory pointed by register pair DE in 8085 processor.

    a) STA D   b) STA E   c) STAX D   d) STAX E

12. The number of 8085 machine cycles required to execute the instruction LDA 2000H is

    a) 2T   b) 3T   c) 4T   d) 6T

13. Which of the following is a two byte 8085 instruction?

    a) STA 1200H   b) LDA 3000H   c) MVI A,12H   d) MOV A,B

14. In which of the following 8085 instructions, the operand is specified by the instruction itself?

    a) CMA   b) SPHL   c) XCHG   d) all the three

15. Which of the following is an invalid 8085 instruction?

    a) STAX D   b) STAX H   c) STA 2000H   d) LDAX D

16. The content of HL register and DE register are 1234_H and 2567_H respectively. A single byte 8085 instruction which is used to add the two 16 bit data is

    a) DAA   b) DAD D   c) DAD H   d) ADD D

17. Which of the following 8085 instructions change the program control to new location?

    a) SPHL   b) PCHL   c) LDA   d) STA

18. The _____ instruction of 8085 is used to implement the 8-bit full adder.

    a) ADD B   b) ACI 35H   c) ADC B   d) all the three

19. The following set of 8085 instructions are used to multiply the content of HL register pair by three

    a) DAA   b) DAD H   c) DAD H   d) ADD H

                              DAD H   ADD H

20. The content of accumulator, A = FF$_H$ and carry flag CY = 0. What would be the content of A register and carry flag after executing the 8085 instructions (i) INR A and (ii) ADD 01H respectively?

    a) (i) A=00H ; CY = 1   b) (i) A=00H ; CY = 0   c) (i) A = 00H ; CY = 1   d) (i) A=00H ; CY = 0

      (ii) A=00H ; CY = 1   (ii) A=00H ; CY = 1   (ii) A = 00H ; CY = 0   (ii) A=00H ; CY = 0

21. Which of the following 8085 instruction is used to clear the accumulator?

    a) MOV A, 00H   b) XRA A   c) ANI A,00H   d) all the three

22. Which of the following 8085 instructions use the auxiliary carry flag for its execution?

    a) ADC   b) SBB   c) CMP   d) DAA

23. The two single byte 8085 instructions which are used (i) to load SP register with HL pair and (ii) to exchange the content of HL pair and top of the stack respectively are

    a) SPHL and XTHL   b) XTHL and SPHL   c) SPHL and XCHG   d) XTHL and XCHG

24. Which of the following 8085 instruction sequences are used to swap the nibbles of A register?

    a) RAL   b) RAR   c) RRC   d) RLC
        RAL   RAR   RRC   RLC
                                  RAR   RRC
                                  RAR   RRC

25. Given the sequence of 8085 instructions

Address	Instructions	Address	Instructions
2000:	MVI A, 12H	3000:	INR A
2002:	CPI 05H		RET
2004:	CZ 3000H		
2007:	ADI A,05H		
2009:	HLT		

What would be the content of A register after the execution of the 8085 program given above.

    a) 17$_H$ b) 18$_H$ c) 12$_H$ d) 05$_H$

26. The following 8085 instruction forces the bus to idle state

    a) NOP   b) HLT   c) RET   d) all the three

27. An 8085 instruction used for generating software delay is

    a) HLT   b) RET   c) NOP   d) EI

28. The order of 8085 machine cycle executed in executing the instruction STA 1200H are

    a) opcode fetch, memory read, memory write

b) opcode fetch, memory write, memory read

c) opcode fetch, memory read, memory read, memory read

d) opcode fetch, memory read, memory read, memory write

29. *Which of the following instruction is used for data transfer between IO device and the processor*

    a) MOV   *b)* IN   *c)* XCHG   *d)* all the above

30. *Which of the following pin is used to differentiate the memory access and I/O access?*

    a) $M/\overline{IO}$ *b)* $\overline{RD}$ *c)* $\overline{WR}$ *d)* $DT/\overline{R}$

31. *What will be the content of SP register of 8085 after a writing operation into stack? SP = 4218H before write operation.*

    a) 4219_H b) 4217_H c) $421A_H$ d) 4216_H

32. *Identify the operations performed by the following 8085 program.*

```
LXI   H,2000H
MVI   M,28H
MOV   A,M
XRI   0FFH
INR   A
MOV   M,A
```

    a) Inverts the content of 2000_H b) complements the content of 2000_H

    c) change the sign of content of 2000_H d) Increment content of 2000_H by one

33. *what will be the content of SP register and DE register pair after executing the following 8085 instruction?*

```
2000: LXI  SP,3020H
2003: CALL 2006
2006: NOP
2007: POP  H
```

    a) $SP = 3020_H$; $HL = 2003_H$ b) $SP = 3022_H$; $HL = 2006_H$

    c) $SP = 3020_H$; $HL = 2006_H$ d) $SP = 3020_H$; $HL = 2003_H$

34. *Consider the following 8085 main program and subroutine. What will be the value of the accumulator after executing the line 4,5 and 6?*

```
1.   MVI   A,27H        SUBROUTINE: INR A
2.   ANI   0FH                      ORI 00H
3.   PUSH  A                        RET
4.   CALL  SUBROUTINE
5.   ADD   A
6.   POP   A
7.   HLT
```

    a) 08_H, 10_H and 07_H respectively   b) 07_H, 08_H and 10_H respectively

    c) 10_H, 07_H and 08_H respectively   d) 08_H, 07_H and 10_H respectively

Answers

1. b	6. a	11. c	16. b	20. b	24. c	28. d	32. c
2. d	7. b	12. c	17. b	21. d	25. a	29. b	33. c
3. b	8. b	13. c	18. c	22. d	26. b	30. a	34. a
4. c	9. d	14. d	19. b	23. a	27. c	31. c	
5. d	10. b	15. b					

IV. Answer the following questions.

E2.1 How many T-states are required to execute interrupt acknowledge machine cycle of 8085? Justify the answer

E2.2 Identify the operations performed and operand in the following 8085 instructions?

     a) MVI A, 12H   b) CMA   c) RLC

E2.3 Write the number of bytes and machine code of the following 8085 instructions.

     a) LDAX B   b) JC 2000H   c) RST 4   d) NOP

E2.4 How fast the 8085 processor with clock frequency 1 MHz executes the following instructions

     a) MVI A, 00H   b) XRA A. Comment on the operation.

E2.5 Identify the machine cycles involved in the following 8085 instructions

     a) SHLD 4000H   b) OUT 80H   c) INX B   d) DAD B

E2.6 In an 8085 microprocessor based system, it is desired to increment the contents of memory location whose address is available in DE register pair and store the result in same location. Write the sequence of instruction to perform the above mentioned action.

E2.7 In an 8085 microprocessor, what would be the contents of accumulator after the following instructions are executed?

     XRA A
     MVI B, F0H
     SUB B

E2.8 Write the sequence of 8085 instructions to add two decimal numbers 12 and 39 which are stored in A and B registers respectively.

E2.9 The content of register B and accumulator A of 8085 microprocessor are 49_H and $3A_H$ respectively. What would be the content of A and the status of all the flags after execution of SUB B instruction?

E2.10 Assume that the accumulator contains the data 65_H and the 8085 instruction MOV C,A ($4F_H$) is fetched. List the operations performed in various T-states while executing the instruction.

E2.11 Write an 8085 assembly language program without using MOV instruction to load the accumulator with content of memory location 3600_H.

E2.12 Write an 8085 assembly language program to transfer 08 bytes of data from memory location which starts at 2050_H to another memory location which starts at 2100_H.

E2.13 Write an 8085 assembly language program to exchange 08 bytes of data from memory location which starts at 2050_H to another memory location which starts at 2100_H.

E2.14 Write an 8085 assembly language program to square the given data stored at 3000H and store the results at 3001_H.

E2.15 Write an 8085 ALP to calculate the average room temperature . Use 5 different readings noted down at 5 different instances throughout the day. Assume that five readings are stored in memory 2500_H onwards

E2.16 It is desired to multiply the numbers $0A_H$ and $0B_H$ and store the result in the accumulator. The numbers are available in registers B and C respectively. A part of 8085 program for this purpose is given below.

```
        MVI   A,00H
LOOP:
        HLT
        END
```

Complete the program by writing the missing sequence of instructions

E2.17 The question i) and ii) are linked questioins. Consider an 8085 microprocessor based system.

i) The following program starts at location 0100_H.

```
        LXI   SP,00FFH
        LXI   H,0701H
        MVI   A,20H
        SUB   M
```

What will be the content of accumulator when the program counter reaches 0109_H?

ii) If in addition the following code exists from 0109_H onwards

```
        ORI 40H
        ADD M
```

What will be the result in the accumulator after the last instruction is executed?

E2.18   Consider the following sequence of instructions for an 8085 microprocessor based system.

Memory Address	Instructions
FF00	MVI A,FFH
FF02	INR A
FF03	JC FF0CH
FF06	ORI A8H
FF08	JM FF15H
FF0B	XRA A
FF0C	OUT PORT1
FF0E	HLT
FF0F	NOP
FF10	XRI FFH
FF12	OUT PORT2
FF14	HLT
FF15	MVI A,FFH

Memory Address	Instructions
FF17	ADI 02H
FF19	RAL
FF1A	JZ FF23H
FF1D	JC FF12H
FF20	JNC FF12II
FF23	CMA
FF24	OUT PORT3
FF26	HLT

a)   If the program execution begins at location $FF00_H$, write down the sequence of instruction which are actually executed till a HLT instruction. (Assume all flags are initially RESET).

b)   Which of the three ports (port-1, port-2, and port-3) will be loaded with data. What is the bit pattern of the data?

E2.19 The program and machine code for an 8085 microprocessor are given below.

```
        3E        MVI   A,C3
        C3
        00        NOP
        80        ADD   B
        3D        DEC   A
        C2        JNZ   800A
```

```
0A
80
C3        JMP    800C
0C
80
D3        OUT    10
10
76        HLT
```

The starting address of the above program is $7FFF_H$. What would happen if it is executed from 8000_H?

E2.20 An 8085 assembly language program is given below

```
        MVI    C,03H
        LXI    H,2000H
        MOV    A,M
        DCR    C

LOOP1:  INX    H
        MOV    B,M

        CMP    B

        JNC    LOOP2

        MOV    A,B
LOOP2:  DCR    C
        JNZ    LOOP1
        STA    2100H
        HLT
```

The contents of memory locations are

$$2000_H = 18_H \; ; \qquad 2001_H = 10_H \; ; \qquad 2002_H = 2B_H$$

a) What does the above program do?

b) At the end of the program what will be

    i) the content of the register A, B, C, H and L?

    ii) the condition of the carry and zero flags.

    iii) the content of the memory locations 2000_H, 2001_H, 2002_H and 2100_H

E2.21 Write an assembly language program without using any arithmetic instruction to store hexadecimal $5D_H$ in the flag register of 8085 microprocessor. Data in other registers of the processor must not alter upon executing this program.

E2.22 An 8085 assembly language program is given below.

```
Line 1:  MVI   A,B5H
Line 2:  MVI   B,0EH
Line 3:  XRI   69H
Line 4:  ADD   B
Line 5:  ANI   9FH
Line 6:  CPI   9FH
Line 7:  STA   3010H
Line 8:  HLT
```

a) What will be the content of the accumulator just after execution of the ADD instruction in line 4?

b) What will be the status of the cy and z flag after executing line 7 of the program.

E2.23 Following is the segment of a 8085 assembly language program.

```
        LXI    SP,EFFFH
        CALL   3000H
                .
                .
                .
                .
```

```
3000H:    LXI    H,3CF4H
          PUSH   PSW
          SPHL
          POP    PSW
          RET
```

What will be the contents of SP on completion of RET execution

E2.24 Write an 8085 ALP to find out how many positive integers and negative integers are there in a six byte array which starts at 2000_H.

E2.25 The following program is written to find whether the given signed numbers located at 2100_H is positive or negative and stores 00_H if the data is positive and 01_H is negative at memory location 2101_H. Identify the logical error if any.

```
              LDA    2100H
              RRC    A
              JNC    POSITIVE
              MVI    A,01H
              JMP    HALT
POSITIVE:     MVI    A,00H
HALT:         HLT
```

E2.26 In 8085 microprocessor the following program is executed. What will be the accumulator content at the end of program?

```
              MVI    A,04H
              MVI    B,04H
HERE:         ADD    B
              DCR    B
              JNZ    HERE
              ADI    A,06H
              HLT
```

E2.27 Give the content of 8085 register A and the memory locations 2000_H and 2100_H after executing the following program?

```
              LXI    D,2000H
              LXI    H,2100H
              XRA    A
              INR    A
              MOV    M,A
              XCHG
              MOV    M,A
              MOV    A,M
              DCR    A
              XCHG
              DCR    M
```

E2.28 What will be the content of SP, A and B registers after executing the lines 6, 7, 8 and 9 in the following program? Assume content of 2000_H is 08_H.

```
1.    MVI    C,00H
2.    LDA    2000H          2800H:    MVI    A,099H
3.    ADI    A,02H                    MVI    B,03H
4.    MOV    B,A                      SUB    B
5.    INR    B                        RET
6.    LXI    SP,2500H
7.    PUSH   B
8.    CALL   2800H
9.    POP    B
```

CHAPTER 3

8051 MICROCONTROLLER

3.1 INTRODUCTION TO MICROCONTROLLERS

Since the invention of microprocessors, different companies have started manufacturing more and more sophisticated processors with improved features such as large data bus, large address bus, sophisticated memory management techniques and instruction set, capability of handling a wide range of integer and floating-point data, parallel processing of instructions, etc. But these sophisticated processors are not necessary for small applications such as controlling a motor, monitoring/controlling temperature, switching ON/OFF traffic lights, etc. In the 1980s, the manufacturers of microprocessors realized that there is a need for low cost, compact, single-chip programmable systems for small dedicated applications and so started manufacturing another class of programmable ICs called microcontrollers.

3.1.1 COMPARISON OF MICROPROCESSORS AND MICROCONTROLLERS

(AU, Nov/Dec' 19, 5 Marks)

Microcontrollers are similar to microprocessors, but they are designed to work as a true single-chip system by integrating all the devices needed for a system on a single chip. The basic functional units of a microprocessor will be ALU, a set of registers, timing and control unit. The microcontroller will have these functional blocks and in addition may have IO ports, a programmable timer, RAM memory and EPROM/EEPROM memory. Some microcontrollers may even have internal ADC and/or DAC.

3.1.2 FUNCTIONAL BUILDING BLOCKS OF A MICROCONTROLLER

(AU, Nov/Dec' 19, 13 Marks)

The microcontroller is a programmable IC manufactured by VLSI (**V**ery **L**arge **S**cale **I**ntegration) technique, and capable of performing arithmetic and logical operations. The various functional blocks of a typical microcontroller are shown in Fig. 3.1.

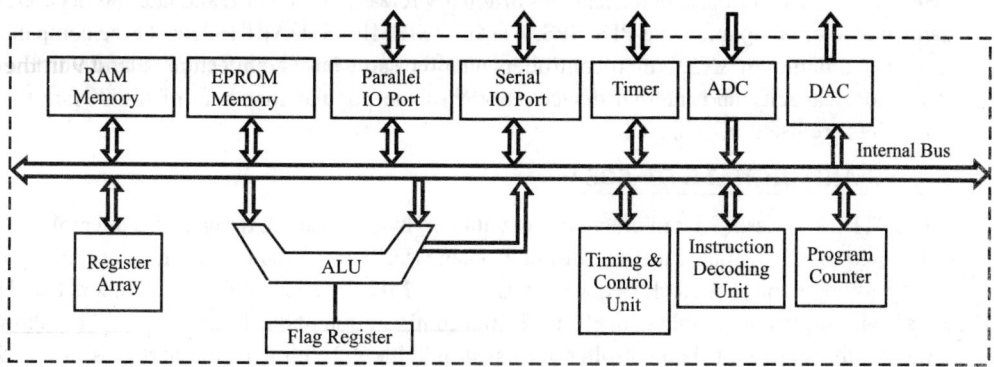

Fig. 3.1: Functional block diagram of a microcontroller.

The basic functional blocks of a microcontroller are the ALU, Flag register, Register array, Program Counter (PC), Instruction Decoding Unit, Timing and control unit, RAM memory, EPROM/EEPROM memory, Parallel IO port, Serial IO port, Programmable timer, ADC and DAC. All microcontrollers may not have all the blocks shown in Fig. 3.1. Some of the functional blocks shown in Fig. 3.1, may not be available in certain microcontrollers.

The ALU is the computational unit of the microcontroller which performs arithmetic and logical operations. The various conditions of the result are stored as status bits called flags in the flag register. The register array and internal RAM memory are used as a temporary storage device for storing temporary data during execution of a program.

The program codes and permanent data are stored in EPROM/EEPROM. In microcontroller-based systems, an external memory is provided only when the internal memory is not sufficient and so in most of the microcontroller-based systems, the program and data are stored in the internal memory of the microcontroller itself.

The program counter generates the address of the instructions to be fetched from the memory and sends it through the internal bus to the memory. (If the instruction to be fetched is stored in the external memory then the address is sent through IO ports to the external memory. Because the microcontrollers communicate with the external world only through IO ports.) The memory will send the instruction codes, which are decoded by the instruction decoding unit and send the information to the timing and control unit. The timing and control unit will generate the necessary control signals for internal and external operation of the microcontroller.

The parallel and serial IO ports are used for interfacing IO devices like switches, keyboard, LCD/ LED, ADC, DAC, etc., and also for any other input/output operations.

Microcontrollers do not have a dedicated external address and data bus. Therefore, for interfacing any additional peripheral devices, the external address and data buses are formed only by using port lines.

Microcontrollers with internal ADC can directly accept analog signals for processing. Likewise, microcontrollers with internal DAC can directly generate analog signals for controlling analog devices. A programmable timer can be used for time-based operations and it can also be used as a counter.

3.2 HARDWARE ARCHITECTURE AND PINOUTS

The 8051 family of microcontrollers was originally released by INTEL and later on licensed to many semiconductor companies like PHILIPS, ATMEL, SIEMENS, HARRIS, etc. These companies have developed a family of 8X5X microcontrollers with the same base architecture but with different internal memory capacity and internal devices. Some of the popular members of 8X5X family of microcontrollers are listed in Table 3.1.

3.2.1 PINS AND SIGNALS OF 8051

The INTEL 8051 is an 8-bit microcontroller with 128 byte internal RAM and 4 kB internal ROM. The 8051 is a 40-pin IC available in Dual In-line Package (DIP) and it requires a single power supply of +5 V. Its maximum internal clock frequency rating is 12 MHz.The 8X5X family members listed in Table 3.1 are pin-to-pin compatible with 8051. The pin configuration of 8051 microcontroller is shown in Fig. 3.2 and the signals of the controller are listed in Table 3.2. Some of the port pins of a 8051 microcontroller have alternate functions and they are listed in Table 3.3.

The 8051 microcontroller has 32 IO pins and they are organized as four numbers of an 8-bit parallel port. The ports are denoted as port-0, port-1, port-2 and port-3. Each port can be used either as an 8-bit parallel port or 8 numbers of 1-bit port (i.e., individual pins of each port can be used as 1-bit IO line independently). When used as 1-bit port, the port pins are denoted as PX.Y, where X can take values 0 to 3 and Y can take values 0 to 7. For example, the bit-0 of port-1 is denoted as P1.0. The ports behave as latches during output operation and as buffers during input operation. Except port-0 all other ports are provided with internal pull up. Hence, while using port-0 for IO operation, external pull up should be provided.

Table 3.1: Some Members of the 8X5X Family of Microcontrollers

Microcontroller	Internal memory capacity				Number of timers
	RAM	ROM	EPROM	FLASH EPROM	
8031	128 bytes	-	-	-	2
8032	256 bytes	-	-	-	3
8051	128 bytes	4 kB	-	-	2
8052	256 bytes	8 kB	-	-	3
8751	128 bytes	-	4 kB	-	2
8752	256 bytes	-	8 kB	-	3
8951	128 bytes	-	-	4kB	2
8952	256 bytes	-	-	8 kB	3

Table 3.2: Signals of a 8051 Microcontrollers

Pins/Signal	Description
P0.7 - P0.0	Port-0 input/output pins.
P1.7 - P1.0	Port-1 input/output pins.
P2.7 - P2.0	Port-2 input/output pins.
P3.7 - P3.0	Port-3 input/output pins.
RST	Reset input
X1, X2	Pins for crystal connection. The signal at X2 can be used as clock signal for peripherals.
PSEN	Program store enable. Used as read control or enable for external program memory.
ALE/PROG	Address Latch Enable or program pulse input during EPROM/ ROM programming.
EA/V_{PP}	External Access or Programming voltage.
V_{cc}	Power supply (+5 V)
V_{ss}	Power supply ground (0 V)

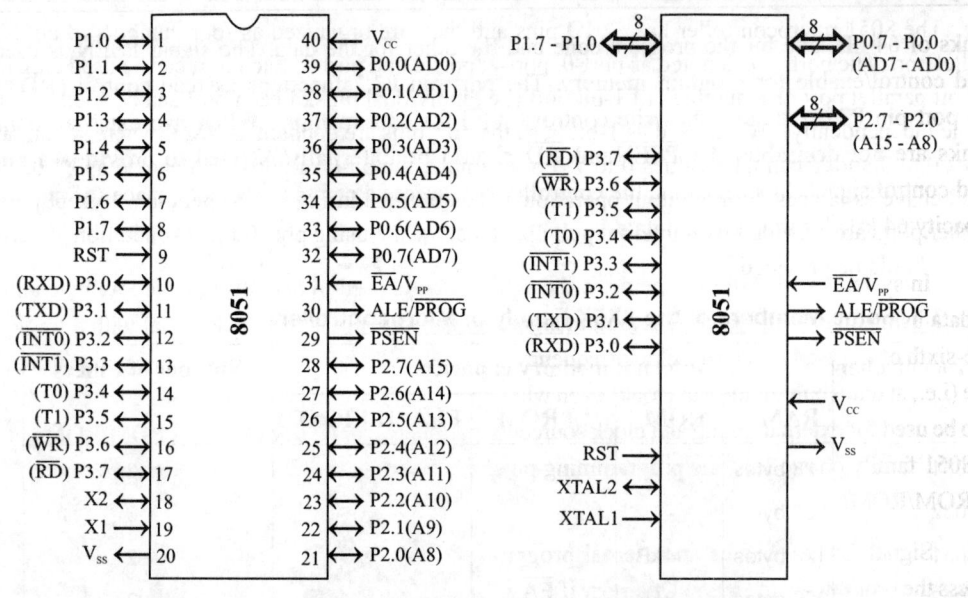

Note: *The signals shown within brackets are alternate functions of port pins.*

Fig. 3.2: Pin configuration of an INTEL 8051 microcontroller.

Table 3.3: Alternate Functions of Port Pins

Port pins	Alternate signal	Description
P0.7-P0.0	AD7-AD0	Multiplexed low byte address/data.
P2.7-P2.0	A15-A8	High byte address
P3.7	$\overline{RD}$	External memory read control signal
P3.6	$\overline{WR}$	External memory write control signal
P3.5	T1	External input to timer 1
P3.4	T0	External input to timer 0
P3.3	$\overline{INT1}$	External interrupt 1
P3.2	$\overline{INT0}$	External interrupt 0
P3.1	TxD	Serial data output
P3.0	RxD	Serial data input

Except port-1 all other ports have alternate functions. (Port-1 can be used only for IO operation.) When external memory is employed, port-0 functions as multiplexed low byte address or data lines, and port-2 functions as high byte address lines. Therefore, for accessing external memory the microcontroller uses a 16-bit address and accesses the memory in bytes. Hence, the addressable memory space is 64 kB ($2^{16} = 64$k). The 8051 allows the external memory to be organized as two banks of 64 kB, one for the program/code and the other for the data. The signal $\overline{PSEN}$ is used as read control/enable for program memory. The port pin P3.7 functions as read control ($\overline{RD}$) and the port pin P3.6 functions as write control ($\overline{WR}$) for data memory. When two external memory banks are not desirable, the $\overline{PSEN}$ and $\overline{RD}$ should be externally ANDed to provided a single read control signal. In such cases, the controller will access a common memory space (of maximum capacity 64 kB) for program and data.

In systems with external memory, the signal ALE is used to demultiplex the low byte address or data using an external latch. The output signal on the ALE pin is the clock signal with a frequency one-sixth of a crystal or internal clock frequency. The controller will output the ALE signal at a constant rate (i.e., at one-sixth of internal clock) even when there is no external memory. Therefore, the ALE can also be used for external timing and clock source for peripherals or IO devices. In EPROM/ROM version of 8051 family controllers, the programming pulse can be input through ALE during programming of EPROM/ROM.

Signal $\overline{EA}$ is used as an external program memory access control. The microcontroller will access the program from external memory if $\overline{EA}$ pin is grounded. During programming mode of internal EPROM/ROM, this pin is used to supply the programming voltage (+12 V).

> Note: *For programming the internal EPROM/ROM of 8051 family of microcontroller, a separate programmer should be employed. The controllers listed in Table 3.1, do not have ISP (in-system programmable) facility.*

The X1 and X2 pins are provided for external quartz crystal connection, in order to generate the required clock for the microcontroller. The maximum frequency of a quartz crystal that can be connected to an 8051 microcontroller is 12 MHz. (There are higher clock versions of the 8051 family of microcontrollers. For details please refer to manufacturers data sheet.) Alternatively, the external clock can be supplied through an X1 pin. The internal clock frequency of an 8051 microcontroller is same as a crystal frequency or externally supplied clock frequency. When a crystal is connected between X1 and X2, the controller will output a clock signal through X2 whose frequency is same as crystal frequency and this clock signal can be used for peripheral or IO devices.

The RST signal is used to reset the microcontroller in order to bring the controller to a known state. For proper reset the RST pin should be held **low** for at least two machine cycles. When the 8051 controller is reset, all the internal registers are cleared except the port latches, stack pointer and SBUF register. The internal RAM is not affected by reset. The content of the various registers of 8051 after a reset are listed in Table 3.4.

Table 3.4: Contents of Registers After a Reset

Register	Content after reset	Register	Content after reset
PC	00_H	SP	07_H
ACC	00_H	TCON	00_H
B	00_H	TH0	00_H
PSW	00_H	TL0	00_H
DPTR	0000_H	TH1	00_H
P0-P3	FF_H	TL1	00_H
IP	$xxx00000_B$	SCON	00_H
IE	$0xx00000_B$	SBUF	Indeterminate
TMOD	00_H	PCON	$0xxx00000_B$

3.2.2 ARCHITECTURE OF 8051

The architecture of 8051 is shown in Fig. 3.3. The various functional blocks of 8051 are the ALU, Special Function Registers (SFRs) listed in Table 3.5, Instruction Register (IR), Program Counter (PC), 128 bytes RAM, 4 kB ROM, Port latches and drivers, Oscillator, Timing and Control Unit.

The 8051 has **Harvard architecture** in which the same address in different memory devices or banks is used for program (or code) and data. Therefore, the architecture has two dedicated 16-bit address pointers, namely, **Program Counter (PC)** and **Data Pointer (DPTR)**. The PC is used as the address pointer to access program instructions and it is automatically incremented after every byte of instruction fetch. The DPTR is used as address pointers to read/write data in data memory and it is programmable using instructions.

Since, the size of the address pointers are 16-bit they can address up to $2^{16} = 64k$ memory locations. Hence, 8051 supports two memory banks of 64 kB each, one for program and the other for data. In 8051, when the $\overline{EA}$ pin is tied to the V_{cc} (logic-1), the first 4 kB of program memory address space refers to 4 kB internal ROM and the remaining 60 kB refer to external (EPROM/RAM) memory. In 8051, when the $\overline{EA}$ pin is tied to the ground (logic-0), the entire 64 kB of program address space refers to external (EPROM/RAM) memory.

The 8051 has separate 256 bytes internal RAM accessed by using an 8-bit address. In this 256 bytes address space, the first 128 addresses are allotted to internal RAM and the next 128 bytes are allotted to SFR. The internal RAM/SFR can be accessed by using MOV instructions and external data memory (RAM) can be accessed by using MOVX instruction.

The 8051 has four 8-bit ports, namely, port-0, port-1, port-2 and port-3. Each port has a latch and driver (or buffer). When external memory is employed the port-0 lines will function as multiplexed low byte address/data lines and port-2 lines will function as high byte address lines. Also, the port pins P3.7 and P3.6 are used to output read and write control signals respectively. In fact, each pin of port-3 has an alternate function which are listed in Table 3.3.

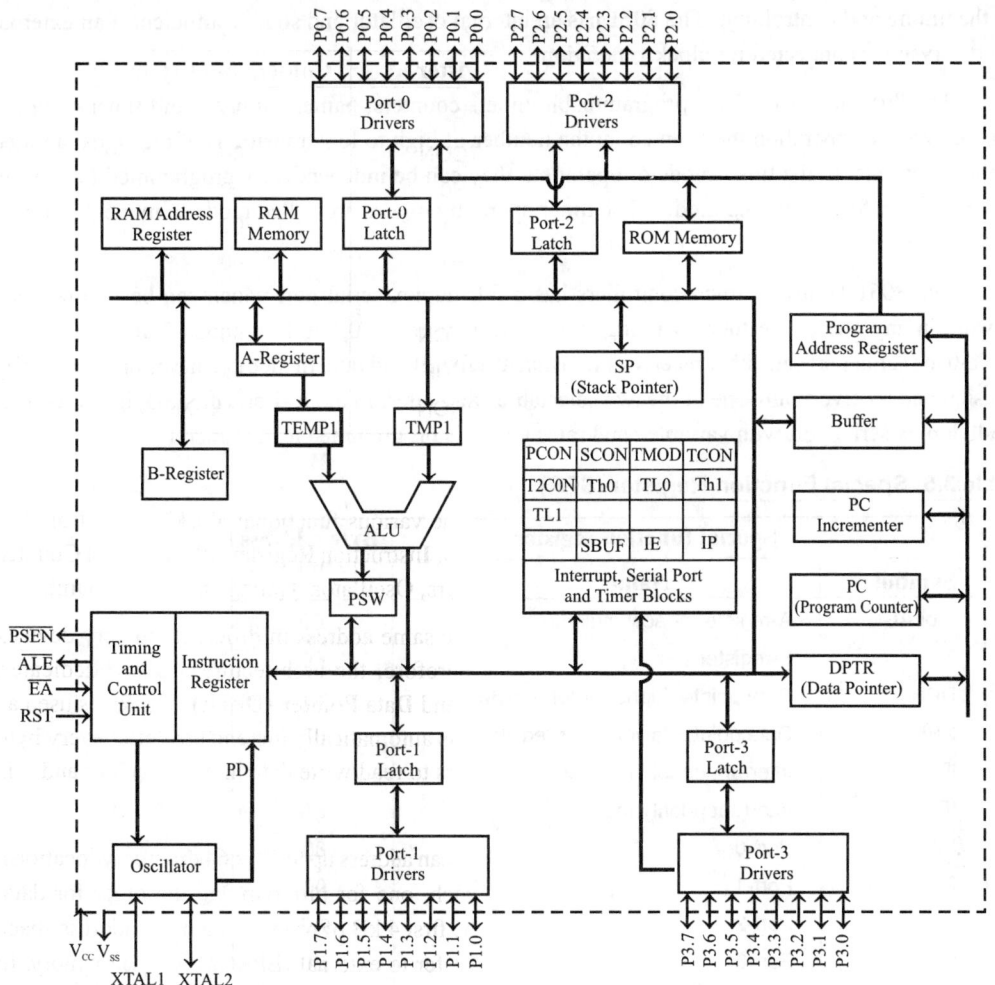

Fig. 3.3: Architecture of a 8051 microcontroller.

Port-1 is a dedicated IO port and does not have any alternate function. The ports are also mapped as internal memory in the controller and so they can be addressed as memory locations for 8-bit operation.

The SFRs include 21 internal registers listed in Table 3.6. (A detailed explanation of SFRs is presented in the Section 3.3.) The SFRs are mapped as internal data memory. The data memory address space 80_H to FF_H are reserved for SFRs. Each register of SFR has one-byte address. Some of the registers are both byte and bit-addressable. (The registers whose address ends with 0_H or 8_H are bit-addressable.)

The 8051 has an 8-bit ALU which performs arithmetic and logical operations on binary data. The A and B registers are used to hold the input data and the result of the ALU operation. Starting from the address stored in the PC, the controller will fetch the instructions one by one, and store in IR, which decodes the instructions and give information to the timing and control unit. Using the information supplied by the IR unit, the control signals necessary for internal and external operations are generated by the timing and control unit. The 8051 has an internal oscillator and so it is sufficient if an external quartz crystal is connected for clock generation.

The 8051 has two 16-bit programmable timers/counters, namely, timer-1 and timer-0. In the counter mode of operation they can count the number of high to low transitions of the signal applied to the timer pins. In the timer mode of operation, they can be independently programmed to work in any one of the four operating modes. The timer operating modes are called mode-0, mode-1, mode-2 and mode-3.

The 8051 family of microcontrollers has a full duplex serial port which can be programmed to work in any one of the four operating modes, namely, mode-0, mode-1, mode-2 and mode-3. In mode-0 the serial port can either receive or transmit at a fixed baud rate. In mode-2, it can simultaneously transmit and receive at any one of the two selectable baud rates. In mode-1 and mode-3, it can work as a full duplex serial port with variable baud rates which is programmed using timer-1.

Table 3.5: Special Function Register (SFR)

Special function register		Byte address in hexa
Symbol	Name	
A or ACC	A-register or accumulator	E0
B	B-register	F0
DPH	Data pointer higher order register	83
DPL	Data pointer lower order register	82
IE	Interrupt enable register	A8
IP	Interrupt priority register	B8
P0	Port-0	80
P1	Port-1	90
P2	Port-2	A0
P3	Port-3	B0
PCON	Power control register	87
PSW	Program status word	D0
SCON	Serial port control register	98
SBUF	Serial port data buffer	99
SP	Stack pointer	81
TMOD	Timer/Counter mode control register	89
TCON	Timer/Counter control register	88
TL0	Timer-0 low order register	8A
TH0	Timer-0 high order register	8C
TL1	Timer-1 low order register	8B
TH1	Timer-1 high order register	8D

Note: The registers whose address ends with 0_H or 8_H are bit addressable.

3.2.3 PROGRAMMING MODEL OF 8051

The programming model of the 8051 microcontroller is shown in Fig. 3.4. The model will show the programmable internal devices of a processor or controller. During execution of a program the contents of the registers and RAM memory locations shown in the programming model are altered.

The programming model of an 8051 microcontroller includes a 128 byte internal RAM and all the Special Function Registers (SFRs). In addition, the programming model of an 8051 includes a 4 kB internal ROM. The internal RAM locations are separate data memory address space and accessed either by direct or indirect addressing using an 8-bit address. The SFRs are also separate data memory address spaces and are accessed by direct addressing using an 8-bit address. The 4 kB internal ROM in an 8051 controller is mapped as program memory and accessed using a 16-bit address.

The first 32 bytes of internal RAM are organized as four groups of eight registers. Each group is called a **register bank** and denoted as bank-0, bank-1, bank-2 and bank-3. The eight registers of a bank are denoted as R_n where n takes values from 0 to 7. At any one time, the controller can use any one of the register bank as general purpose registers (or scratch pad registers). The selection of the register bank depends on the value of bits RS0 and RS1 in the PSW register. After a reset the PSW register is cleared and so the controller works with register bank-0.

The internal RAM locations in the address range 20_H to $2F_H$ are bit addressable. The internal RAM locations in the address range 30_H to $7F_H$ can be used as general purpose RAM. The hexa address of the bit-addressable RAM are listed in Table 3.6. The 16 RAM locations in the address range 20_H to $2F_H$ has a capacity of 128 bits ($16 \times 8 = 128$). Each bit in this RAM area can be addressed by an 8-bit address in the range 00_H to $7F_H$, as shown in Table 3.6.

Table 3.6: Byte Address of Bit-Addressable RAM

Byte address	Hexa address of bit position							
	B_7	B_6	B_5	B_4	B_3	B_2	B_1	B_0
20	07	06	05	04	03	02	01	00
21	0F	0E	0D	0C	0B	0A	09	08
22	17	16	15	14	13	12	11	10
23	1F	1E	1D	1C	1B	1A	19	18
24	27	26	25	24	23	22	21	20
25	2F	2E	2D	2C	2B	2A	29	28
26	37	36	35	34	33	32	31	30
27	3F	3E	3D	3C	3B	3A	39	38
28	47	46	45	44	43	42	41	40
29	4F	4E	4D	4C	4B	4A	49	48
2A	57	56	55	54	53	52	51	50
2B	5F	5E	5D	5C	5B	5A	59	58
2C	67	66	65	64	63	62	61	60
2D	6F	6E	6D	6C	6B	6A	69	68
2E	77	76	75	74	73	72	71	70
2F	7F	7E	7D	7C	7B	7A	79	78

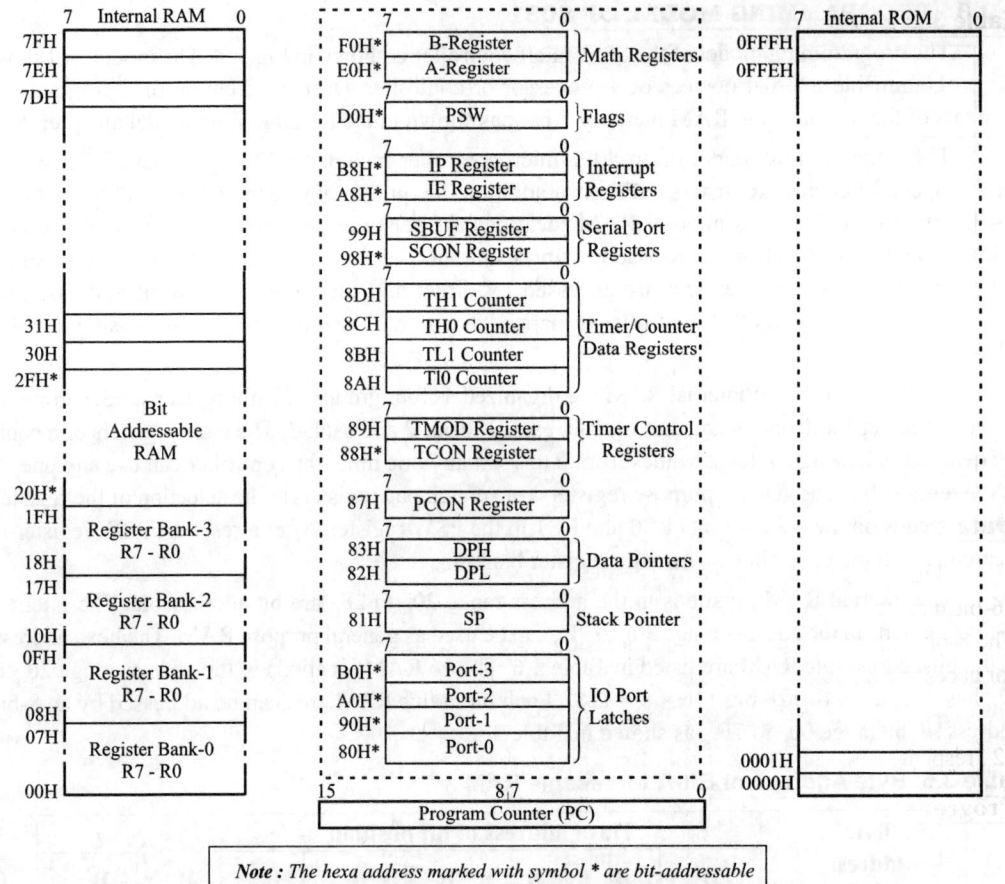

Note : *The hexa address marked with symbol * are bit-addressable*

Fig. 3.4: Programming model of an 8051 microcontroller.

3.2.4   SPECIAL FUNCTION REGISTER (SFR) OF 8051

The 8051 is provided with 21 special function registers and they are used for selecting various programmable features of the microcontroller. The special functions of most of the SFR are distinguishable. Each SFR has an internal one-byte address assigned to it.

Some of the registers are both byte and bit-addressable. The SFR along with their byte address are listed in Table 3.5. The bit-addressable register along with address for each bit are listed in Table 3.7.

A and B Registers

The A and B registers are called CPU registers. They are used to hold the data for most of the CPU (ALU) operations. The sizes of A and B registers are 8-bit and they are mapped as on-chip data memory with byte address $E0_H$ and $F0_H$ respectively. These registers are also bit-addressable.

In most of the ALU operations, the result is stored in the A-register and so, it is also known as the **accumulator.**

Table 3.7: Bit Address of SFR

SFR	Hexa address of bit position							
	B_7	B_6	B_5	B_4	B_3	B_2	B_1	B_0
B	F7	F6	F5	F4	F3	F2	F1	F0
A or ACC	E7	E6	E5	E4	E3	E2	E1	E0
PSW	D7	D6	D5	D4	D3	D2	D1	D0
IP	BF	BE	BD	BC	BB	BA	B9	B8
P3	B7	B6	B5	B4	B3	B2	B1	B0
IE	AF	AE	AD	AC	AB	AA	A9	A8
P2	A7	A6	A5	A4	A3	A2	A1	A0
SCON	9F	9E	9D	9C	9B	9A	99	98
P1	97	96	95	94	93	92	91	90
TCON	8F	8E	8D	8C	8B	8A	89	88
P0	87	86	85	84	83	82	81	80

Data Pointer (DPTR)

The data pointer is a 16-bit register used to hold the 16-bit address of data memory. The 16-bit data pointer can also be used as two numbers of 8-bit data pointer, namely, DPH and DPL. The 8-bit data pointers are used for accessing internal RAM and SFR. The 16-bit data pointer is used for accessing external data memory.

The 8-bit data pointer DPH and DPL are mapped as internal memory with byte address 83_H and 82_H respectively. The contents of the data pointers are programmable using instructions.

Program Status Word (PSW)　　　　　　　　　　　　　　　　*(AU, Nov/Dec' 19, 2 Marks)*

The program status word stores the status of the result of the ALU operations and some of the status of the processor by means of a 1-bit status called flag. The PSW is also known as a **flag register**. The flags are useful for the programmer to test the condition of the result and make decisions.

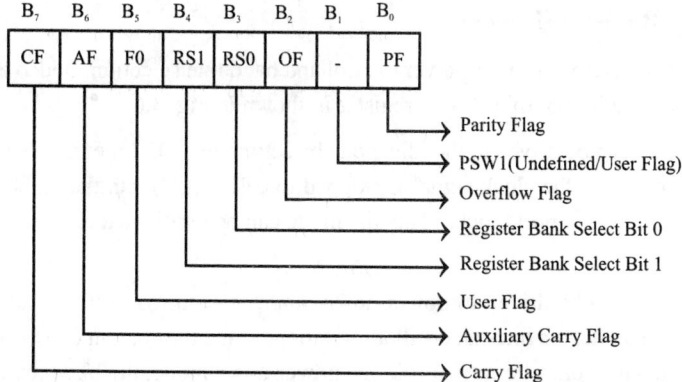

Fig. 3.5: Format of PSW of an 8051 family of microcontrollers.

The format of the PSW of an 8051 microcontroller is shown in Fig. 3.5. The PSW consists of four math flags, two register bank select bits and two user flag bits. The math flags are carry, auxiliary carry, overflow and parity flags. These flags are altered after arithmetic and logical operations depending on the result. The carry flag is set when the result has a carry. When there is a carry from the lower nibble to the upper nibble, the auxiliary carry is set. When the result has even parity, the parity flag is set. In signed mathematical operations if the size of the result exceeds the maximum range then the overflow flag is set.

Table 3.8: Selection of a Register Bank

The register bank select bits RS1 and RS0 are used to select any one of the four register banks of the internal RAM. At any one time, the microcontroller can work with (or access) only one register bank selected by these bits. The bank select bits are programmable and after reset the controller defaults to bank-0. The selection

Bank select bits		Selected	Range of hexa
RS1	RS0	register bank	address of the selected bank
0	0	Bank-0	00_H - 07_H
0	1	Bank-1	08_H - $0F_H$
1	0	Bank-2	10_H - 17_H
1	1	Bank-3	08_H - $1F_H$

of a register bank using the RS1 and RS0 bits are listed in Table 3.8. The user flag bits can be used by the programmer to indicate the status of certain events during program execution.

Stack Pointer (SP)

The stack pointer always holds the 8-bit address at the top of stack. The 8051 microcontroller supports the LIFO (**L**ast-**I**n-**F**irst-**O**ut) stack, and the stack may reside anywhere in on-chip RAM (i.e., the programmer can reserve any portion of on-chip RAM as stack.) After a reset, the stack pointer is initialized to 07_H. The stack can be accessed using PUSH and POP instructions. During a PUSH operation, the stack pointer is automatically incremented by one and during POP operation the stack pointer is automatically decremented by one.

Power Control Register (PCON)

The PCON register is used for power control and baud rate selection. It also consists of general purpose user flags. The format of a PCON register is shown in Fig. 3.6.

The controller can be driven to the idle mode by setting the IDL bit of the PCON register. When the idle mode is activated, the clock signal is stopped to CPU(ALU), but the clock signal is supplied to interrupt, timer and serial port blocks. The idle mode can be terminated either by an interrupt or by hardware reset.

The controller can be driven to power-down mode by setting the PD bit of a PCON register. During power-down mode, the internal oscillator is stopped, and the content of SFR and internal RAM are preserved. When the controller remains in the power down mode, the V_{CC} (power supply voltage) can be reduced to 2 V. The power-down mode can be terminated only by a hardware reset and during hardware reset the content of SFR are altered but the content of internal RAM are preserved.

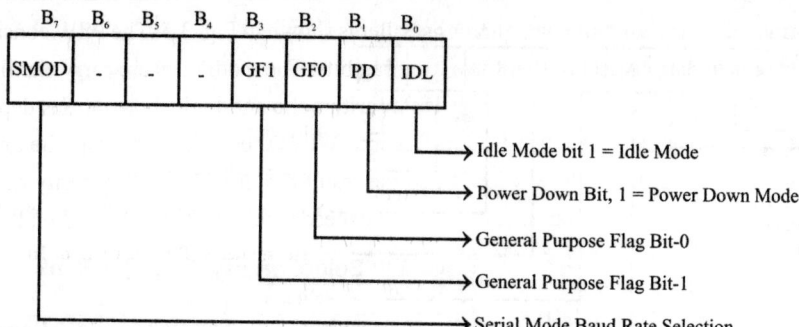

Fig. 3.6: Format of a PCON register of 8051 family of microcontrollers.

Note: The idle and power-down mode are not available in NMOS version of 8051.

The SMOD bit is used to decide the baud rate in serial port operating modes 1, 2 or 3. In mode 2, if SMOD = 0, then the baud rate is 1/64 of oscillator frequency and if SMOD = 1 then the baud rate is 1/32 of oscillator frequency.

The general purpose flag bits GF1 and GF0 can be used by the programmer to indicate the status of certain events during program execution.

Serial Data Buffer (SBUF) Register

The SBUF register is used to hold the parallel data during transmission and reception. During serial reception, the serial data is received via RxD pin and converted to parallel data and stored in the receive buffer. During serial transmission, the parallel data is stored in the transmit buffer and then converted to serial data to transmit via TxD pin.

The transmit and receive buffers are assigned the same internal address 99_H but the transmit buffer can be accessed only for write operation and the receive buffer can be accessed only for read operation. When data is written to SBUF, it goes to transmit buffer and when data is read from SBUF it comes from the receive buffer.

Serial Port Control Register (SCON)

The format of a SCON register is shown in Fig. 3.7. The SCON register consists of mode selection bits, the 9th data bit (bit-B_8) for transmit and receive, and the serial port interrupt bits TI and RI. The bits SM0 and SM1 are used to select any one of the four operating modes for serial transmission and reception. The four modes of a serial port are mode-0, mode-1, mode-2 and mode-3.

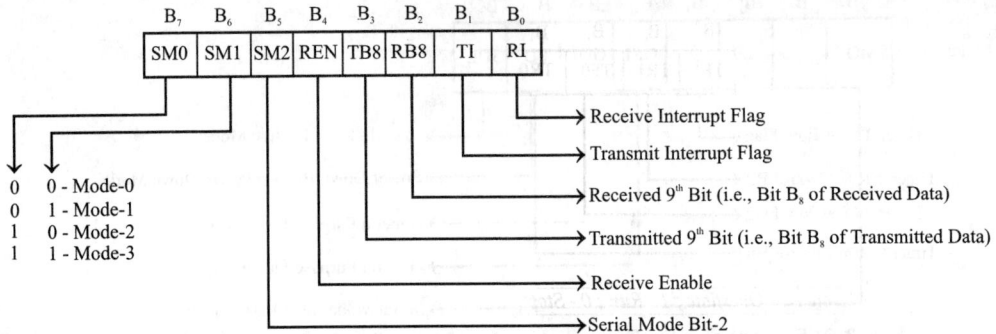

Fig. 3.7: Format of a SCON register of the 8051 family of microcontrollers.

Timer Mode Control (TMOD) Register

The TMOD register is used to select the operating mode and the timer/counter operation of the timers. The format of a TMOD register is shown in Fig. 3.8. The lower four bits of the TMOD register is used to control timer-0 and the upper four bits are used to control timer-1. The two timers can be independently programmed to operate in various modes. The register has two separate two bit field M0 and M1 to program the operating mode of timers. The operating modes of timers are mode-0, mode-1, mode-2 and mode-3. In all these operating modes, the oscillator clock is divided by 12 and applied as the input clock to the timer.

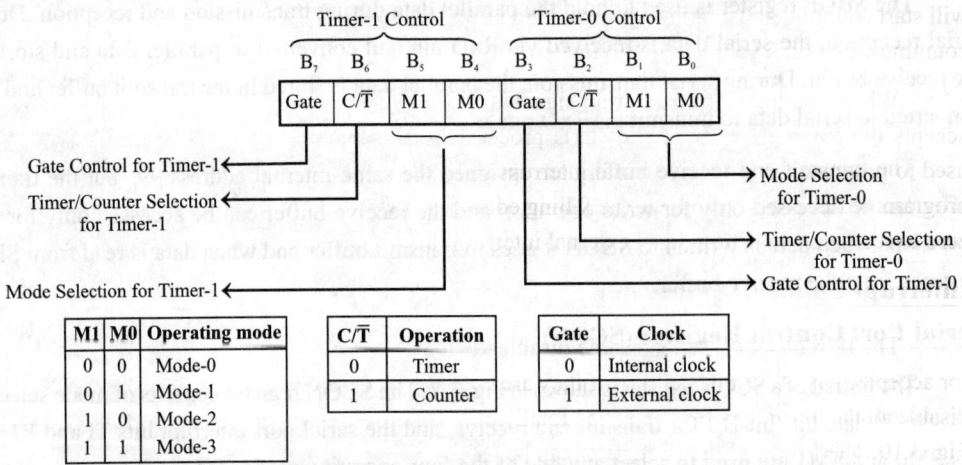

Fig. 3.8: Format of the TMOD register of the 8051 family of microcontroller.

Timer Control (TCON) Register

The TCON register consists of timer overflow flags, timer run control bits, external interrupt flags and external interrupt-type control bits. The format of a TCON register is shown in Fig. 3.9.

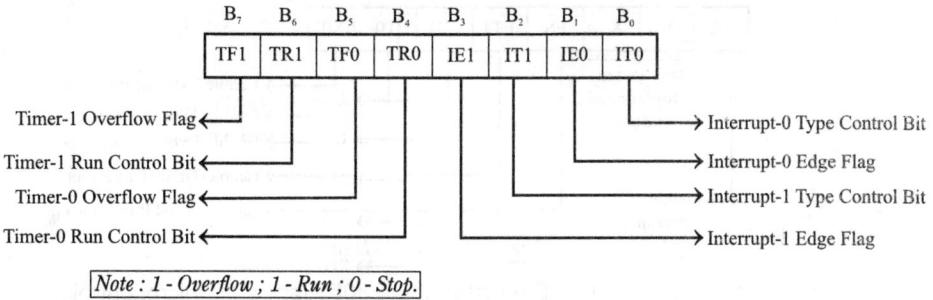

Fig. 3.9: Format of a TCON register of the 8051 family of microcontroller.

The timers in an 8051 microcontroller are upcounters and keep on incrementing as long as the clock is applied. Therefore, when the clock is applied after reaching the maximum value (i.e., the content of the counter is all 1s), the content of the counter will become zero (i.e., all 0 s). This condition is called timer overflow and it is also the end of timing which a program wants to maintain by using the timer. The TCON register has a 1-bit flag, TF for each timer to indicate the timer overflow or end of timing. Whenever the timer/counter overflows, the TF flag is set to one. The TF flag is also used as an interrupt signal to initiate the execution of a subroutine. When the controller vectors to subroutine, the TF flag is cleared.

The TR bit is used to start/stop the timer/counter. When the TR bit is set to one, the timer/counter will start counting and continue the counting as long as the TR bit is one. The timer/counter will stop counting when the TR bit is cleared to zero.

When a valid external interrupt signal is detected, the IE flag is set to one. When the controller accepts the external interrupt and starts processing it, the IE flag is cleared to zero. The IT bit is used to program the type of external interrupt signal to be recognized by the controller. The IT bit is programmed as one to recognize the falling edge-triggered external interrupt and it is programmed as zero to recognize logic **low** level external interrupt.

Interrupt Enable (IE) Register

The IE register is used to enable/disable the interrupts of an 8051. The interrupts are recognized (or accepted) by the controller only if they are enabled. The IE register can be programmed to enable/ disable all the five interrupts of an 8051 totally or individually. The format of an IE-register is shown in Fig. 3.10. The EA bit of the IE register can be programmed as zero, to disable all the five interrupts of 8051. When EA bit is programmed as one, the interrupts are enabled provided their individual enable bits are programmed as one. (The EA bit is also called global enable.)

Each interrupt has one-bit field to enable or disable it individually. When EA = 1, if the enable bit of a particular interrupt is programmed as one then it is enabled and if the enable bit is programmed as zero, then it is disabled.

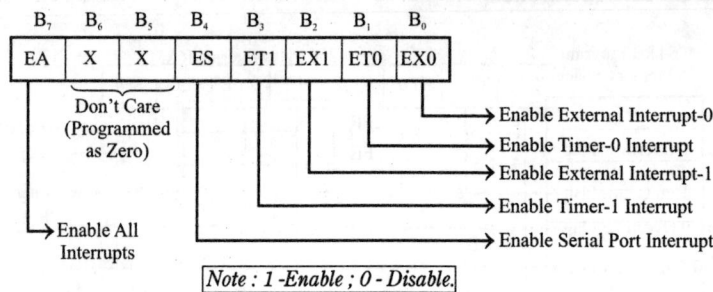

Fig. 3.10: Format of the IE register of the 8051 family of microcontrollers.

Interrupt Priority (IP) Register

The 8051 has five interrupts and the normal priority of these interrupts from highest to lowest are external interrupt-0, Timer-0 interrupt, External interrupt-1, Timer-1 interrupt and serial port interrupt.

The IP register can be programmed to make the priority of any of the interrupt as highest. The format of an IP register is shown in Fig. 3.11. The IP register has one-bit field for the priority of each interrupt. When the priority bit of a particular bit is programmed as one then its priority will be highest. In 8051, while servicing a lower priority interrupt a higher priority interrupt will be recognized but another lower priority interrupt will not be recognized.

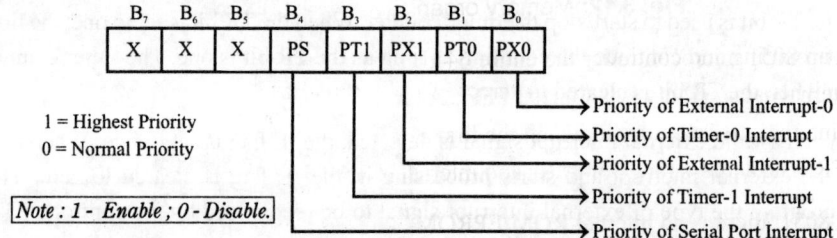

Fig. 3.11: Format of the IP register of an 8051 family of microcontrollers.

3.3 MEMORY ORGANIZATION   *(AU, Nov/Dec' 19, 5 Marks)*

A microcontroller-based system requires both EPROM and RAM. The EPROM is required for permanent program and permanent data storage. The RAM is required for temporary data storage and stack. The 8051 has 64 kB program memory address space and 64 kB data memory address space.

The microcontroller can only read from the program memory and the signal PSEN is used as read control for reading the program memory. Therefore read only memories like ROM/EPROM/EEPROM can be employed as the program memory. The microcontroller can read and write with data memory. It has a separate read control signal, $\overline{RD}$ and write control signal, $\overline{WR}$ for reading and writing with data memory respectively. Hence read-write memories like static RAM can be employed as data memory. The interfacing of external memory to the 8051 microcontroller are shown in te Fig. 3.12.

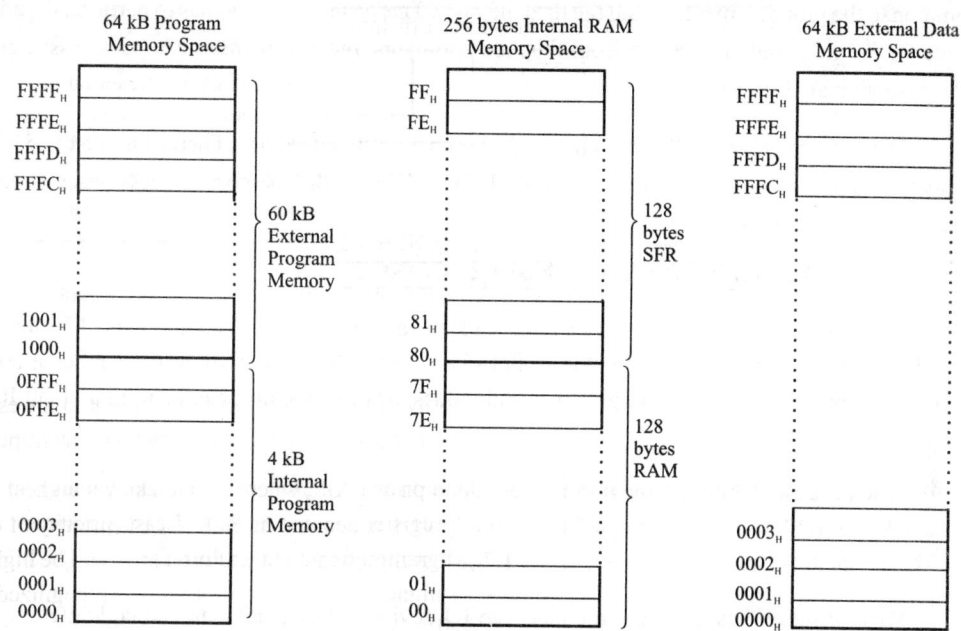

Fig. 3.12: Memory organization in the 8051 microcontroller.

In an 8051 microcontroller the entire 64 kB data memory space is external. The address range of the external data memory is 0000_H to $FFFF_H$. Apart from the external data memory the 8051 has 256 bytes of internal data memory in which the first 128 bytes are called RAM and the next 128 byte are called SFR. The address range of SFRs and internal RAM are 00_H to FF_H.

In 8051, there is no internal ROM/EPROM and so the entire 64 kB program memory space in the range 0000_H to $FFFF_H$ are external. Therefore in an 8031/8051-based system the pin EA is always tied **low** or grounded (0V).

The 8051, has a 4 kB internal ROM which can be mapped to the first 4 kB address space of the program memory if the EA pin is tied **high** or tied to V_{CC} (+5V). This means that the usage of internal ROM in the 8051 is optional. When EA is tied **high** or tied to V_{CC} (+5V) the internal 4 kB ROM is mapped as the first 4 kB of the program memory address space and when EA is tied **low** or grounded (0V), the internal ROM is ignored or cannot be accessed. When EA is tied **high**, the internal 4 kB ROM will be mapped as the program memory in the address range 0000_H to $0FFF_H$ and the external program memory 60 kB will have the address range 1000_H to $FFFF_H$. When EA is tied to the ground, the entire 64 kB program memory address space is external with an address range 0000_H to $FFFF_H$.

In an 8051-based system, it is possible to have a single memory bank for both program and data, to minimize the cost of the system where memory requirement is less. When a single memory bank is provided, the combined read control signal is generated by logically ANDing the $\overline{PSEN}$ and $\overline{RD}$ signals. When a single bank is provided the total memory capacity is 64 kB and this address space is common to the program memory and the data memory. The system designer has to partition the address space for program and data. Apart from this, the 256 bytes internal memory can be accessed as data memory using an 8-bit address.

The 8051 microcontroller does not provide separate IO addresses. Therefore in an 8051 based system, only memory-mapped IO is possible. Hence some of the memory address space should be reserved for IO devices.

3.4 IO PORTS AND DATA TRANSFER CONCEPTS

The 8051 microcontroller has 32 IO pins and they are organized as four numbers of 8-bit parallel port. The ports are denoted as Port-0, Port-1, Port-2 and Port-3. Each port can be used either as an 8-bit parallel port or 8 numbers of 1-bit port (i.e., individual pins of each port can be used as 1-bit IO line independently).

When used as 1-bit port, the port pins are denoted as PX.Y, where X can take values 0 to 3 and Y can take values 0 to 7. For example, the bit-0 of port-1 is denoted as P1.0 (Least Significant Bit of Port-1) and the bit-7 of port-1 is denoted as P1.7 (Most Significant Bit of Port-1).

Most of these pins are used to connect to IO devices or external data and code memory. The ports behave as latches during output operation and as buffers during input operation. Except port-0 all other ports are provided with internal pull up. Hence, while using port-0 for IO operation, external pull up should be provided.

In order to set a port pin as output pin, the corresponding data bit must be cleared in the port register and in order to set a port pin as input pin, the corresponding data bit must be set high in the port register. For example, when port pin P1.0 has to be set as output, then initialize the port pin by writing "0" to pin P1.0, and when port pin P1.0 has to be set as input, then initialize the port pin by writing "1" to pin P1.0.

General Internal Hardware Structure (or circuit) of an IO Port Pin of 8051

Each bit of an IO port of 8051 has the following components to read/write a binary bit.

1. An internal data bus line for data transfer between the port pin and internal bus.

2. A D-latch to store the value of the data bit during write operation and is controlled by "Write to latch" control signal. When "Write to latch" = 1, the data on the internal bus line is latched into the output of D-latch.

3. Two tristate buffers, are employed for read operation. One buffer is used to read the data on the output of D-latch and it is controlled by "Read latch" control signal. Another buffer is used to read the data on the port pin and it is controlled by "Read pin" control signal. When "Read pin" = 1, the data present at the port pin is read and when "Read latch" = 1, the data in the output of D-latch is read.

4. FET transistor to drive signal at the port pin. When the signal at the gate of transistor is "0", the transistor will be OFF and so the drain-source path is open. When the signal at the gate of the transistor is "1", the transistor will be ON and so the drain-source path is short.

PORT-0 PIN INTERNAL HARDWARE STRUCTURE (OR CIRCUIT)

The Port-0 serves as input, output or as a bi-directional lower order address and data bus (AD_0-AD_7) for external memory. The internal hardware structure (or circuit) of each pin in Port-0 is shown in Fig. 3.13. Port-0 has two FET driver transistors at the port pin.

When a port-0 pin is used as an input, the port pin is initialized by writing "1" to the D-latch, which will make the signal at transistor gate as "0". Therefore, both the transistors will turn OFF, which in turn causes the pin to "float" in a high impedance state, and so the port pin is connected only to the input buffer.

When a port-0 pin is used as an output, the port pin is initialized by writing "0" to D-latch, which will make the signal at transistor gate as "1". Therefore, both the transistors will turn ON, which in turn causes the port pin to be connected to ground. Now, during a write operation when a "0" is written the same condition exists, but when a "1" is written then transistors are OFF and the port pin floats, and so an external pull-up resistor should be connected to maintain logic **high** at the port pin.

When port-0 is used as an address bus to external memory, internal control signal switch the address lines to the transistor gates. Logic "1" on an address bit will turn the upper transistor ON and the lower transistor OFF to provide logic **high** at the port pin. When the address bit is a "0", the lower transistor is ON and the upper transistor is OFF to provide logic **low** at the port pin. After the address output, the port-0 pins are internally initialized for data transfer. Hence, for normal address/data interfacing (or for external memory access), no pull-up resistors are required.

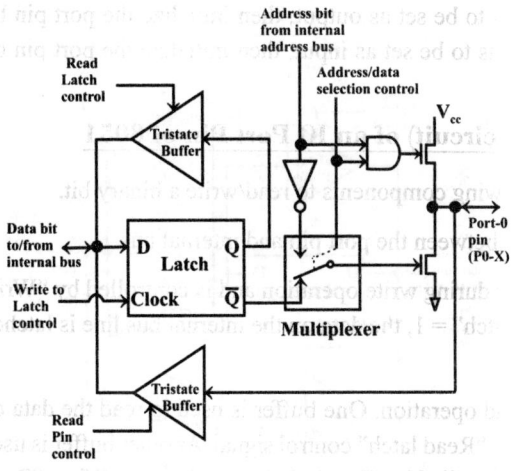

Fig. 3.13: Circuit of port-0 bit.

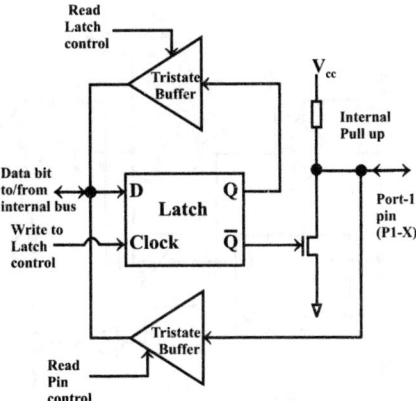

Fig. 3.14: Circuit of port-1 bit.

PORT-1 PIN INTERNAL HARDWARE STRUCTURE (OR CIRCUIT)

The Port-1 can serve as either input or output and does not have any alternate function. The internal hardware structure (or circuit) of each pin in Port-1 is shown in Fig. 3.14. Since the port-1 does not have alternate function, the output of D-latch is connected directly to the gate of the transistor which has an internal pull-up resistor as shown in Fig. 3.14. This internal pull-up resistor help to maintain a stable logic level when used as output port.

When a port-1 pin is used as an input, the port pin is initialized by writing "1" to the D-latch, which will make the signal at transistor gate as "0". Therefore, the transistor will turn OFF, which in turn causes the port pin and input of buffer to be pulled high by the internal pull up. An external circuit can overcome the high impedance pull-up and drive the pin **low** or **high** to input '0' or '1' respectively.

When a port-1 pin is used as an output, the port pin is initialized by writing "0" to D-latch, which will make the signal at transistor gate as "1". Therefore, the transistor will turn ON, which in turn causes the port pin to be connected to ground. Now, during a write operation, when a "0" is written the same condition exists, but when a "1" is written then the transistor will be OFF, and so the port pin is pulled **high** by the internal pull up.

PORT-2 PIN INTERNAL HARDWARE STRUCTURE (OR CIRCUIT)

Port-2 serves as input, output or as a higher order address bus ($A_8 - A_{15}$) for external memory. The internal hardware structure (or circuit) of each pin in port-2 is shown in Fig. 3.15. The port-2 pins are also provided with internal pull up and so the input/output operation of port-2 pins are similar to that of port-1.

When port-2 is used as an address bus to external memory, the internal control signal switches the address lines to the transistor gates.

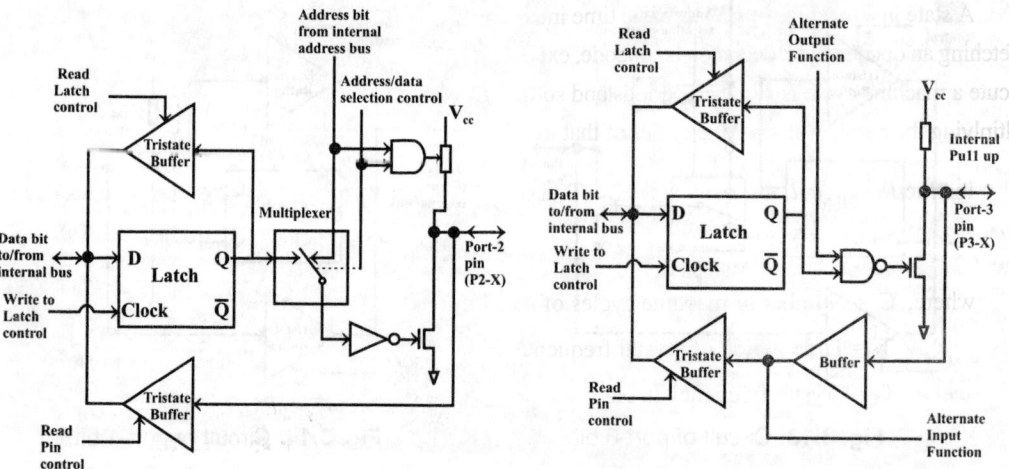

Fig. 3.15: Circuit of port-2 bit.

Fig 3.16: Circuit of port-3 bit.

PORT-3 PIN INTERNAL HARDWARE STRUCTURE (OR CIRCUIT)

Port-3 serves as input or output and also each pin of port-3 has an alternate function. The internal hardware structure (or circuit) of each pin in Port-3 is shown in Fig. 3.16. Port-3 pins are also provided with internal pull up and so the input/output operation of port-3 pins are similar to that of port-1.

In order to select the alternate functions of port-3 pins, the port pin has to be initialized by writing "1" to it and the alternate function of port-3 pins are controlled by various other special function registers.

3.5 MACHINE CYCLES AND TIMING DIAGRAM

The timing diagram provides information about the various conditions of the signal while a machine cycle is executed.

The external basic operations performed by a microcontroller are called machine cycles. The executive of an instruction involves execution of one or more machine cycles in a specified order. The 8051 microcontroller takes one to four machine cycles to execute an instruction. The basic timing of the 8051 machine cycle is shown in Fig. 3.17.

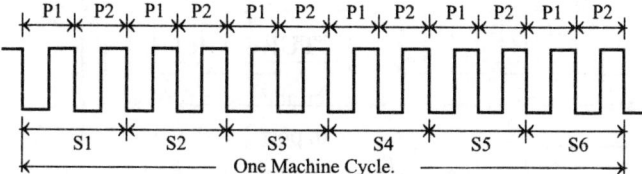

Fig. 3.17: Basic timing of a machine cycle.

The entire timing of a machine cycle of 8051 is divided into 6 states and they are denoted as S1, S2, S3, S4, S5 and S6. The timing of each state is two clock periods and they are denoted as P1 and P2.

A state in a machine cycle is a basic time interval for discrete operation of the microcontroller such as fetching an opcode byte, decoding an opcode, executing an opcode, writing a data, etc. The time taken to execute a machine cycle is 12 clock periods and so the time taken to execute an instruction is obtained by multiplying the number of machine cycles of that instruction by 12 clock periods.

Instruction execution time $= C \times 12 \times T$

$$= C \times 12 \times \frac{1}{f}$$

where, C = Number of machine cycles of an instruction

T = Time period of crystal frequency in seconds

f = Crystal frequency in Hz

The 8051 microcontroller has four machine cycles. These are:

1. **External program memory fetch cycle**

2. **External data memory read cycle**

3. **External data memory write cycle**

4. **Port operation cycle**

3.5.1 EXTERNAL PROGRAM MEMORY FETCH CYCLE

The External program memory fetch machine cycle is executed by the 8051 to fetch the opcode and subsequent instruction bytes from the memory. The timing diagram of an external memory fetch cycle is shown in the Fig. 3.18. During one machine cycle (6 states), two consecutive bytes of program memory are read. In one-byte instruction, the second byte is discarded. The timing of various signals involved in the fetch operation are shown in the timing diagram in Fig. 3.18.

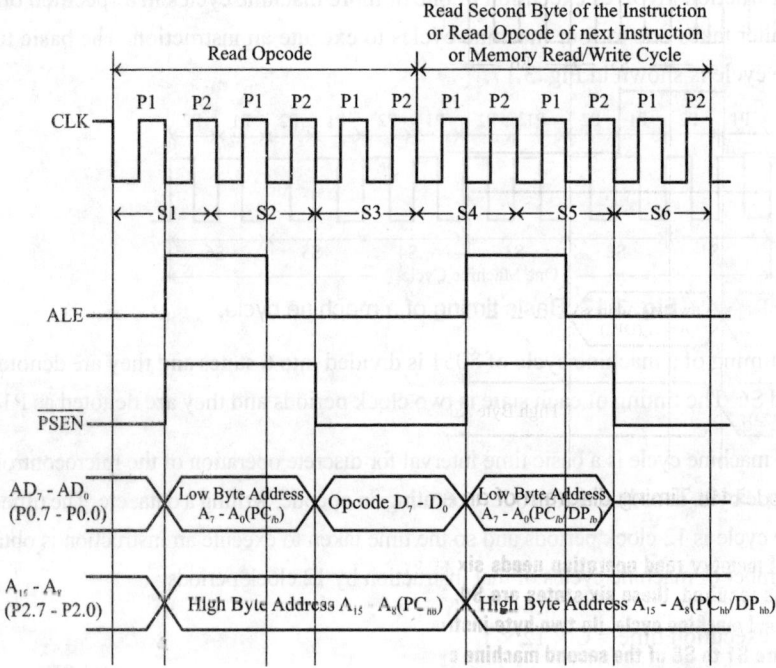

Fig. 3.18: Timing diagram of an external program memory fetch cycle.

1. At the falling edge of phase P2 of first state S1, the microcontroller outputs the low byte address on AD$_7$-AD$_0$ lines and high byte address on A$_{15}$-A$_8$ lines. For program memory fetch, the content of the Program Counter (PC) is the address of the program code. The ALE is asserted high to enable the address Latch.

2. At the middle of state S2, the ALE is asserted low and this enables the Latch to take low byte of the address and keep on its output lines.

3. The program store enable $\overline{PSEN}$ is asserted low to fetch the opcode from the memory and load on the AD$_7$-AD$_0$ lines at the state S3.

4. The microcontroller utilizes the first three states (S1, S2 and S3) to fetch the opcode from the memory.

5. During the remaining three states of one machine cycle (S4, S5 and S6), the microcontroller fetches the second byte of the same instruction or opcode of the next instruction or executes the external memory read/ write cycle.

6. When executing the one-byte instruction, the states S4, S5 and S6 are used by the processor for internal operations to decode the instructions and for completing the task specified by the one-byte instruction.

3.5.2 EXTERNAL DATA MEMORY READ CYCLE

The memory read cycle is executed by the 8051 to read a data from the external data memory. The data memory read cycle is executed immediately after an opcode fetch if the instruction execution requires an external data memory access. The timing diagram of the external memory read cycle is shown in Fig. 3.19. The timing of various signals involved in read operation are shown in the timing diagram.

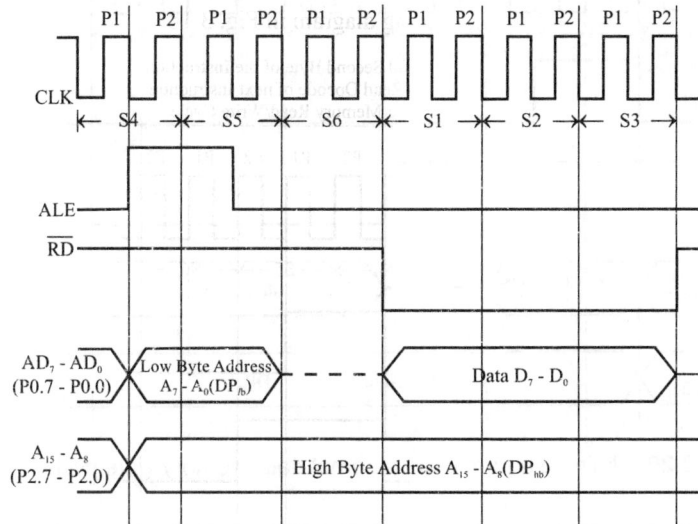

Fig. 3.19: Timing diagram of an external data memory read cycle.

1. External memory read operation needs six states. In one-byte instruction, when the external memory access is required, these six states are S4, S5 and S6 of the first machine cycle and S1, S2 and S3 of the second machine cycle. (In two-byte instruction, when external memory access is required these six states are S1 to S6 of the second machine cycle.)

2. At the first half of the state S4, the ALE is asserted high to enable the address latch. In this state, the microcontroller outputs the content of the Data Pointer low (DP$_{lb}$) on AD$_7$- AD$_0$ lines and the content of the Data Pointer high (DP$_{hb}$) on A$_{15}$-A$_8$ lines.

3. At the S1 state of the second machine cycle, the memory read signal is asserted low and the data in the specified address can be read and placed on the AD$_7$-AD$_0$ lines. The read signal is asserted low for the three states S1, S2 and S3 of the second machine cycle.

4. At the end of the S3 state of the second cycle, the RD signal is asserted low and at this time the data is latched into the microcontroller.

5.   The high byte of address A_{15}-A_8 is valid for six states, i.e., from the S4 of the first machine cycle to S3 of second machine cycle.

6.   At the last three states of the second machine cycle, the microcontroller reads the next byte of program memory and discards it.

3.5.3 EXTERNAL DATA MEMORY WRITE CYCLE

The memory write cycle is executed by the 8051 to store the data to the external data memory. The timing diagram of the memory write cycle is shown in Fig. 3.20. The timings of various signals involved in write operation are shown in the timing diagram.

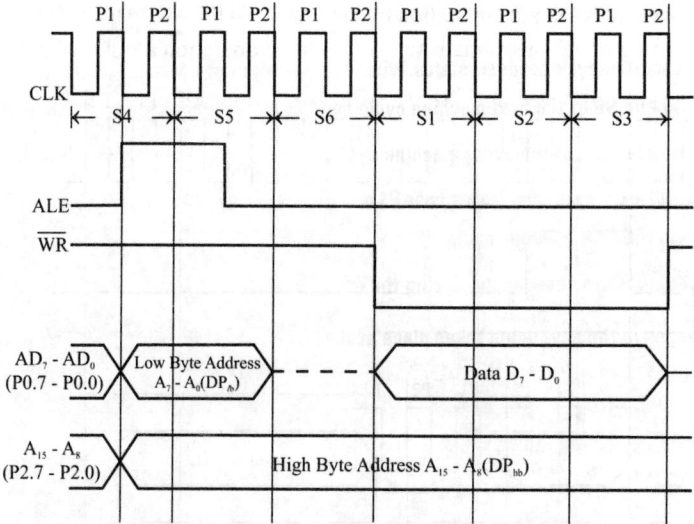

Fig. 3.20: Timing diagram of an external data memory write cycle.

1.   This write operation needs six states. In one-byte instruction, when the external memory access is required, these six states are S4, S5 and S6 of the first machine cycle and S1, S2 and S3 of the second machine cycle. (In two-byte instruction, when external memory access is required these six states are S1 to S6 of the second machine cycle.)

2.   At the first half of the state S4, the ALE is asserted high to enable the address latch. In this state, the microcontroller outputs the content of the Data Pointer low (DP_{lb}) on the AD_7- AD_0 lines and the content of Data Pointer high (DP_{hb}) on the A_{15}-A_8 lines.

3.   At the S1 state of the second machine cycle, the memory write signal is asserted low and the data is output on the AD_7-AD_0 lines. The write signal is asserted low by the microcontroller for three states S1, S2 and S3 of the second machine cycle.

4.   At the end of the S3 state of the second cycle, the WR signal is asserted low and at this time the data is latched into the external memory.

5.   The high byte of address A_8-A_{15} valid for six states, i.e., from S4 of the first machine cycle to S3 of the second machine cycle.

6.   At the last three states of the second machine cycle, the microcontroller reads the next byte of the program memory and discards it.

3.5.4 PORT OPERATION CYCLE

The port value can be changed during the port operation machine cycle. The timing diagram of the port operation cycle is shown in the Fig. 3.21. The various signals during the port operation cycle are shown in the timing diagram. By using some instructions we can change the port value immediately. For this, the 8051 executes the port operation cycle.

1. The port operation cycle needs six states. When port operation is required in 1-byte instruction, these six states are S4, S5 and S6 of the first machine cycle and S1, S2 and S3 of the second machine cycle.

2. In the fifth state S5 of the every machine cycle, all the port values are sampled.

3. The ports P0 and P1 are sampled at phase P1 of state S5. The ports P2, P3, RST are sampled at phase P2 of the state S5 in the first machine cycle.

4. The port values are changed by placing the new value on the specified port.

5. The changes in the new value takes place at the S1 state of the second machine cycle.

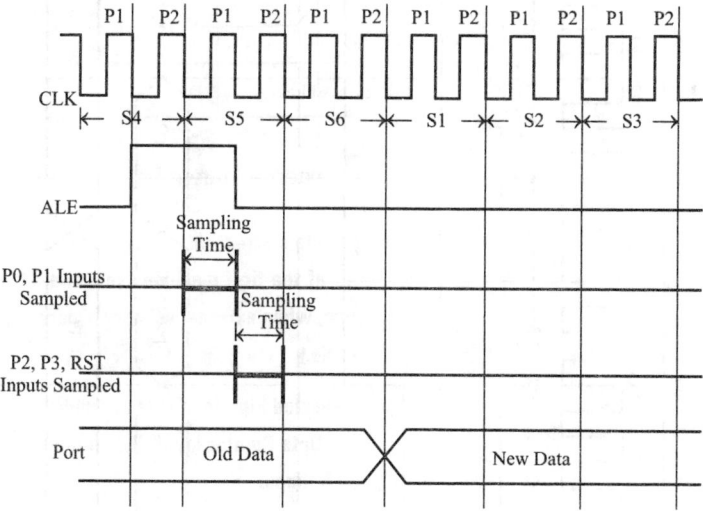

Fig. 3.21: Timing diagram of a port operation cycle.

3.5.5 TIMING DIAGRAM OF 8051 INSTRUCTIONS

The size of an 8051 instruction is one to three bytes. The first byte is opcode and the subsequent bytes are address or data. The 8051 microcontroller executes the instructions in one to four machine cycles.

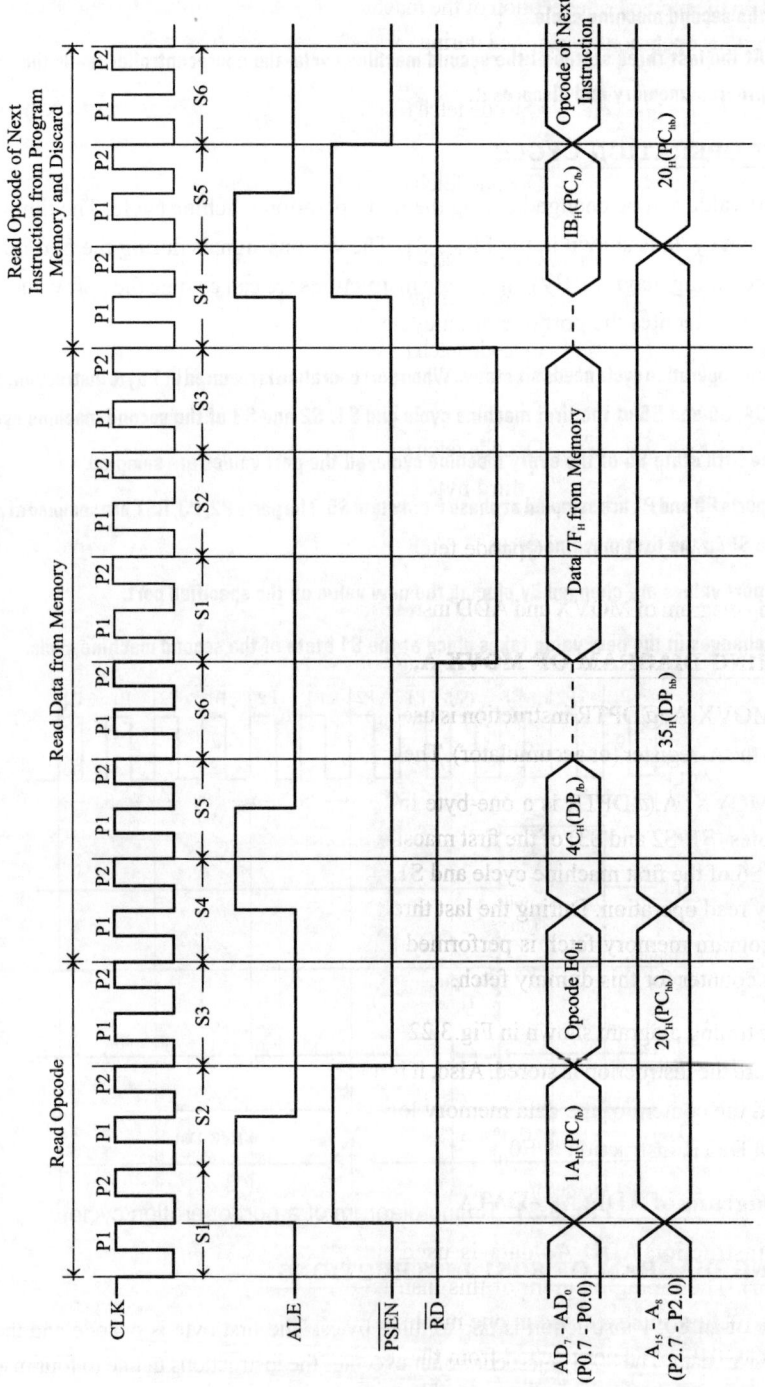

Fig. 3.22: Timing diagram of MOVX A, @DPTR.

Based on the method of execution of the machine cycles, the instructions can be classified as shown below. The various operations performed during execution is also shown below.

Case (i) : *1-byte, 1-cycle* - Opcode fetch (3 states) + Dummy program memory fetch (3 states)

Case (ii) : *2-byte, 1-cycle* - Opcode fetch (3 states) + Fetch second byte of instruction (3 states)

Case (iii) : *1-byte, 2-cycle* - Opcode fetch (3 states) + Data memory read/write (6 states) + dummy program memory fetch (3 states)

Case (iv) : *2-byte, 2-cycle* - Opcode fetch (3 states) + Fetch second byte of instruction (3 states) + Data memory read/write (6 states)

Case (v) : *3-byte, 2-cycle* - Opcode fetch (3 states) + Fetch second byte (3 states) + fetch third byte (3 states) + dummy program memory fetch (3 states)

Case (vi) : *1-byte, 4-cycle* - Opcode fetch (3 states) +Dummy fetches (7 × 3 states)

The timing diagram of MOVX and ADD instructions are shown in Fig. 3.22 and Fig. 3.23 respectively:

3.5.6 TIMING DIAGRAM OF MOVX A,@DPTR

The MOVX A,@DPTR instruction is used to move the content of the data memory addressed by the DPTR to the A-register (or accumulator). The timing diagram of this instruction is shown in Fig. 3.22.

The MOVX A,@DPTR is a one-byte instruction and executed in two machine cycles. In the first three states (S1, S2 and S3) of the first macshine cycle, the opcode is fetched and the next six states (S4, S5 and S6 of the first machine cycle and S1, S2 and S3 of the second machine cycle) are used for data memory read operation. During the last three states (S4, S5 and S6) of the second machine cycle, a dummy program memory fetch is performed and it is discarded. The controller will not increment the program counter for this dummy fetch.

In the timing diagram shown in Fig.3.22 it is assumed that $201A_H$ is the address of the program memory where the instruction is stored. Also, it is assumed that the content of the DPTR (**Data P**ointer) is $354C_H$ and the content of the data memory location with address $354C_H$ be $7F_H$. The opcode of the MOVX A,@ DPTR instruction is $E0_H$.

Timing diagram of ADD A, #DATA

The instruction ADD A,#data is used to add an 8-bit immediate data to the A-register (accumulator). The timing diagram of this instruction is shown in Fig. 3.23. The ADD A,#data is a two-byte instruction and executed in one machine cycle. In the first three states (S1, S2 and S3) of the machine cycle, the opcode is fetched from the program memory. In the next three states (S4, S5 and S6), the immediate data (which is the second byte of instruction) is fetched from the program memory.

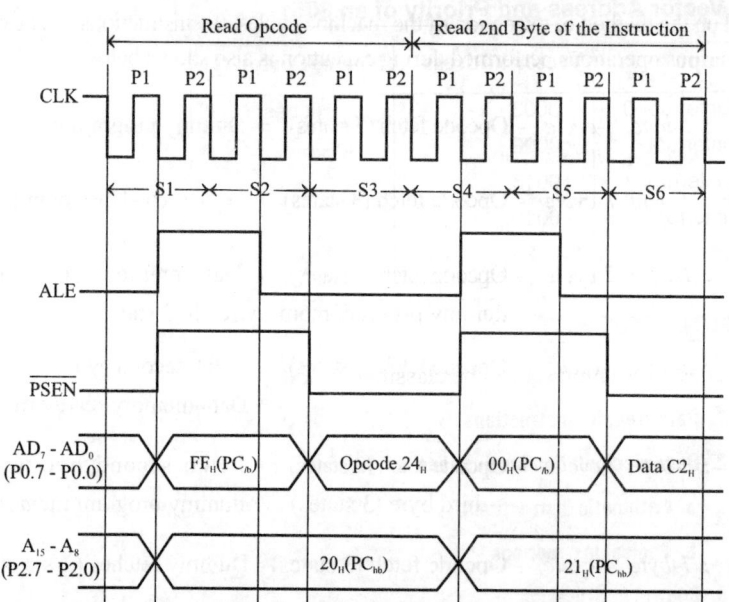

Fig. 3.23: Timing diagram of ADD A, #10$_H$.

In the timing diagram shown in Fig. 3.23, it is assumed that 20FF$_H$ and 2100$_H$ are the address of the program memory where the two bytes of the instruction are stored. Also, it is assumed that the immediate data in the instruction is C2$_H$. The opcode of ADD A,#data instruction is 24$_H$.

3.6   INTERRUPTS IN 8051

The interrupts are signals that are generated to interrupt the normal program execution of the microcontroller, in order to carry out a specific task/work. The specific task/work to be executed by an interrupt will be written as a program and stored in code memory as a subroutine program called the interrupt service subroutine.

The 8051 has five interrupts. In this, two interrupts are external interrupts and the remaining three are internal interrupts. The two external interrupts are interrupts initiated by applying appropriate signals through the pins $\overline{INT0}$ and $\overline{INT1}$, and they are called external interrupt-0 and external interrupt-1 respectively. The internal interrupts are initiated by timer-0, timer-1 and the serial port. All the interrupts of 8051 are maskable and vectored interrupts. The vector address and the priorities of the interrupts of 8051 are listed in Table 3.10. (The priorities of the interrupts can also be altered by programming the IP register.)

Table 3.10: Vector Address and Priority of an 8051

Interrupt	Vector address	Normal priority
External interrupt-0	0003_H	Highest
Timer-0 interrupt	$000B_H$	
External interrupt-1	0013_H	
Timer-1 interrupt	$001B_H$	
Serial port interrupt	0023_H	Lowest

3.7 INSTRUCTION SET

The 8051 instruction set can be classified as follows:

1. **Data transfer instructions**

2. **Data Manipulation instructions**

    a. **Arithmetic instructions**

    b. **Logical instructions**

3. **Control instructions**

    a. **Branching instructions**

    b. **Boolean instructions**

4. **IO instructions**

A brief explanation about each instruction is given in the following section. The list of symbols/abbreviations used in the instruction set are listed below:

Symbols/Abbreviations Used in the Instruction Set

Rn	Register R_7 to R_0 of currently selected register bank	bit	Address of bit-addressable RAM/SFR
direct	8-bit address of internal RAM/SFR	&	Logical AND
A	Accumulator	|	Logical OR
@A+	Memory addressed indirectly through the accumulator	~	Complement/Logical NOT
		^	Logical Exclusive-OR
@Ri	Internal RAM addressed indirectly through R_0 or R1	CF	Carry flag
		B_n	n^{th} bit of register/memory
@DPTR	External data memory addressed indirectly through DPTR	B_{n+1}	$(n + 1)^{th}$ bit of register/memory
		$(A)_{3-0}$	Lower nibble of accumulator
#data	8-bit immediate data/constant	$(A)_{7-4}$	Upper nibble of accumulator
#data16	16-bit immediate data/constant	$(PC)_{7-0}$	Lower byte of program counter
addr11	11-bit address	$(PC)_{15-8}$	Upper byte of program counter
addr16	16-bit address	$(PC)_{10-0}$	Lower 11 bits of program counter
offset	8-bit signed offset value in the range -128_{10} to $+127_{10}$	$(RAM)_{3-0}$	Lower nibble of RAM.

3.8 ADDRESSING MODES

Every instruction of a program has to operate on a data. The method of specifying the data to be operated by the instruction is called addressing. The 8051 has the following types of addressing:

    1. Immediate addressing

    2. Direct addressing

    3. Register addressing

    4. Register indirect addressing

    5. Implied addressing

    6. Relative addressing

Immediate Addressing

In immediate addressing mode, an 8/16-bit immediate data/constant is specified in the instruction itself.

> ### Example:
>
> *MOV A, #6CH*
>
> Move the immediate data $6C_H$ given in the instruction to the A-register. (Accumulator).

> ### Example:
>
> *MOV DPTR, #0100H*
>
> Load the immediate 16-bit constant given in the instruction in the DPTR (Data pointer). This constant will be an address of the data memory location.

Direct Addressing

In direct addressing mode, the address of the data is directly specified in the instruction. The direct address can be the address of an internal data RAM location (00_H to $7F_H$) or the address of a special function register (80_H to FF_H).

> ### Example:
>
> *MOV A,07 H*
>
> The address of the R7-register of bank-0 is 07. This instruction will move the content of the R7-register to the A-register (Accumulator).

Register Addressing

In register addressing mode, the instruction will specify the name of the register in which the data is available.

> ### Example:
>
> *MOV R2,A*
>
> The content of the A-register (accumulator) is moved to register R2 of the currently selected memory bank.

Register Indirect Addressing

In this mode, the instruction specifies the name of the register in which the address of the data is available. The internal data RAM locations (00_H to $7F_H$) can be addressed indirectly through registers R1 and R0. The registers R3 to R7 cannot be used for register indirect addressing. The external RAM can be addressed indirectly through DPTR.

Example:

MOV A,@R0

The internal RAM location R0 holds the address of the data. The content of the RAM location addressed by R0 is moved to the A-register (Accumulator).

Implied Addressing

In implied addressing mode, the instruction itself specifies the data to be operated by the instruction.

Example:

CPL C

Complement carry flag.

Relative Addressing

In relative addressing mode, the instruction specifies the address relative to the program counter. The instruction will carry an offset whose range is -128_{10} to $+ 127_{10}$. The offset is added to the PC to generate the 16-bit physical address.

Example:

JC Offset

If carry is one, the program control jumps to an address obtained by adding the content of the program counter and offset value in the instruction.

3.9 DATA TRANSFER INSTRUCTIONS

The instruction set of the 8051 microcontroller includes a variety of instructions for data transfer between the registers and the memory locations. The various mnemonics used for data transfer instructions are MOV, MOVC, MOVX, PUSH, POP, XCH and XCHD, and they perform any one of the following operations:

1. Copy the content of an SFR to the internal memory or vice versa.

2. Load an immediate operand to the SFR/internal memory.

3. Exchange the content of the SFR/internal memory with the accumulator.

4. Copy the content of the program memory to the accumulator.

5. Copy the content of the data memory to the accumulator or vice versa.

The data transfer instructions of 8051 are listed in Table 3.11 with a brief explanation about each instruction.

Table 3.11: Data Transfer Instructions

S.No.	Instruction	Symbolic representation	Explanation
1.	MOV A,Rn	(A) ← (Rn)	The content of register Rn is moved to the accumulator (A-register). The Rn can be any one of the 8 registers of the currently selected bank.
2.	MOV A,direct	direct = 8-bit address of internal RAM/SFR (A) ← (RAM/SFR)	The content of internal RAM/SFR (whose address is specified directly in the instruction) is moved to the accumulator (A-register).
3.	MOV A,@Ri	(Ri) = Internal RAM address (A) ← (RAM)	The content of internal RAM memory (whose address is specified by the Ri-register is moved to the accumulator (A-register)). The register Ri can be either R0 or R1 or the currently selected register bank.
4.	MOV A,#data	(A) ← data	The data given in the instruction is moved to the accumulator (A-register).
5.	MOV Rn,A	(Rn) ← (A)	The content of the accumulator is moved to register Rn, where Rn is any one of the 8 registers of the currently selected register bank.
6.	MOV Rn,direct	direct = 8-bit address of internal RAM/SFR (Rn)← (RAM/SFR)	The content of internal RAM/SFR (whose address is directly specified in the instruction) is moved to register Rn, where Rn is any one of the 8 registers of the currently selected register bank.
7.	MOV Rn,#data	(Rn) ← data	The immediate data given in the instruction is moved to register Rn, where Rn is any one of the 8 registers of the currently selected register bank.
8.	MOV direct,A	direct = 8-bit address of internal RAM/SFR (RAM/SFR) ← (A)	The content of the accumulator is moved to internal RAM/SFR whose address is directly specified in the instruction.
9.	MOV direct,Rn	direct = 8-bit address of internal RAM/SFR (RAM/SFR) ← (Rn)	The content of register Rn is moved to internal RAM/SFR whose address is directly specified in the instruction.
10.	MOV direct,direct	direct = 8-bit address of internal RAM/SFR (RAM/SFR) ←(RAM/SFR)	The content of one internal RAM/SFR is moved to another internal RAM/SFR. The address of the source and destination are directly specified in the instruction.
11.	MOV direct,@Ri	(Ri) = Internal RAM address of source operand direct = Internal RAM/SFR address of destination operand (RAM/SFR) ← (RAM)	The content of internal RAM whose address is specified by Ri is moved to another internal RAM/SFR whose address is directly specified in the instruction. The register Ri can be either R0 or R1.

Table 3.11: continued...

S.No.	Instruction	Symbolic representation	Explanation
12.	MOV direct,#data	direct = Address of internal RAM/SF (RAM/SFR) ← data	The immediate data given in the instruction is moved to the internal RAM/SFR, whose address is directly specified in the instruction.
13.	MOV @Ri,A	(Ri) = Internal RAM address (RAM) ← (A)	The content of the accumulator is moved to an internal RAM location whose address is specified by the Ri-register. The register Ri can be either R0 or R1.
14.	MOV @Ri,direct	direct = Internal RAM/SFR address of source operand (Ri) = Internal RAM address of destination operand. (RAM) ← (RAM/SFR)	The content of the internal RAM/SFR whose address is directly specified in the instruction is moved to another internal RAM location whose address is specified by the Ri-register. The register Ri can be either R0 or R1.
15.	MOV @Ri,#data	(Ri) = Internal RAM address (RAM) ← data	The immediate data given in the instruction is moved to an internal RAM location, whose address is specified by the Ri-register. The register Ri can be R0 or R1.
16.	MOV DPTR,#data16	(DPTR) ← data16	The 16-bit constant (data16) given in the instruction is moved to the DPTR. (The content of the DPTR is used as address of external data memory in the subsequent instruction.)
17.	MOVC A,@A+DPTR	(A) + (DPTR) = Address of program memory (A) ← (program memory)	This instruction will copy a byte from the code/program memory to the accumulator. The address of the program memory is given by the sum of the content of the DPTR and accumulator before the move operation.
18.	MOVC A,@A+PC	(PC) ← (PC)+1 (A) + (PC) = Address of program memory (A) ← (program memory)	This instruction will copy a byte from the code/program memory to the accumulator. The address of the program memory is given by the sum of the PC and the accumulator. Here, the content of the PC is incremented before adding to A to get the address of the code memory.
19.	MOVX A,@Ri	(Ri) = 8-bit address external data RAM (A) ← (RAM)	The content of external data RAM is moved to the accumulator. The content of register Ri is the 8-bit address of the external memory. The register Ri can be either R0 or R1 of the currently selected register bank.
20.	MOVX A,@DPTR	(DPTR) = 16-bit address of external data RAM (A) ← (RAM)	The content of external data RAM is moved to the accumulator. The content of DPTR is the 16-bit address of the external RAM.

Table 3.11: continued...

S.No.	Instruction	Symbolic representation	Explanation
21.	MOVX @Ri,A	(Ri) = 8-bit address of external data RAM (RAM) ← (A)	The content of the accumulator is moved to the external data RAM. The content of the Ri is the 8-bit address of external RAM. The register Ri can be either R0 or R1 of the currently selected register bank.
22.	MOVX @DPTR,A	(DPTR) = 16-bit address of external data RAM (RAM) ← (A)	The content of the accumulator is moved to the external data RAM. The content of the DPTR is the 16-bit address of the external RAM.
23.	PUSH direct	(SP) ← (SP) + 1 direct = 8-bit address of internal RAM/SFR ((SP)) ← (RAM/SFR)	The stack pointer is incremented by one. The content of the internal RAM/SFR (whose address is directly specified in the instruction) is moved to the internal RAM memory pointed by the SP.
24.	POP direct	direct = 8-bit address of internal RAM/SFR (RAM/SFR) ← ((SP)) (SP) ← (SP) − 1	The content of the internal RAM memory pointed by the SP is moved to the internal RAM/SFR (whose address is directly specified in the instruction). Then the stack pointer is decremented by one.
25.	XCH A,Rn	(A) $\rightleftarrows$ (Rn)	The content of the register Rn is exchanged with the accumulator. The register Rn can be any one of the eight registers of the currently selected register bank.
26.	XCH A,direct	direct = 8-bit address of internal RAM/SFR (A) $\rightleftarrows$ (RAM/SFR)	The content of the internal RAM /SFR whose address is directly specified in the instruction is exchanged with the accumulator.
27.	XCH A,@Ri	(Ri) = 8-bit address of internal RAM (A) $\rightleftarrows$ (RAM)	The content of the internal RAM whose address is specified by the Ri-register is exchanged with the accumulator. The register Ri can be either R0 or R1 of the currently selected register bank.
28.	XCHD A,@Ri	(Ri) = 8-bit address of internal RAM (A)$_{3-0}$ $\rightleftarrows$ (RAM)$_{3-0}$	The lower nibble of the internal RAM addressed by Ri-register is exchanged with the lower nibble of the accumulator. The content of the upper nibble of RAM and accumulator are not altered. The Register Ri can be either R0 or R1 of the currently selected register bank.

3.10 DATA MANIPULATION INSTRUCTIONS

The instructions that manipulate data either by arithmetic or logical instructions can be called data manipulating instructions. These instructions can be further classified into arithmetic instructions and logical instructions.

3.10.1 ARITHMETIC INSTRUCTIONS

The arithmetic group includes instructions for performing addition, subtraction, multiplication, division, increment and decrement operation on the binary data. The mnemonics used in arithmetic instructions are ADD, ADDC, SUBB, INC, DEC, MUL, DIV and DA. The results of most of the arithmetic operations are stored in the accumulator except a few decrement and increment operations. The arithmetic instructions except increment and decrement instructions modify flags of 8051. The arithmetic instructions of 8051 are listed in Table 3.12.

Table 3.12: Arithmetic Instructions

S.No.	Instruction	Symbolic representation	Explanation
29.	ADD A,Rn	$(A) \leftarrow (A) + (Rn)$	The content of the register Rn and the accumulator are added. The result is stored in the accumulator. The register Rn can be any one of the eight registers of the currently selected register bank.
30.	ADD A,direct	direct = 8-bit address of internal RAM/SFR $(A) \leftarrow (A) + (RAM/SFR)$	The content of the internal RAM/SFR and the accumulator are added. The result is stored in the accumulator. The address of internal RAM/SFR is directly specified in the instruction.
31.	ADD A,@Ri	(Ri) = Address of internal RAM $(A) \leftarrow (A) + (RAM)$	The content of the internal RAM and the accumulator are added. The result is stored in the accumulator. The register Ri holds the address of the internal RAM and Ri can be either R0 or R1 of the currently selected register bank.
32.	ADD A,#data	$(A) \leftarrow (A) + data$	The immediate data given in the instruction is added to accumulator.
33.	ADDC A,Rn	$(A) \leftarrow (A) + CF + (Rn)$	This instruction is same as **ADD A, Rn** except that the current value of carry flag (i.e., previous carry) is also added to the sum.
34.	ADDC A,direct	direct = address of internal RAM/SFR $(A) \leftarrow (A) + CF + (RAM/SFR)$	This instruction is same as **ADD A, direct** except that the current value of carry flag (i.e., previous carry) is also added to the sum.
35.	ADDC A,@Ri	(Ri) = address of internal RAM $(A) \leftarrow (A) + CF + (RAM)$	This instruction is same as **ADD A, @Ri** except that the current value of the carry flag (i.e., previous carry) is also added to sum.

Table 3.12: *continued...*

S.No.	Instruction	Symbolic representation	Explanation
36.	ADDC A,#data	(A) ← (A) + CF + data	The immediate data given in the instruction, the carry flag and the content of the accumulator are added. The result is stored in the accumulator.
37.	SUBB A,Rn	(A) ← (A) – CF – (Rn)	The carry flag and the content of the Rn-register are subtracted from the content of the accumulator. The result is stored in the accumulator. The register Rn can be any one of the 8 registers of the currently selected register bank. The 8051 perform 2's complement subtraction and then complement carry.
38.	SUBB A,direct	direct = Address of internal RAM/SFR (A) ← (A) – CF– (RAM/SFR)	The carry flag and the content of RAM/SFR (specified by direct address) are subtracted from the content of the accumulator. The result is stored in the accumulator.
39.	SUBB A,@Ri	(Ri) = Address of internal RAM (A) ← (A) – CF – (RAM)	The carry flag and the content of RAM (specified by the Ri-register) are subtracted from the content of the accumulator. The result is stored in the accumulator. The register Ri can be either R0 or R1 of the currently selected register bank.
40.	SUBB A,#data	(A)← (A) – CF – data	The carry flag and the data given in the instruction are subtracted from the accumulator. The result is stored in the accumulator.
41.	INC A	(A) ← (A) + 1	The content of the accumulator is incremented by one.
42.	INC Rn	(Rn) ← (Rn) +1	The content of the register Rn is incremented by one. The Rn can be any one of the eight registers of the currently selected register bank.
43.	INC direct	direct = Address of internal RAM/SFR (RAM/SFR) ← (RAM/SFR) +1	The content of RAM/SFR (whose address is directly given in the instruction) is incremented by one.
44.	INC @Ri	(Ri) = Address of internal RAM/SFR (RAM) ← (RAM) +1	The content of RAM (whose address is specified by Ri) is incremented by one. The Ri can be either R0 or R1 of the currently selected register bank.
45.	DEC A	(A) ← (A) –1	The content of the accumulator is decremented by one.
46.	DEC Rn	(Rn) ← (Rn) – 1	The content of register Rn is decremented by one. The Rn can be any one of the eight registers of the currently selected register bank.

Table 3.12: continued...

S.No.	Instruction	Symbolic representation	Explanation
47.	DEC direct	direct = Address of internal RAM/SFR (RAM/SFR) ← (RAM/SFR) − 1	The content of RAM/SFR (whose address is directly given in the instruction) is decremented by one.
48.	DEC @Ri	(Ri) = Address of internal RAM (RAM) ← (RAM) −1	The content of RAM (whose address is specified by Ri) is decremented by one. The Ri can be either R0 or R1 of the currently selected register bank.
49.	INC DPTR	(DPTR) ← (DPTR) + 1	The 16-bit content of the DPTR (Data Pointer) is incremented by one.
50.	MUL AB	(B) (A) ← (A) × (B) high low byte byte	The contents of A and B registers are multiplied. The low byte of the product is stored in the A-register and the high byte of the product is stored in the B-register.
51.	DIV AB	(A) ← (A) ÷ (B) Quotient (B) ← (A) MOD (B) Remainder	The content of the A-register is divided by the content of the B-register. The quotient is stored in the A-register and the remainder is stored in B-register.
52.	DAA	i) If $(A)_{3-0}$ > 9 or AF = 1 then $(A)_{3-0}$ ← $(A)_{3-0}$ + 06 ii)If $(A)_{7-4}$ > 9 or CF = 1 then $(A)_{7-4}$ ← $(A)_{7-4}$ + 06	This instruction is executed after the addition of two packed BCD data, to convert the result in the accumulator to the packed BCD data. If the lower nibble of the accumulator is greater than 09 or the AF is set, then it is corrected by adding 06. If the upper nibble of the accumulator is greater than 09 or the CF is set, then it is corrected by adding 06.

3.10.2 LOGICAL INSTRUCTIONS

The logical group includes instructions for performing logical AND, OR, Exclusive-OR, and Complement operations, and instructions for right and left rotation. The mnemonics used in logical operations are ANL, ORL, XRL, CLR, CPL, RL, RLC, RR, RRC and SWAP. The logical operations except rotate through carry do not modify the flags of 8051. In rotate through carry, the carry flag alone is modified. In most of the logical instructions, the result is stored in the accumulator and in some instructions the result is stored in the internal RAM/SFR. The logical instructions of 8051 are listed in Table 3.13 with a brief explanation about each instruction.

Table 3.13: Logical Instructions

S.No.	Instruction	Symbolic representation	Explanation
53.	ANL A,Rn	(A) ← (A) & (Rn)	The content of the register Rn and the accumulator are bit by bit logically ANDed, and the result is stored in the accumulator. The register Rn can be any one of the 8 registers of the currently selected register bank.
54.	ANL A,direct	direct = Address of internal RAM/SFR (A) ← (RAM/SFR) & (A)	The content of the RAM/SFR (whose address is directly given in the instruction) and the accumulator are bit by bit logically ANDed, and the result is stored in the accumulator.
55.	ANL A,@Ri	(Ri) = Address of internal RAM (A) ← (RAM) & (A)	The content of the RAM (whose address is specified by Ri) and the accumulator are bit by bit logically ANDed, and the result is stored in the accumulator. The register Ri can be either R0 or R1 of the currently selected register bank.
56.	ANL A,#data	(A) ← (A) & data	The data given in the instruction and the content of the accumulator are bit by bit logically ANDed, and the result is stored in the accumulator.
57.	ANL direct, A	direct = Address of internal RAM/SFR (RAM/SFR) ← (RAM/SFR) & (A)	The content of the accumulator and the RAM/SFR are bit by bit logically ANDed, and the result is stored in the RAM/SFR. The address of the RAM/SFR is directly specified in the instruction.
58.	ANL direct,#data	direct = Address of internal RAM/SFR (RAM/SFR) ← (RAM/SFR) & data	The data given in the instruction and the content of RAM/SFR are bit by bit logically ANDed, and the result is stored in RAM/SFR. The address of the RAM/SFR is directlyspecified in the instruction.
59.	ORL A,Rn	(A) ← (A)	(Rn)	The content of the register Rn and the accumulator are bit by bit logically ORed, and the result is stored in the accumulator. The register Rn can be any one of the 8 registers of the currently selected register bank.
60.	ORL A,direct	direct = Address of internal RAM/SFR (A) ← (RAM/SFR)	(A)	The content of the RAM/SFR (whose address is directly given in the instruction) and the accumulator are bit by bit logically ORed, and the result is stored in the accumulator.
61.	ORL A,@Ri	(Ri) = Address of internal RAM (A) ← (RAM)	(A)	The contents of the RAM (whose address is specified by Ri) and the accumulator are bit by bit logically ORed, and the result is stored in the accumulator. The register Ri can be either R0 or R1 of the currently selected register bank.

Table 3.13: continued...

S.No.	Instruction	Symbolic representation	Explanation
62.	ORL A,#data	(A) ← (A) | data	The data given in the instruction and the content of the accumulator are bit by bit logically ORed, and the result is stored in the accumulator.
63.	ORL direct,A	direct = Address of internal RAM/SFR (RAM/SFR) ← (RAM/SFR) | (A)	The content of the accumulator and the RAM/SFR are bit by bit logically ORed, and the result is stored in the RAM/SFR. The address of the RAM/SFR is directly specified in the instruction.
64.	ORL direct,#data	direct = Address of internal RAM/SFR (RAM/SFR) ← (RAM/SFR) | data	The data given in the instruction and the content of the RAM/SFR are bit by bit logically ORed, and the result is stored in the RAM/SFR. The address of the RAM/SFR is directly specified in the instruction.
65.	XRL A,Rn	(A) ← (A) ^ (Rn)	The contents of the register Rn and the accumulator are bit by bit logically exclusive-ORed, and the result is stored in the accumulator. The register Rn can be anyone of the 8 registers of the currently selected register bank.
66.	XRL A,direct	direct = Address of internal RAM/SFR (A) ← (RAM/SFR) ^ (A)	The content of the RAM/SFR (whose address is directly given in and instruction) and the accumulator are bit by bit logically exclusive-ORed, and the result is stored in the accumulator.
67.	XRL A,@Ri	(Ri) = Address of internal RAM (A) ← (RAM) ^ (A)	The content of the RAM (whose address is specified by Ri) and the accumulator are bit by bit logically exclusive-ORed, and the result is stored in the accumulator. The register Ri can be either R0 or R1 of the currently selected register bank.
68.	XRL A,#data	(A) ← (A) ^ data	The data given in the instruction and the content of the accumulator are bit by bit logically exclusive-ORed, and the result is stored in the accumulator.
69.	XRL direct,A	direct = Address of internal RAM/SFR (RAM/SFR) ← (RAM/SFR)^ (A)	The content of the accumulator and RAM/SFR are bit by bit logically exclusive-ORed, and the result is stored in the RAM/SFR. The address of the RAM/SFR is directly specified in the instruction.
70.	XRL direct,#data	direct = Address of internal RAM/SFR (RAM/SFR) ← (RAM/SFR)^ data	The data given in the instruction and the content of RAM/SFR are bit by bit logically exclusive-ORed, and the result is stored in the RAM/SFR. The address of the RAM/SFR is directly specified in the instruction.

Table 3.13: continued...

S.No.	Instruction	Symbolic representation	Explanation
71.	CLR A	$(A) \leftarrow 0$	The content of the accumulator is cleared.
72.	CPL A	$(A) \leftarrow \sim (A)$	The content of the accumulator is complemented.
73.	RL A		The content of the accumulator is rotated left by one bit. The most significant digit (B_7) is moved to the least significant digit (B_0) position.
74.	RLC A		The content of the accumulator along with the carry is rotated left by one bit. The carry is moved to the least significant digit position and the most significant digit is moved to the carry.
75.	RR A		The content of the accumulator is rotated right by one bit. The least significant digit (B_0) is moved to the most significant digit (B_7) position.
76.	RRC A		The content of the accumulator along with the carry is rotated right by one bit. The carry is moved to the most significant digit (position) and the least significant digit (B_0) is moved to carry.
77.	SWAP A	$(A)_{3-0} \rightarrow (A)_{7-4}$	The higher nibble of the accumulator is exchanged with the lower nibble of the accumulator.

3.11 CONTROL INSTRUCTIONS

The instructions that changes the proram flow are called control instructions. Normally, program instructions are executed sequencially so that instructions are executed one by one. But when a control instruction is encountered the sequential execution is modified depending on the condition specified by the control instruction. The control instructions can be classified into program branching and boolean instructions.

3.11.1 PROGRAM BRANCHING INSTRUCTIONS

Normally, a program is executed sequentially and the PC (Program Counter) keeps track of the address of the instructions and it is incremented appropriately after each fetch operation. The program branching instructions will modify the content of the PC so that, the program control branches to a new address. The program branching instructions of 8051 includes conditional and unconditional branching instructions. In conditional branching instructions, the content of the PC is modified, only if the condition specified in the instruction is true, whereas in unconditional branching instruction, the PC is always modified. The instructions like ACALL and LCALL will save the previous value of the PC in the stack before modifying the PC. The program branching instructions of 8051 are listed in Table 3.14 with a brief explanation about each instruction.

Table 3.14: Program Branching Instructions

S.No.	Instruction	Symbolic representation	Explanation
78.	ACALL addr11	$(PC) \leftarrow (PC) + 2$ $(SP) \leftarrow (SP) + 1$ $((SP)) \leftarrow (PC)_{7-0}$ $(SP) \leftarrow (SP) + 1$ $((SP)) \leftarrow (PC)15-8$ $(PC)_{10-0} \leftarrow addr11$	This instruction is used to unconditionally call a subroutine which resides within the same 2k block of the program memory in which the instruction following ACALL is stored. This instruction first increments the PC by two, to point to the address of the instruction next to ACALL. Next, the content of the SP is incremented by one and the low byte of PC is saved in the stack memory pointed by SP. Again, content of the SP is incremented by one and then, high byte of PC is saved in the stack memory pointed by the SP. Then, the 11-bit address given in the instruction is moved to the lower 11-bit position of the PC. (The 11- bit address is the second byte of the instruction and upper 3 bits of the opcode.) Now, the controller starts fetching the instructions from this new address.
79.	LCALL addr16	$(PC) \leftarrow (PC) + 3$ $(SP) \leftarrow (SP) + 1$ $((SP)) \leftarrow (PC)_{7-0}$ $(SP) \leftarrow (SP) + 1$ $((SP)) \leftarrow (PC)_{15-8}$ $(PC) \leftarrow addr16$	This instruction is used to unconditionally call a subroutine anywhere in the 64 k space. First, the PC is incremented by three to point to the next instruction. Next, the SP is incremented and the content of the PC is saved in the stack memory pointed by the SP. Then, the SP is incremented and the content of the PC is saved in the stack memory pointed by the SP. Then, the 16-bit address given in the instruction is moved to the PC and so the controller fetching the instruction from this new address. starts
80.	RET	$(PC)_{15-8} \leftarrow ((SP))$ $(SP) \leftarrow (SP) - 1$ $(PC)_{7-0} \leftarrow ((SP))$ $(SP) \leftarrow (SP) - 1$	This instruction is used to terminate a subroutine. On execution of this instruction, the content of the stack memory pointed by the SP is moved to the high byte of PC and SP is decremented by one. Then, the content of the stack memory pointed by SP is moved to the low byte of PC and again the SP is decremented by one.
81.	RETI	$(PC)_{15-8} \leftarrow ((SP))$ $(SP) \leftarrow (SP) - 1$ $(PC)_{7-0} \leftarrow ((SP))$ $(SP) \leftarrow (SP) - 1$	This RETI instruction is used to terminate an interrupt service subroutine. This instruction moves the top of the stack to the PC similar to that of RET instruction and in addition restores the interrupt logic to accept additional interrupts of the same priority level as the one just processed.

Table 3.14: continued....

S.No	Instruction	Symbolic representation	Explanation
82.	AJMP addr11	(PC) ← (PC) + 2 (PC)$_{10-0}$ ← addr11	This instruction is used to unconditionally jump to a memory location within the same 2k block of program memory in which the instruction following AJMP is stored. First, the PC is incremented by two to point to the address of next instruction and then the 11-bit address given in the instruction is moved to the lower 11-bit position of the PC. The 11-bit address is the second byte of the instruction and upper 3 bits of opcode
83.	LJMP addr16	(PC) ← addr16	This instruction is used to unconditionally jump to any location in the 64 k memory space. Upon execution of this instruction, the 16-bit address given in the instruction is moved to the PC, and so the controller starts fetching the instruction from this new address.
84.	SJMP offset	(PC) ← (PC) + 2 (PC) ← (PC) + offset	This instruction is used to unconditionally transfer the program control to a new address obtained by adding the 8-bit signed offset to the content of the PC. The offset will be in the range of -128_{10} to $+127_{10}$.
85.	JMP @A+DPTR	(A) + (DPTR) = Address (PC) ← Address	This instruction computes the address to which the program control has to be transferred and loads this address in the PC. The address is given by the sum of the signed 8-bit in the accumulator and the 16-bit content of the DPTR.
86.	JZ offset	(PC) ← (PC) + 2 If (A) = 0 then (PC) ← (PC) + offset	First, the content of the PC is incremented by two. Next, the content of accumulator is checked. If the content of the accumulator is zero, then the 8-bit signed offset given in the instruction is added to the PC, so that the program control branches to new address. If the accumulator is not zero then PC is not modified, so that the next instruction of the program is fetched and executed.
87.	JNZ offset	(PC) ← (PC) + 2 If (A) ≠ 0 then (PC) ← (PC) + offset	First, the content of the PC is incremented by two. Next, the content of the accumulator is checked. If the content of the accumulator is not zero, then the 8-bit signed offset given in the instruction is added to the PC, so that the program control branches to a new address. If the accumulator is zero, then PC is not modified so that the next instruction of the program is fetched and executed.

Table 3.14: continued...

S.No.	Instruction	Symbolic representation	Explanation

88.	CJNE A,direct,offset	(PC) ← (PC) + 3
		direct = Address of internal RAM/SFR
		If (A) ≠ (RAM/SFR) then
		(PC) ← (PC) + offset
		If (A) < (RAM/SFR) then, CF ← 1
		If (A) > (RAM/SFR) then, CF ← 0
			First, the PC is incremented by three to point to the next instruction. The content of the accumulator and the internal RAM/SFR (whose address is directly specified in the instruction) are compared. If the contents are not equal then the program control is transferred to a new address. The new address is the sum of the PC and offset given in the instruction. Also, if the content of the accumulator is less than RAM/SFR, then the carry flag is set, otherwise it is cleared.

89.	CJNE A,#data,offset	(PC) ← (PC) + 3
		If (A) ≠ data then
		(PC) ← (PC) + offset
		If (A) < data then, CF ← 1
		If (A) > data then, CF ← 0
			This instruction is same as **CJNE A,direct,offset** except that the comparison is performed with the immediate data given in the instruction and the accumulator.

90.	CJNE Rn,#data,offset	(PC) ← (PC) + 3
		If (Rn) ≠ data then
		(PC) ← (PC) + offset
		If (Rn) < data then, CF ← 1
		If (Rn) > data then, CF ← 0
			This instruction is same as **CJNE A,direct,offset** except that the comparison is performed with the content of the Rn and the immediate data. The Rn can be any one of the eight registers of the currently selected register bank.

91.	CJNE @Ri,#data,offset	(PC) ← (PC) + 3
		(Ri) = Address of internal RAM
		If (RAM) ≠ data then
		(PC) ← (PC) + offset
		If (Ri) < data then, CF ← 1
		If (Ri) > data then, CF ← 0
			This instruction is same as **CJNE A,direct,offset** except that the comparison is performed between the RAM (whose address is specified by Ri) and the immediate data given in the instruction. The register Ri can be either R0 or R1 of the currently selected register bank

92.	DJNZ Rn,offset *[AU May'15, 2 marks]*	(PC) ← (PC) + 2 (Rn) ← (Rn) − 1 If (Rn) ≠ 0 then (PC) ← (PC) + offset	First, the PC is incremented by two to point to the address of the next instruction. Then, the content of register Rn is decremented by one. If the content of Rn (after decrement) is not equal to zero, the the offset given in the instruction is added to the PC so that, the program control branches to a new address. The register Rn can be any one of the eight registers of the currently selected register bank

Table 3.14: continued...

S.No.	Instruction	Symbolic representation	Explanation
93.	DJNZ direct,offset	(PC) ← (PC) + 2 direct = Address of RAM/SFR (RAM/SFR) ← (RAM/SFR) −1 If (RAM/SFR) ≠ 0 then (PC) ← (PC) + offset	This instruction is same as **DJNZ Rn, offset** except that the content of the RAM/SFR is decremented and compared. The address of the RAM/SFR is directly specified in the instruction.
94.	NOP	(PC) ← (PC) + 1	This instruction will not perform any operation, except that the PC is incremented by one to point to the next instruction. Execution of NOP will produce a delay of one machine cycle time and so, this instruction can be used to create small delays in multiples of machine cycle time.

3.11.2 BOOLEAN INSTRUCTIONS

The boolean instructions operate on a particular bit of a data. This group includes instructions which clear, complement or move a particular bit of bit-addressable RAM/SFR or carry flag. It also include jump instructions, which transfers the program control to a new address, if a particular bit is set or cleared. The boolean instructions of the 8051 are listed in Table 3.15 with a brief explanation about each instruction.

Table 3.15: Boolean Instructions

S.No.	Instruction	Symbolic representation	Explanation
95.	CLR C	CF ← 0	Clear carry flag.
96.	CLR bit	bit = Address of particular bit of RAM/SFR (bit) ← 0	The particular bit of RAM/SFR whose address is specified in the instruction is cleared to zero.
97.	SETB C	CF ← 1	The carry flag is set to one.

Table 3.15: continued ....

S.No.	Instruction	Symbolic representation	Explanation
98.	SETB bit	bit = Address of particular bit of RAM/SFR (bit) ← 1	The particular bit of RAM/SFR whose address is specified in the instruction is set to one.
99.	CPL C	CF ← ~ CF	The carry flag is complemented.
100.	CPL bit	bit = Address of particular bit of RAM/SFR (bit) ← ~ (bit)	The particular bit of RAM/SFR whose address is specified in the instruction is complemented.
101.	ANL C,bit	bit = Address of particular bit of RAM/SFR CF ← CF & (bit)	The particular bit of RAM/SFR is logically ANDed with the carry flag and the result is stored in the carry flag.
102.	ANL C,/bit	bit = Address of particular bit of RAM/SFR (bit) ← ~ (bit) CF ← CF & (bit)	The complement of the particular bit of RAM/SFR is logically ANDed with the carry flag and the result is stored in the carry flag.
103.	ORL C,bit	bit = Address of particular bit of RAM/SFR CF ← CF | (bit)	The particular bit of RAM/SFR is logically ORed with the carry flag and the result is stored in the carry flag.
104.	ORL C,/bit	bit = Address of particular bit of RAM/SFR (bit) ← ~ (bit) CF ← CF | (bit)	The complement of the particular bit of RAM/SFR is logically ORed with the carry flag and the result is stored in the carry flag.
105.	MOV C,bit	bit = Address of particular bit of RAM/SFR CF ← (bit)	The particular bit of RAM/SFR is moved to the carry flag. The address of the bit is directly given in the instruction.
106.	MOV bit,C	bit = Address of particular bit of RAM/SFR (bit) ← CF	The carry flag is moved to a particular bit of RAM/SFR whose address is directly specified in the instruction.

Table 3.15: continued...

S.No.	Instruction	Symbolic representation	Explanation
107.	JC offset	(PC) ← (PC) + 2 If CF = 1 then, (PC) ← (PC) + offset	The content of the PC is incremented by two to point to the next instruction. Then the carry flag is checked. If the carry flag is one then the offset given in the instruction is added to the PC so that the program control is transferred to a new address.
108.	JNC offset	(PC) ← (PC) + 2 If CF = 0 then, (PC) ← (PC) + offset	Same as JC offset, except that the branching will take place if the carry flag is zero.
109.	JB bit, offset	bit = Address of particular bit of RAM/SFR (PC) ← (PC) + 3 If If (bit¹ = 1, then (PC) ← (PC) + offset	The content of the PC is incremented by three to point to the next instruction. The particular bit of RAM/SFR is tested. If the bit is one, then the offset given in the instruction is added to PC so that the program control is transferred to a new address.
110.	JNB bit, offset	bit = Address of particular bit of RAM/SFR (PC) ← (PC) + 3 If (bit) = 0, then (PC) ← (PC) + offset	The content of the PC is incremented by three to point to the next instruction. The particular bit of RAM/SFR is tested. If the bit is zero then the offset given in the instruction is added to the PC so that the program control is transferred to a new address.
111.	JBC bit, offset	bit = Address of particular bit of RAM/SFR (PC) ← (PC) + 3 If (bit) = 1, then bit ← 0 ; (PC) ← (PC) +offset	The content of the PC is incremented by three to point to the next instruction. The particular bit of RAM/SFR is tested. If the bit is one, then clear the bit and the offset given in the instruction is added to the PC, so that the program control is transferred to a new address.

3.12 IO INSTRUCTIONS

The IO ports of 8051 are mapped as RAM memory and so that content of port can be accessed by direct addressing. Therefore the instructions that employs direct addressing can be called IO instructions.

Example: Mov A, diorect; Mov cirect, A; Add A, direct; DJNZ direcr,offiset; CJNE A, direct,offset; ORL direct,#data.

3.13 COMPARISON OF 8085 AND 8051 ASSEMBLY LANGUAGE PROGRAMMING

The 8085 has **von Neumann architecture** in which there will be a single memory bank with a common address space for program and data. The system designer has to divide or allot the space for program and data depending on the need. The 8051 has **Harvard architecture** which support two different memory banks, one for program and another for data, with independent or separate address space. The 8051 microcontrollers have more advanced features than 8085 microprocessor and some of the features are listed here.

1. Program codes are stored in program address space and accessed using program address pointer called program counter (PC).

2. Data can be stored in either in data memory space or program memory space. When data is stored in external data memory space it is accessed using data memory pointers R0, R1 and DPTR. When data is stored in internal memory space it can be accessed by direct addressing or using data memory pointer DPTR. When data is stored in program memory space it can be accessed using program memory pointer PC.

3. Jump address can be 3 types: 16-bit address, 11-bit address, 8-bit signed offset address

4. Subtraction is always with previous borrow and so carry should be set or cleared depending upon program logic before executing the subtract instruction.

5. In 8051 combined decrement, compare and jump instructions using zero flag are available which makes programming task easier.

6. 8051 has number of bit manipulating instructions. So that individual bits of data can modified more efficiently and also can be used for decision making.

7. In 8051, 8-bit multiplication can be performed using single instruction MUL. Similarly, 8-bit division can be performed using single instruction DIV.

8. Microcontrollers are designed to run continuously and so program execution cannot be halted by any instruction. Alternatively, program execution can be made to remain in an idle loop or simple loop without doing any work, until reset or interrupt.

3.14 SHORT-ANSWER QUESTIONS

Q3.1 What is a microcontroller?

A microcontroller is a programmable semiconductor device available as IC and capable of performing arithmetic and logical operations.

Q3.2 What are the basic units of a microcontroller?

The basic units of a microcontroller are the ALU, a set of registers, IO ports, memory, timing and control unit. In addition some of the controllers may have timers, ADC and DAC.

Q3.3   List the features of an 8051 microcontroller.

The features of an 8051 microcontroller are,

1. 8-bit controller operating on bit and byte operand

2. Provides separate code and data memory address space

3. 256 bytes internal RAM and 4 kB internal ROM

4. 64/60 kB external program memory address space

5. 64 kB external data memory address space

6. Four numbers of 8-bit parallel ports

7. One number of programmable serial port

8. Two numbers of programmable timers

9. Five members of vectored interrupts

Q3.4   List the alternate functions of the ports of an 8051 microcontroller.

Port pins	Alternate signal	Description
P0.7 - P0.0	$AD_7 - AD_0$	Multiplexed low byte address/data.
P2.7 - P2.0	$A_{15} - A_8$	High byte address
P3.7	RD	External memory read control signal
P3.6	WR	External memory write control signal
P3.5	T1	External input to timer 1
P3.4	T0	External input to timer 0
P3.3	INT1	External interrupt 1
P3.2	INT0	External interrupt 0
P3.1	TxD	Serial data output
P3.0	RxD	Serial data input

Q3.5   List the interrupts of an 8051 microcontroller.

The 8051 microcontroller has five interrupts and they are (in the order of higher to lower priority) External interrupt-0, Timer-0 interrupt, External interrupt-1, Timer-1 interrupt and Serial port interrupt.

Q3.6   What are dedicated address pointers in an 8051?

The 8051 has two dedicated address pointers: Program Counter (PC) and Data Pointer (DPTR). The PC is used as an address pointer for programs and DPTR is used as an address pointer for data.

Q3.7   What are SFRs?

SFRs (Special Function Registers) are internal registers of a microcontroller dedicated for specific functions. These registers can be used only for their specified/defined functions and cannot be used for any other function. In microcontrollers, the SFRs are mapped as internal data memory and can be accessed by direct addressing.

Q3.8   What are register banks in an 8051?

The register banks are internal RAM locations of 8051 which can be used as general-purpose registers or scratch pad registers. The first 32 bytes of internal RAM of 8051 are organized as four register banks with each bank consisting of eight locations. At any one time, the processor can work with only one register bank depending on the value of bits RS0 and RS1 in the PSW register.

Q3.9   What is PSW in an 8051?

The flag register of an 8051 is called PSW (**P**rogram **S**tatus **W**ord). The PSW consists of four math flags, two register bank select bits and two user flags. The math flags are carry, auxiliary carry, overflow and parity flag. The register bank select bits are RS0 and RS1. The user flags are F0 and PSW1.

Q3.10 How is stack implemented in an 8051?

The 8051 supports a LIFO (**L**ast-**I**n-**F**irst-**O**ut) stack and the stack can reside anywhere in the internal RAM. The 8051 has an 8-bit **S**tack **P**ointer (SP) to indicate the top of stack. The stack can be accessed using PUSH and POP instructions. During PUSH, the SP is automatically incremented by one and during POP, the SP is automatically decremented by one.

Q3.11 What are the operating modes of the serial port of an 8051?

The operating modes of the serial port of an 8051 are mode-0, mode-1, mode-2 and mode-3. In mode-0, the serial port functions as a half-duplex serial port at fixed baud rate and one data character is framed as 8 bits. In modes 1, 2 and 3, the serial port can function as full-duplex serial port. In modes 1 and 3, the baud rate is variable and in mode-1, the baud rate is either 1/32 or 1/64 of oscillator frequency. In mode-1, one data character should be framed as 10 bits, and in modes 2 and 3, one data character is framed as 11 bits.

Q3.12 How is the baud rate decided in modes 1 and 3 of serial transmission in an 8051?

In serial transmission modes 1 and 3 of 8051, the baud rate depends on the SMOD bit of PCON register and the timer-1 overflow rate as shown below:

$$\text{The baud rate in mode 1 or 3} = \frac{2^{\text{SMOD}}}{32} \times (\text{Timer-1 overflow rate})$$
(in modes 1 and 3)

Q3.13 What are the operating modes of the timer of an 8051?

The operating modes of the timers of an 8051 are mode-0, mode-1, mode-2 and mode-3. In mode-0 the timers will function as 13-bit timers and in mode-1, the timers will function 16-bit timers. In mode-2, the timers will function as 8-bit timers with auto reload feature. Timer-0 alone can work in mode-3 and in this mode the TL0 will function as an 8-bit timer controller by standard timer-0 control bits and TH0 will function as 8-bit timer controlled by timer-1 control bits.

Q3.14 What is state in an 8051 microcontroller?

The state is the basic time unit for discrete operation of the controller such as fetching an opcode, executing an opcode, writing a data, etc. A machine cycle consists of six states and the timing of each state is 2 oscillator clock periods.

Q3.15 How can the time taken to execute an instruction be estimated in an 8051 controller?

The time taken to execute an instruction by an 8051 controller is obtained by multiplying the time to execute a machine cycle by the number of machine cycles of the instruction. The time to execute a machine cycle is 12 clock periods.

$\therefore$ Time to execute an instruction $= C \times 12 \times T = C \times 12 \times \dfrac{1}{f}$

where, C = Number of machine cycles of an instruction.

T = Time period of crystal frequency in seconds.

f = Crystal frequency in Hz.

Q3.16 What is the size of 8051 instructions?

The size of 8051 instructions is one to three bytes. The first byte is an opcode and the subsequent bytes are the address or data.

Q3.17 How does an 8051 microcontroller differentiate between external program memory access and data memory access?

With external program memory, the controller can perform only read operations but with external data memory, the controller can perform both read and write operations. For reading program memory, the controller asserts $\overline{\text{PSEN}}$ as **low**, for reading data memory the controller asserts $\overline{\text{RD}}$ as **low**, and for writing data memory the controller asserts $\overline{\text{WR}}$ as **low**.

Q3.18 What are the addressing modes available in an 8051 controller?

The addressing modes available in the 8051 microcontroller are

1.	Immediate addressing	4.	Register indirect addressing
2.	Direct addressing	5.	Implied addressing
3.	Register addressing	6.	Relative addressing

Q3.19 Explain register indirect addressing in an 8051.

In register indirect addressing, the instruction specifies the name of the register in which the address of the data is available. The internal data RAM locations can be addressed indirectly through registers R1 and R0. The external RAM can be addressed indirectly through the DPTR (Data pointer).

Example : **MOV A,@R0** - The content of the RAM location addressed by the R0 is moved to the A-register.

Q3.20 Explain relative addressing in an 8051.

In relative addressing mode, the instruction specifies the address relative to the Program Counter (PC). The instruction will carry an offset whose range is -128_{10} to $+127_{10}$. The offset is added to the PC to generate the 16-bit physical address.

Example : **JC offset** - If carry is one, then the program control jumps to an address obtained by adding the content of the PC and the offset value in the instruction.

Q3.21 How can the 8051 instructions be classified ?

The 8051 instructions can be classified into the following five groups:

1. Data transfer instructions
2. Arithmetic instructions
3. Logical instructions
4. Branching instructions
5. Boolean instructions

Q3.22 List the instructions of 8051 that affect all the flags of 8051.

The 8051 instructions that affect all the flags are ADD, ADDC, and SUBB.

Q3.23 List the instructions of 8051 that affect the overflow flag in 8051.

The 8051 instructions that affect the overflow flag are ADD, ADDC, DIV, MUL and SUBB.

Q3.24 List the instructions of 8051 that affect only the carry flag.

The 8051 instructions that affect only the carry flag are,

ANL C,bit	CPL C	RRC A
ANL C,/bit	MOV C,bit	RLC A
CJNE	ORL C,bit	SETB C
CLR C	ORL C,/bit	

Q3.25 List the instructions of 8051 that always clear the carry flag.

The instructions that always clear the carry flag are CLR C, DIV and MUL.

Q3.26 What are the operations performed by the Boolean variable instructions of an 8051?

The Boolean variable instructions can clear or complement or move a particular bit of bit-addressable RAM/SFR or carry flag. They can also transfer the program control to a new address if a particular bit is set or cleared.

Q3.27 Can single bit of a port be accessed in 8051? If yes, how? Give an example. *(AU, Nov/Dec' 19, 2 Marks)*

The single bit of ports can be accessed by using the syntex PX.Y in the operand field of bit manipulating instruction, where X can take values 0 to 3 (port number) and Y can take values 0 to 7 (bit number).

Example

MOV C,bit   ;The value of bit is moved to carry flag.

ANL C,bit   ;The value of bit and carry flag are logically ANDed and result is
                   ;stored in carry flag.

3.15 EXERCISES

I. Fill in the blanks with appropriate words

1. One machine cycle of 8051 consists of _____ number of clock pulses.

2. _____ machine cycle of 8051 is executed to load the data from external data memory into the internal register.

3. The 8051 instruction MOVX A, @ DPTR is executed in _____ number of machine cycles.

4. The _____ addressing mode is used in the 8051 instruction ADD R1, 20H.

5. The 8051 instruction that copies the data of external RAM pointed by R_1 register into A register is _____.

6. The _____ instruction of 8051 is used to exchange the lower nibbles of two operands.

7. The borrow which results from subtraction is always reflected on _____ flag.

8. The 8051 instruction SUB A, R1 is _____ instruction.

9. The quotient and remainder are stored in _____ and _____ registers respectively after executing the 8051 instruction DIV AB.

10. The _____ logical instruction of 8051 is used to mask the bits.

11. The instructions that change the sequence of execution are called _____ instructions.

12. The _____ instruction of 8051 is executed to jump to a location within the same page of 2K memory space unconditionally.

13. The _____ instruction of 8051 is used to clear the 0th bit of port 1.

14. The JNB C, offset instructioin of 8051 is equivalent to _____.

15. The result of 8051 instruction RL A, if A = 01_H is _____.

16. The 8051 instruction CJNE stands for _____.

17. The _____ architecture uses separate address for code and data.

18. The size of the two external memory banks of 8051 is _____.

19. The interrupt vector address of second highest priority interrupt of 8051 is _____.

20. The _____ and _____ registers of 8051 are used to access the program memory and external data memory respectively.

21. The special function registers(SFRs) of 8051 include _____ number of internal registers.

22. The _____ and _____ bits of PSW registers are used to select the register bank of 8051.

23. The address range of 16 RAM locations of internal 128 byte RAM of 8051 are from _____ to _____.

24. After the reset, the stack pointer of 8051 is initialized to the address _____.

25. The _____ bit of SCON register is used to enable multiprocessor communication using 8051 controller.

26. The _____ bit of TCON register is used to start/stop the timer/counter in 8051.

27. The _____ bit of Interrupt Enable (IE) register of 8051 is called as global enable.

28. The size of the 8051 external program memory when $\overline{EA} = 1$ is _____ bytes and mapped in the address range _____ to _____.

Answers

1. 12	8. invalid	15. 02_H	22. RS_0, RS_1
2. data memory read	9. A, B	16. compare and jump if not equal	23. 20_H, $2F_H$
3. two	10. ANL	17. Harvard	24. 07_H
4. direct	11. control transfer	18. 64 KB	25. SM2
5. MOV A, @R1	12. AJMP	19. $000B_H$	26. TR
6. XCHD	13. CLR P1.0	20. Program Counter(PC) and Data Pointer(DPTR)	27. EA
7. carry	14. JNC offset	21. 21	28. 60k, 1000_H, $FFFF_H$

II. State whether the following statements are True/False.

1. One state of 8051 consists of two clock pulses.

2. The 8051 microcontroller utilizes all the six states to fetch the opcode from code memory.

3. The 8051 special function registers can be accessed using direct addressing mode.

4. MOV A, 07H instruction of 8051 uses immediate addressing mode.

5. Data transfer between two internal RAM/SFR is not possible in 8051.

6. External data memory of 8051 controller can be read/written only through A register.

7. Internal RAM memory of 8051 can be pointed only by R_0 and R_1 register.

8. The 8051 instructions INC and DEC do not affect carry flag.

9. The destination register is always the A register in 8051 addition and subtraction.

10. The 8051 instruction CPL A changes the magnitude of A register.

11. The execution of 8051 instruction ANL A does not affect carry flag.

12. The 8051 always consider previous carry (borrow) while performing subtraction.

13. The rotate instructions of 8051 are used to transmit data serially.

14. The program counter(PC) of 8051 is always modified by unconditional jump instruction.

15. The 8051 instruction ACALL is a three byte instruction.

16. All the 8051 conditional jumps are short jumps.

17. The 8051 instruction "JNB bit, offset " jumps to the specified relative address if specified bit is 1.

18. Microcontrollers are general purpose systems.

19. Microcontrollers can have on-chip timers.

20. The 8031 has 128 byte internal RAM and 4 kB internal ROM.

21. The port-1 of 8051 is a dedicated I/O port.

22. The 8051 family of microcontrollers are built with Harvard architecture.

23. The 8051 family microcontrollers come with in system programmers.

24. The internal clock frequency of 8051 is same as crystal frequency.

25. All the interrupts of 8051 are maskable and vectored interrupts.

26. The 8051 special function registers(SFRs) whose address ends with 0_H (or) 8_H are bit addressable.

27. The Data Pointer (DPTR) register of 8051 is automatically incremented.

28. All SFR's of 8051 are bit addressable.

29. After the reset, the 8051 always works with register bank-0.

30. The stack memory of 8051 is always placed in external data memory.

31. The 8051 timer is an up-counter.

32. In 8051 the TF flag of TCON register is reset to zero whenver the timer/counter overflows.

33. The priorities of 8051 interrupts are fixed.

34. The internal 4 kB EPROM of 8051 is not used when $\overline{EA} = 0$.

35. It is not possible to have a single memory bank for both data and program in 8051.

Answers

1. True	7. True	13. True	19. True	25. True	31. True
2. False	8. True	14. True	20. False	26. True	32. False
3. True	9. True	15. False	21. True	27. False	33. False
4. False	10. False	16. True	22. True	28. False	34. True
5. False	11. True	17. False	23. False	29. True	35. False
6. True	12. True	18. False	24. True	30. False	

III. Choose the right answer for the following questions.

1. *The time required to execute one machine cycle of 8051 is*

    a) 6 states   b) 12 clock pulses   c) a or b   d) none

2. *Which of the following machine cycles are executed by 8051 to execute the instruction CLR A?*

    a) External program memory fetch cycle

    b) External data memory read cycle

    c) External data memory write cycle

    d) Port operation cycle

3. *Identify the 8051 instruction which uses immediate addressing mode*

    a) ADD A, 20H   b) MOV R1, #20H   c) both a and b   d) none

4. *Identify the invalid 8051 instruction*

    a) MOV A, #75H   b) ADD A, 80H   c) MOV A, R1   d) MOV A, @R4

5. *This 8051 instruction copies the data from code memory into accumulator*

    a) MOVX A, @PC   b) MOVX A, @A+DPTR   c) MOVC A, @PC   d) MOVC A, @A+PC

6. *The 8051 instruction used to perform multibyte addition is*

    a) ADD   b) ADDC   c) both a and b   d) neither a nor b

7. *Which of the following 8051 instruction is used to multiply the contents of internal RAM E0$_H$ and F0$_H$?*

    a) MUL F0H   b) MUL E0H   c) MUL AB   d) MUL B

8. *What is the status of CY and overflow flag after executing the 8051 instruction SUBB A, 09H? Assume A = 12$_H$ and CY = 0 before execution of instruction.*

    a) CY = 0 ; OF = 0   b) CY = 0 ; OF = 1   c) CY = 1 ; OF = 0   d) CY = 1 ; OF = 1

9. *Identify the invalid 8051 instruction.*

    a) ADD A, 12H   b) ADD A, #12H   c) ADD A, R1   d) ADD R1, A

10. The _____ instruction of 8051 is used to set the least significant bit of the register 'A'

   a) ANL A, #01H   *b)* ANL A, 01H   *c)* ORL A, 01H   *d)* ORL A, #01H

11. Which of the following 8051 instruction is equivalent to MOV A, #00H?

   a) CPL A   *b)* CLR A   *c)* both a and b   *d)* neither a nor b

12. The content of A register is 95_H. What is the status of A after executing the 8051 instruction SWAP A?

   a) 90H   *b)* 05H   *c)* 00H   *d)* 59H.

13. Which of the following is not a valid jump instruction of 8051?

   a) JMP 2000H   *b)* JNZ 12H   *c)* JC 1000H   *d)* JNC F2H

14. The _____ instruction of 8051 is used to terminate an interrupt service subroutine

   a) RET   *b)* RETI   *c)* either a or b   *d)* neither a nor b

15. Which of the following instruction is used to set the carry flag when the port-1 pin P1.4 is 0? Assume carry flag is intially reset.

   a) ANL C, P1.4   *b)* ANL C, /P1.4   *c)* ORL C, P1.4   *d)* ORL C, /P1.4

16. What is the status of CY flag after execution of the following 8051 codes

   MOV A, #82H

   CJNE A, #20H, NEXT

   a) CY = 0   *b)* CY = 1   *c)* carry does not change   *d)* CY = $\overline{CY}$

17. Assume CY = 1 and P1.2 = 0. What is the status of CY flag after the execution of the 8051 instruction ANL C, / P1.2?

   a) CY = 0   *b)* CY = 1   *c)* CY = $\overline{CY}$ *d)* CY is not altered

18. In 8051 short jump instruction, if MSB of relative address = 1 then it is

   a) forward jump   *b)* backward jump   *c)* a or b   *d)* neither a nor b

19. Which of the following is called as ROM less version of 8051?

   a) 8031   *b)* 8032   *c)* 8751   *d)* 8752

20. Which of the following pin of 8051 is externally AND with $\overline{RD}$ pin, to produce read control signal when there is only one external memory?

   a) ALE   *b)* $\overline{EA}$ *c)* $\overline{PSEN}$ *d)* RxD

21. The following 8051 pin is used as programming pulse for programming 8051

   a) $\overline{EA}$ *b)* ALE   *c)* $\overline{WR}$ *d)* TxD

22. Which of the following interrupts of 8051 has been assigned the lowest priority?

   a) Timer-0 interrupt *b)* Timer-1 interrupt c) External interrupt-0 *d)* Serial port interrrupt

23. Which of the following condition is required to store a 8051 program of size 15 kB in external code memory without ignoring internal 4 kB ROM?

   a) $\overline{EA}$ = 1   *b)* $\overline{EA}$ = 0   *c)* $\overline{PSEN}$ = 1   *d)* $\overline{PSEN}$ = 0

24. Which of the following data memory address space are alloted for 8051 SFR's?

a) 00_H to $7F_H$ b) 80_H to FF_H c) 00_H to FF_H d) 20_H to $2F_H$

25. The following register of 8051 is used to select the operating modes of timer/counter.

a) TCON   b) TMOD   c) PCON   d) SCON

26. Which of the following mode of serial port function with fixed baud rate in 8051 controller?

a) Mode-0   b) Mode-1   c) Mode-2   d) Mode-3

27. The mode-2 of timer/counter in 8051 is also called

a) 13-bit timer/counter   b) 16-bit timer/counter

c) autoreload 8-bit timer/counter   d) autoreload 16-bit timer/counter

28. An external pulse is applied to $\overline{INT}$ pin of 8051. What should be the content of TMOD register to count the pulse using timer/counter-1 in mode-0?

a) 00_H b) $C0_H$ c) 80_H d) $0C_H$

29. What should be the content of Interrupt Enabler (IE) register to enable the serial port interrupt in 8051 controller?

a) 80_H b) 81_H c) 01_H d) 82_H

30. Which of the following is internal interrupt of 8051?

a) Timer-0 interrupt b) Timer-1 interrupt c) Serial port interrupt d) all the three

31. The following register of 8051 is a bit-addressable register.

a) TCON   b) PCON   c) SP   d) TMOD

32. Which of the following flag is not supported by 8051?

a) carry   b) auxilary carry   c) sign   d) parity

33. The program counter in 8051 is a _____ bit register.

a) 8-bit   b) 16-bit   c) 32-bit   d) 64-bit

Answers

1. c	7. c	13. c	19. a	25. b	31. a
2. a	8. a	14. b	20. c	26. a	32. c
3. a	9. d	15. d	21. b	27. c	33. b
4. d	10. d	16. a	22. d	28. b	
5. d	11. b	17. b	23. a	29. b	
6. b	12. d	18. b	24. b	30. d	

IV. Answer the following questions.

E3.1 Compare 8051 and 8052 microcontrollers

E3.2 How is the external program memory accessed in 8051? Explain with diagram.

E3.3 How are the interrupts of 8051 classified?

E3.4 List the vector address and priority of 8051 interrupts.

E3.5 Why is 8031 called as ROM less version of 8051?

E3.6 Find the contents of the condition flags in the 8051 after adding the binary numbers $A2_H$ and $C8_H$.

E3.7 Write the binary word to be written into the PSW of the 8051, to select the register bank-3 in the internal RAM.

E3.8 Specify the size of various memory supported by 8051.

E3.9 How many ports are bit-addressable in 8051?

E3.10 What happens in power down mode of 8051?

E3.11 What happens in idle down mode of 8051?

E3.12 Compare power down and idle mode of 8051.

E3.13 What is the significance of $\overline{EA}$ line of 8051?

E3.14 Draw the format of PSW register of 8051. Explain each flag

E3.15 What is the significance of SMOD bit of PCON register in 8051?

E3.16 Draw the format of SCON register of 8051. Explain each bit.

E3.17 How is multiprocessor communication implemented using 8051?

E3.18 What is meant by autoreload counter? How is it implemented in 8051?

E3.19 How is the timer/counter used to count the internal and external pulses in 8051?

E3.20 When does the timer/counter start and stop counting in 8051?

E3.21 How are the interrupts of 8051 enabled?

E3.22 How are the priority of interrupts of 8051 changed?

E3.23 The 8051 microcontroller works at 10 MHz. How much time is taken by 8051 to execute the instruction ADD A, #12H?

E3.24 Where the program control is transferred when the instruction JMP 06H is executed by 8051? Assume PC = 2100_H.

E3.25 Identify the addressing mode for the following 8051 instructions

a) MOV A,R2   b) MOVX A,@DPTR   c) MUL AB

d) XRL A,20H   e) CJNE A,#04H,07H

E3.26 Write down the steps involved in executing the 8051 instruction PUSH 07H. Assume SP = 3070_H and the $R_7 = 25_H$.

E3.27 Differentiate XCH and XCHD instructions of 8051.

E3.28 Write down the 8051 instsruction sequences to transfer the data stored at internal RAM location 45_H into external data memory 36_H.

E3.29 Name the 8051 instruction used to increment the external data memory pointer by one

E3.30 Identify the syntax errors if any in the following 8051 instructions

    a.   MOV A,@R4   b.   INC PC   c. MUL A,#25H   d. SUB R2,R4

E3.31 The content of the register A = 78_H. What do the 8051 instructions RL A and RLC A produce? Assume CF = 1.

E3.32 Calculate the physical jumping address after the execution of the instruction SJMP F2H if it is stored in the address 2536_H.

E3.33 Write down the single equivalent 8051 instruction of the following

    DEC R1
    JNZ 06_H

E3.34 Find the status of CY flag after execution of the following 8051 instructions sequentially

    CLR   C
    SETB C
    SETB P1.4
    CPL   C
    ANL   C,P1.4

E3.35 Write down the 8051 instructions used to read the status of LED which is connected to P1.6. If the LED is ON switch it OFF and jump three steps forward in the program.

E3.36 Show the complete interface of 16 kB EPROM IC27128 and 8 kB RAM IC6264 with the 8051. Ignore the internal 4 kB ROM.

CHAPTER 4

PERIPHERAL INTERFACING

4.1 INTRODUCTION

Programmable peripheral devices are designed to perform various input/output functions and specific routine activities. Every programmable device will have one or more control registers. The programmable devices can be set up to perform specific functions by writing control words into the control registers. The control word is an instruction which informs the peripheral about various functions it has to perform. The format of the control word will be specified by the manufacturer of the peripheral devices.

INTEL have developed a number of peripheral devices that can be used with 8085/8086/8088 based systems. Some of the peripheral devices developed by INTEL for 8085/8086/8088 based system are Parallel peripheral interface-8255, Keyboard/Display controller-8279 and Programmable Timer 8254. A brief discussion about these devices and their interfacing with an 8085 processor are presented in this chapter.

The parallel peripheral interface-8255 is used to interface a slow IO device to the fast processor and to achieve an efficient data transfer between them. The 8279 is used to relieve the processor from time-consuming routine activities like keyboard scanning and display refreshing. The programmable timers are used to maintain various timings and to initiate time-based activities.

4.2 PROGRAMMABLE PERIPHERAL INTERFACE - INTEL 8255

(AU, Nov/Dec' 19, 8 Marks)

The INTEL 8255 is a device used to implement parallel data transfer between processor and slow peripheral devices like ADC, DAC, keyboard, 7-segment display, LCD, etc.

The 8255 has three ports: Port-A, Port-B and Port-C. The ports A and B are 8-bit parallel ports. Port-A can be programmed to work in any one of the three operating modes as input or output port. The three operating modes are:

Mode - 0 → Simple IO port

Mode - 1 → Handshake IO port

Mode - 2 → Bidirectional IO port

The port-B can be programmed to work either in mode-0 or mode-1 as input or output port. The port-C pins (8 pins) have different assignments depending on the mode of ports A and B. If ports A and B are programmed in mode-0, then the port-C can perform any one of the following function:

1. **As 8-bit parallel port in mode-0 for input or output.**

2. **As two numbers of 4-bit parallel port in mode-0 for input or output.**

3. **The individual pins of port-C can be set or reset for various control applications.**

If port-A is programmed in mode-1/mode-2 and port-2 is programmed in mode-1 then some of the pins of port-C are used for handshake signals and the remaining pins can be used as input/output lines or individually set/reset for control applications.

IO Modes of 8255

Mode-0 :   In this mode, all the three ports can be programmed either as the input or the output port. In mode-0, the outputs are latched and the inputs are not latched. The ports do not have handshake or interrupt capability. The ports in mode-0 can be used to interface DIP switches, hexa-keypad, LEDs and 7-segment LEDs to the processor.

Mode-1 :   In this mode, only ports A and B can be programmed either as the input or output port. In mode-1, handshake signals are exchanged between the processor and peripherals prior to data transfer. The port-C pins are used for handshake signals. Input and output data are latched. Interrupt driven data transfer scheme is possible.

Mode-2 :   In this mode, the port will be a bidirectional port. (i.e., the processor can perform both read and write operations with an IO device connected to a port in mode-2.) Only port-A can be programmed to work in mode-2. Five pins of port-C are used for handshake signals. This mode is used primarily in applications such as data transfer between two computers or floppy disk controller interface.

Pins, Signals and Internal Block Diagram of 8255

The pin description of 8255 is shown in Fig. 4.1. It has 40 pins and requires a single +5V supply. The internal block diagram of 8255 is shown in Fig. 4.2

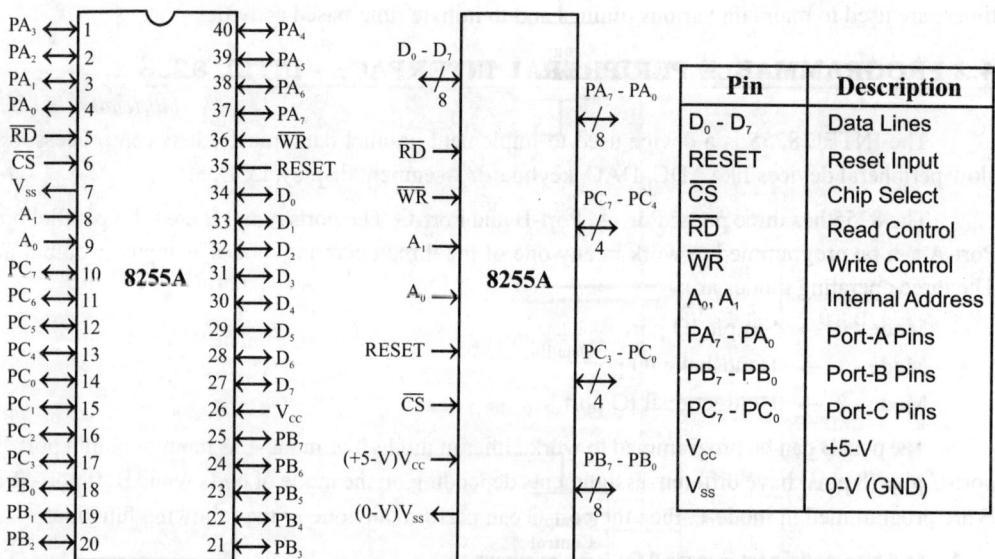

Fig. 4.1: Pin description of 8255.

The ports are grouped as Group A and Group B. The group A has port-A, port-C upper and its control circuit. The group B comprises port-B, port-C lower and its control circuit. The read/write control logic requires six control signals. These signals are:

$\overline{\text{RD}}$ (Read) : This control signal enables the read operation. When this signal is **low**, the microprocessor reads data from a selected IO port of the 8255A.

$\overline{\text{WR}}$ (Write) : This control signal enables the write operation. When this signal goes **low**, the microprocessor writes into a selected IO port or the control register.

RESET : This is an active **high** signal. It clears the control register and sets all ports in the input mode.

$\overline{\text{CS}}$, A_0 and A_1 : These are device select signals. The address lines A_0 and A_1 of 8255 can be connected to any two address lines of the processor to provide internal addresses. The A_0 and A_1 selects any one of the 4 internal devices as shown in Table 4.1. The 8255 will remain in **high impedance** state if the signal input to $\overline{\text{CS}}$ is **high** and the device can be brought to normal logic by making the signal input to $\overline{\text{CS}}$ as logic **low**.

Table 4.1:

Internal address		Device selected
A_1	A_0	
0	0	Port-A
0	1	Port-B
1	0	Port-C
1	1	Control Register

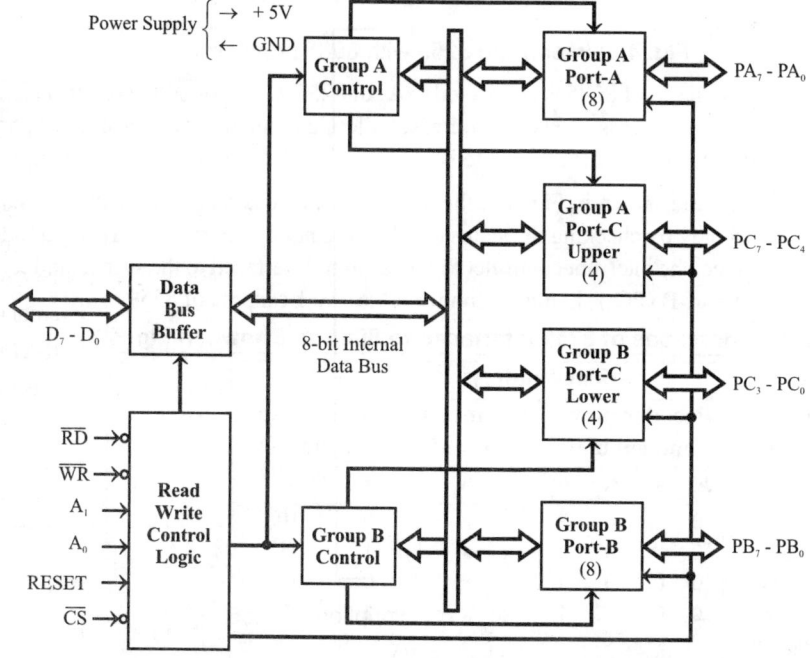

Fig. 4.2: Internal block diagram of 8255.

4.2.1 INTERFACING OF 8255 WITH 8085 PROCESSOR *(AU, Nov/Dec' 19, 5 Marks)*

A simple schematic for interfacing the 8255 with an 8085 processor is shown in Fig. 4.3. The 8255 can be either memory-mapped or IO-mapped in the system. In the schematic shown in Fig. 4.3, the 8255 is IO-mapped in the system. The chip select signals for IO-mapped devices are generated by using a 3-to-8 decoder. The address lines A_4, A_5 and A_6 are decoded to generate eight chip select signals (IOCS-0 to IOCS-7) and in this, the chip select IOCS-1 is used to select 8255. The address line A_7 and the control signal IO/$\overline{M}$ are used as enable for the decoder.

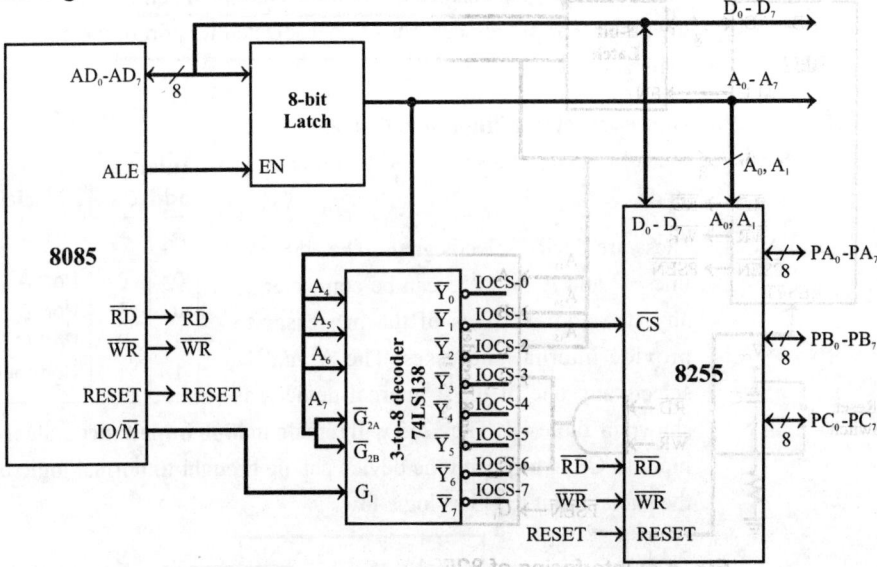

Fig. 4.3: Interfacing 8255 with 8085 processor.

The address line A_0 of 8085 is connected to A_0 of 8255 and A_1 of 8085 is connected to A_1 of 8255 to provide the internal addresses. The IO addresses allotted to the internal devices of 8255 are listed in Table 4.2. The data lines D_0-D_7 are connected to D_0-D_7 of the processor to achieve parallel data transfer.

In the schematic shown in Fig. 4.3, the interrupt scheme is not included and so the data transfer can be performed only by checking the status of 8255 and not by interrupt method. For interrupt driven data transfer scheme, the interrupt controller 8259 has to be interfaced to the system and the interrupts of port-A (PC_3) and port B (PC_0) should be connected to two IR inputs of 8259.

Table 4.2: IO Addresses of 8255 Interfaced to 8085 as Shown in Fig. 4.3

Internal device	Binary address								Hexa address
	Decoder input and enable				Input to address pins of 8255				
	A_7	A_6	A_5	A_4	A_3	A_2	A_1	A_0	
Port-A	0	0	0	1	x	x	0	0	10
Port-B	0	0	0	1	x	x	0	1	11
Port-C	0	0	0	1	x	x	1	0	12
Control Register	0	0	0	1	x	x	1	1	13

Note : Don't care "x" is considered as zero.

4.2.2 INTERFACING OF 8255 WITH 8051 MICROCONTROLLER

The INTEL 8255 can be interfaced to an 8051 microcontroller for additional port requirement. A simple schematic for interfacing the 8255 with 8051 microcontroller is shown in Fig. 4.4. The 8255 can be mapped only as a memory device in an 8051 microcontroller. Moreover, both the read and write operation is possible only if the devices are mapped in the data memory address space. (If the devices are mapped in the program memory address space, then only read operation is possible.)

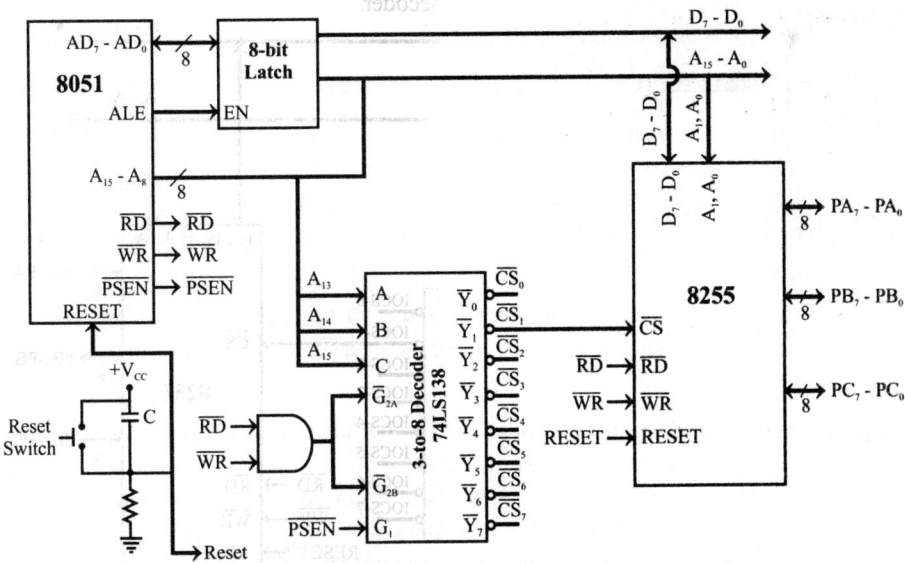

Fig. 4.4: Interfacing of 8255 with 8051 Microcontroller.

Table 4.3: Address Allotted to 8255 Interfaced to 8051 as Shown in Fig. 4.4

Internal device	Decoder input			Input to address pins of 8255														Hexa address	Comment
	A_{15}	A_{14}	A_{13}	A_{12}	A_{11}	A_{10}	A_9	A_8	A_7	A_6	A_5	A_4	A_3	A_2	A_1	A_0		
Port-A	0	0	1	x	x	x	x	x	x	x	x	x	x	x	0	0	2000	External
Port-B	0	0	1	x	x	x	x	x	x	x	x	x	x	x	0	1	2001	data
Port-C	0	0	1	x	x	x	x	x	x	x	x	x	x	x	1	0	2002	memory
Control Register	0	0	1	x	x	x	x	x	x	x	x	x	x	x	1	1	2003	address space

Note : Don't care "x" is considered as zero.

The chip select signals needed for devices mapped in the data memory address space are generated by using a 3-to-8 decoder. The address lines A_{13}, A_{14} and A_{15} are applied as input to decoder and so they are decoded to generate 8-chip select signals CS_0 to CS_7, and in this, the signal CS_1 is used to select the 8255. The control signals $\overline{RD}$ and $\overline{WR}$ are logically ANDed and applied as logic **low** enable for the decoder. The $\overline{PSEN}$ is connected to logic **high** enable of the decoder. $\overline{PSEN}$ will be asserted **low** while accessing program memory and so, the data memory decoder will be disabled. When program memory is not accessed, $\overline{PSEN}$ will be **high** and so the data memory decoder will be enabled.

The address line A_0 of 8031 is connected to A_0 of 8255 and the address line A_1 of 8031 is connected to A_1 of 8255 to provide internal addresses. The addresses allotted to the internal devices of 8255 are listed in Table 4.3. The data lines D_7- D_0 are connected to D_7 - D_0 of the processor to achieve parallel data transfer.

4.2.3 PROGRAMMING (OR INITIALIZING) 8255   *(AU, Nov/Dec' 19, 8 Marks)*

The 8255 has two control words: IO **M**ode **S**et control **W**ord (MSW) and **B**it **S**et/**R**eset (BSR) control word. The MSW is used to specify IO functions and BSR word is used to set/reset individual pins of port-C. Both the control words are written in the same control register. The control register differentiates them by the value of bit B_7. The BSR control word does not affect the functions of ports A and B.

Bit B_7 of the control register specifies either the IO function or the bit set/reset function. If $B_7 = 1$, then the bits B_6- B_0 determine the IO functions in various modes. If bit $B_7 = 0$, then the bits B_6- B_0 determine the pin of port-C to be set or reset.

The 8255 ports are programmed (or initialized) by writing a control word in the control register. For setting IO functions and mode of operation, the IO mode set control word is sent to the control register. For setting/resetting a pin of port-C, the bit set/reset control word is sent to the control register. The format of the IO mode set control word is shown in Fig. 4.5 and the format of bit set/reset control word is shown in Fig. 4.6. The various functions (assignments) of port-C pins during the different operating modes of ports A and B are listed in Table 4.4.

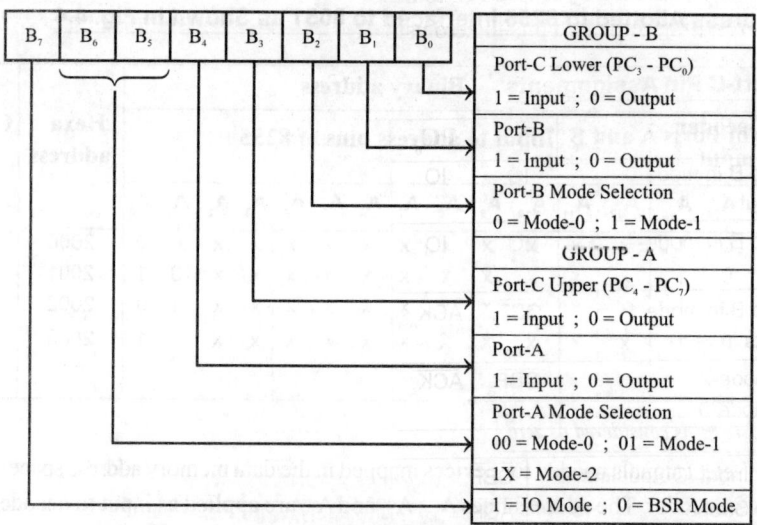

Fig. 4.5: Format of IO mode set control word of 8255.

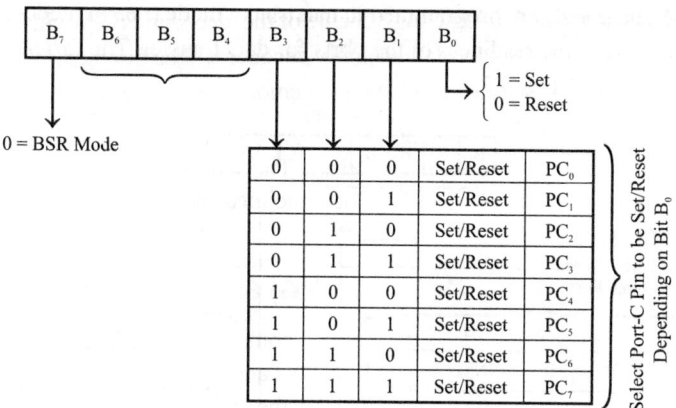

Fig. 4.6: Format of Bit Set/Reset control word of 8255.

In handshake mode (i.e., in mode-1 and mode-2), the data transfer between the processor and the port can be implemented either by the interrupt method or by checking the status of the 8255 ports. In an interrupt-driven data transfer scheme, when the port is ready it interrupts the 8086 processor through the NMI/INTR pin for a read/write operation. In status check technique, the 8086 processor can check the status of ports A and B by reading port-C. When the port is ready for data transfer, the processor executes a read/write cycle.

The 8255 has two internal flip-flops as interrupt enables ($INTE_A$ and $INTE_B$) for port-A and port-B interrupt signals. In interrupt driven data transfer scheme, the 8255 generates an interrupt signal, only if these flip-flops are enabled by using BSR control word. The $INTE_A$ is enabled by setting PC_4 to **high** and $INTE_B$ is enabled by setting PC_2 to **high** using BSR control word. The interrupt signal can be disabled by resetting these two bits to zero using BSR control word.

Table 4.4: Port-C Pin Assignments

Functions of Ports A and B	PC_7	PC_6	PC_5	PC_4	PC_3	PC_2	PC_1	PC_0
Ports A and B in mode-0 Input/Output	IO	IO	IO	IO	IO	IO	IO	IO
Ports A and B in mode-1 Input ports	IO	IO	IBF_A	$\overline{STB}_A$	$INTR_A$	$\overline{STB}_B$	IBF_B	$INTR_B$
Ports A and B in mode-1 Output ports	$\overline{OBF}_A$	$\overline{ACK}_A$	IO	IO	$INTR_A$	$\overline{ACK}_B$	$\overline{OBF}_B$	$INTR_B$
Port-A in mode-2 Port-B in mode-0	$\overline{OBF}_B$	$\overline{ACK}_A$	IBF_A	$\overline{STB}_A$	$INTR_A$	IO	IO	IO

IO	-	Input /Output line	OBF	-	Output Buffer Full
$\overline{STB}$	-	Strobe	$\overline{ACK}$	-	Acknowledge
IBF	-	Input Buffer Full	The subscript A denotes port-A signal.		
INTR	-	Interrupt Request	The subscript B denotes port-B signal.		

When port-A and port-B are programmed in handshake mode (i.e., in mode-1 and mode-2), the port-C can be read to know the readiness of the ports for data transfer. The format of the status word read from port-C is shown in Fig. 4.7.

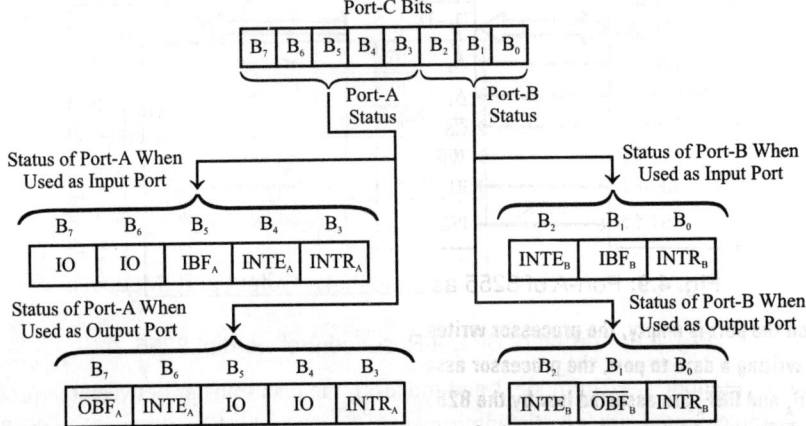

Fig. 4.7: Format of status word of 8255 for handshake input and output operation.

8255 Handshake Input Port (Mode-1)

The signals used for data transfer between input device and 8085 microprocessor using port-A of 8255 as handshake input port (Mode-1) are shown in Fig. 4.8.

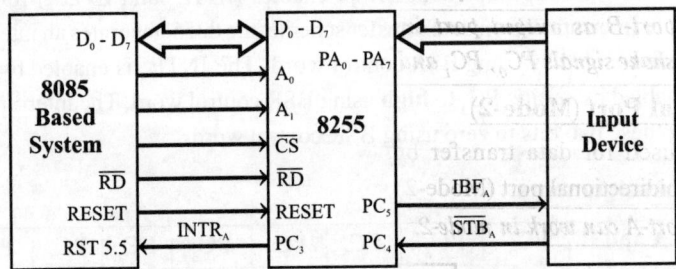

Fig. 4.8: Port-A of 8255 as handshake input port (Mode-1).

1. The input device checks the IBF_A signal. If it is low, then the input device places the data on the port lines PA_0-PA_7 and asserts $\overline{STB}_A$ low and after a delay time $\overline{STB}_A$ is asserted high.

2. When $\overline{STB}_A$ is low, the 8255 asserts IBF signal high and at the rising edge of $\overline{STB}_A$ the data is latched to the port and $INTR_A$ is set high.

3. When $INTR_A$ goes high, the processor is interrupted through the RST 5.5 input pin to execute a subroutine for reading the data from the port. For a read operation, the processor asserts $\overline{RD}$ low and then high.

4. When $\overline{RD}$ is low, the $INTR_A$ is reset (asserted low) by 8255 and at the rising edge of $\overline{RD}$, the IBF is asserted low and the input device can send the next data.

> Note : For port-B as input port in mode-1, same operations are performed, but for handshake signals PC_0, PC_1 and PC_2 are used.

8255 Handshake Output Port (Mode-1)

The signals used for data transfer between output device and an 8085 microprocessor using port-A of 8255 as the handshake output port (Mode-1) are shown in Fig. 4.9.

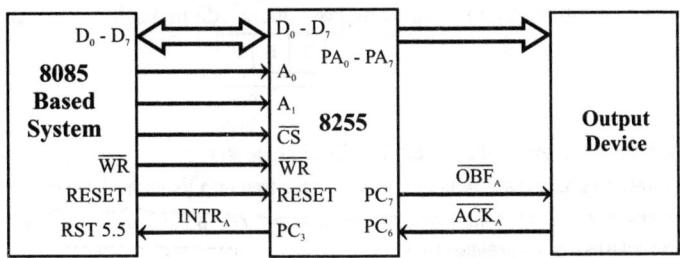

Fig. 4.9: Port-A of 8255 as handshake Output port (Mode-1).

1. When the port is empty, the processor writes a byte in the port.
2. For writing a data to port, the processor asserts $\overline{WR}$ low and then high. At the rising edge of $\overline{WR}$, both the $INTR_A$ and $\overline{OBF}_A$ are asserted low by the 8255.
3. The $\overline{OBF}_A$ signal informs the output device that the data is ready. If the output device accepts the data then it sends an acknowledgement signal by asserting $\overline{ACK}_A$ low and then high.
4. When $\overline{ACK}_A$ is low, the OBF_A is asserted high by the 8255. When $\overline{ACK}_A$ is high, the $INTR_A$ is set (asserted high), to interrupt the processor.
5. When $INTR_A$ goes high, the processor is interrupted through RST 5.5 input pin to execute an interrupt service routine to load the next data in the output port.

> *Note :* *For port-B as output port in mode-1, same operations are performed, but for handshake signals PC_0, PC_1 and PC_2 are used.*

8255 Bidirectional Port (Mode-2)

The signals used for data transfer between IO device and 8085 microprocessor using port-A of 8255 as a bidirectional port (Mode-2) are shown in Fig. 4.10.

> *Note : Only port-A can work in mode-2.*

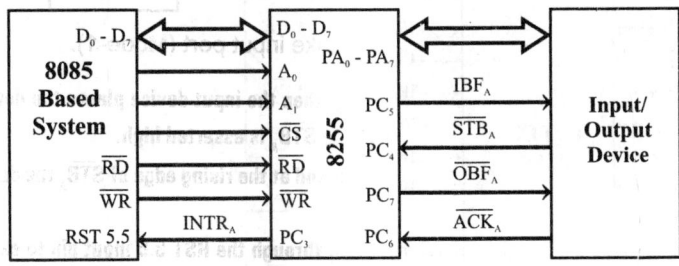

Fig. 4.10: Port-A of 8255 as bidirectional port (Mode-2).

In mode-2, the port can be used either as an input port or as an output port. At any one time, the processor will perform either read or write operation. In mode-2, the read operation can be followed by write, or write operation can be followed by read. The signals involved and the operations performed for read operation are similar to mode-1 input port. The signals involved and the operations performed for write operation are similar to mode-1 output port.

4.3   PROGRAMMABLE INTERRUPT CONTROLLER - INTEL 8259

The 8259 is a programmable interrupt controller. It is used to expand the interrupts of an 8085 or 8086 processor. One 8259 can accept eight interrupt requests and allow one by one to the processor INTR pin. The interrupt controller can be used in cascaded mode in a system to expand the interrupts up to 64.

Features of an 8259

1. It is programmed to work with either an 8085 or 8086 processor.
2. It manages 8 interrupts according to the instructions written into its control registers.
3. In an 8085 processor-based system, it vectors an interrupt request anywhere in the memory map and the interrupt vector address is programmable.
4. The priorities of the interrupts are programmable. The different operating modes which decides the priorites are automatic rotation mode, specific rotation mode and fully nested mode.
5. The interrupts can be masked or unmasked individually.
6. The 8259 is programmed to accept either level triggered interrupt signals or edge triggered interrupt signals.
7. The 8259 provides the status of the pending interrupts, masked interrupts and interrupt being serviced.
8. The 8259s can be cascaded to accept a maximum of 64 interrupts.

4.3.1 INTERFACING 8259 WITH 8085 MICROPROCESSOR

The 8259 is a 28-pin IC packed in DIP. The various pins of an 8259 are shown in Fig. 4.11. It requires two internal address and they are $A_0 = 0$ or $A_0 = 1$. It can be either memory-mapped or IO-mapped in the system. The interfacing of an 8259 to 8085 is shown in Fig. 4.12. In Fig. 4.12, the 8259 is IO-mapped in the system. The low order data bus lines D_0-D_7 are connected to D_0-D_7 of the 8259. The address line A_0 of the 8085 processor is connected to A_0 of the 8259 to provide the internal address. The 8259 requires one chip select signal. The chip select signal for 8259 is generated by using 3-to-8 decoder.

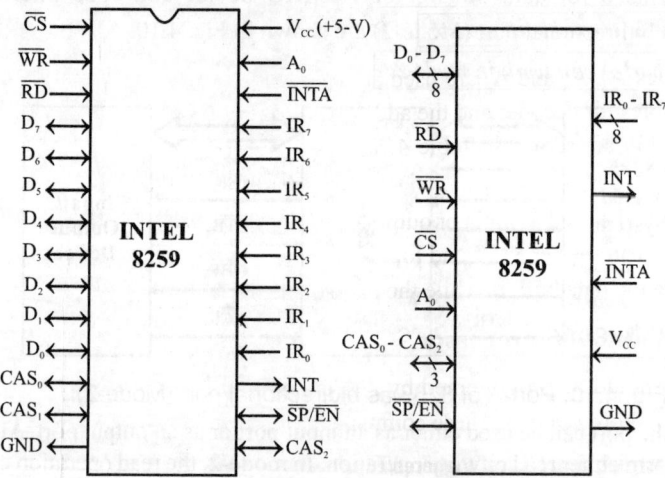

Fig. 4.11: Pin details of 8259.

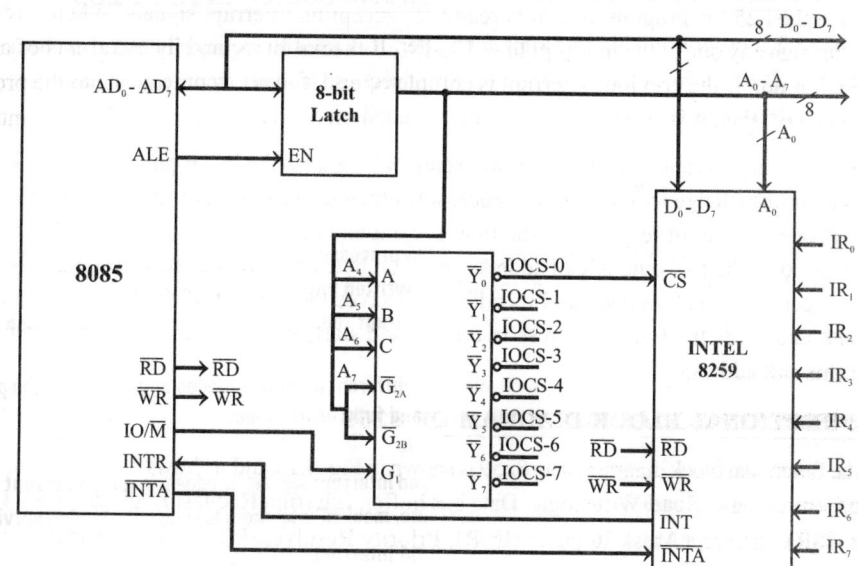

Fig. 4.12: Interfacing 8259 to 8085 microprocessor.

Table 4.5: IO Address of 8259 Interfaced to 8085 as Shown in Fig. 4.12

	Binary Address								
	Decoder input/ enable				Input to address pin of 8259				Hexa address
	A_7	A_6	A_5	A_4	A_3	A_2	A_1	A_0	
For A_0 of 8259 to be zero	0	0	0	0	x	x	x	0	00
For A_0 of 8259 to be one	0	0	0	0	x	x	x	1	01

Note : Don't care "x" is considered as zero.

The address lines A_4, A_5 and A_6 are used as input to the decoder. The control signal $IO/\overline{M}$ is used as logic **high** enables for decoder and the address line A_7 is used as logic **low** enable for decoder. The IO addresses of 8259 are shown in Table 4.5. The signals CAS_0-CAS_2 are used only in cascade operations of 8259s.

The $\overline{SP/EN}$ pin can be used as input or output signal. In non-buffered mode, it is used as an input signal and tied to logic-1 in master 8259 and logic-0 in slave 8259. In buffered mode it is used as an output signal to disable the data buffers while the data is transferred from 8259A to the CPU.

Working of 8259 With 8085 processor

First, the 8259 should be programmed by sending Initialization Command Word (ICW) and Operational Command Word (OCW). These command words will inform 8259 about the following,

- **Type of interrupt signal (Level triggered/Edge triggered).**
- **Type of processor (8085/8086).**
- **Call address and its interval (4 or 8).**

- **Masking of interrupts.**
- **Priority of interrupts.**
- **Type of end of interrupt.**

Once the 8259 is programmed, it is ready for accepting interrupt signals. When it receives an interrupt through any one of the interrupt lines IR_0-IR_7, it checks for its priority and also checks whether it is masked or not. If the previous interrupt is completed and if the current request has highest priority and is unmasked, then it is serviced.

For servicing this interrupt, the 8259 will send INT signal to the INTR pin of the 8085. In response it expects an acknowledge $\overline{INTA}$ from the processor. When the processor accepts the interrupt, it sends three $\overline{INTA}$ one by one. In response to the first, second and third $\overline{INTA}$ signals, the 8259 will supply the CALL opcode, the low byte of call address and high byte of the call address respectively. Once the processor receives the call opcode and its address, it saves the content of the Program Counter (PC) in the stack and loads the CALL address in the PC and starts executing the interrupt service routine stored in this call address.

4.3.2 FUNCTIONAL BLOCK DIAGRAM OF 8259   *(AU, Nov/Dec' 19, 13 Marks)*

The functional block diagram of an 8259 is shown in Fig. 4.13 and it shows eight functional blocks. They are Control logic, Read/Write logic, Data bus buffer, Interrupt Request Register (IRR), In-Service Register (ISR), Interrupt Mask Register (IMR), Priority Resolver (PR), and Cascade buffer.

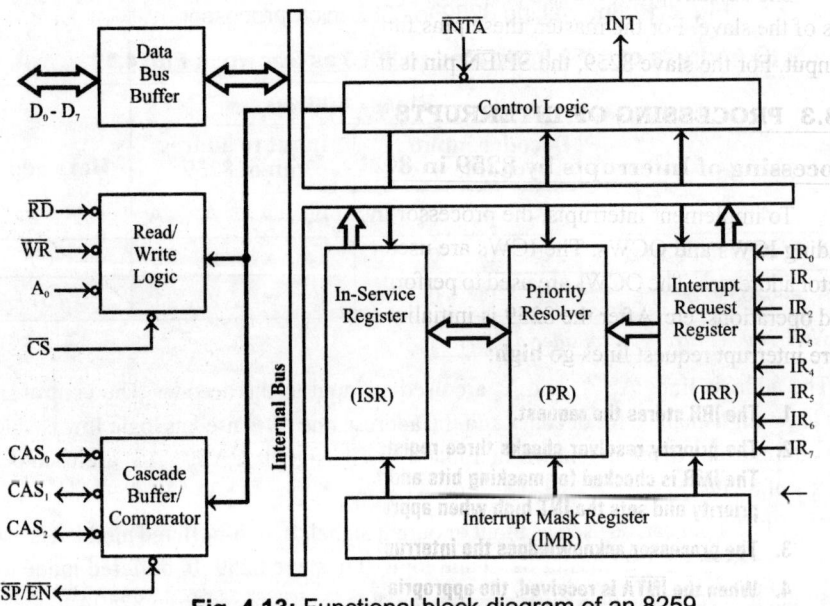

Fig. 4.13: Functional block diagram of an 8259.

The data bus and its buffer are used for the following activities:

1. **The processor sends the control word to the data bus buffer through D_0- D_7.**

2. **The processor reads the status word from the data bus buffer through D_0- D_7.**

3. **From the data bus buffer the 8259 call opcode and address (in case of an 8085) through D_0- D_7 to the processor.**

The processor uses the $\overline{RD}$, $\overline{WR}$ and A_0 to read or write the 8259. The 8259 is selected by the $\overline{CS}$. The IRR has eight input lines $(IR_0\text{-}IR_7)$ for interrupts. When these lines go **high**, the requests are stored in the IRR. It registers a request only if the interrupt is unmasked. Normally IR_0 has the highest priority and IR_7 the lowest. The priorities of an interrupt request input are also programmable.

The interrupt mask register stores the masking bits of the interrupt lines to be masked. The relevant information is sent by the processor through the OCW1. The in-service register keeps track of which interrupt input is currently being serviced. For each input that is currently being serviced, the corresponding bit will be set in the in-service register. The priority resolver examines the interrupt request, mask and in-service registers and determines whether the INT signal should be sent to the processor or not.

The cascade buffer/comparator is used to expand the interrupts of the 8259. Figure 4.14 is an example of 8259s in cascade connection. It is called the master 8259. To each interrupt request input of the master 8259 $(IR_0\text{-}IR_7)$ one slave 8259 can be connected. The 8259s interrupting the master 8259 are called slave 8259s.

Each 8259 has its own addresses so that each 8259 can be programmed independently by sending the command words and the status bytes can be read from it independently.

The cascade pins $(CAS_0, CAS_1$ and $CAS_2)$ from the master are connected to the corresponding pins of the slave. For the master, these pins function as output, and for the slave device they function as input. For the slave 8259, the $\overline{SP}/\overline{EN}$ pin is tied **low** to let the device know that it is a slave.

4.3.3 PROCESSING OF INTERRUPTS BY AN 8259

Processing of Interrupts by 8259 in 8085

To implement interrupts, the processor interrupt should be enabled and 8259 be initialized by sending ICWs and OCWs. The ICWs are used to set up the proper conditions and specify the CALL vector addresses. The OCWs are used to perform functions such as masking interrupts, setting up status, read operations, etc. After the 8259 is initialized, the following sequence of events occur when one or more interrupt request lines go **high**:

1. **The IRR stores the request.**
2. **The priority resolver checks three registers (IRR, IMR, ISR). The IRR is checked for interrupt request. The IMR is checked for masking bits and the ISR for the interrupt request being served. It resolves the priority and sets the INT high when appropriate.**
3. **The processor acknowledges the interrupt by sending the $\overline{INTA}$ signal.**
4. **When the $\overline{INTA}$ is received, the appropriate priority bit in the ISR is set to indicate which interrupt level is being served, and the corresponding bit in the IRR is reset to indicate that the request is accepted. Then, the opcode of the CALL instruction is placed on the data bus.**
5. **When the processor decodes the CALL instruction, it places two more $\overline{INTA}$ signals on the data bus.**
6. **When the 8259 receives the second $\overline{INTA}$, it outputs the low-order byte of the CALL address on the data bus. When the third $\overline{INTA}$ signal is received, the 8259 outputs high-order byte of CALL address on the data bus. The CALL address is the vector memory location for the interrupt. (this address is programmed by sending ICW1 and ICW2 to the control register during the initialization.)**

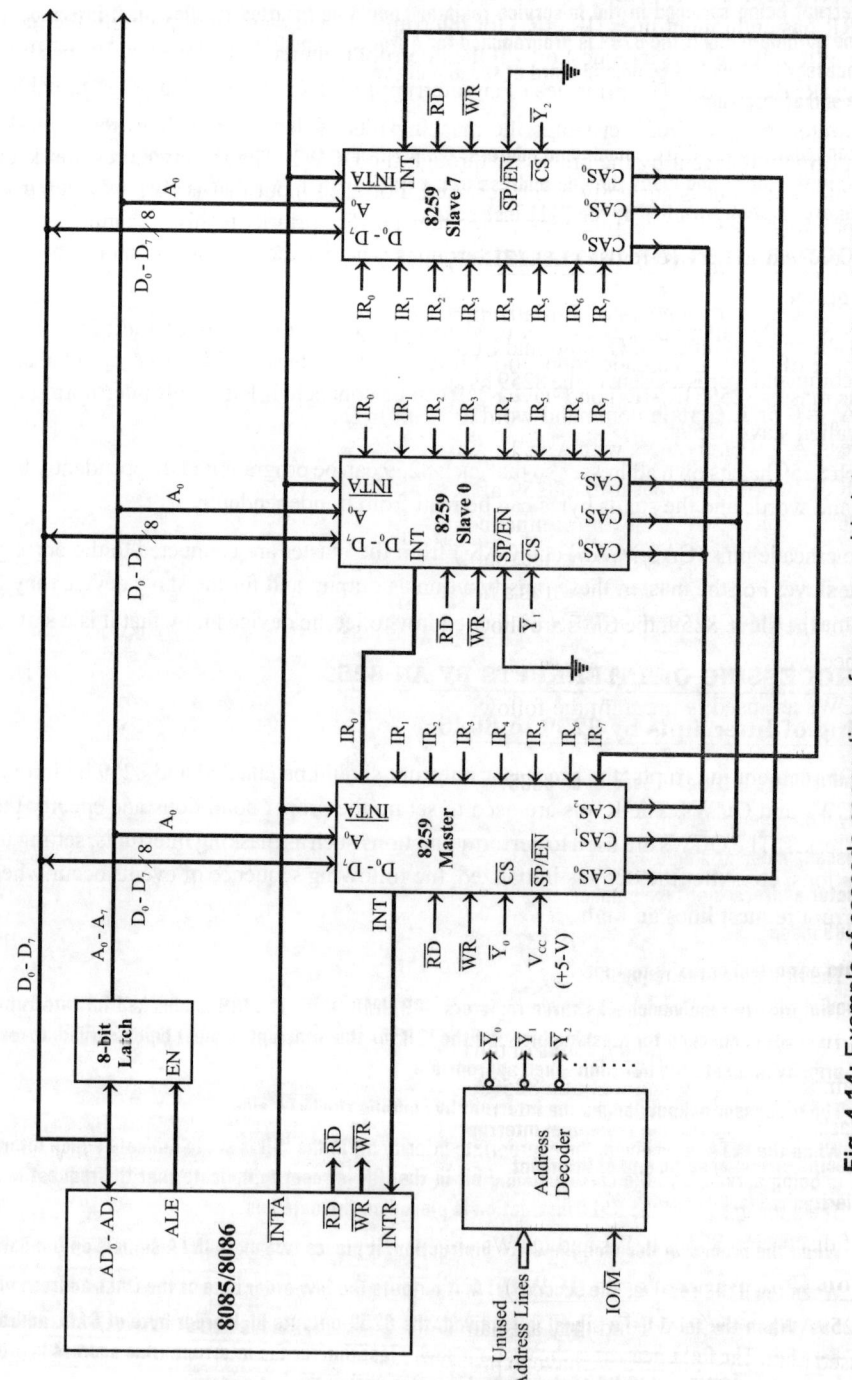

Fig. 4.14: Example of cascade connection of programmable interrupt controllers-8259.

7. Once the processor reads the CALL opcode and address from the 8259 the bit corresponds to the current interrupt being serviced in the in-service register should be resetted to allow next interrupt. This is done automatically if the 8259 is programmed for Automatic End Of Interrupt (AEOI). Alternatively, the processor can sends a command word at the end of the interrupt service routine to inform 8259 about the end of interrupt.

8. After receiving the CALL opcode and address, the processor saves the content of the Program Counter (PC) in the stack and loads the call address in the PC. Thus the program control is transferred to the memory location specified by the CALL instruction.

4.3.4 PROGRAMMING (OR INITIALIZING) 8259

The 8259 has four numbers of Initialization Command Word (ICW) and three numbers of Operational Command Word (OCW). The command words are sent to the 8259 by selecting it by $\overline{CS} = 0$ and $A_0 = 0$ or 1. Certain command word are sent to the internal address, $A_0 = 0$ and others with $A_0 = 1$.

The OCW1 should be sent to the 8259 after sending the ICWs. The OCW2 can be sent at any time (either before servicing a interrupt or at the end of the interrupt service routine). The order of sending the ICWs and OCWs are shown as a flowchart in Fig. 4.15. The format of the ICWs and OCWs are shown in Figs. 4.16 and 4.17 respectively.

The ICWs are used to program the following features of an 8259.

- Call address interval in case of an 8085
- Level or edge triggered
- Cascade mode or single
- Vector addresses or type number
- 8085 mode
- Auto or normal end of interrupt
- Special fully nested mode

The OCWs are used to read the status of the interrupts and also to program the following features of a 8259:

- Masking or unmasking of individual interrupts.
- Specific or non-specific end of interrupt.
- Priority modes.

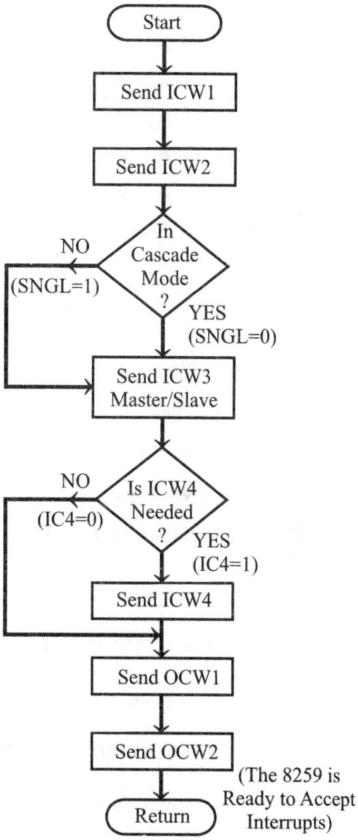

Fig. 4.15: Sending order of ICWs and OCWs.

A brief discussion about ICWs and OCWs are presented in the following sections.

Initialization Command Words (ICWs)

The 8259A has four ICWs and they are named as ICW1, ICW2, ICW3 and ICW4. When only one 8259 is used in the system, then we have to program the 8259 by sending ICW1, ICW2 and ICW4. When a number of 8259s are used in the system, then we have to program each 8259 by sending all the four ICWs. The format of the ICW3 for a master and slave 8259 are different.

ICW1 :   The ICW1 programs the basic operations of 8259. In 8085 based-system, the bit ADI is used to program a call address interval of 4 or 8 and the upper three bits (B_5, B_6 and B_7) of ICW1 are used to program the upper three bits of low byte of the call address. The lower five bits of the low byte of the call address are automatically inserted by the 8259 as shown in Table 4.6. The single or cascade mode of operation is selected by programming the "SNGL" bit. The LTIM bit determines whether the interrupt request input is positive edge-triggered or level-triggered.

Table 4.6: Low Byte CALL Address

Interrupt input	Low byte call address														
	Interval = 4								Interval = 8							
	B_7	B_6	B_5	B_4	B_3	B_2	B_1	B_0	B_7	B_6	B_5	B_4	B_3	B_2	B_1	B_0
IR0	A_7	A_6	A_5	0	0	0	0	0	A_7	A_6	0	0	0	0	0	0
IR1	A_7	A_6	A_5	0	0	1	0	0	A_7	A_6	0	0	1	0	0	0
IR2	A_7	A_6	A_5	0	1	0	0	0	A_7	A_6	0	1	0	0	0	0
IR3	A_7	A_6	A_5	0	1	1	0	0	A_7	A_6	0	1	1	0	0	0
IR4	A_7	A_6	A_5	1	0	0	0	0	A_7	A_6	1	0	0	0	0	0
IR5	A_7	A_6	A_5	1	0	1	0	0	A_7	A_6	1	0	1	0	0	0
IR6	A_7	A_6	A_5	1	1	0	0	0	A_7	A_6	1	1	0	0	0	0
IR7	A_7	A_6	A_5	1	1	1	0	0	A_7	A_6	1	1	1	0	0	0

ICW2 :   In 8085, the ICW2 is used to program the high byte of call address. The lower three bits of type number are automatically inserted by the 8259 and the upper five bits are programmable. The binary code inserted in the lower three bits for interrupt request IR_0 to IR_7 are 000 to 111.

For example, if the bits T_3 to T_7 are chosen as 10010, then the following interrupt type numbers are associated with IR_0 to IR_7. For any interrupt request input through IR_0-IR_7 lines, the associated interrupt type is executed by the processor.

IR_0 is associated with type-90_H interrupt (90_H = 1001 0000)

IR_1 is associated with type-91_H interrupt (91_H = 1001 0001)

IR_2 is associated with type-92_H interrupt (92_H = 1001 0010)

IR_3 is associated with type-93_H interrupt (93_H = 1001 0011)

IR_4 is associated with type-94_H interr upt (94_H = 1001 0100)

IR_5 is associated with type-95_H interrupt (95_H = 1001 0101)

IR_6 is associated with type-96_H interrupt (96_H = 1001 0110)

IR_7 is associated with type-97_H interrupt (97_H = 1001 0111)

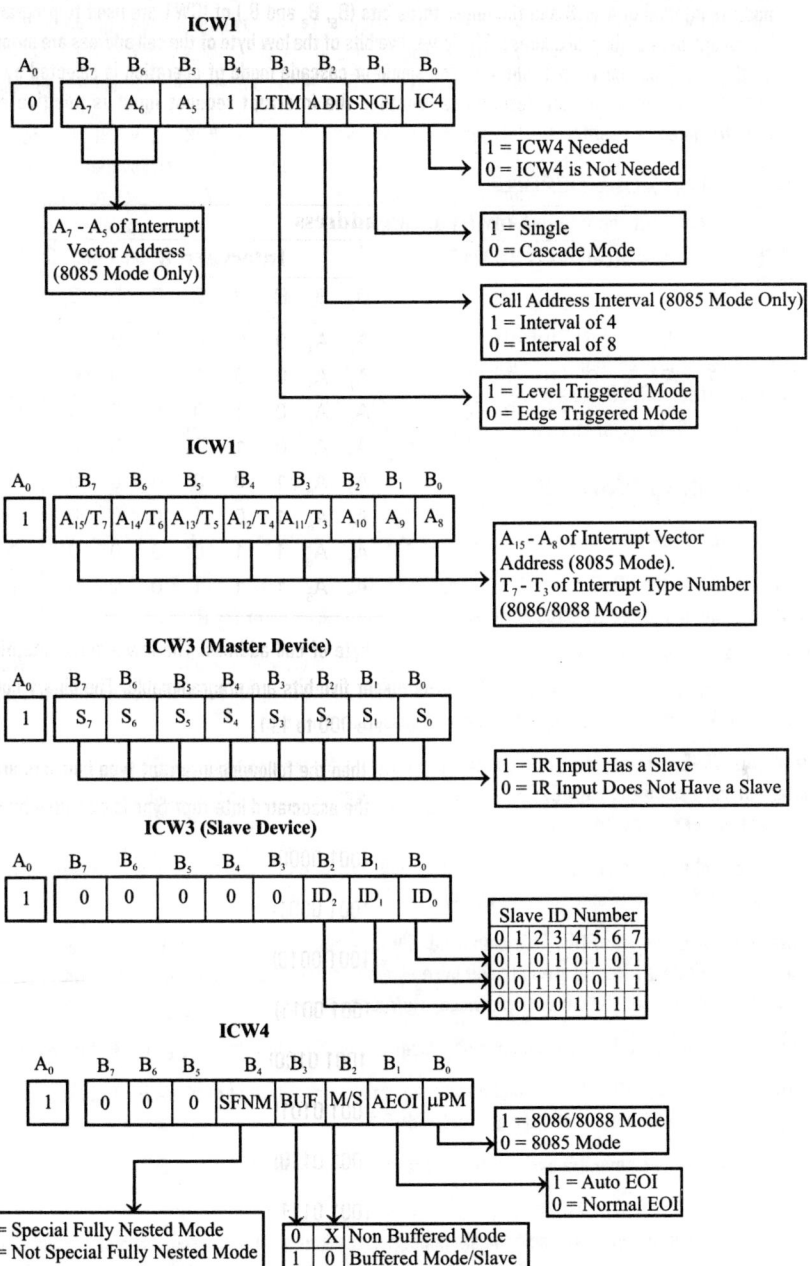

Fig. 4.16: Format of ICWs.

ICW3 :   The ICW3 should be sent to 8259s in cascade operations. Separate formats are provided for master and slave 8259s. In the cascade mode, slave 8259s are connected to one or more IR inputs of the master 8259 and each slave is provided with a slave ID number. The connection of slave 8259s to the IR inputs of the master are informed to the master through ICW3. For slave 8259s, the ID numbers are informed through the ICW3.

ICW4 :   The ICW4 is used to inform 8259 whether it is connected to an 8085 or an 8086-based system. For an 8085- based system the right most bit is set to zero, whereas in an 8086-based system it is set to one. The AEOI bit is used to program the method of terminating the interrupt. If AEOI is set to one, then the 8259 will automatically reset the interrupt request bit in the in-service register after supplying the type number to the processor. If the AEOI bit is programmed as zero, then the processor has to send the OCW2 to terminate the interrupt.

The BUF and M/S bits are used to select the buffered or non-buffered operation of a master/slave 8259. The SFNM bit is used to nest or include the priorities of the slave IR input with the master IR input. For example if IR_4 of a master 8259 has a slave 8259 connected to it and they are programmed for SFNM operation. Now the priorities of IR_0 to IR_7 of slave 8259 will be higher than IR_5 to IR_7 of master 8259.

Operation Command Words (OCWs)

The 8259 has three Operation Command Words (OCWs): OCW1, OCW2 and OCW3.

OCW1 :   OCW1 is sent to the 8259 to mask or unmask the IR inputs of the 8259. At any time, the mask status of the interrupts can be read by the processor by using the same address of the OCW1.

OCW2 :   The OCW2 is sent to the 8259A only when the AEOI mode (in ICW4) is not selected. The OCW2 is sent by the processor to decide on the type of End-Of-Interrupt (EOI) and to program the priorities of the interrupt (i.e., IR inputs of 8259A). The different methods of EOI are discussed as follows.

1. **Non-specific End-of-Interrupt :** This command is sent by the processor to the 8259 to terminate the current interrupt being serviced by the 8259. It resets the corresponding bit in the in-service register of the 8259 and allows the next higher priority interrupt.

2. **Specific End-of-Interrupt :** This command is sent by the processor to reset or terminate a specific interrupt request, decided by the lower 3-bits of OCW2.

3. **Rotate on Non-specific EOI :** This command will take action same as that of a non-specific EOI except that it rotates the priorities after resetting the bit in-service register. In this case, the interrupts will have rotating priority, in which the priority of the currently serviced interrupt becomes the least.

4. **Rotate on Automatic EOI :** This command is sent to the 8259 to select automatic EOI with rotating priority.

5. **Rotate on Specific EOI :** This command will take action similar to that of a specific EOI except that it rotates the priorities of the interrupts after they are serviced.

6. **Set priority :** The command is sent to set the priority of the interrupt level specified by the lower three bits of the OCW2 as the least.

OCW3 :   The OCW3 is used to set any special mask mode, poll the active interrupt request and to read the in-service and interrupt request registers. In special mask mode, the mask status are negated to allow the interrupts masked by and interrupt mask register.

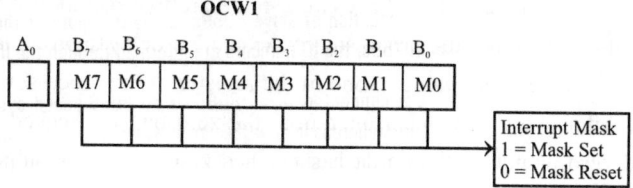

OCW1

A₀	B₇	B₆	B₅	B₄	B₃	B₂	B₁	B₀
1	M7	M6	M5	M4	M3	M2	M1	M0

Interrupt Mask
1 = Mask Set
0 = Mask Reset

OCW2

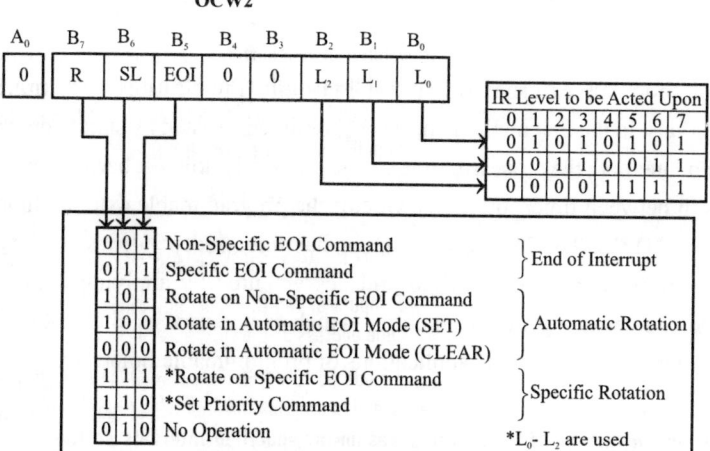

A₀	B₇	B₆	B₅	B₄	B₃	B₂	B₁	B₀
0	R	SL	EOI	0	0	L₂	L₁	L₀

IR Level to be Acted Upon

0	1	2	3	4	5	6	7
0	1	0	1	0	1	0	1
0	0	1	1	0	0	1	1
0	0	0	0	1	1	1	1

0	0	1	Non-Specific EOI Command	End of Interrupt
0	1	1	Specific EOI Command	
1	0	1	Rotate on Non-Specific EOI Command	Automatic Rotation
1	0	0	Rotate in Automatic EOI Mode (SET)	
0	0	0	Rotate in Automatic EOI Mode (CLEAR)	
1	1	1	*Rotate on Specific EOI Command	Specific Rotation
1	1	0	*Set Priority Command	
0	1	0	No Operation	*L₀- L₂ are used

OCW3

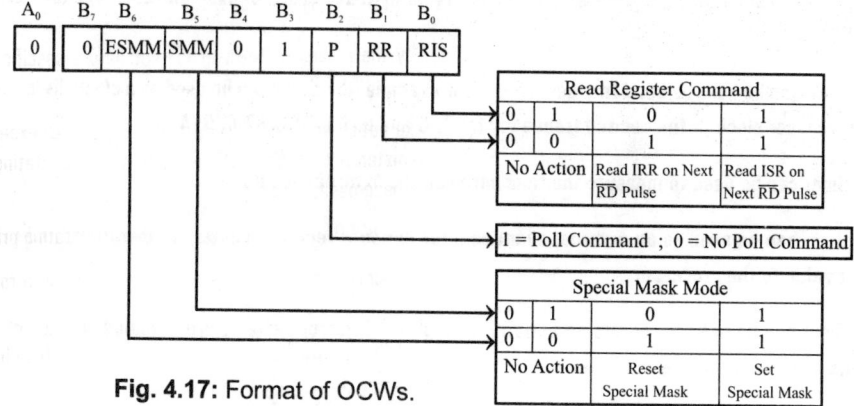

A₀	B₇	B₆	B₅	B₄	B₃	B₂	B₁	B₀
0	0	ESMM	SMM	0	1	P	RR	RIS

Read Register Command			
0	1	0	1
0	0	1	1
No Action		Read IRR on Next $\overline{RD}$ Pulse	Read ISR on Next $\overline{RD}$ Pulse

1 = Poll Command ; 0 = No Poll Command

Special Mask Mode			
0	1	0	1
0	0	1	1
No Action		Reset Special Mask	Set Special Mask

Fig. 4.17: Format of OCWs.

4.4 PROGRAMMABLE TIMER - INTEL 8254

When the processor has to perform time-based activities, there are two methods to maintain the timings of the operations. In one method, the processor can execute a delay subroutine. In this method, the delay subroutine will load a count value in one of the register of the processor and start decrementing the count value. After every decrement operation, the zero flag is checked to verify whether the count has reached zero or not. If the count has reached zero, the delay subroutine is terminated. Now the desired time will be elapsed and the processor can perform the desired time-based task. In this method, the time is estimated in terms of processor clock periods needed to execute the delay subroutine.

In the second method, an external timer can maintain the timings and interrupt the processor at periodic intervals. In the first method, the processor time is wasted by simply decrementing a register. But in the second method, the processor time can be efficiently utilized, because the processor can perform other tasks in between timer interrupts. One of the programmable external timer device is the 8254 developed by INTEL. The INTEL 8254 timer has three independent counters. In each counter, a count value can be incremented loaded and decremented by applying a clock signal. At the end of count, each counter will generate an output which can be used as interrupt to processor to initiate the time-based activity. Some of the applications of programmable timers are given below:

1. The timer can interrupt a time-sharing operating system at specified intervals so that it can switch programs.

2. The timer can send timing signals at periodic intervals to IO devices. (For example, start of conversion signal to ADC.)

3. The timer can be used as baud rate generator. (For example, the timer can be used as a clock divider to divide the processor clock to the desired frequency for TxC and RxC of USART-8251A.)

4. The timer can be used to measure the time between the external events.

5. The timer can be used as an external event counter to count repetitive external operations and inform the count value to the processor.

6. The timer can be used to initiate an activity through the interrupt after a programmed number of external events have occured.

The 8254 is a 24-pin IC packed in DIP and requires a single +5 V supply. The pin configuration of 8254 is shown in Fig. 4.18. The functional block diagram of 8254 is shown in Fig. 4.19.

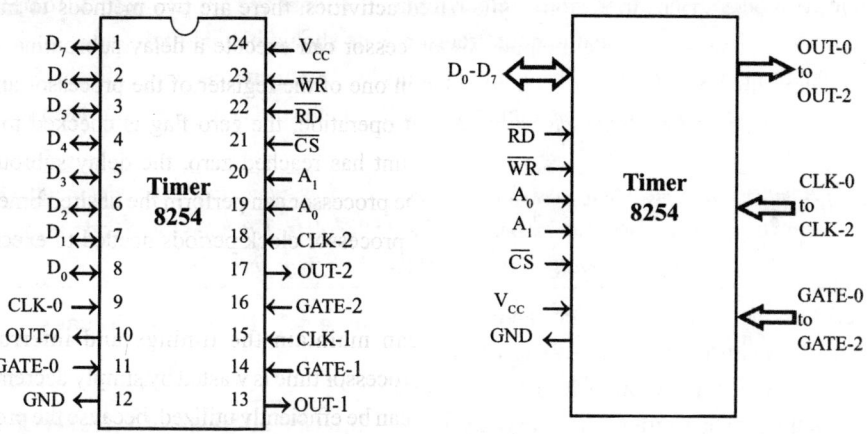

Pin	Description
$D_0 - D_7$	Bidirectional data lines
$\overline{CS}$	Chip select
$\overline{RD}$	Read control
WR	Write control
A_0, A_1	Internal address
CLK-0 to CLK-2	Clock input to counters
GATE-0 to GATE-2	Gate control input to counters
OUT-0 to OUT-2	Output of counters

Fig. 4.18: Pin configuration of an 8254 timer.

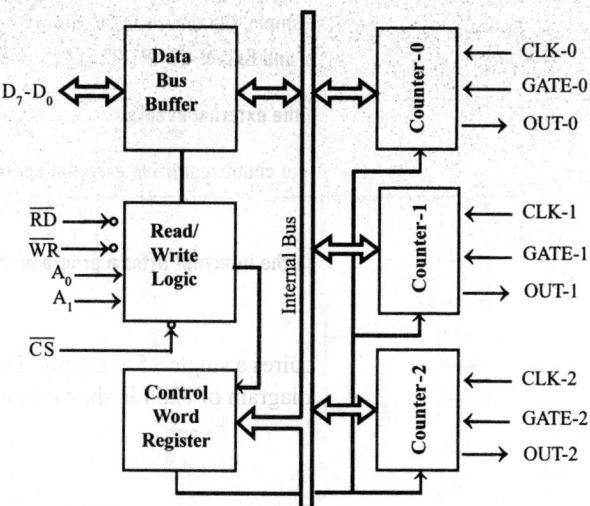

Fig. 4.19: Functional block diagram of an 8254 timer.

The 8254 has three independent 16-bit counters which can be programmed to work in any one of the possible six modes. Each counter has a clock input, gate input and counter output. To operate a counter, a count value has to be loaded in count register, gate should be tied **high** and a clock signal should be applied through clock input. The counter counts by decrementing the count value by one in each cycle of clock signal and generates an output depending on the mode of operation. The maximum input clock frequency for 8254 is 10 MHz.

> *Note :*   *Another timer released by INTEL is the 8253 which is a low clock version of the 8254. The maximum input clock frequency to the 8253 is 2.6 MHz. The 8253 and 8254 are pin to pin compatible and functionally same except the clock frequency.*

The 8254 has eight data lines which can be used for communication with the processor. The control words and count values are written into the 8254 registers through the data bus buffer. The CS is used to select the chip. The address lines A_0 and A_1 are used to select any one of the four internal devices as shown in Table 4.7. The control signals $\overline{RD}$ and $\overline{WR}$ are used by the processor to perform read/write operation. The processor can read the count value in the count register with/without stopping the counter at any time.

Table 4.7: Internal Addresses of 8254

Internal address		Device selected
A_1	A_0	
0	0	Counter-0
0	1	Counter-1
1	0	Counter-2
1	1	Control Register

4.4.1 INTERFACING 8254 WITH 8085 PROCESSOR

A simple schematic for interfacing the 8254 with an 8085 processor is shown in Fig. 4.20. The 8254 can be either memory-mapped or IO-mapped in the system. In the schematic shown in Fig. 4.20, the 8254 is IO-mapped in the system. The chip select signals for IO-mapped devices are generated by using a 3-to-8 decoder. The address lines A_4, A_5 and A_6 are decoded to generate eight chip select signals (IOCS-0 to IOCS-7) and in this, the chip select IOCS-5 is used to select the 8254. The address line A_7 and the control signal $IO/\overline{M}$ are used to enable the decoder.

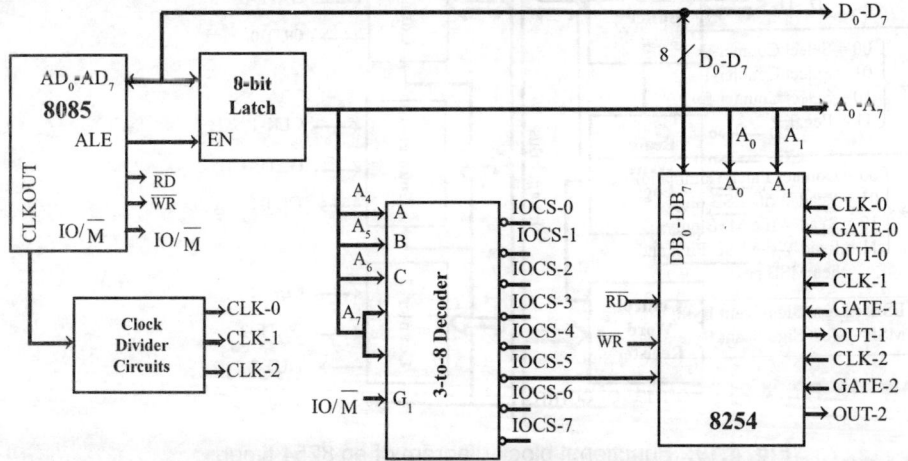

Fig. 4.20: Interfacing of 8254 with 8085 processor.

The address lines A_0 and A_1 of the 8085 are connected to A_0 and A_1 of the 8254 to provide the internal addresses. The IO addresses allotted to the internal devices of 8254 are listed in Table 4.8. The data lines D_0-D_7, $\overline{RD}$ and $\overline{WR}$ signals of the 8254 are connected to the D_0-D_7, $\overline{RD}$ and $\overline{WR}$ of the processor respectively to achieve parallel data transfer.

Table 4.8: IO Addresses of 8254 Interfaced to 8085 as Shown in Fig. 4.21

| Internal device | Binary address | | | | | | | | Hexa address |
| | Decoder input and enable | | | | Input to address pins of 8254 | | | | |
	A_7	A_6	A_5	A_4	A_3	A_2	A_1	A_0	
Counter - 0	0	1	0	1	x	x	0	0	50
Counter - 1	0	1	0	1	x	x	0	1	51
Counter - 2	0	1	0	1	x	x	1	0	52
Control Register	0	1	0	1	x	x	1	1	53

Note : Don't care "x" is considered as zero.

The clock signals required for the counters can be obtained either from the processor clock output or from an external clock source. The clock signal from a 8085 can also be divided to lower values by using clock divider circuits and then applied to clock input of counters.

4.4.2 PROGRAMMING 8254

Each counter of 8254 can be individually programmed by writing a control word followed by the count value. The format of the control word is shown in Fig. 4.21.

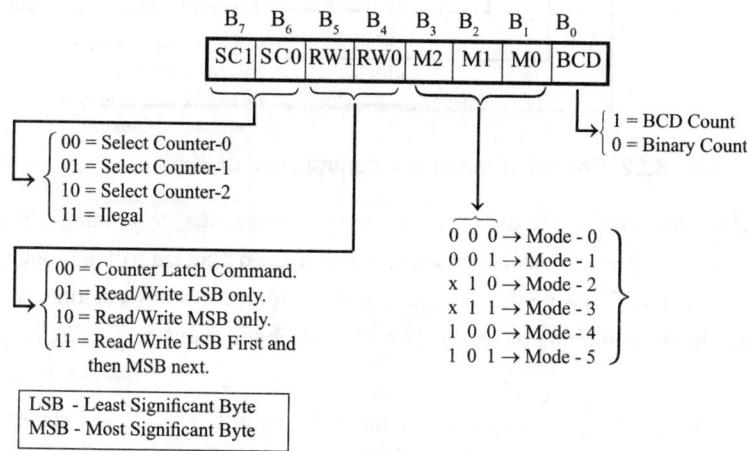

Fig. 4.21: Format of control word for timer 8254.

The bit B_0 (BCD) of control word is used to select BCD or binary count and the bits B_1 to B_3 (M0, M1 and M2) are used to select the mode of operation for the counter specified by bits B_6 and B_7 of control word. Please remember that for each counter separate control word has to be sent to the same control register address. The 8254 identifies the control word for a particular counter from bits B_6 and B_7 of the control word.

The bits B_4 and B_5 are used for read/write command. These bits are programmed for reading/ writing the 16-bit count value in the proper order. If the count value is read without stopping the counter, then the count value may change between reading the LSB and MSB. To avoid this, the counter latch command can be used to latch the count value to an internal latch available at the output of each counter before the read operation.

Alternatively, a separate read-back control word is available for latching the count value in the 8254. (This control word is not available in 8253.) The format of read-back control word of 8254 is shown in Fig. 4.22. This control word has to be send to the same control register address before the read operation to latch the count value. The control register identifies this control word from the value of bits B_6 and B_7.

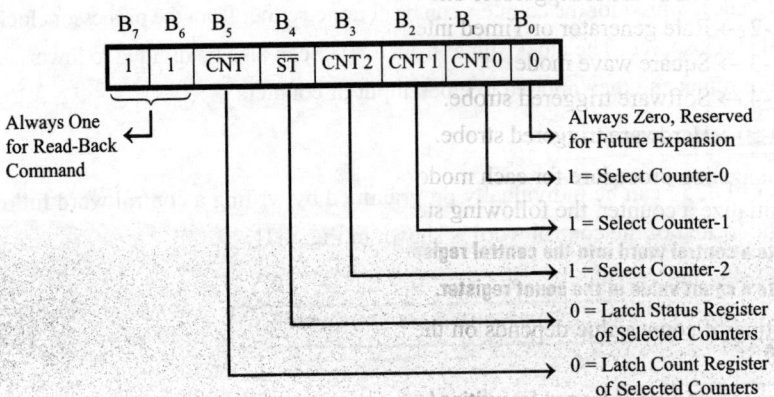

Fig. 4.22: Format of read-back control word of 8254.

The read-back control word can be used to latch one or all the counters by sending a single control word. This control word is also used to latch the status register to the output latch of the counters, so that the status registers can be read by using the respective counter address. At any one time we can latch either the count value by programming the bit B_5 as zero or latch the status register by programming the bit B_4 as zero.

The format of the status register of each counter is shown in Fig. 4.23. The status word of a counter can be read to check the programmed status of the counter and also to verify whether the count value has reached terminal count i.e., zero or not.

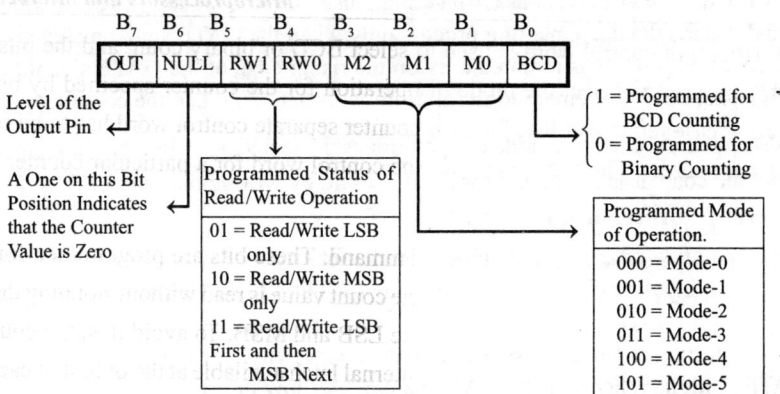

Fig. 4.23: Format of status word of each counter of 8254.

4.4.3 OPERATING MODES OF 8254

The 8254 has six modes of operation. Each counter of 8254 can be independently programmed to work in one of the possible six operating modes. The six modes are:

Mode-0 → Interrupt on terminal count.

Mode-1 → Hardware retriggerable one shot.

Mode-2 → Rate generator or Timed interrupt generator.

Mode-3 → Square wave mode.

Mode-4 → Software triggered strobe.

Mode-5 → Hardware triggered strobe.

The initialization procedure for each mode is almost same, but the output of each mode will be different. To initialize a counter, the following steps are necessary:

1. **Write a control word into the control register.**

2. **Write a count value in the count register.**

The writing of count value depends on the control word. There may be three possible choice. These are:

i) **If the control word is framed for writing LSB only, then write LSB alone.**

ii) **If the control word is framed for writing MSB only, then write MSB alone.**

iii) **If the control word is framed for writing LSB first and MSB next, then write LSB first and write MSB next.**

> *Note :* LSB - Significant Byte (Low order byte).
>
> MSB - Most Significant Byte (High order byte).

In all the modes the GATE signal acts as a control signal to start, stop or maintain the counting process. In modes 0, 2, 3 and 4 once the count value is loaded in the counter, the timer starts decrementing the count value if the GATE is **high**. Whenever the GATE signal goes **low**, the counter stops counting and will resume counting only when the GATE is made **high** again.

In modes 1 and 5 the GATE act as a triggering pulse. In these modes, the count value is loaded in the counter and it starts the decrementing process only when the GATE signal makes a low-to-high transition (i.e., the count process is initiated only on the rising edge of the GATE signal.) In modes 1 and 5 the GATE signal need not remain **high** (after initiation), to maintain the counting process.

A brief description about each mode of operation is presented here. In the following discussions it is assumed that the counter is initialized for binary count, by writing only the LSB of the count.

Mode-0 : Interrupt on terminal count

In mode-0 operation when a count value is loaded in a counter it starts decrementing the count value by one for each input clock pulse (provided the GATE is **high**) and asserts the output as **high** when the count value is zero (i.e., on terminal count). This low-to-high transition of the counter output can be used as an interrupt to the processor to initiate any activity. In mode-0 the 8254 will count as long as the GATE is **high**. Whenever the GATE signal goes **low** the counter stops counting and will resume counting only when the GATE is made **high** again.

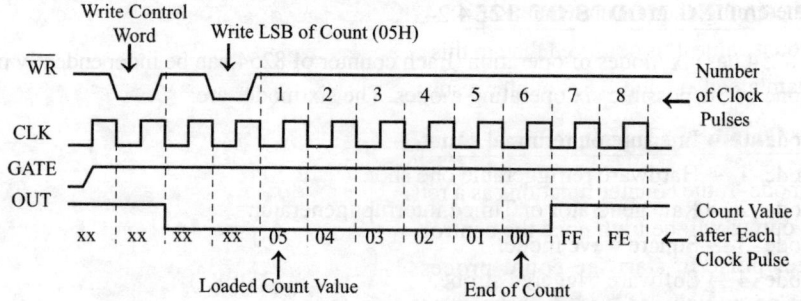

Fig. a: Timing diagram of Mode-0 with GATE always **high**.

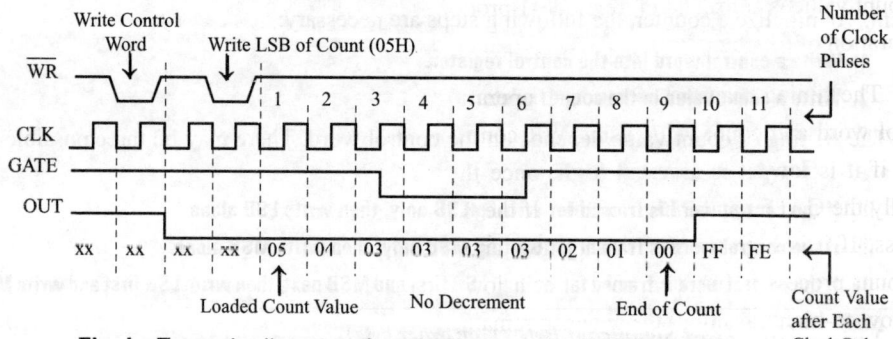

Fig. b: Example diagram of mode-0 when the GATE is made **low** for small duration before the terminal count.

Note : "xx" represents undefined count value.

Fig. 4.24: Timing diagram of mode-0 of 8254.

The timing diagram for mode-0 operation is shown in Fig. 4.24. In the timing diagram of Fig. 4.24(a) initially the counter output remains **high**, and it is assumed that the GATE is always **high**. The processor writes the control word and count value using write control signal ($\overline{WR}$). Once the control word is written into the control register, the output goes **low**. After the write operation of count value by the processor, the 8254 requires one clock pulse to load the count value in the respective count register. Therefore in the first clock pulse after $\overline{\text{WR}}$ goes **high**, the 8254 loads the count value in the count register and in each subsequent clock pulse, the count value is decremented by one. When the count value becomes zero, the output of the counter is asserted **high**.

Figure 4.24 shows the timing diagram for a count value of 05H initially loaded in the count register. Here the output goes **high** after 6 (5 + 1 = 6) clock pulse. In general, if a count value of N is loaded in the count register then the output goes **high** after N + 1 clock pulses. Please remember that the counter continues to decrement the count value even after zero (00 → FF ; FF → FE and so on) as the long as the GATE is **high** and the clock signal is supplied. The output of the counter remains **high** until a new count or command is sent to the counter.

In the timing diagram shown in Fig. 4.24 (b), the GATE is made **low** for a small period before the terminal count value. It is observed that, in this period, the count value is not decremented and previous value is maintained as such. The counter resumes operation only when the GATE is made **high** again.

Mode-1 : Hardware Retriggerable One Shot

In mode-1, the counter functions as a retriggerable monostable multivibrator (one shot). In this mode, the output will be **high** once the control word is sent to the control register. The GATE acts as a trigger pulse to start the count process. When a low-to-high transition of GATE signal occurs, the count value is loaded in the counter and the count is decremented by one for each clock pulse. When the count value is loaded in the counter the OUTPUT goes **low** and it becomes **high** when the count value is zero. Therefore mode-1 produces a logic **low** pulse output whose width is equal to the duration of the count.

The timing diagram of mode-1 operation is shown in Fig. 4.25. The processor writes the control word and count value using $\overline{\text{WR}}$ control signal. Initially the output is assumed to be **high**. Even if it is **low**, it is asserted **high**, once the control word is written into the control register. Initially the GATE can be **high** or **low**. If the GATE is **low** then it is made **high** to initiate the count process. If it is **high** then it is made **low** and after a small delay it is made again **high**, because the count process is initiated only after a low-to-high transition of GATE. After the trigger pulse (i.e., low-to-high transition) the gate can remain either in **high** state or in **low** state.

The first clock pulse after a low-to-high transition of gate is used to load the count value in the counter and for each subsequent clock the count value is decremented by one. Once the count value is loaded in the counter the output is asserted **low** and at the end of the count, when the count value is zero, the output is asserted **high**. In the timing diagram shown in Fig. 4.25(a), a count value of 05H is loaded and so the output remains **low** for 5 clock periods. In general, if a count value of N is loaded in the counter then the output will remain **low** for N clock periods. Therefore the output low pulse width will be N times the clock period.

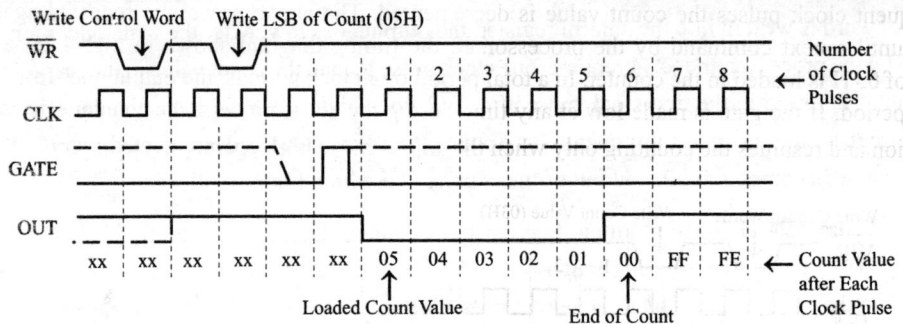

Fig. a: Timing diagram of mode-1.

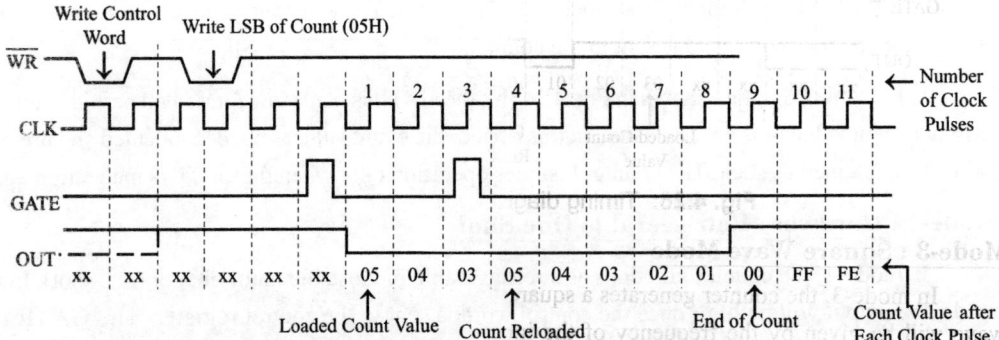

Fig. b: Timing diagram of mode-1 with GATE retriggering before end of count.

Note: "xx" represents undefined count value.

Fig. 4.25: Timing diagram of mode-1 of 8254.

In the timing diagram of Fig. 4.25(b), the GATE is retriggered before the end of the count. In this case, the original count value is reloaded again in the clock pulse after gate retriggering and the count value is decremented by one in each subsequent clock pulse.

Mode-2 : Rate Generator or Timed Interrupt Generator

The mode-2 is used to generate a periodic low pulse of width equal to one clock period. If a count value of N is loaded in the counter then the output will go **low** once in N clock periods. Therefore the frequency of low pulse generated will be equal to the input clock frequency divided by N. For mode-2 operation GATE should be always **high**.

The timing diagram of mode-2 operation is shown in Fig. 4.26. The processor writes the control word and count value using $\overline{WR}$ control signal. Initially the output is assumed to be **high**. Even if it is low, it is asserted **high**, once the control word is written into the control register. The GATE input is permanently tied to logic **high**. In the first clock pulse after the $\overline{WR}$ signal goes **high**, the count value is loaded in the counter and the count value is decremented by one for each subsequent clock pulse.

Initially the output is **high**. When the count reaches one, the output is asserted **low**. In the next clock pulse, the output is asserted **high** and the original count value is reloaded. In the subsequent clock pulses the count value is decremented. The above process is repeated again and again until a next command by the processor. In the timing diagram shown in Fig. 4.26, a count value of 03 H is loaded in the counter. In a total period of 3 clock periods, the output goes **low** for one clock period. If the gate is made **low** at any time during the count process, the counter will stop the operation and resumes the counting only when the gate is made **high** again.

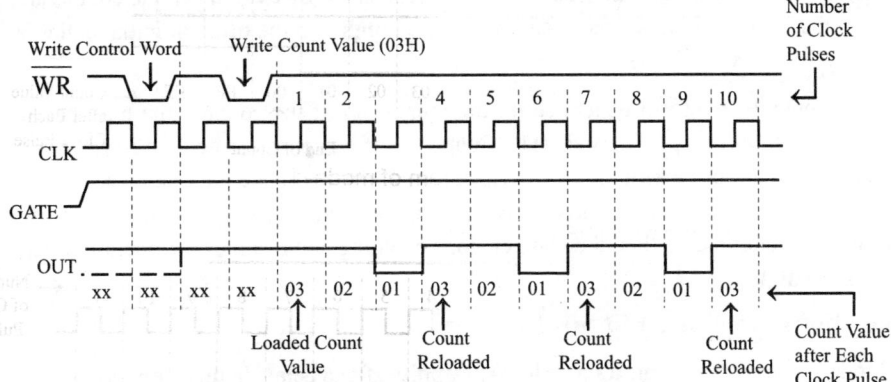

Fig. 4.26: Timing diagram of mode-2 of 8254.

Mode-3 : Square Wave Mode

In mode-3, the counter generates a square wave at the output pin. The frequency of the square wave will be given by the frequency of the input clock signal divided by the count value loaded in the count register. If the count value N is an even number, then the output will be alternatively **high** for N/2 clock periods and **low** for N/2 clock periods. If the count value is an odd number, then the output will be alternatively **high** for $\dfrac{N+1}{2}$ clock periods and **low** for clock periods (i.e., when the count value is odd, then the output **high** period will be more than low period by one clock period.) The timing diagram of mode-3 is shown in Fig. 4.27.

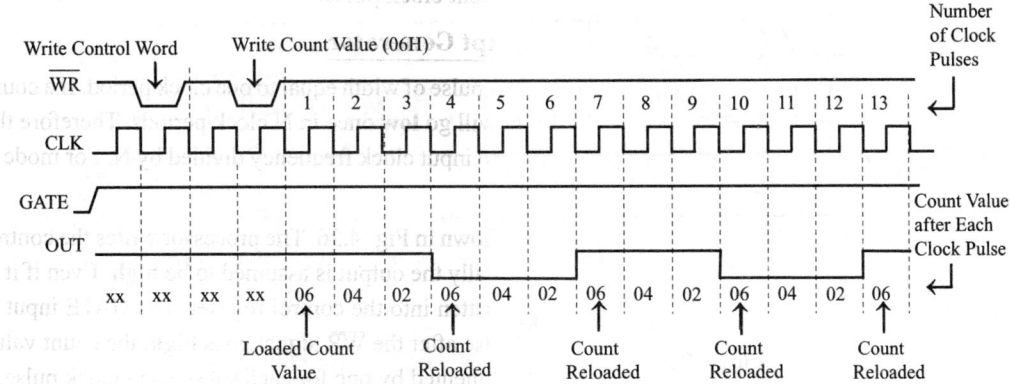

Fig. 4.27: Timing diagram of mode-3 of 8254.

In the timing diagram shown in Fig. 4.27 a count value of 06H is loaded in the counter. The count value is loaded in the counter in the first clock pulse after the $\overline{WR}$ signal goes **high**. Then for each subsequent clock pulse the count is decremented by two. When the count value reaches two then in the next clock pulse, the output is asserted **low** and original/initial count is reloaded in the counter and for each subsequent clock pulse the count is decremented by two. When the count value reaches two then in the next clock pulse, the output is asserted **high** and the original/initial count is reloaded and the above process is repeated again and again.

In the output waveform generated on the output pin of the counter, the high period and low period are equal to three clock periods. The frequency of waveform generated is given by the clock signal divided by six, because six clock periods are required to generate one cycle of output wave. Throughout the mode-3 operation the GATE input signal should be maintained as **high**. If it is made **low** during the count process then the counter stops counting and resumes the operation only after the GATE is made **high**.

Mode-4 : Software Triggered Strobe

Mode-4 is used to generate a single logic **low** pulse after a delay. In this mode when a count value, N is loaded in the counter, a logic **low** pulse of width equal to one clock period is generated in the (N + 1)th clock pulse. Here the delay time is N clock periods. This signal is often used as strobe signal in parallel data transfer scheme. Mode-4 is called a software triggered strobe because the counter starts its operation once the count value is written into the count register by a software instruction. However the GATE input signal should remain **high** throughout the mode-4 operation.

The timing diagram of the mode-4 operation is shown in Fig. 4.28. The GATE is permanently tied to **high**. The processor writes the control word and count value using the write control signal. In the first clock pulse after $\overline{WR}$ signal goes **high**, the count value is loaded in the counter and in each subsequent clock pulses the count value is decremented by one. When the count value reaches zero, the output is asserted **low** for one clock period and then it is made **high**.

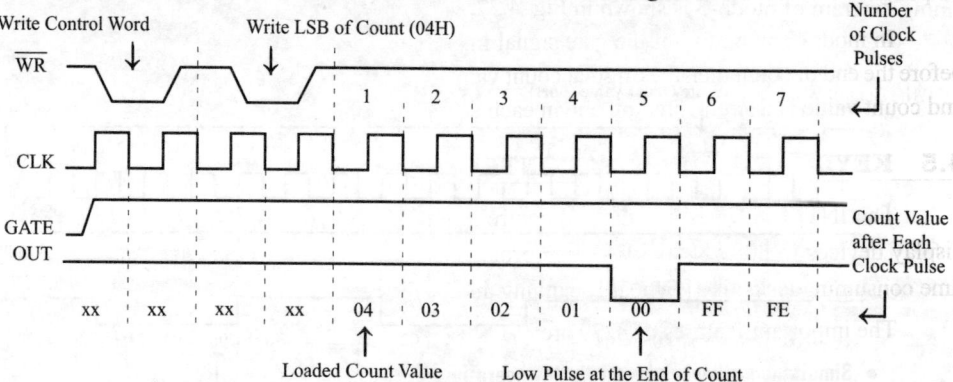

Fig. 4.28: Timing diagram of mode-4 operation of 8254.

Here a count value of 04H is loaded in the counter. Initially the output remains **high** and in the fifth clock pulse the output goes **low** for one clock period. In mode-4 operation, if the GATE is made **low** during count process then the counter stops counting and resumes the operation only when the GATE is made **high**.

Mode-5 : Hardware Triggered Strobe

The mode-5 is same as that of mode-4, except that the counter is initiated by a low-to-high transition of the GATE signal. In mode-4, the counter will start decrementing the count value immediately after the write operation of count value by the processor. But in mode-5, the counter will wait for a low-to-high transition of GATE signal after the write operation of count value by the processor.

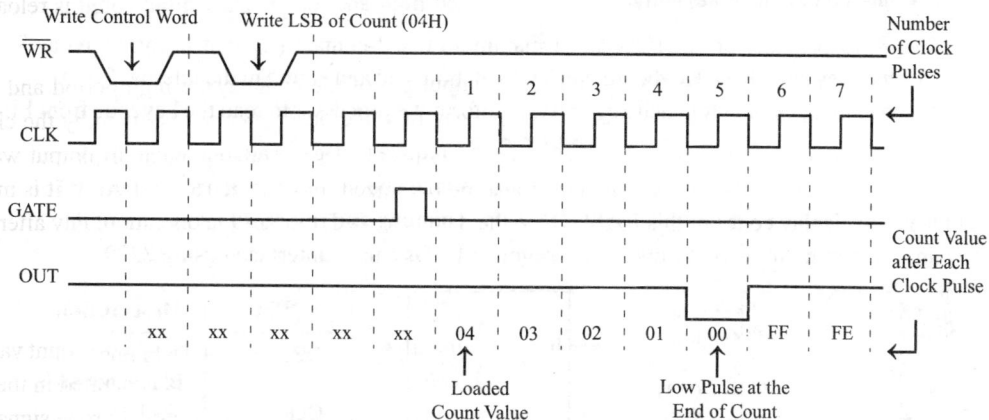

Fig. 4.29: Timing diagram of mode-5 operation of 8254.

The timing diagram of a mode-5 operation is shown in Fig. 4.29. In the first clock pulse after a low-to-high transition of GATE, the count value is loaded in the counter and for each subsequent clock pulse the count value is decremented by one. When the count value reaches zero, the output is asserted **low** for one clock period and then it is made **high**. Here a count value of 04 H is loaded in the counter. Initially the output remains **high** and the counter wait for a low-to-high transition of the GATE signal. In the fifth clock pulse after a low-to-high trasition of GATE signal, the output goes **low** for one clock period.

In mode-5 operation, if the gate signal makes another low-to-high transition (i.e., retriggered) before the end of count then the original count value is reloaded in the clock pulse after gate retriggering and count value is decremented by one in each subsequent clock pulse.

4.5 KEYBOARD / DISPLAY CONTROLLER - INTEL 8279

The INTEL 8279 is a dedicated controller specially developed for interfacing keyboard and display devices to 8085/8086/8088 microprocessor-based system. It relieves the processor from the time consuming tasks like keyboard scanning and display refreshing.

The important features of 8279 are:

- **Simultaneous keyboard and display operations.**
- **2-key lockout or N-key rollover with contact debounce.**
- **Scanned keyboard mode.**
- **Scanned sensor mode.**
- **Strobed input entry mode.**

- **8-character keyboard FIFO.**
- **16-character display.**
- **Right or left entry 16-byte display RAM.**
- **Mode programmable from CPU.**
- **Programmable scan timing.**
- **Interrupt output on key entry.**

The 8279 provides an interface for a maximum of 64-contact key matrix (arranged as 8×8 matrix array of key switches). Keyboard entries are debounced and stored in the internal FIFO RAM. It generates an interrupt signal for each key entry, to inform the processor to read the keycode from FIFO.

The 8279 provides a multiplexed interface for 7-segment LEDs and other popular display devices. It consist of a 16×8 display RAM which can also be organized into a dual 16×4 RAM. The CPU has to load the display codes in this RAM. Once the data is loaded, the 8279 takes care of display and refreshing. A maximum of 16 numbers of 7-segment LEDs can be interfaced using 8279.

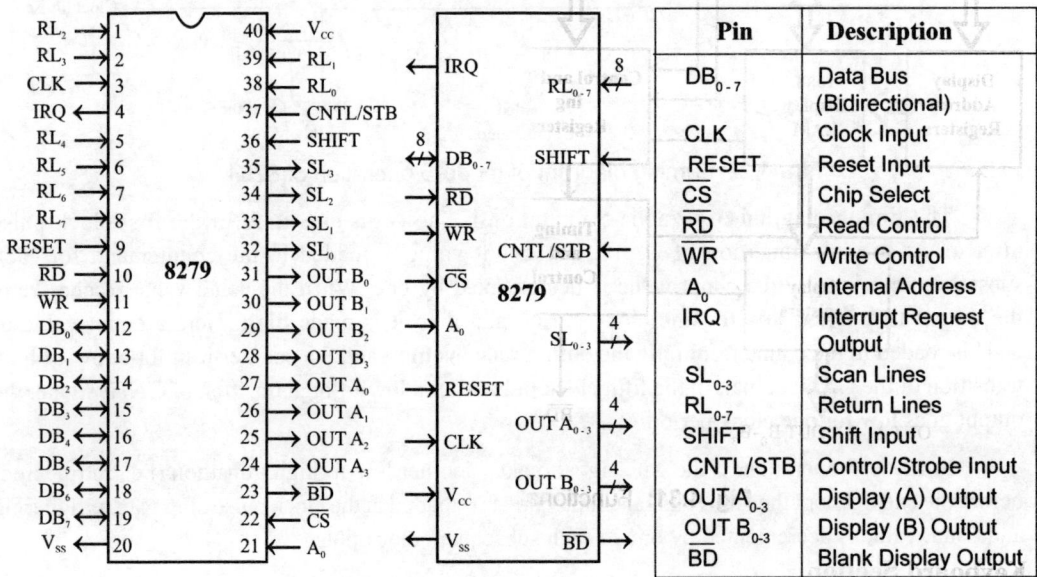

Fig. 4.30: Pin description of 8279.

The 8279 is a 40-pin IC available in DIP (**D**ual **I**n-line **P**ackage). The pin configuration of 8279 is shown in Fig. 4.30. The 8279 has two internal addresses decided by the logic level of A_0. If A_0 is **low,** then the processor can read or write to data register of 8279. If A_0 is **high,** then the processor can write to the control register or read the status register. The 8279 can be either IO-mapped or memory-mapped in the system.

4.5.1 BLOCK DIAGRAM OF 8279

The functional block diagram of an 8279 is shown in Fig. 4.31. The four major sections of 8279 are keyboard, scan, display and CPU interface.

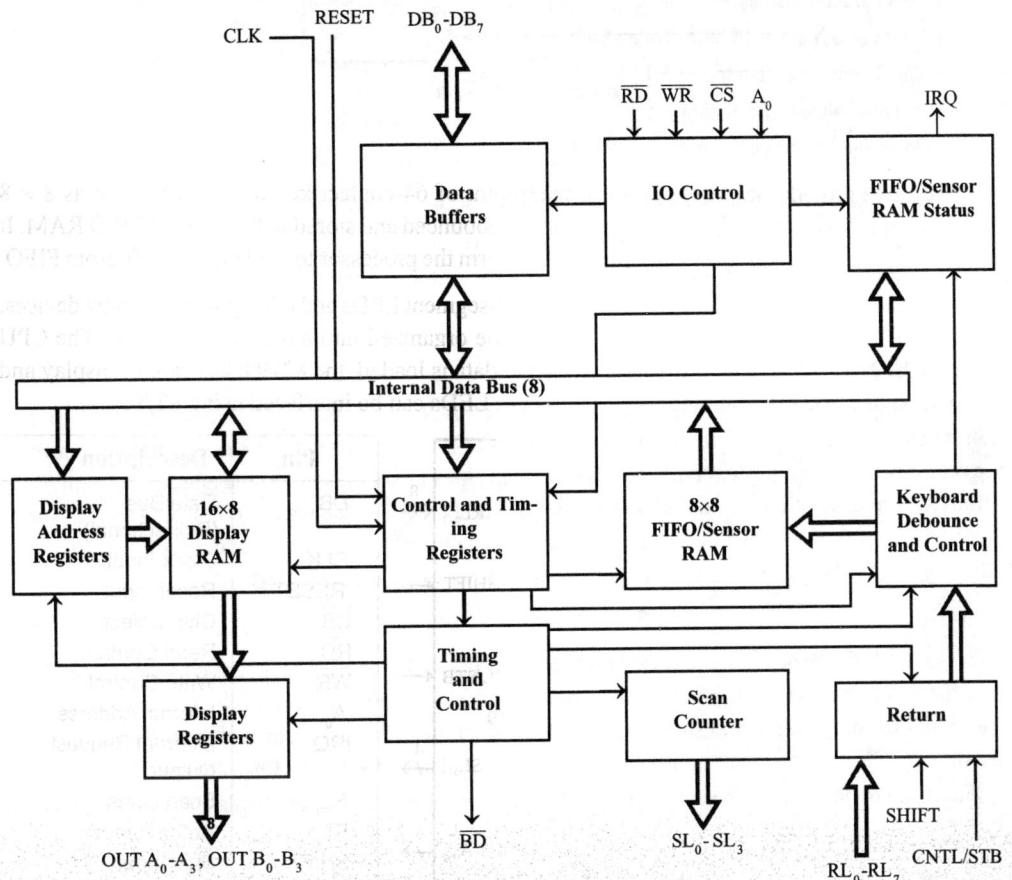

Fig. 4.31: Functional block diagram of 8279.

Keyboard Section

The keyboard section consists of eight return lines RL_0-RL_7 that can be used to form the columns of a keyboard matrix. It has two additional inputs : shift and control/strobe. The keys are automatically debounced. The two operating modes of the keyboard section are 2-key lockout and N-key rollover. In the 2-key lockout mode, if two keys are pressed simultaneously, only the first key is recognized. In the N-key rollover mode, simultaneous keys are recognized and their codes are stored in FIFO.

The keyboard section also has an 8×8 FIFO (**F**irst-**I**n-**F**irst-**O**ut) RAM. The FIFO can store eight keycodes in the scan keyboard mode. The status of the shift key and control key are also stored along with the keycode. The 8279 generates an interrupt signal when there is an entry in the FIFO. The format of keycode entry in the FIFO for scan keyboard mode is shown in Fig. 4.32.

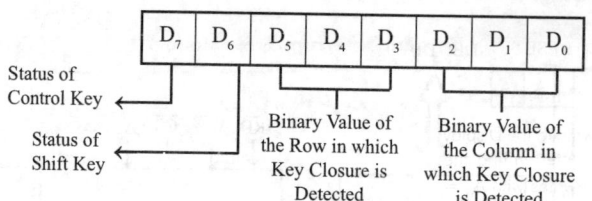

Fig. 4.32: Keycode entry in FIFO for scan keyboard mode.

In sensor matrix mode, the condition (i.e., open/close status) of 64 switches is stored in the FIFO RAM. If the condition of any of the switches change, then the 8279 asserts IRQ as **high** to interrupt the processor.

Display Section

The display section has eight output lines divided into two groups A_0-A_3 and B_0-B_3. The output lines can be used either as a single group of eight lines or as two group of four lines, in conjunction with the scan lines for a multiplexed display. The output lines are connected to the anodes through the driver transistor in case of common cathode 7-segment LEDs. The cathodes are connected to the scan lines through driver transistors. The display can be blanked by $\overline{BD}$ line. The display section consist of 16×8 display RAM. The CPU can read from or write into any location of the display RAM.

Scan Section

The scan section has a scan counter and four scan lines, SL_0 to SL_3. In decoded scan mode, the output of the scan lines will be similar to a 2-to-4 decoder. In encoded scan mode, the output of the scan lines will be binary count, and so an external decoder should be used to convert the binary count to decoded output. The scan lines are common for keyboard and display. These lines are used to form the rows of a matrix keyboard and also connected to digit drivers of a multiplexed display, to turn ON/OFF.

CPU Interface Section

The CPU interface section takes care of the data transfer between 8279 and the processor. This section has eight bidirectional data lines DB_0-DB_7 for data transfer between 8279 and the CPU. It requires two internal addresses $A_0 = 0$ or 1 for selecting either data buffer or control register of 8279. The control signals $\overline{WR}, \overline{RD}, \overline{CS}$ and A_0 are used for read/write to 8279. It has an interrupt request line IRQ, for interrupt driven data transfer with processor.

The 8279 requires an internal clock frequency of 100 kHz. This can be obtained by dividing the input clock by an internal prescaler. The prescaler can take a value from 2 to 31, which is programmable. The RESET signal sets the 8279 in the 16-character display with two-key lockout keyboard mode. Also, the reset will set the clock prescaler to 31.

4.5.2 PROGRAMMING THE 8279

The 8279 can be programmed to perform various functions through eight command words. The formats of the command words and a brief explanation are presented in Fig. 4.33.

Write Display RAM

Code : | 1 | 0 | 0 | AI | A | A | A | A |

The CPU sets up the 8279 for a write to the Display RAM by first writing this command. After writing the command with $A_0 = 1$, all subsequent writes with $A_0 = 0$ will be to the Display RAM. The addressing and auto increment functions are identical to those for the Read Display RAM.

Display Write Inhibit/Blanking

$$\quad\quad\quad\quad\quad\quad A\quad B\quad A\quad B$$

Code : | 1 | 0 | 1 | X | IW | IW | BL | BL |

The IW Bits can be used to mask nibble A and nibble B in application requiring separate 4-bit display ports. By setting the IW flag (IW=1) for one of the ports, the port becomes masked.

The BL flags are available for each nibble. The last Clear command issued determines the code to be used as a blank.

Clear

Code : | 1 | 1 | 0 | C_D | C_D | C_D | C_F | C_A |

The CD bits are available in this command to clear all rows of the Display RAM to a selectable blanking code as follows:

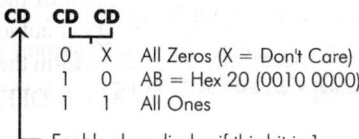

CD	CD	CD	
0	X		All Zeros (X = Don't Care)
1	0		AB = Hex 20 (0010 0000)
1	1		All Ones

Enable clear display if this bit is 1

If the C_F bit is asserted ($C_F = 1$), the FIFO status is cleared and the interrupt output line is reset.

C_A, the clear all bit, has the combined effect of C_D and C_F; it uses the C_D clearing code on the Display RAM and also clears FIFO status. Furthermore, it resynchronizes the internal timing chain.

Read FIFO/Sensor RAM

Code : | 0 | 1 | 0 | AI | X | A | A | A |

X = Don't care

The CPU sets up the 8279 for a read of the FIFO/Sensor RAM by first writing this command. In the Scan keyboard Mode, the Auto-Increment flag (AI) and the Ram address bits (AAA) are irrelevant.

In the Sensor Matrix Mode, the RAM address bits AAA select one of the 8 rows of the Sensor RAM. If the AI flag is set (AI = 1), each successive read will be from the subsequent row of the sensor RAM.

End Interrupt/Error Mode Set

Code : | 1 | 1 | 1 | E | X | X | X | X |

X = Don't care

For the sensor matrix mode this command lowers the IRQ line and enables further writing into RAM. For the N-Key rollover mode , if the E bit is programmed to 1, the chip will operate in the special Error mode.

Keyboard/Display Mode Set

Code : | 0 | 0 | 0 | D | D | K | K | K |

DD

0	0	Eight No.of 8-bit character display	-Left entry
0	1	Sixteen No.of 8-bit character display	-Left entry
1	0	Eight No.of 8-bit character display	-Right entry
1	1	Sixteen No.of 8-bit character display	-Right entry

KKK

0	0	0	Encoded Scan Keyboard - 2-Key Lockout
0	0	1	Decoded Scan Keyboard- 2-Key Lockout
0	1	0	Encoded Scan Keyboard - N-Key Rollover
0	1	1	Decoded Scan Keyboard - N-Key Rollover
1	0	0	Encoded Scan Sensor Matrix
1	0	1	Decoded Scan Sensor Matrix
1	1	0	Strobed Input, Encoded Display Scan
1	1	1	Strobed Input, Decoded Display Scan.

Program Clock

Code : | 0 | 0 | 1 | P | P | P | P | P |

All timing and multiplexing signals for the 8279 are generated by an internal prescaler. This prescaler divides the external clock (pin 3) by a programmable integer. Bits PPPPP determine the value of this integer which ranges from 2 to 31. Choosing a divisor that yields 100 kHz will give the specified scan and debounce times.

Read Display RAM

Code : | 0 | 1 | 1 | AI | A | A | A | A |

The CPU sets up the 8279 for a read of the Display RAM by first writing this command. The address bits AAAA select one of the 16 rows of the Display RAM. If the AI flag is set (AI=1), this row address will be incremented after each read or write to the Display RAM.

Fig. 4.33: 8279 Command word formats.

4.5.3 KEYBOARD AND DISPLAY INTERFACE USING 8279

In a microprocessor-based system, when the keyboard and the 7-segment LED display are interfaced using ports or latches then the processor has to carryout the following tasks:

- **Keyboard scanning**
- **Key debouncing**
- **Keycode generation**
- **Sending display code to LED**
- **Display refreshing**

The above functions has to be performed continuously in specified time intervals. Hence most of the processor time will be utilized for the above task. To overcome this problem, the dedicated keyboard/display controller such as INTEL 8279 can be employed in the system. The 8279 provides a hardware solution for keyboard and display interfacing in microprocessor-based system.

When 8279 is employed, the processor task is to program the 8279 by sending the control words and load the display code in display RAM of 8279. Once the 8279 is programmed it takes care of keyboard scanning, debouncing, keycode generation and display refreshing. Whenever 8279 detects a key press, it informs the processor through interrupt so that the processor can read the keycode from FIFO of 8279.

4.5.4 INTERFACING 8279 WITH 8085 PROCESSOR

A typical hexa keyboard and 7-segment LED display interfacing circuit using 8279 is shown in Fig. 4.34. The circuit can be used in an 8085 microprocessor system and consists of 16 numbers of hexa-keys and 6 numbers of 7-segment LEDs. The 7-segment LEDs can be used to display a six digit alphanumeric character.

The 8279 can be either memory-mapped or IO-mapped in the system. In the circuit of Fig. 4.34, the 8279 is IO-mapped. The address line A_0 of the system is used as A_0 of 8279. The clock signal for 8279 is obtained by dividing the output clock signal of 8085 by a clock divider circuit.

The chip select signal CS is obtained from the IO address decoder of the 8085 system. The chip select signals for IO-mapped devices are generated by using a 3-to-8 decoder. The address lines A_4, A_5 and A_6 are used as input to the decoder. The address line A_7 and the control signal IO/$\overline{M}$ are used as enable for the decoder. The chip select signal IOCS-3 is used to select 8279. The IO address of the internal devices of 8279 are shown in Table 4.9.

Table 4.9: IO Addresses of 8279 Interfaced to 8085 as Shown in Fig. 4.34

Internal device	Binary address								Hexa address
	Decoder input and enable				Input to address line of 8279				
	A_7	A_6	A_5	A_4	A_3	A_2	A_1	A_0	
Data register	0	0	1	1	x	x	x	0	30
Control register	0	0	1	1	x	x	x	1	31

Note : Don't care "x" is considered as zero.

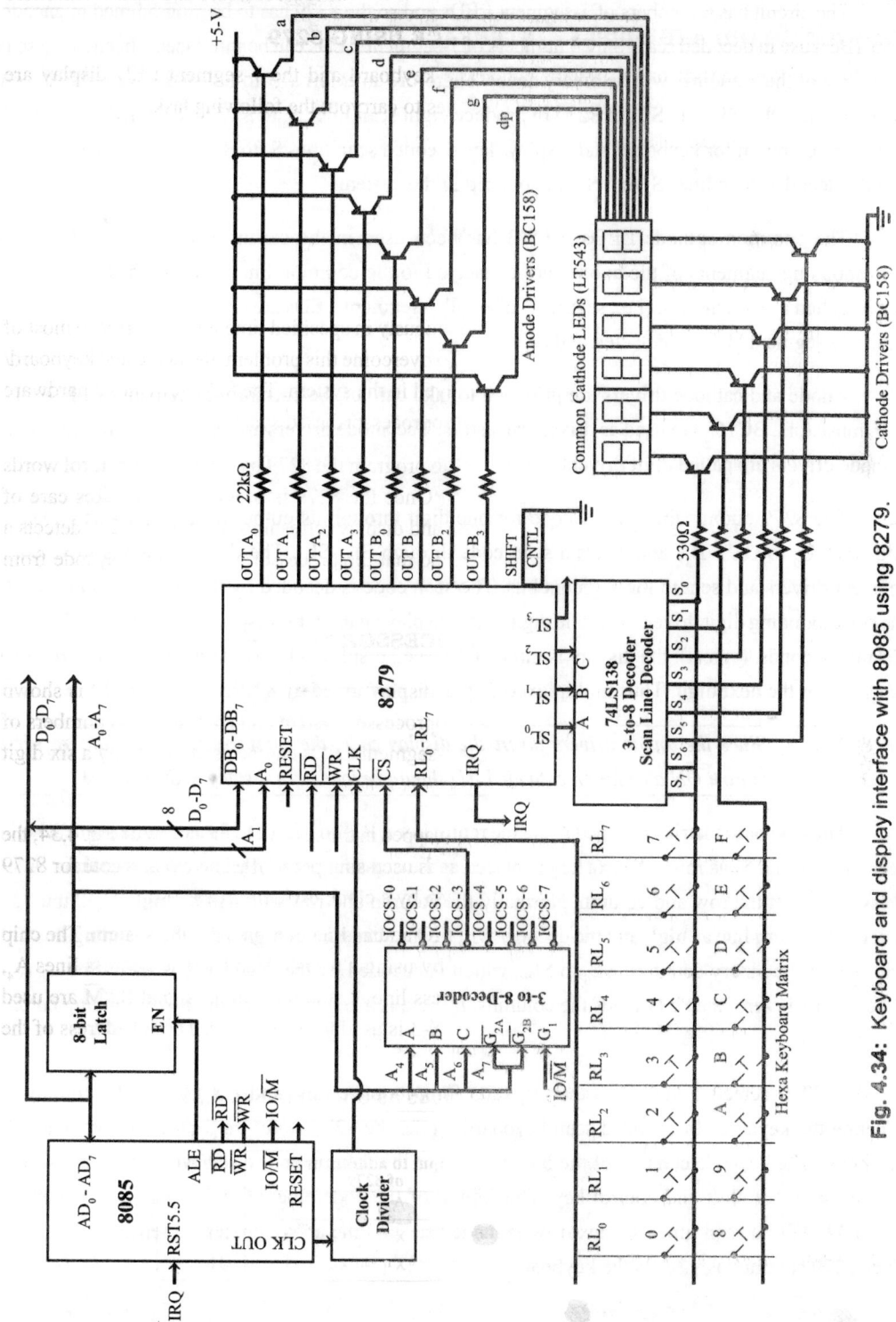

Fig. 4.34: Keyboard and display interface with 8085 using 8279.

The circuit has 6 numbers of 7-segment LEDs and so the 8279 has to be programmed in encoded scan. (Because in decoded scan, only 4 numbers of 7-segment LEDs can be interfaced) In encoded scan, the output of the scan lines will be binary count. Therefore an external 3-to-8 decoder is used to decode the scan lines SL_0, SL_1 and SL_2 of 8279 to produce eight scan lines S_0 to S_7. The decoded scan lines S_0 and S_1 are common for keyboard and display. The decoded scan lines S_2 to S_5 are used only for display and the decoded scan lines S_6 and S_7 are not used in the system.

The common cathode LEDs, LT543 has been used in the circuit shown in Fig. 4.34. The corresponding segments of the anodes are connected to the common line to form a bus and this bus can be called a segment bus. (i.e., segment "a" of all 7-segment LEDs are connected to a common line, similarly segment "b" is connected and so on.)

Anode and cathode drivers are provided to take care of the current requirement of LEDs. The pnp transistors, BC158 are used as driver transistors. The anode drivers are called segment drivers and cathode drivers are called digit drivers.

The 8279 outputs the display code for one digit through its output lines (OUT A_0 to OUT A_3 and OUT B_0 to OUT B_3) and sends a scan code through SL_0-SL_3. The display code is inverted by segment drivers and sent to the segment bus. The scan code is decoded by the decoder and turns ON the corresponding digit driver. Now one digit of the display character is displayed. After a small interval (10 milliseconds, typical), the display is turned OFF (i.e., display is blanked) and the above process is repeated for the next digit. Thus multiplexed display is performed by 8279.

Note :   *Since the anode drivers invert the display code, the complement of the data required to turn ON a common cathode LED should be loaded in display RAM of 8279 .*

The keyboard matrix is formed using the return lines, RL_0 to RL_7 of 8279 as columns and decoded scan lines S_0 and S_1 as rows. A hexa key is placed at the crossing point of each row and column. A key press will short the row and column. Normally the column and row line will be **high**. (i.e., the 8279 will tie the return line as **high** and the decoder will tie the scan line as **high**.) During scanning the 8279 will output the binary count on SL_0 to SL_3, which is decoded by the decoder to make a row as zero. When a row is zero, the 8279 reads the columns. If there is a key press then the corresponding column will be zero.

If 8279 detects a key press then it waits for debounce time and again reads the columns to generate the keycode. In encoded scan keyboard mode, the 8279 stores an 8-bit code for each valid key press. The keycode consists of the binary value of the column and row in which the key is found and the status of shift and control key. The format of the code entered in FIFO RAM is shown in Fig. 4.32. After a scan time, the next row is made zero and the above process is repeated and so on. Thus 8279 continuously scans the keyboard.

4.5.5 INTERFACING 8279 WITH 8051 MICROCONTROLLER

The 8279 can be interfaced to an 8051 controller to provide the keyboard and 7-segment LED display. A simple schematic for interfacing 8279 with 8051 microcontroller is shown in Fig. 4.35. In this scheme the display drivers, scan line decoders and keyboard matrix are same as that in Fig. 4.34. The working of this keyboard and display interface is also similar to the interface shown in Fig. 4.34.

In the schematic shown in Fig. 4.35, the 8279 is mapped as data memory, because the 8051 controller supports only memory mapping of IO devices, and read and write operations are possible only with data memory.

The data lines D_7- D_0 of 8051 are connected to DB_7 - DB_0 lines of 8279. The address line A_0 of 8031/8051 is connected to A_0 of 8279 to provide the internal address. The address lines A_{13}, A_{14} and A_{15} are decoded to generate eight chip select signals, and in this CS_2 is used as chip select for 8279. The signals $\overline{RD}$ and $\overline{WR}$ are logically ANDed and used as logic **low** enable for the decoder and the signal $\overline{PSEN}$ is used as logic **high** enable for the decoder. The addresses allotted to 8279 are listed in Table 4.10.

The clock frequency at X_2 pin of 8051 can be divided using any clock divider circuit to generate 3 MHz clock for 8279. A RC circuit can be employed to generate the reset signal for 8051 as well as for 8279.

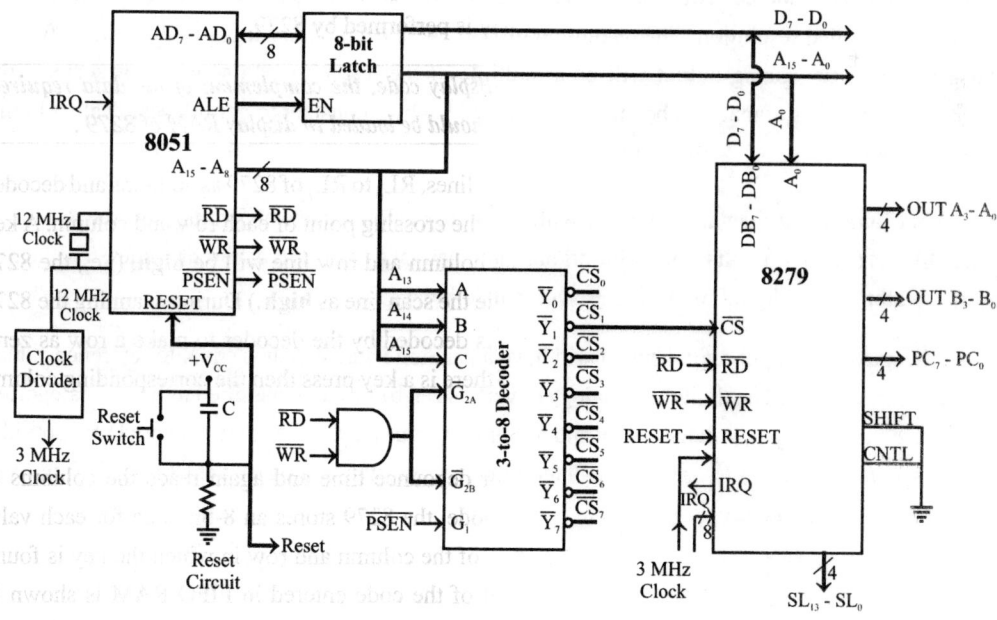

Fig. 4.35: Interfacing of 8279 with 8051 Microcontroller.

Table 4.10: Address Alloted 8279 Interfaced to 8051 as Shown in Fig. 4.35

Internal device	Binary address		Hexa address	Comment
	Decoder input A_{15} A_{14} A_{13}	Input to Address pin of 8279 A_{12} A_{11} A_{10} A_9 A_8 A_7 A_6 A_5 A_4 A_3 A_2 A_1 A_0		
Data Register	0 1 0	x x x x x x x x x x x x 0	4000	External data memory address space
Control Register	0 1 0	x x x x x x x x x x x x 1	4001	

Note: Don't care "x" is considered as zero.

4.6 DAC INTERFACE

In many applications, the microprocessor has to produce analog signals for controlling certain analog devices. Basically, the microprocessor system can produce only digital signals. In order to convert the digital signal to analog signal, a **D**igital-to-**A**nalog **C**onverter (DAC) has to be employed.

The DAC will accept a digital (binary) input and convert to analog voltage or current. Every DAC will have "n" input lines and an analog output. The DAC requires a reference analog voltage (V_{ref}) or current (I_{ref}) source. The smallest possible analog value that can be represented by the n-bit binary code is called resolution. The resolution of DAC with n-bit binary input is $\frac{1}{2^n}$ of the reference analog value. Every analog output will be a multiple of the resolution. In some converters, the input reference analog signal will be multiplied or divided by a constant to get full scale value. In this case the resolution will be $\frac{1}{2^n}$ of the full scale value.

For example, consider an 8-bit DAC with reference analog voltage of 5 volts. Now the resolution of the DAC is $(1/2^8) \times 5$ volts. The 8-bit digital input can take, $2^8 = 256$ different values. The analog values for all possible digital inputs are as shown in Table 4.11.

The maximum input digital signal will have an analog value which is equal to reference analog value minus resolution. The digital-to-analog converters can be broadly classified into three categories, and they are current output, voltage output and multiplying type DAC. The current output DAC provides an analog current as output signal. In voltage output DAC, the analog current signal is internally converted to voltage signal.

TABLE 4.11:

Digital input	Analog output
0000 0000	$\frac{0}{2^8} \times 5$ Volts
0000 0001	$\frac{1}{2^8} \times 5$ Volts
0000 0010	$\frac{2}{2^8} \times 5$ Volts
0000 0011	$\frac{3}{2^8} \times 5$ Volts
.	.
.	.
.	.
1111 1111	$\frac{255}{2^8} \times 5$ Volts

In multiplying type DAC, the output is given by the product of the input signal and the reference source and the product is linear over a broad range. Basically, there is not much difference between these three types and any DAC can be viewed as multiplying DAC.

The basic components of a DAC are resistive network with appropriate values, switches, a reference source and a current to voltage converter as shown in Fig. 4.36.

The switches in the circuit of Fig. 4.36 can be transistors which connect the resistance either to ground or V_{ref}. The resistors are connected in such a way that for any possible binary input, the total current I_T is in binary proportion. The operational amplifier converts the current I_T to a voltage signal V_0, which can be calculated from the following equation.

$$V_0 = V_{ref} \frac{R_f}{R} \left(\frac{D_2}{2^1} + \frac{D_1}{2^2} + \frac{D_0}{2^3} \right)$$

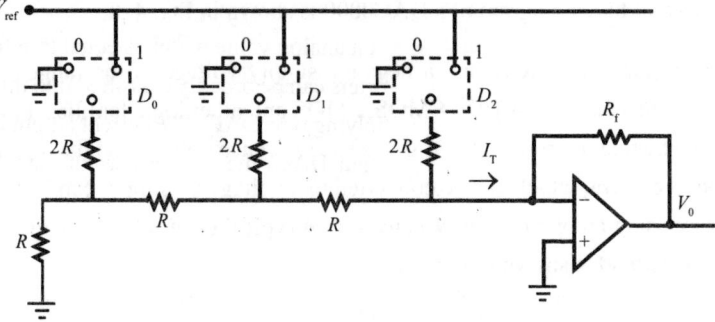

Fig. 4.36: A typical R/2R ladder resistive network as DAC.

The circuit of Fig. 4.36 can be modified as 8-bit DAC by increasing the number of R/2R ladder. For an 8-bit DAC the output voltage is given by,

$$V_0 = V_{ref} \frac{R_f}{R} \left(\frac{D_7}{2^1} + \frac{D_6}{2^2} + \frac{D_5}{2^3} + \frac{D_4}{2^4} + \frac{D_3}{2^5} + \frac{D_2}{2^6} + \frac{D_1}{2^7} + \frac{D_0}{2^8} \right)$$

The time required for converting the digital signal to analog signal is called **conversion time**. It depends on the response time of the switching transistors and the output amplifier. If the DAC is interfaced to the microprocessor, then the digital data (signal) should remain at the input of DAC, until the conversion is complete. Hence, to hold the data a latch is provided at the input of DAC.

The Digital-to-Analog converters compatible to the microprocessors are available with or without internal latch and I to V converting amplifier. The AD558 of the Analog Device is an example of an 8-bit DAC with an internal latch and I to V converting amplifier. The output of AD558 is an analog voltage signal.

The AD558 can be directly interfaced to 8086 microprocessor bus and it requires only two control signals: **Chip Select** ($\overline{CS}$) and **Chip Enable** ($\overline{CE}$). [No handshake signals are necessary for interfacing a DAC. The time between loading two digital data to the DAC is controlled by software time delay.]

The DAC0800 of the National Semiconductor Corporation is an example of an 8-bit DAC without internal latch and I to V converting amplifier. The DAC0800 can be interfaced to the microprocessor using either a port device or a latch.

4.6.1 DAC0800

The DAC0800 is an 8-bit, high speed, current output DAC with a typical settling time (conversion time) of 100 ns. It produces complementary current output which can be converted to voltage by using a simple resistor load.

The DAC0800 is available as a 16-pin IC in DIP. The pin configuration of DAC0800 is shown in Fig. 4.37 and the internal block diagram of a DAC0800 is shown in Fig. 4.38.

The DAC0800 requires a positive and a negative supply voltage in the range of ± 5 V to ± 18 V. It can be directly interfaced with TTL, CMOS, PMOS and other logic families. For TTL input, the threshold pin should be tied to ground ($V_{LC} = 0$ V). The reference voltage and the digital input will decide the analog output current, which can be converted to a voltage by simply connecting a resistor to output terminal or by using an op-amp I to V converter. A typical example of generating a positive voltage output using DAC0800 is shown in Fig. 4.39.

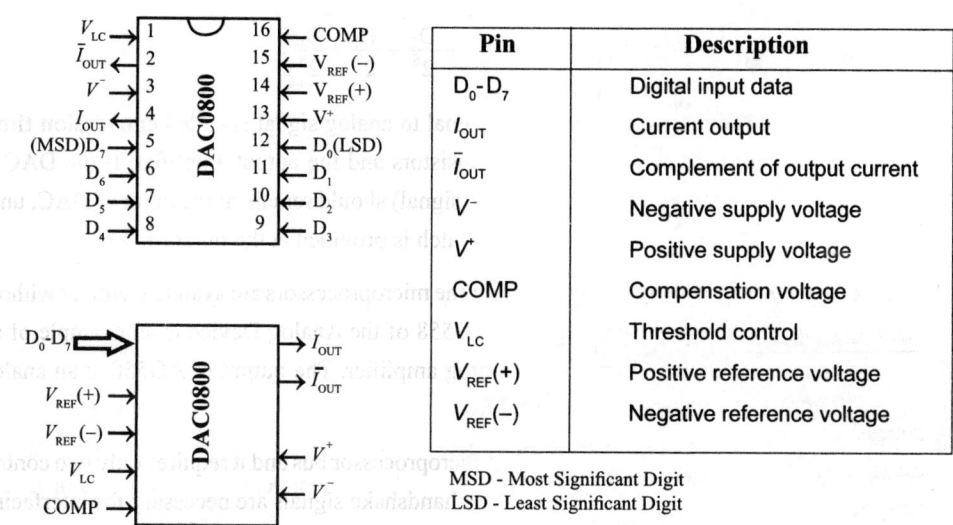

Fig. 4.37: Pin description of DAC0800.

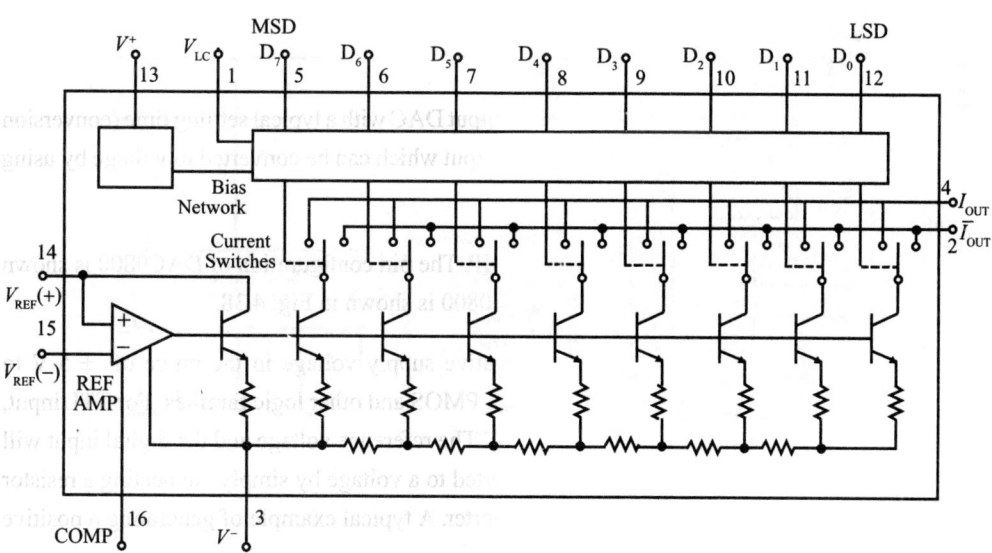

Fig. 4.38: Block diagram of DAC0800.

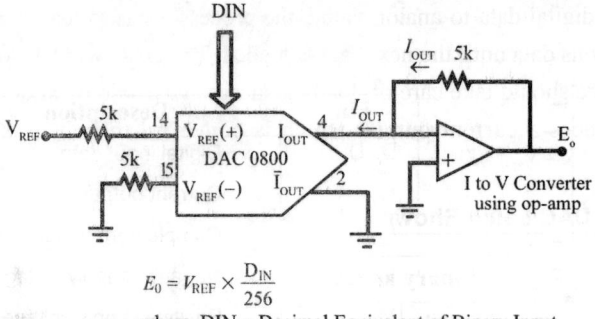

$$E_0 = V_{REF} \times \frac{D_{IN}}{256}$$

where, DIN = Decimal Equivalent of Binary Input

Fig. 4.39: DAC 0800 with I to V converter to produce positive output voltage.

4.6.2 INTERFACING DAC0800 WITH 8085

(AU, Nov/Dec' 19, 15 Marks)

The DAC0800 can be interfaced to an 8085 system bus by using an 8-bit latch and the latch can be enabled by using one of the chip select signals generated for IO devices. A simple schematic for interfacing DAC0800 with 8085 is shown in Fig. 4.40. In this schematic, the DAC0800 is interfaced using an 8-bit latch 74LS273 to the system bus. The 3-to-8 decoder 74LS138 is used to generate chip select signals for IO devices. The address lines A_4, A_5 and A_6 are used as input to decoder. The address line A_7 and the control signal $IO/\overline{M}$ are used as enable for the decoder. The decoder will generate eight chip select signals and in this, the signal IOCS-7 is used as enable for latch of the DAC. The IO address of the DAC is shown in Table 4.12.

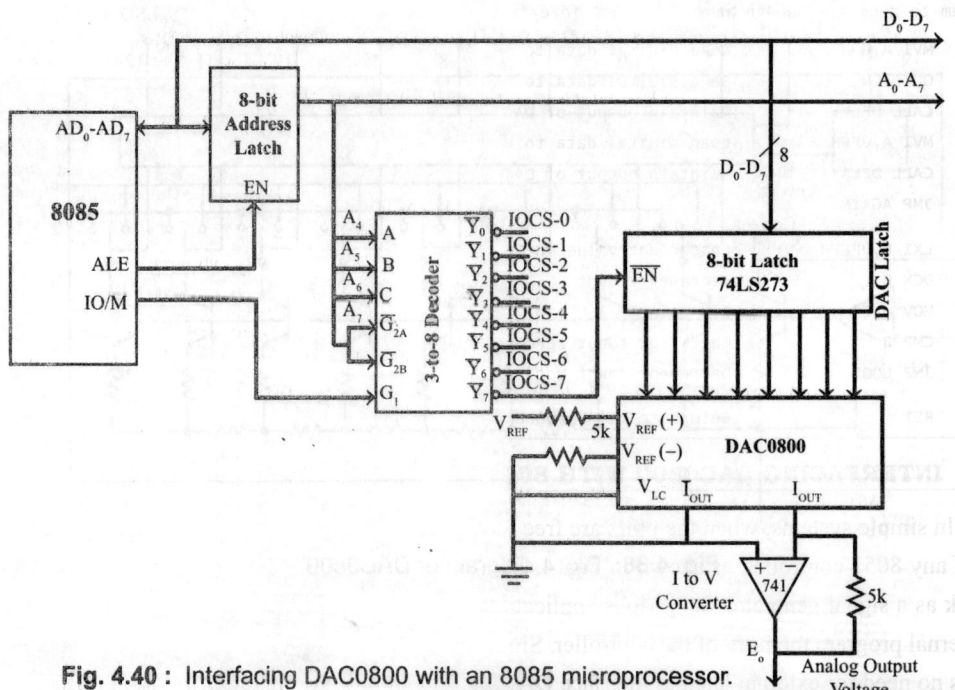

Fig. 4.40 : Interfacing DAC0800 with an 8085 microprocessor.

In order to convert a digital data to analog value, the processor has to load the data to latch. The latch will hold the previous data until the next data is loaded. The DAC will take definite time to convert the data. The software should take care of loading successive data only after the conversion time. The DAC 0800 produces a current output, which is converted to voltage output using a I to V converter.

Table 4.12: IO Address of DAC Latch Shown in Fig. 4.40

Device	Binary address								Hexa address
	Decoder Input and enable				Unused address lines				
	A_7	A_6	A_5	A_4	A_3	A_2	A_1	A_0	
DAC Latch 74LS273	0	1	1	1	x	x	x	x	70

EXAMPLE PROGRAM 1

(AU, Nov/Dec' 19, 15 Marks)

This example program is developed for generation of square waveform using the DAC interfaced to the 8085 system bus as shown in Fig. 4.40. The DAC is interfaced to the system bus with IO address 70_H. In order to generate a square wave, first the digital data corresponding to negative maximum is send to DAC and then after a time delay, the digital data corresponding to positive maximum is send to DAC and again after a time delay, the process is repeated continuously. The frequency of the square wave is decided by the amount of time delay introduced between two voltage levels.

Assembly language program

```
;Program to generate square wave using DAC interfaced to 8085 system bus

AGAIN:   MVI A,00H       ;Load digital data to generate negative maximum
         OUT 70H         ;Send digital data to DAC
         CALL DELAY      ;Maintain output of DAC at negative maximum for half the time period
         MVI A,0FFH      ;Load digital data to generate positive maximum
         CALL DELAY      ;Maintain output of DAC at positive maximum for half the time period
         JMP AGAIN

DELAY:   LXI B,0FFFH     ;Load count value in BC pair (Assume count as 0FFFH)
LOOP:    DCX B           ;Decrement count
         MOV A,C         ;Get C in A
         CMP B           ;Check for count zero
         JNZ LOOP        ;Decrement count until zero

         RET             ;Return to main program
```

4.6.3 INTERFACING DAC0800 WITH 8051 MICROCONTROLLER

In simple systems, when the ports are free the DAC0800 can be directly interfaced to an 8-bit port of any 8051 controller, as shown in Fig. 4.41. In this system, the controller can be programmed to work as a signal generator for various applications and the program can be permanently stored in the internal program memory of the controller. Since the 8x5x ports are internally provided with latch there is no need for external latch to interface DAC0800.

The DAC0800 can also be interfaced to an 8051 microcontroller as memory-mapped IO, as shown in Fig. 4.41. In this case an 8-bit latch such as 74LS273 is interfaced to the system bus and mapped in the data memory address space with 16-bit address. The DAC0800 is connected to output lines of the latch. The controller will load the digital data to the latch and it will hold the data on its output lines. The next data will be loaded to the latch only when previous data has been converted to analog value. The loading of consecutive data to the latch of DAC is controlled by software time delay.

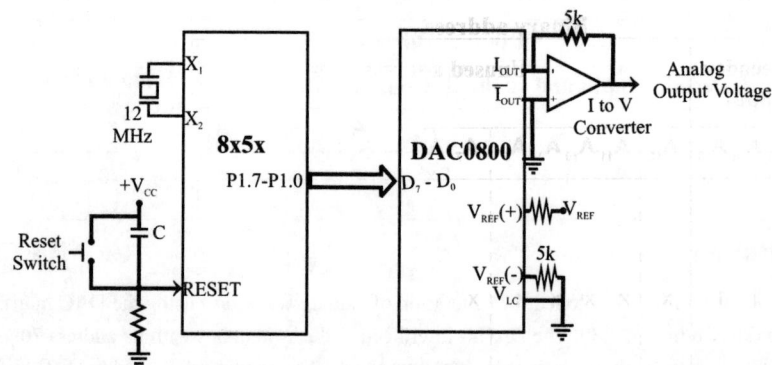

Fig. 4.41: Interfacing DAC0800 to a port of 8051 as MicroController.

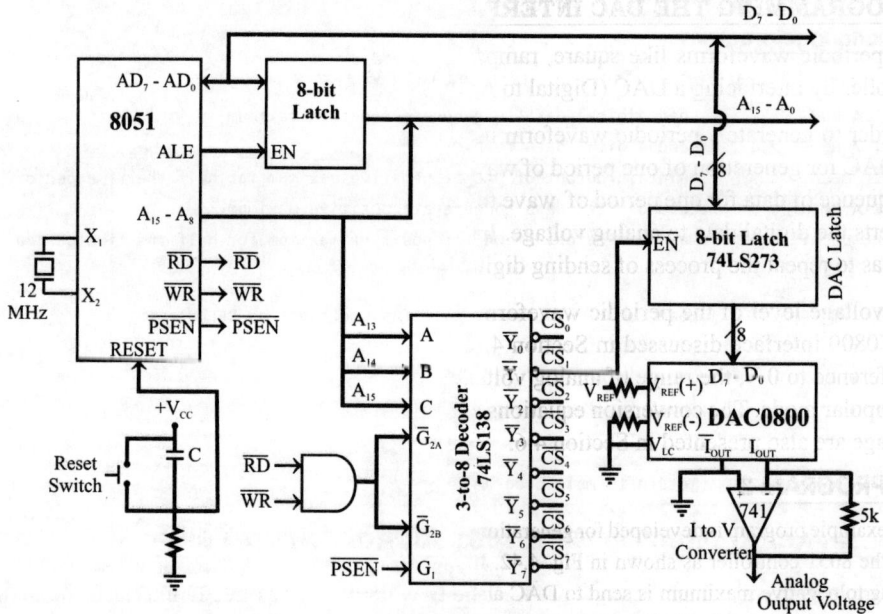

Fig. 4.42: Interfacing DAC0800 to 8051 as memory-mapped IO.

The address lines A_{13}, A_{14} and A_{15} are decoded to generate 8 chip select signals and in this the signal CS_3 is used as logic **low** enable for the DAC latch 74LS273. The signals $\overline{RD}$ and $\overline{WR}$ are logically ANDed and used as logic **low** enable for the decoder and the signal PSEN is used as logic **high** enable for the decoder. The address allotted to DAC latch is shown in Table 4.13.

Table 4.13: Address Allocation to DAC Latch Interfaced to Controller as Shown in Fig. 4.42

Device	Binary address				Hexa address	Comment
	Decoder input	Unused address lines				
	$A_{15}\,A_{14}\,A_{13}$	A_{12} $A_{11}\,A_{10}\,A_9\,A_8$	$A_7\,A_6\,A_5\,A_4$	$A_3\,A_2\,A_1\,A_0$		
DAC External Latch 74LS273	0 1 1	x x x x x	x x x x	x x x x	6000	data memory address space

4.6.4 PROGRAMMING THE DAC INTERFACED (DAC0800) WITH 8051

The periodic waveforms like square, ramp, triangular, sine can be generated using the 8051 microcontroller by interfacing a DAC (Digital to Analog Controller).

In order to generate a periodic waveform using DAC, the sequence of digital data that has to be sent to DAC for generation of one period of waveform has to be determined. The controller has to send the sequence of data for one period of wave to DAC one by one with or without delay, so that the DAC converts the digital data to analog voltage. In order to generate the waveform continuously, the controller has to repeat the process of sending digital data of one period continuously.

The voltage level of the periodic waveform is decided by the reference voltage of the DAC. In the DAC0800 interface discussed in Section 4.6, when the positive reference is tied to +5 V and negative reference to 0 V, the range of analog voltage will be 0 to 5 volts in unipolar mode and –5 V to +5 V in bipolar mode. The conversion equations and relation between the digital data and converted analog voltage are also presented in Section 4.6.

EXAMPLE PROGRAM 2

This example program is developed for generation of a square waveform using the DAC directly interfaced to port-1 of the 8051 controller as shown in Fig. 4.42. In order to generate a square wave, first the digital data corresponding to negative maximum is send to DAC and then after a time delay, the digital data corresponding to positive maximum is send to DAC and again after a time delay, the process is repeated continuously. The frequency of the square wave is decided by the amount of time delay introduced between two voltage levels.

Flowchart

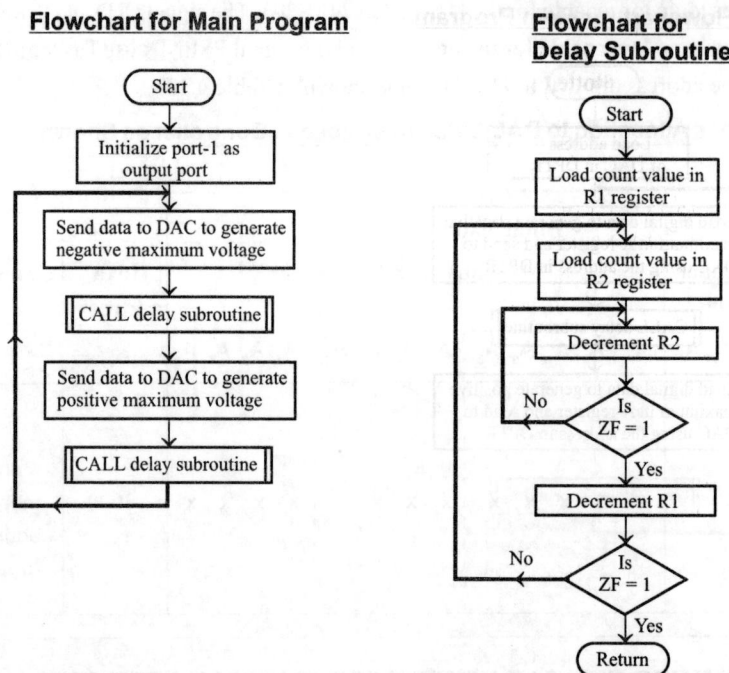

Flowchart for Main Program

Flowchart for Delay Subroutine

Assembly language program

```
;Program to generate square wave using DAC directly interfaced to port of 8051
;Program to generate square wave using DAC directly interfaced to port of 8051

        MOV     P1,#00H    ;Initialize port-1 as output port

AGAIN:  MOV     P1,#00H    ;Send digital data to DAC to generate negative maximum
        ACALL   DELAY      ;Maintain output of DAC at negative maximum for half the time period

        MOV     P1,#FFH    ;Send digital data to DAC to generate positive maximum
        ACALL   DELAY      ;Maintain output of DAC at positive maximum for half the time period

        SJMP    AGAIN

DELAY:  MOV     R1,#FFH    ;Load count value in R1 register
        MOV     R2,#FFH    ;Load count value in R2 register
LOOP1:  DJNZ    R2,LOOP1   ;Decrement count in R2 one by one until zero
LOOP2:  DJNZ    R1,LOOP2   ;Decrement count in R1 ono by one until zero
        RET

        END
```

EXAMPLE PROGRAM 3

This example program is developed for generation of square waveforms using the DAC interfaced to the 8051 system bus as shown in Fig. 4.42. The DAC is interfaced to the system bus with data memory address 6000H. The logic of square-wave generation is same as Chapter 5, Example Program 13.

Flowchart

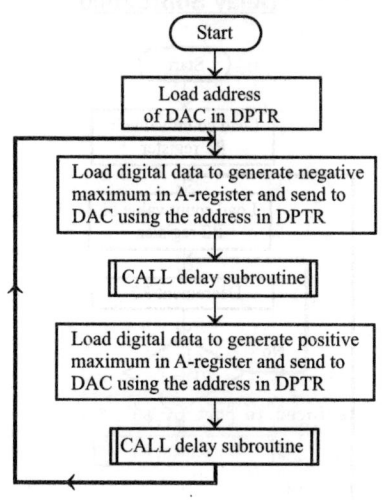

Flowchart for Main Program

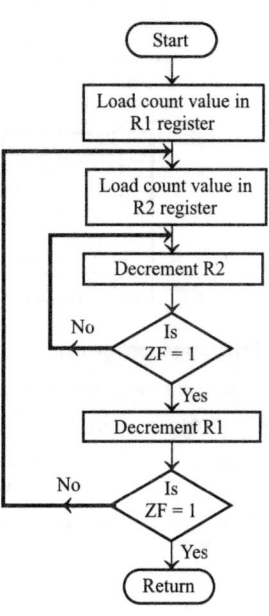

Flowchart for Delay Subroutine

Assembly language program

```
;Program to generate square wave using DAC interfaced to 8051 system bus

AGAIN:  MOV     DPTR,#6000H  ;Load address of DAC in DPTR
        MOV     A,#00H       ;Load digital data to generate negative maximum
        MOVX    @DPTR,A      ;Send digital data to DAC
        ACALL   DELAY        ;Maintain output of DAC at negative maximum for half the time period

        MOV     A,#FFH       ;Load digital data to generate positive maximum
        MOVX    @DPTR,A      ;Send digital data to DAC
        ACALL   DELAY        ;Maintain output of DAC at positive maximum for half the time period

        SJMP    AGAIN

DELAY:  MOV     R1,#FFH      ;Load count value in R1 register
        MOV     R2,#FFH      ;Load count value in R2 register
LOOP1:  DJNZ    R2,LOOP1     ;Decrement count in R2 one by one until zero
LOOP2:  DJNZ    R1,LOOP2     ;Decrement count in R1 one by one until zero
        RET

        END
```

EXAMPLE PROGRAM 4

This example program is developed for generation of ramp waveforms using the DAC directly interfaced to port-1 of the 8051 controller as shown in Fig. 4.42. In order to generate one period of a ramp wave, the digital data is send continuously from minimum to maximum (00H to FFH) one by one with or without delay. Then the process is repeated continuously to generate the ramp waveform. The time delay and the total time taken by the program to generate the all possible 256 values of digital data from 00H to FFH will decide the frequency of the ramp waveform.

Flowchart

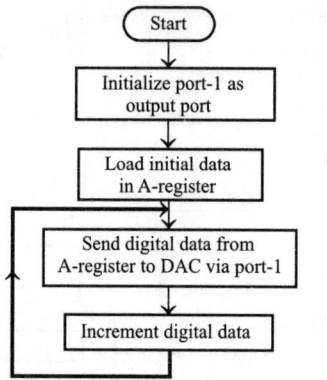

Assembly language program

```
;Program to generate ramp wave using the DAC directly interfaced to port of 8051
        MOV   P1,00H    ;Initialize port-1 as output port

        MOV   A,#00H    ;Load initial digital data in A register
CONTIN: MOV   P1,A      ;Send digital data to DAC

        INC   A         ;Increment digital data one by one
        SJMP  CONTIN    ;and send to DAC continuously

        END
```

EXAMPLE PROGRAM 5

This example program is developed for generation of ramp waveform using the DAC interfaced to the 8051 system bus as shown in Fig. 4.42. The DAC is interfaced to the system bus with data memory address 6000H. The logic of ramp wave generation is same as Chapter 5, Example Program 15.

Flowchart

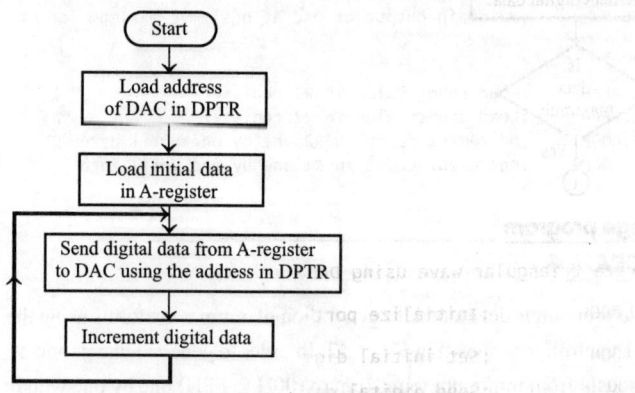

Assembly language program

```
;Program to generate ramp wave using the DAC interfaced to 8051 system bus
        MOV    DPTR,#6000H  ;Load address of DAC in DPTR
        MOV    A,#00H       ;Load initial digital data in A register
CONTIN: MOVX   @DPTR,A      ;Send digital data to DAC
        INC    A            ;Increment digital data one by one
        SJMP   CONTIN       ;and send to DAC continuously
        END
```

EXAMPLE PROGRAM 6

This example program is developed for generation of triangular waveform using the DAC directly interfaced to port-1 of the 8051 controller as shown in Fig. 4.42. In order to generate the raising edge of one period of a triangular wave, the digital data is send continuously from minimum to maximum (00H to FFH) one by one with or without delay, and then to generate the falling edge of one period of triangular wave the digital data is send continuously from maximum to minimum (FFH to 00H) one by one with or without delay. Then the process is repeated continuously to generate the triangular waveform. The time delay and the total time taken by the program to generate the all possible 256 values of digital data from 00H to FFH for raising edge and the all possible values of digital data from FFH to 00H for falling edge will decide the frequency of the triangular waveform.

Flowchart

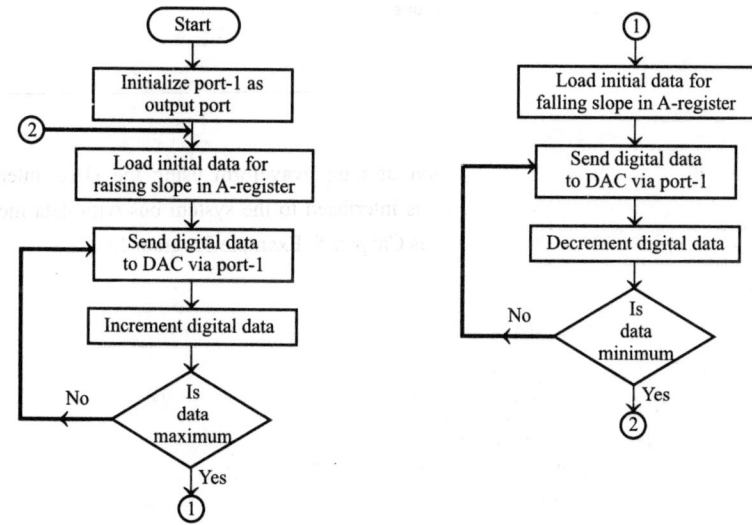

Assembly language program

```
;Program to generate triangular wave using DAC interfaced directly to port of 8051
        MOV    P1,#00H        ;Initialize port-1 as output port
AGAIN:  MOV    A,#00H         ;Set initial digital data for raising slope as 00H
RAISE:  MOV    P1,A           ;Send digital data to DAC
        INC    A              ;Increment the digital data
        CJNE   A,#FFH,RAISE   ;Repeat until the data reaches upper limit
        MOV    A,#FFH         ;Set initial digital data for falling slope as FFH
```

```
FALL:    MOV    P1,A          ;Send digital data to DAC
         DEC    A             ;Decrement the digital data
         CJNE   A,#00H,FALL   ;Repeat until the data reaches lower limit

         JMP    AGAIN         ;Repeat generation of next cycle
         END
```

EXAMPLE PROGRAM 7

This example program is developed for generation of triangular waveform using the DAC interfaced to the 8051 system bus as shown in Fig. 4.42. The DAC is interfaced to the system bus with data memory address 6000H. The logic of the triangular wave generation is same as Example Program 19.

Flowchart

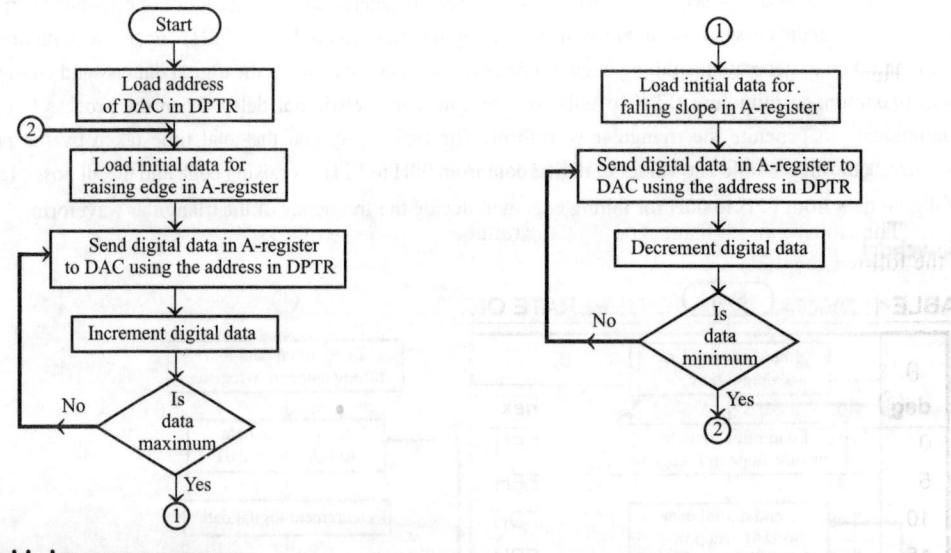

Assembly language program

```
;Program to generate triangular wave using the DAC interfaced to 8051 system bus

AGAIN:   MOV    DPTR,#6000H   ;Load address of DAC in DPTR
         MOV    A,#00H        ;Set initial digital data for raising slope as 00H
RAISE:   MOVX   @DPTR,A       ;Send digital data to DAC
         INC    A             ;Increment the digital data
         CJNE   A,#FFH,RAISE  ;Repeat until the data reaches upper limit

         MOV    A,#FFH        ;Set initial digital data for falling slope as FFH
FALL:    MOVX   @DPTR,A       ;Send digital data to DAC
         DEC    A             ;Decrement the digital data
         CJNE   A,#00H,FALL   ;Repeat until the data reaches lower limit

         JMP    AGAIN         ;Repeat generation of next cycle
         END
```

EXAMPLE PROGRAM 8

This example program is developed for generation of sinusoidal waveform using the DAC directly interfaced to port-1 of 8051 controller as shown in Fig. 4.42. In order to generate of one period of a sine wave, the digital data has to be calculated at equal time intervals of a period and stored as a data base. The controller program has to take digital data from this data base one by one and sent to DAC with or without delay. Then the process is repeated continuously to generate the sine waveform. The time delay and the total time taken by the program to send the digital values of one period will decide the frequency of the sine waveform. Here, one period of the sine wave is divided into 72 equal intervals and the digital data for each interval is calculated and tabulated in the following table.

Calculation of Digital Data For One Period of Sine Wave

In bipolar mode DAC the analog voltage, E_o for a digital data X is given by,

$$E_o = V_{REF} \times \left(\frac{-255 + 2X}{256} \right)$$

From the above equation, the equation of digital data, X can be obtained as,

$$X = \frac{1}{2} \left[\frac{E_o \times 256}{V_{REF}} + 255 \right]$$

Here, $E_o = 5 \times \sin \theta$

$V_{REF} = 5$

$$\therefore X = \frac{1}{2} \left[\frac{5 \times \sin \theta \times 256}{5} + 255 \right] = 128 \times \sin \theta + 127.5$$

The value of digital data X for θ in the range 0° to 355°, in steps of 5°, are calculated and tabulated in the following table.

TABLE 1: DIGITAL DATA TO GENERATE ONE CYCLE OF SINEWAVE

θ deg	X dec	X hex	θ deg	X dec	X hex	θ deg	X dec	X hex	θ deg	X dec	X hex
0	127	7FH	90	255	FFH	180	127	7FH	270	0	00H
5	138	8AH	95	255	FFH	185	116	74H	275	0	00H
10	149	95H	100	253	FDH	190	105	69H	280	1	01H
15	160	A0H	105	251	FBH	195	94	5EH	285	3	03H
20	171	ABH	110	247	F7H	200	83	53H	290	7	07H
25	181	B5H	115	243	F3H	205	73	49H	295	11	0BH
30	191	BFH	120	238	EEH	210	63	3FH	300	16	10H
35	200	C8H	125	232	E8H	215	54	36H	305	22	16H
40	209	D1H	130	225	E1H	220	45	2DH	310	29	1DH
45	218	DAH	135	218	DAH	225	36	24H	315	36	24H
50	225	E1H	140	209	D1H	230	29	1DH	320	45	2DH
55	232	E8H	145	200	C8H	235	22	16H	325	54	36H
60	238	EEH	150	191	BFH	240	16	10H	330	63	3FH
65	243	F3H	155	181	B5H	245	11	0BH	335	73	49H
70	247	F7H	160	171	ABH	250	7	07H	340	83	53H
75	251	FBH	165	160	A0H	255	3	03H	345	94	5EH
80	253	FDH	170	149	95H	260	1	01H	350	105	69H
85	255	FFH	175	138	8AH	265	0	00H	355	116	74H

Flowchart

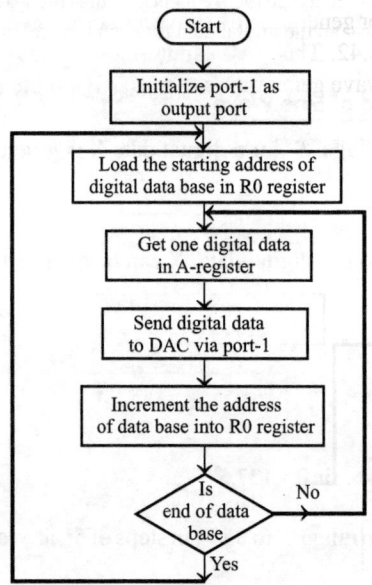

Assembly language program

```
;Program to generate sinewave using DAC directly interfaced to port of 8051

        MOV   P1,#00H        ;Initialize port-1 as output port

AGAIN   MOV   R0,#10H        ;Load starting address of table in R0 register

CONTIN: MOV   A,@R0          ;Get the digital data in A register
        MOV   P1,A           ;Send digital data to DAC
        INC   R0             ;Increment the address
        CJNE  R0,#58H,CONTIN ;Continue output of digital data one by one, until end of table

        SJMP  AGAIN          ;Repeat generation of next cycle

;Digital data to generate one cycle of sinewave

        ORG 10H

TABLE:  DB 7FH,8AH,95H,A0H,ABH,B5H,BFH,C8H
        DB D1H,DAH,E1H,E8H,EEH,F3H,F7H,FBH
        DB FDH,FFH,FFH,FFH,FDH,FBH,F7H,F3H
        DB EEH,E8H,E1H,DAH,D1H,C8H,BFH,B5H
        DB ABH,A0H,95H,8AH,7FH,74H,69H,5EH
        DB 53H,49H,3FH,36H,2DH,24H,1DH,16H
        DB 10H,0BH,07H,03H,01H,00H,00H,00H
        DB 01H,03H,07H,0BH,10H,16H,1DH,24H
        DB 2DH,36H,3FH,49H,53H,5EH,69H,74H

        END
```

EXAMPLE PROGRAM 9

This example program is developed for generation of sine waveforms using the DAC interfaced to the 8051 system bus as shown in Fig. 4.42. The DAC is interfaced to the system bus with data memory address 6000H. The logic of sine-wave generation is same as Example Program 14.

Flowchart

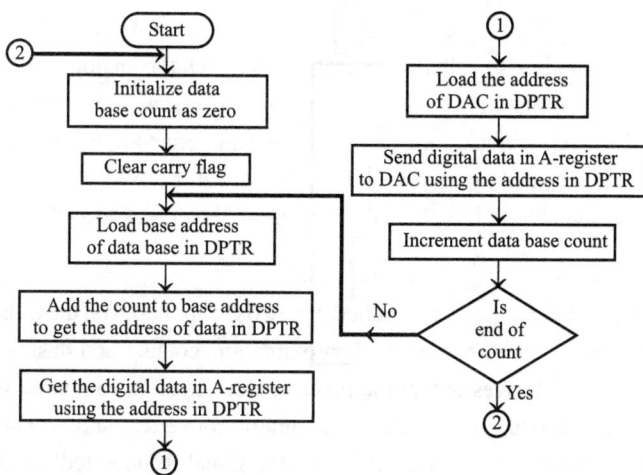

Assembly language program

```
;Program to generate sinewave using DAC interfaced to 8051 system bus

AGAIN:  MOV   R1,#00H        ;Initialize count as zero
        CLR   C              ;Clear carry flag

CONTIN: MOV   DPTR,#TABLE    ;Load base address of table in DPTR
        MOV   A,R1           ;Get the count in A register
        ADDC  A,DPL          ;Add the count to low byte of DPTR
        MOV   DPL,A          ;to get the address of digital data in DPTR

        MOVX  A,@DPTR        ;Get the digital data in A register
        MOV   DPTR,#6000H    ;Load the address of DAC in DPTR
        MOVX  @DPTR,A        ;Send digital data to DAC

        INC   R1             ;Increment count
        MOV   A,R1           ;Get the count in A register
        CJNE  A,#48H,CONTIN  ;Continue output of digital data one by one, until end of count

        SJMP  AGAIN          ;Repeat generation of next cycle

;Digital data to generate one cycle of sinewave
TABLE:  DB  7FH,8AH,95H,A0H,ABH,B5H,BFH,C8H
        DB  D1H,DAH,E1H,E8H,EEH,F3H,F7H,FBH
        DB  FDH,FFH,FFH,FFH,FDH,FBH,F7H,F3H
        DB  EEH,E8H,E1H,DAH,D1H,C8H,BFH,B5H
        DB  ABH,A0H,95H,8AH,7FH,74H,69H,5EH
        DB  53H,49H,3FH,36H,2DH,24H,1DH,16H
        DB  10H,0BH,07H,03H,01H,00H,00H,00H
        DB  01H,03H,07H,0BH,10H,16H,1DH,24H
        DB  2DH,36H,3FH,49H,53H,5EH,69H,74H

        END
```

4.7 ADC INTERFACE

In many applications, an analog device has to be interfaced to the digital system. But the digital devices cannot accept the analog signals directly and so the analog signals are converted to equivalent digital signals (data) using an **A**nalog-to-**D**igital **C**onverter (ADC).

The **A**nalog-to-**D**igital (A/D) conversion is the reverse process of **D**igital to **A**nalog (D/A) conversion. The A/D conversion is also called quantization, in which the analog signal is represented by an equivalent binary data. The analog signals vary continuously and are defined for any interval of time. The digital signals (or data) can take only finite values and defined only for discrete instant of time. If the digital data is represented by an n-bit binary then it can have 2^n different values. In A/D conversion the given analog signal has to be divided into steps of 2^n values, and each step is represented by one of the 2^n values.

The analog-to-digital converters can be classified into two groups based on the technique involved for conversion. The first group includes successive approximation, counter and flash-type converters. The technique involved in these devices is that the given analog signal is compared with internally generated analog signal. The second group includes integrator converters and voltage to frequency converters. In the devices of the second group, the given analog signal is converted to time or frequency and the new parameters (time or frequency) is compared with the known values to produce digital signal.

The trade-off between the two techniques is based on Accuracy vs Speed. The successive approximation and the flash type are faster but generally less accurate than the integrator and the voltage-to-frequency type converters. Also, the flash type is costlier. The successive approximation type converters are used for high speed conversion and the integrating type converters are used for high accuracy.

The resolution of the converter is the minimum analog value that can be represented by the digital data. If the ADC gives n-bit digital output and the full scale analog input is X volts, then the resolution is $\frac{1}{2^n} \times X$ volts. In an ADC, another critical parameter is conversion time. The conversion time is defind as the total time required to convert an analog signal into its digital equivalent. It depends on the conversion technique and the propagation delay in various circuits.

Successive Approximation ADC

A successive approximation ADC consists of D/A converter, successive approximation register and comparator. Figure 4.43 shows the functional blocks of a typical successive approximation A/D converter.

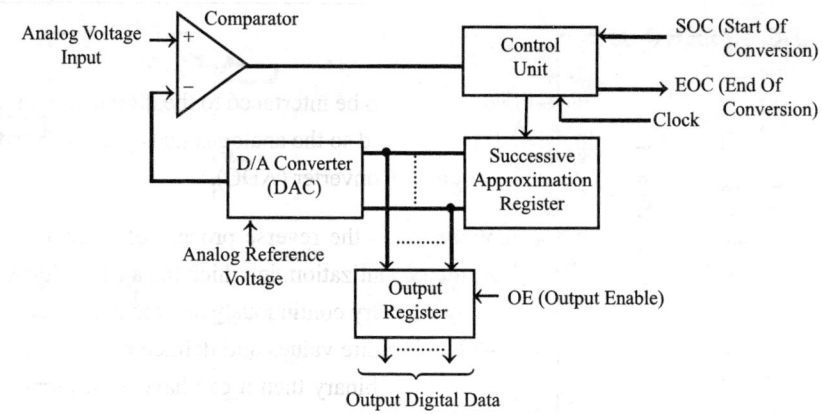

Fig. 4.43: Successive approximation A/D converter.

The conversion process is initiated by a **S**tart **O**f **C**onversion (SOC) signal from the processor to the ADC. On receiving the SOC, the control unit of the ADC will give a start command to the successive approximation register and it starts generating digital signal by the successive approximation method. The generated digital data is converted to analog signal by the D/A converter and then compared with the given analog signal. When the analog signals are equal, the comparator output informs the control unit to stop generation of digital signal. The digital data available at this instant is given as output through output register. Also, the control unit generates a signal to indicate the **E**nd **O**f **C**onversion (EOC) process to the processor.

Successive Approximation Method of Conversion

In this method, the MSD (**M**ost **S**ignificant **D**igit) is first set to **"1"** and all other digits are reset to **"0"**. The analog signal generated for this digital data is compared with the given analog signal. (Initially, the comparator output will be **high**. After comparison the output of the comparator remains in **high** state if the given analog signal is higher than the generated analog signal. Otherwise, if the given signal is less than the generated signal then the output of the comparator changes from **high** to **low** state.) If the output state of the comparator changes then the MSD is reset to **"0"** otherwise it is retained as **"1"**. Then the above process is repeated by setting the next higher order bit to **"1"**. The process is continued for each bit starting from MSD to LSD. (During a process, the higher order bits are the bits determined in earlier steps and the lower order bits are reset to "0".) After one complete cycle through MSD to LSD, the data available on the successive approximation register will be the digital equivalent of the given analog signal.

4.7.1 ADC0809

The ADC0809 is an 8-bit successive approximation type ADC with inbuilt 8-channel multiplexer. The ADC0809 is suitable for interface with 8086 microprocessor. The ADC0809 is available as a 28-pin IC in DIP (**D**ual **I**n-line **P**ackage). The ADC0809 has a total unadjusted error of ±1 LSD (**L**east **S**ignificant **D**igit). The ADC0808 is also same as ADC0809 except the error. The total unadjusted error in ADC0808 is $\pm\frac{1}{2}$ LSD. The pin configuration of ADC0809/ADC0808 is shown in Fig. 4.44.

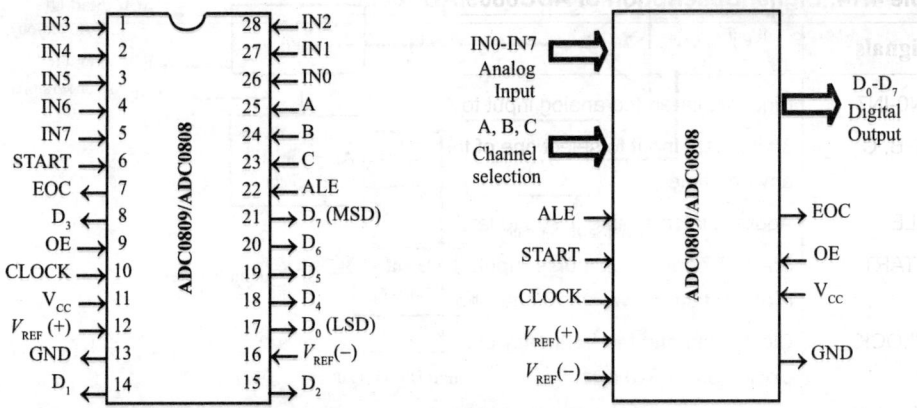

LSD = Least Significant Digit, MSD = Most Significant Digit

Fig. 4.44: Pin configuration of ADC0809/ADC0808.

The internal block diagram of ADC0809/ADC0808 is shown in Fig. 4.45. The various functional blocks of ADC are 8-channel multiplexer, comparator, 256R resistor ladder, switch tree, successive approximation register, output buffer, address latch and decoder.

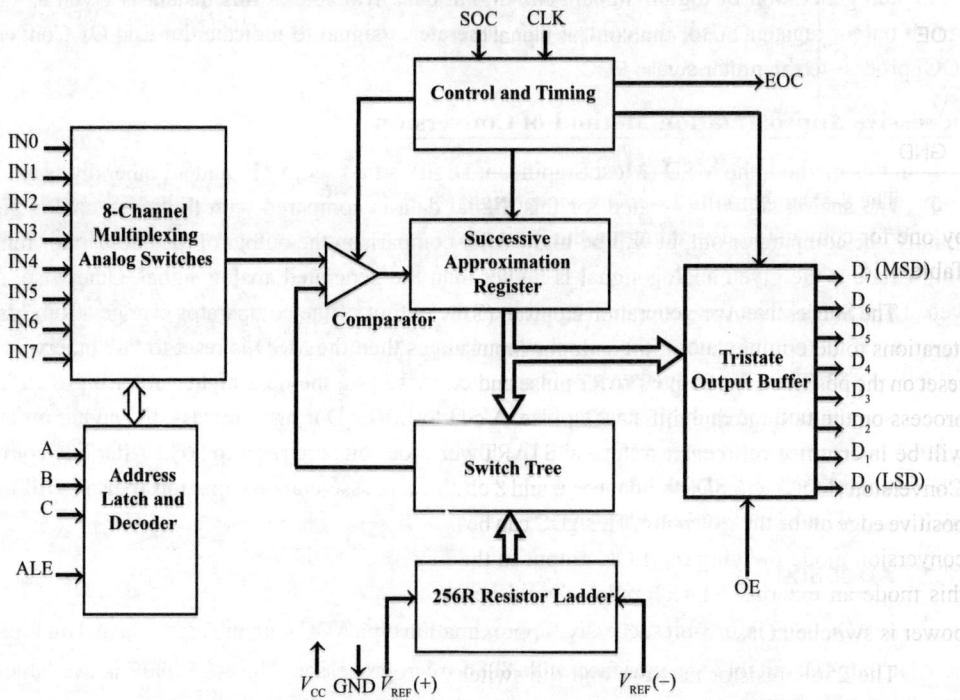

Fig. 4.45: Functional block diagram of ADC0809/ADC0808.

Table 4.14: Signal Description of ADC0809/ADC0808

Signals	Description
IN0-IN7	Eight single-ended analog input to ADC.
A, B, C	3-bit binary input to select one of the eight analog signals for conversion at any one time.
ALE	Address latch enable. Used to latch the 3-bit address input to an internal latch.
START	Start of conversion pulse input. To start ADC process this signal should be asserted **high** and then **low**. This signal should remain **high** for atleast 100 ns.
CLOCK	Clock input and the frequency of clock can be in the range of 10 kHz to 1280 kHz. Typical clock input is 640 kHz.
$V_{REF}(+), V_{REF}(-)$	Reference voltage input. The positive reference voltage can be less than or equal to V_{cc} and the negative reference voltage can be greater than or equal to ground.
D_0-D_7	The 8-bit digital output. The reference voltages will decide the mapping of the analog input to the digital data.
EOC	End of conversion. This signal is asserted **high** by the ADC to indicate the end of conversion process and it can be used as interrupt signal to processor.
OE	Output buffer enable. This signal is used to read the digital data from the output buffer after a valid EOC.
V_{cc}	Power supply, +5 V
GND	Power supply ground, 0 V

The 8-channel multiplexer can accept eight analog inputs in the range of 0 to 5 V and allow one by one for conversion depending on the 3-bit address input. The channel selection logic is shown in Table 4.15.

The **S**uccessive **A**pproximation **R**egister (SAR) performs eight iterations to determine the digital code for input value. The SAR is reset on the positive edge of the START pulse and start the conversion process on the falling edge of START pulse. A conversion processs will be interrupted on receipt of a new START pulse. The **E**nd **O**f **C**onversion (EOC) will go **low** between 0 and 8 clock pulses after the positive edge of the START pulse. The ADC can be used in continuous conversion mode by tying the EOC output to the START input. In this mode an external START pulse should be applied whenever power is switched ON.

The 256R resistor network and the switch tree is shown in Fig. 4.46. The 256R ladder network has been provided instead of conventional R/2R ladder because of its inherent monotonicity, which guarantees no missing digital codes. Also, the 256R resistor network does not cause load variations on the reference voltage.

Table 4.15:

Address input			Selected channel
C	**B**	**A**	
0	0	0	IN0
0	0	1	IN1
0	1	0	IN2
0	1	1	IN3
1	0	0	IN4
1	0	1	IN5
1	1	0	IN6
1	1	1	IN7

The comparator in the ADC0809/ ADC0808 is a chopper-stabilized comparator. It converts the DC input signal into an AC signal and amplifies the AC signal using a high gain AC amplifier. Then it converts AC signal to DC signal. This technique limits the drift component of the amplifier, because the drift is a DC component and it is not amplified/passed by the AC amplifier. This makes the ADC extremely insensitive to temperature, long term drift and input offset errors.

In ADC conversion process, the input analog value is quantized and each quantized analog value will have a unique binary equivalent. The quantization step in ADC0809/ADC0808 is given by,

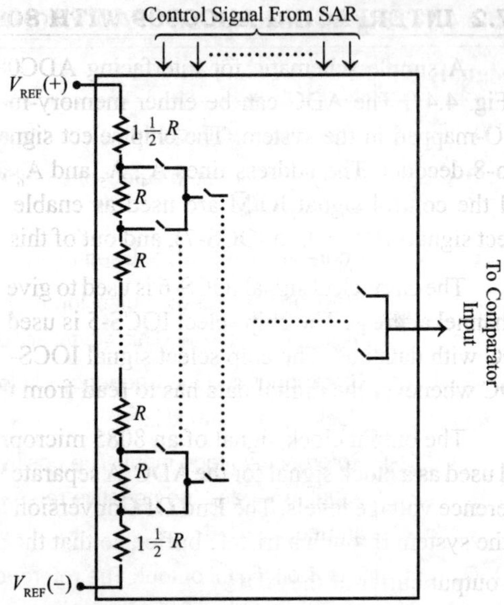

$$Q_{step} = \frac{V_{REF}}{2^8} = \frac{V_{REF}(+) - V_{REF}(-)}{256_{10}}$$

Fig. 4.46: 256R resistor network and switch tree.

The digital data corresponding to an analog input (V_{in}) is given by,

$$\text{Digital data} = \frac{V_{in}}{Q_{step}} \pm \text{Absolute Accuracy} = \left(\frac{V_{in}}{Q_{step}} - 1\right)_{10}$$

EXAMPLE 10

Let, $V_{REF}(+) = 3.84$ V, $V_{REF}(-) = 0$ V

$$\therefore Q_{step} = \frac{V_{REF}(+) - V_{REF}(-)}{256_{10}} = \frac{3.84}{2.56} = 0.015 \text{ V} = 15 \text{ mV}$$

Let the input analog voltage be 2.56 V. Now the digital data corresponding to 2.56 V is given by,

$$\text{Digital data} = \frac{V_{in}}{Q_{step}} - 1 = \frac{2.56}{0.015} - 1 = 169_{10} = A9_H = 1010\ 1001_2$$

EXAMPLE 11

Let $V_{REF}(+) = 5$ V, $V_{REF}(-) = 0$ V

$$\therefore Q_{step} = \frac{V_{REF}(+) - V_{REF}(-)}{256_{10}} = \frac{5}{256} = 0.01953125$$

Let the input analog voltage be 1.25 V. Now the digital data corresponding to 1.25 V is given by,

$$\text{Digital data} = \frac{V_{in}}{Q_{step}} - 1 = \frac{1.25}{0.01953125} - 1 = 63_{10} = 3F_H = 0011\ 1111_2$$

4.7.2 INTERFACING ADC0809 WITH 8085 MICROPROCESSOR

A simple schematic for interfacing ADC0809/ADC0808 with 8085 microprocessor is shown in Fig. 4.47. The ADC can be either memory-mapped or IO-mapped in the system. Here the ADC is IO-mapped in the system. The chip select signals for IO-mapped devices are generated by using a 3-to-8 decoder. The address lines A_4, A_5 and A_6 are used as input to decoder. The address line A_7 and the control signal IO/$\overline{M}$ are used as enable for the decoder. The decoder generates eight chip select signals (IOCS-0 to IOCS-7), and out of this three chip select signals are used for ADC interface.

The chip select signal IOCS-6 is used to give **Start Of** Conversion (SOC) signal to ADC along with a channel address. The chip select IOCS-5 is used to enable the tristate buffer provided for interfacing EOC with data bus. The chip select signal IOCS-7 is inverted and used to enable the output buffer of ADC whenever the digital data has to read from the ADC.

The output clock signal of an 8085 microprocessor is divided by suitable clock divider circuits and used as a clock signal for the ADC. A separate voltage source has to be provided to give an accurate reference voltage levels. The **End Of** Conversion (EOC) signal of ADC is connected to the bus line D_0 of the system through a tristate buffer, so that the processor can check for a valid EOC before reading the output buffer of the ADC.

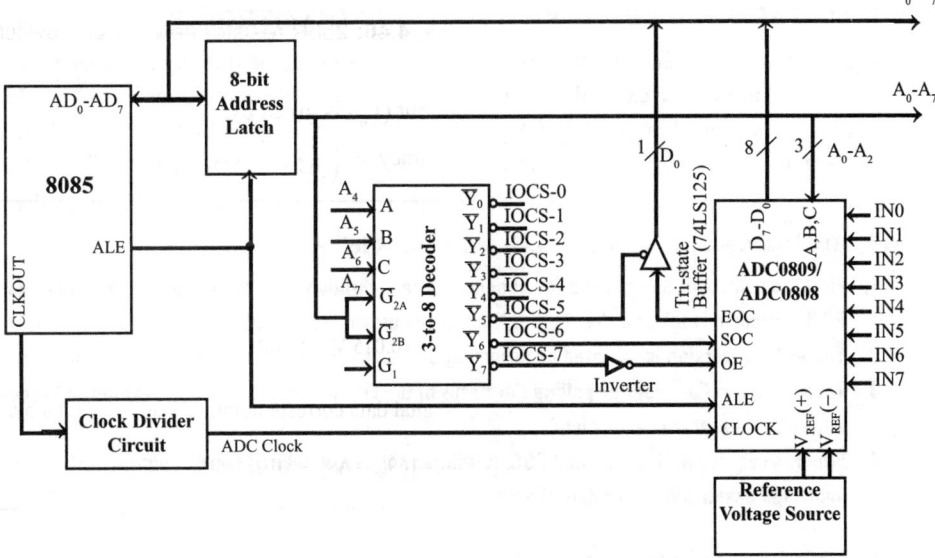

Fig. 4.47: Interfacing ADC0809/ADC0808 with 8085 microprocessor.

The working of ADC 0809 with 8085 will be as follows:

1. First the processor selects a channel by sending an address and SOC pulse is asserted high and low.

2. Once address of the channel and SOC pulse are applied, the ADC will start converting the signal at the selected channel.

3. Then the processor keeps on polling the status of the EOC to verify whether it is set to one. (when the conversion is completed by ADC0809 the EOC is set to one.)

4. When the processor finds a valid EOC, then it will read the digital value from the output buffer of ADC.

Table 4.16: IO Address of ADC0809/ADC0808 Interfaced to 8085 as Shown in Fig. 4.47

Operation performed	Binary address								Hexa address
	Decoder input/enable				Address input to ADC				
	A_7	A_6	A_5	A_4	A_3	A_2	A_1	A_0	
SOC channel-0	0	1	1	0	x	0	0	0	60
SOC channel-1	0	1	1	0	x	0	0	1	61
SOC channel-2	0	1	1	0	x	0	1	0	62
SOC channel-3	0	1	1	0	x	0	1	1	63
SOC channel-4	0	1	1	0	x	1	0	0	64
SOC channel-5	0	1	1	0	x	1	0	1	65
SOC channel-6	0	1	1	0	x	1	1	0	66
SOC channel-7	0	1	1	0	x	1	1	1	67
Read EOC	0	1	0	1	x	x	x	x	50
Read ADC output	0	1	1	1	x	x	x	x	70

4.7.3 INTERFACING ADC0809 WITH 8051 MICROCONTROLLER

In simple systems when the ports are free the ADC0809 can be directly interfaced through the port pins of 8051 controller. An example of a ADC interface with a 8051 controller is shown in Fig. 4.48. In this system the channel address (A, B, C) and the control signals (ALE,SOC, EOC, OE) are applied through port-0 pins. The ADC data (i.e, converted digital data) is read through port 1. The program for ADC conversion can be permanently stored in the internal program memory of the controller.

The ADC interface shown in Fig. 4.48 can work as follows:

1. First the controller has to send the channel address through P0.0 to P0.2 port lines. Then the port pin P0.3 is asserted high and then low to latch the address into the ADC.
2. The ADC conversion is initiated by asserting SOC as high and then low through the P0.4 pin.
3. Then the controller keeps on polling the status of the EOC through the P0.5 pin. (At the end of conversion the ADC will assert EOC as high.)
4. When the controller finds a valid EOC, it will read the digital data from the output buffer by sending a logic high enable signal through the P0.6 pin.

The ADC0809 can also be interfaced to an 8051 microcontroller as memory-mapped IO as shown in Fig. 4.49. The address lines A_0, A_1 and A_2 are used to select the desired channel for conversion. The signals SOC and OE are generated using a decoder. The signal EOC is read by the controller through a tristate buffer. The clock signal at X_2 pin is divided by a suitable clock divider and used as an ADC clock. A separate source is provided for reference voltage. The working of this system is similar to that shown in Fig. 4.49. The addresses allotted to initiate various operations are listed in Table 4.17.

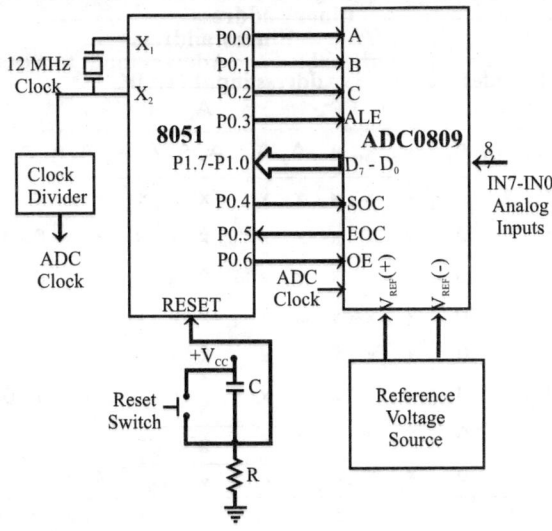

Fig. 4.48: Interfacing ADC0809 through port pins of an 8051 Microcontroller.

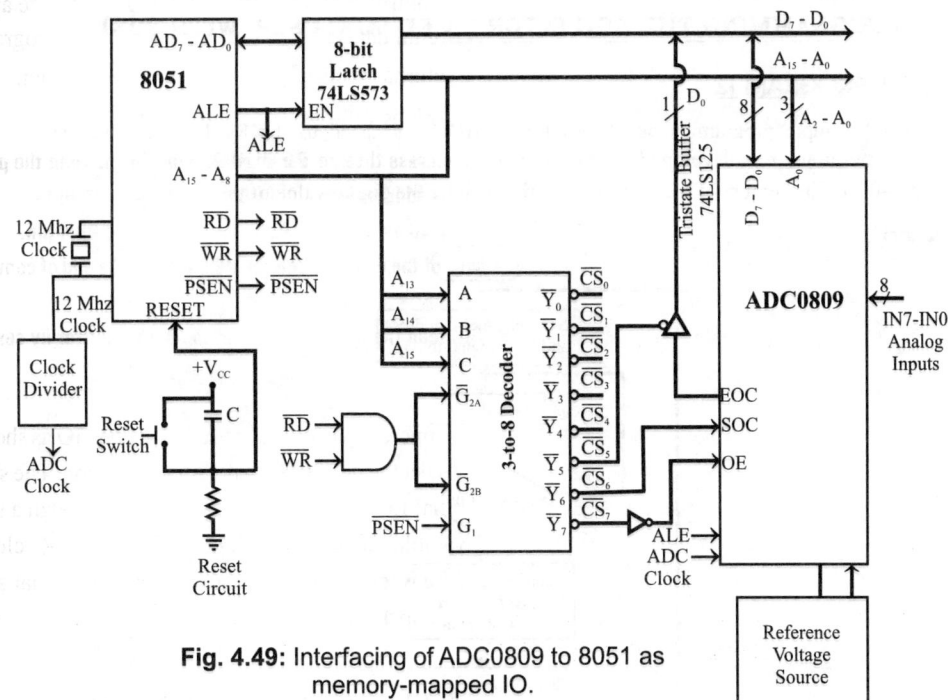

Fig. 4.49: Interfacing of ADC0809 to 8051 as memory-mapped IO.

Table 4.17: Addresss Alloted to ADC0809 Interfaced to 8051 as Shown in Fig. 4.49

Operation performed	Decoder Input A_{15} A_{14} A_{13}	Address input to ADC A_{12} A_{11} A_{10} A_9 A_8 A_7 A_6 A_5 A_4 A_3 A_2 A_1 A_0	Hexa address	Comment
SOC channel-0	1 1 0	x x x x x x x x x x 0 0 0	C000	
SOC channel-1	1 1 0	x x x x x x x x x x 0 0 1	C001	
SOC channel-2	1 1 0	x x x x x x x x x x 0 1 0	C002	
SOC channel-3	1 1 0	x x x x x x x x x x 0 1 1	C003	External
SOC channel-4	1 1 0	x x x x x x x x x x 1 0 0	C004	data
SOC channel-5	1 1 0	x x x x x x x x x x 1 0 1	C005	memory address
SOC channel-6	1 1 0	x x x x x x x x x x 1 1 0	C006	space
SOC channel-7	1 1 0	x x x x x x x x x x 1 1 1	C007	
Read EOC	1 0 1	x x x x x x x x x x x x x	A000	
Read ADC output	1 1 1	x x x x x x x x x x x x x	E000	

Note : Don't care "x" is considered as zero.

4.7.4 PROGRAMMING THE ADC INTERFACED (ADC 0809) WITH 8051

EXAMPLE PROGRAM 12

This example program is developed for an ADC interfaced to the 8051 system bus as shown in Fig. 4.48. The program is developed to read ADC channel-0 and store the digital value in the external RAM location 2400H. The content of the memory location will be the digital value of the last read operation.

Flowchart

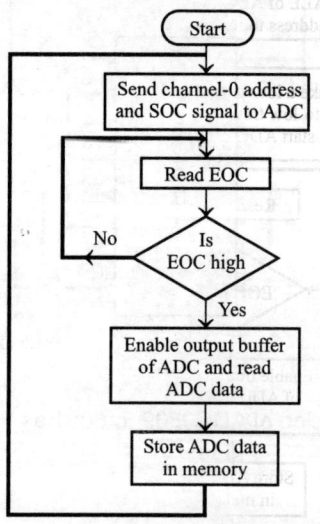

Assembly language program

```
;Program to read ADC interfaced to 8051 system bus

AGAIN:    MOV  DPTR,#C000H    ;Load address of ADC Channel-0 in DPTR
          MOV  A,#00          ;Move a dummy data to A
          MOVX @DPTR,A        ;Send address and SOC to ADC

          MOV  DPTR,#A000H    ;Load address of EOC buffer in DPTR
WAIT:     MOVX A,@DPTR        ;Get the status of EOC in A register
          RLC  A              ;Move EOC status to carry flag
          JNC  WAIT           ;Wait until EOC is high

          MOV  DPTR,#E000H    ;Load address of ADC output buffer in DPTR
          MOVX A,@DPTR        ;Get the ADC data in A register

          MOV  DPTR,#2400H    ;Load address of data memory in DPTR
          MOVX @DPTR,A        ;Store the ADC data in memory location 2400H
          SJMP AGAIN          ;Go to read next ADC data

          END                 ;Assemby end
```

EXAMPLE PROGRAM 13

This example program is developed for ADC interfaced directly to ports of 8051 as shown in Fig. 4.48. The program is developed to read ADC channel-0 and store the digital value in the internal RAM location 7FH. The content of the memory location will be the digital value of the last read operation.

Flowchart

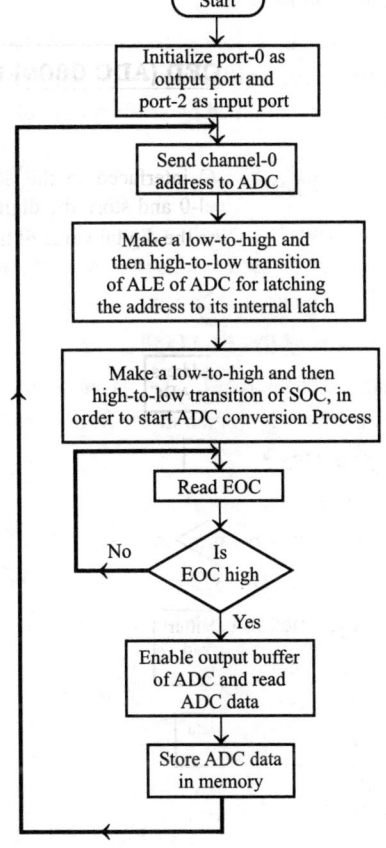

Assembly language program

```
;Program to read ADC interfaced directly to port of 8051

START:  MOV   P0,#00H      ;Initialize port-0 as output port
        MOV   P2,#FFH      ;Initialize port-2 as input port

AGAIN:  MOV   P0,#00H      ;Send channel-0 address to port-0
        SETB  P0.3         ;Set ALE high
        NOP                ;Delay and make ALE low
        NOP
        CLR   P0.3

        SETB  P0.4         ;Set SOC high
        NOP                ;Delay and make SOC low
        NOP

        CLR   P0.4
WAIT:   JNB   P2.7,WAIT    ;Wait until EOC is high

        SETB  P0.5         ;Set OE high, to enable ADC output buffer
        MOV   A,P1         ;Get ADC data in A register
        MOV   7FH,A        ;Store ADC data in internal RAM
        SJMP  AGAIN

        END                ;Assemby end
```

4.8 SHORT-ANSWER QUESTIONS

Q4.1   What is a programmable peripheral device?

If the functions performed by a peripheral device can be altered or changed by a program instruction then the peripheral device is a called programmable device. Usually programmable devices will have control registers. The device can be programmed by sending control word in the prescribed format to the control register.

Q4.2   What are the operating modes of an 8212 ?

The 8212 can be hardwired to work either as a latch or tristate buffer. If mode (MD) pin is tied **high,** then it will work as a latch and so it can be used as an output port. If mode (MD) pin is tied **low,** then it work as tristate buffer and so it can be used as an input port.

Q4.3   What are the various internal devices of INTEL 8155?

The INTEL 8155 is an IC consisting of static RAM, IO ports and a timer. The internal devices of 8155 are 256 bytes of static RAM, three numbers of programmable IO ports and a 14-bit programmable timer.

Q4.4   What are the internal devices of an 8255 ?

The internal devices of an 8255 are port-A, port-B and port-C. The ports can be programmed for either input or output function in different operating modes.

Q4.5   What are the operating modes of port-A of an 8255?

The port-A of an 8255 can be programmed to work in any one of the following operating modes as input or output port:

        Mode-0 : Simple IO port

        Mode-1 : Handshake IO port

        Mode-2 : Bidirectional IO port

Q4.6   What are the functions performed by port-C of an 8255?

   1.   The port-C pins are used for handshake signals.

   2.   Port-C can be used as an 8-bit parallel IO port in mode-0.

   3.   It can be used as two numbers of 4-bit parallel port in mode-0.

   4.   The individual pins of port-C can be set or reset for various control applications.

Q4.7   List some of the features of INTEL 8259 (Programmable Interrupt Controller).

   ● It manage eight interrupt requests   ● The priorities of interrupts are programmable

   ● The interrupt vector addresses are programmable ● The interrupt can be masked or unmasked individually

Q4.8   Write the various functional blocks of INTEL 8259?

   The various functional blocks of 8259 are Control logic, Read/ Write logic, Data bus buffer, Interrupt Request Register (IRR), Interrupt Mask Register (IMR) and In-Service Register (ISR), Priority Resolver (PR) and Cascade buffer.

Q4.9   What is master and slave 8259?

   When 8259s are connected in cascade, one 8259 will be directly interrupting the processor and it is called master 8259. To each interrupt request input of master 8259, one slave 8259 can be connected. The 8259's interrupting the master 8259 are called slave 8259.

Q4.10 How is 8259 programmed?

   The 8259 is programmed by sending Initialization Command Words (ICWs) and Operational Command Words (OCWs).

Q4.11 What are the features of 8259 that are programmed using ICWs?

   The ICWs are used to program the following features of an 8259:

   ● Call address interval (in case of 8085)   ● 8085 or 8086 modes

   ● Cascade mode or single   ● Auto or Normal end of interrupt

   ● Level or Edge triggered   ● Special fully nested mode

   ● Vector address (in case of 8085)
      or Type number (in case of 8086)

Q4.12 What are features of 8259 that can be programmed using OCWs?

   The OCWs are used to program the following features of an 8259:

   ● Masking of individual interrupts.   ● Specific or Non-specific end of interrupt.   ● Priority modes.

Q4.13 What is the difference in programming master 8259 and slave 8259 ?

   The ICW 3 will be different for master 8259 and slave 8259. For master, the ICW3 will inform the IR input that are having slaves. For slave, the ICW3 will inform its slave ID number.

Q4.14 Write a program segment to initialize a single 8259 connected to an 8085 processor.

   Let us assume that 8259 is IO-mapped in the system. The 8259 can be initialized by sending ICW1, ICW2 and OCW1. Let the 8-bit address when $A_0 = 0$ be 00_H and when $A_0 = 1$ be 01_H.

```
MVI A,ICW1 ; Move ICW1 to A-register.
OUT 00H    ; Send ICW1 to 8259.
```

```
MVI A,ICW2 ; Move ICW2 to A-register.
OUT 01H    ; Send ICW2 to 8259.
MVI A,OCW1 ; Move OCW1 to A-register.
OUT 01H    ; Send OCW1 to 8259.
HLT        ; Halt program execution.
```

Q4.15 Write the format of ICW1?

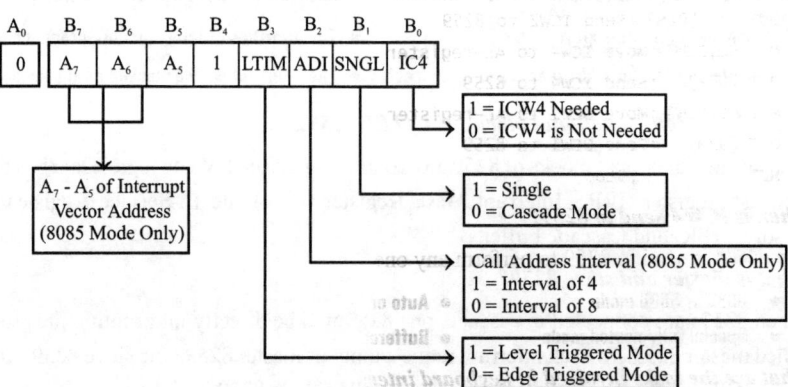

Fig. Q4.15: Format of ICW1.

Q4.16 Frame the Command words ICW1, ICW2 and OCW1 for initializing a single 8259 interfaced to 8085 with the call address interval of 8 and for level triggered interrupt. Also unmask all interrupt inputs. The desired vector address is 5000_H.

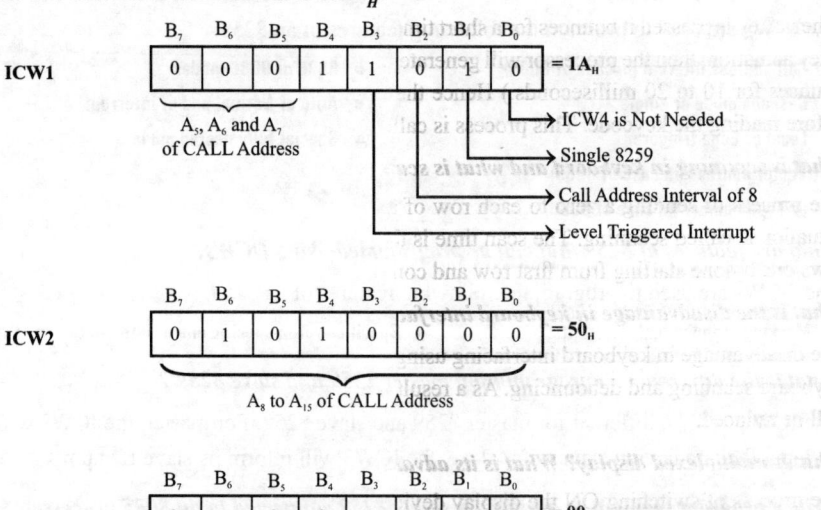

Q4.17 Write a program segment to initialize a single 8259 connected to an 8086 processor.

Let us assume that 8259 is IO-mapped in the system with an even address. The 8259 can be initialized by sending ICW1, ICW2, ICW4 and OCW1. Let the 8-bit address with $A_0 = 0$ be 00_H and when $A_0 = 1$ be 02_H

```
MOVAL,ICW1   ;Move ICW1 to AL-register
OUT [00H]    ;Send ICW1 to 8259
MOV AL,ICW2  ;Move ICW2 to AL-register
OUT    [02H] ;Send ICW2 to 8259
MOV AL,ICW4  ;Move ICW4 to AL-register
OUT [02H]    ;Send ICW4 to 8259
MOV AL,OCW1  ;Move OCW1 to AL-register
OUT [02H]    ;Send OCW1 to 8259
HLT          ;Stop
```

Q4.18 When is ICW4 send to 8259?

The ICW4 is send to 8259 to perform any one of the following features:

- **8085 or 8086 mode**
- **Special fully nested mode**
- **Auto or Normal end of interrupt**
- **Buffered or Non-buffered mode.**

Q4.19 What are the tasks involved in keyboard interface ?

The task involved in keyboard interfacing are Sensing a key actuation, Debouncing the key and Generating keycodes (Decoding the key). These tasks are performed by the software if the keyboard is interfaced through ports and they are performed by hardware if the keyboard is interfaced through 8279.

Q4.20 What is debouncing ?

When a key is pressed it bounces for a short time. If a key code is generated immediately after sensing a key actuation, then the processor will generate the same keycode a number of times. (A key typically bounces for 10 to 20 milliseconds.) Hence the processor has to wait for the key bounces to settle before reading the keycode. This process is called keyboard debouncing.

Q4.21 What is scanning in keyboard and what is scan time?

The process of sending a zero to each row of a keyboard matrix and reading the columns for key actuation is called scanning. The scan time is the time taken by the device/processor to scan all the rows one by one starting from first row and coming back to the first row again.

Q4.22 What is the disadvantage in keyboard interfacing using ports?

The disadvantage in keyboard interfacing using ports is that most of the processor time is utilized in keyboard scanning and debouncing. As a result the computational speed/efficiency of the processor will be reduced.

Q4.23 What is multiplexed display? What is its advantage?

The process of switching ON the display devices one by one for a specified time interval is called multiplexed display. In microprocessor-based systems, six to eight 7-segment LEDs are interfaced to provide multiplexed display. At any one time only one 7-segment LED is made to glow at a time. After a few milliseconds, the next 7-segment LED is made to glow and so on. Due to persistence of vision, it will appear as if the LEDs are glowing continuously. The advantage in multiplexed display is that the power requirement of the display devices is reduced to a very large extent.

Q4.24 What is scanning in display and what is the scan time?

In display devices, the process of sending display codes to 7-segment LEDs to display the LEDs one by one is called scanning (or multiplexed display). The scan time is the time taken to display all the 7-segment LEDs one by one, starting from first LED and coming back to the first LED again.

Q4.25 What is the disadvantage in 7-segment LED interfacing using ports?

The disadvantage in using ports for 7-segment LED interfacing is that most of the processor time is utilized for display refreshing.

Q4.26 What is the advantage in using INTEL 8279 for keyboard and display interfacing?

When 8279 is used for keyboard and display interfacing, it takes care of all the task involved in keyboard scanning and display refreshing. Hence the processor is relieved from the task of keyboard scanning, debouncing, keycode generation and display refreshing, and so the processor time can be more efficiently used for computing.

Q4.27 *List the functions performed by 8279.*

The function performed by 8279 are:

- **Keyboard scanning**   • **Key debouncing**   • **Keycode generation**
- **Informing the key entry to CPU**   • **Storing display codes**   • **Output display codes to LEDs**
- **Display refreshing**

Q4.28 What is the maximum number of keycodes that can be generated by 8279?

In scanned keyboard mode, the maximum size of keyboard matrix array that can be interfaced to 8279 is 8 x 8, which consists of 64 keys. In addition, the 8279 has two control keys called shift and control. For each key press, an 8-bit code is generated and stored in the FIFO (keyboard RAM of 8279). The keycode consists of row and column number of the key in binary along with the status of the shift and control key. Hence with 64 contact keys, shift and control key, a maximum of 256 keycodes can be generated by the 8279.

Q4.29 What are the programmable display features of 8279?

The 8279 can be used for interfacing LEDs or 7 segment LEDs. In decoded scan, 4 numbers of 7-segment LEDs can be interfaced and in encoded scan, a maximum of 16 numbers of 7-segment LEDs can be interfaced. The 8279 can be programmed for left entry or right entry.

Q4.30 What are the different scan modes of of 8279?

The different scan modes of 8279 are decoded scan and encoded scan. In decoded scan mode, the output of the scan lines will be similar to a 2-to-4 decoder. In encoded scan mode, the output of the scan lines will be binary count, and so an external decoder should be used to convert the binary count to the decoded output.

Q4.31 What is the difference in programming the 8279 for encoded scan and decoded scan?

If the 8279 is programmed for decoded scan then the output of scan, lines will be decoded output and if it is programmed for encoded scan then the output of scan, lines will be binary count. In encoded mode, an external decoder should be used to decode the scan lines.

Q4.32 How is a keyboard matrix formed in keyboard interface using 8279?

The return lines, RL_0 to RL_7 of 8279 are used to form the columns of keyboard matrix. In decoded scan, the scan lines SL_0 to SL_3 of 8279 are used to form the rows of keyboard matrix. In encoded scan mode, the scan line SL_0 to SL_3 are connected to input of a decoder and the output lines of decoder are used as rows of keyboard matrix.

Q4.33 What are the operating modes of a timer 8254?

The 8254 timer has six operating modes. These are:

1. Mode-0 $\rightarrow$ Interrupt on terminal count
2. Mode-1 $\rightarrow$ Hardware retriggerable one shot
3. Mode-2 $\rightarrow$ Rate generator or Timed interrupt generator
4. Mode-3 $\rightarrow$ Square wave mode
5. Mode-4 $\rightarrow$ Software triggered strobe
6. Mode-5 $\rightarrow$ Hardware triggered strobe.

Q4.34 What is the function of the GATE signal in timer 8254?

In timer 8254, the GATE signal acts as a control signal to start, stop or maintain the counting process. In modes 0, 2, 3 and 4, the GATE signal should remain **high** to start and maintain the counting process. In modes 1 and 5, the GATE signal has to make a low-to-high transition to start the counting process and need not remain **high** to maintain the counting process.

Q4.35 What will be the frequency of the square wave generated by a 8254 timer in mode-3?

The frequency of the generated square wave is given by the frequency of input clock signal divided by the count value loaded in the count register. If the count value N is an even number then the square wave will be alternatively **high** and **low** for N/2 clock periods. If the count value N is an odd number then the **high** time of square wave will be $\frac{N+1}{2}$ clock periods and **low** time will be $\frac{N-1}{2}$ clock periods.

Q4.36 What is resolution in DAC?

The resolution in DAC is the smallest possible analog value that can be generated by the n-bit binary input. If the reference voltage in n-bit, DAC is V_{REF}, then the resolution is $(1/2^n) \times V_{REF}$ Volts.

Q4.37 What are the internal devices of a typical DAC?

The internal devices of a DAC are R/2R resistive network, an internal latch and current to voltage converting amplifier.

Q4.38 What is settling or conversion time in DAC?

The time taken by the DAC to convert a given digital data to the corresponding analog signal is called conversion time.

Q4.39 What are the different types of ADC?

The different types of ADC are successive approximation ADC, counter type ADC, flash type ADC, integrator converters and voltage-to-frequency converters.

Q4.40 What is resolution and conversion time in ADC?

The resolution in ADC is the minimum analog value that can be represented by the digital data. If the ADC gives n-bit digital output and the analog reference voltage is V_{REF}, then the resolution is $(1/2^n)$ × V_{REF} Volts. The conversion time in ADC is defined as the total time required to convert an analog signal into its digital equivalent.

Q4.41 Write an assembly language program to display 99 in Port A, 1's complement of 99 in Port B and 2's complement of 99 in Port C. Assume the Port addresses are 30_H, 31_H and 32_H for ports A, B and C respectively.   *(AU, Nov/Dec' 19, 5 Marks)*

Solution:

Control word : 1000 0000 = 80H

$\qquad$ ↑ $\quad$ (All ports are output in mode-0)

$\qquad$ IO mode

Program

```
MVI A, 80H        ;Load control word in A
OUT 33H           ;Send to control register (33H is address of control register)
MVI A,99H         ;Load data in A
OUT 30H           ;Send data to port A
CMA               ;Complement accumulator
OUT 31H           ;Send complement data to port B
INR A             ;Get 2's complement in A
OUT 32H           ;Send 2's complement data to port C
HLT
```

4.9 EXERCISES

I. Fill in the blanks with appropriate words

1. In Intel 8255, the port whose individual pins can be set or reset is _____.

2. The internal register which is used to store the command words of 8255 is _____ register.

3. The strobed input/output mode of 8255 is _____.

4. The mode of 8255 which is used for bidirectional data transfer between processor and pheripherals is _____.

5. The _____ control word of Intel 8255 is used to set/reset individual pins of Port-C.

6. The control word which is used to initialize all the ports of 8255 as input ports in mode-0 is _____.

7. The _____ signals of 8255 are used to provide synchronization between transmitter and receiver.

8. In common _____ type LED, all the anode terminals of LEDs are internally shorted.

9. The method by which current requirement for LEDs can be reduced is called _____.

10. The 8254 has _____ numbers of independent 16-bit counters.

11. The maximum input clock frequency of 8254 is _____.

12. The time taken by the DAC to convert a given digital data to a corresponding analog signal is called _____ time.

13. _____ signal of ADC0809 is used to indicate the end of conversion process and can be used as interrupt signal to the processor.

Answers

1. port-C	8. anode
2. control word	9. multiplexed display
3. mode-1	10. three
4. mode-2	11. 2.6 MHz
5. Bit Set/Reset(BSR)	12. conversion
6. $9B_H$	13. End of Conversion (EOC)
7. handshaking	

II. State whether the following statements are True/False.

1. The port-B of Intel 8255 can work either in mode-0 or mode-1.

2. Port C of 8255 can function independently as either input or output port.

3. In BSR mode, all the ports can be used to set and reset individual pins.

4. The mode-2 of Intel 8255 is generally used for data transfer between two computers.

5. In Intel 8255, only port-A can work in mode-2.

6. keyboard interfacing using ports consumes more time than keyboard interfacing using hardware.

7. In common cathode LED, all the cathods of LEDs are connected to logic 1.

8. Maximum of eight 7-segment LED's can only be interfaced with 8086 using 8279.

9. More than one key pressed can be recognized by 8279 in N-key rollover mode.

10. In 8279 sensor matrix mode, the condition of all 64 switches are stored in FIFO RAM.

11. The input clock frequency of 8253 and 8254 are same.

12. The ADC conversion is also called as quantization.

13. ADC 0809 uses successive approximation method to covert analog to digital.

14. The INTEL 8279 keyboard and display controller can be used to generate maximum 256 key-codes.

Answers

1. True	5. True	9. True	13. True
2. True	6. False	10. False	14. True
3. False	7. False	11. True	
4. True	8. True	12. True	

III. Choose the right answer for the following questions.

1. *The Programmable Peripheral Interface(PPI) is also known as*

   a) Serial IO port   b) Parallel IO port   c) Serial Input Port   d) Parallel Output port

2. *Which of the following 8255 port can work in all the three modes?*

   a) port-A   b) port-B   c) port-C   d) all the three

3. *Which of the following port is used to generate handshake signals in mode-1 and mode-2 of 8255?*

   a) Port-A   b) Port-B   c) Port-C lower   d) Port-C upper

4. Which of the following are Group-B port of 8255?

   a) port-A   b) port-B   c) port-C lower   d) b and c

5. Which of the following control word of 8255 is used to set 4th pin of port-C?

   a) 18_H b) 19_H c) 08_H d) 09_H

6. The process of rechecking the keypress in a row of hex keyboard after 10 to 20 ms is called as _____

   a) key actuation   b) key debouncing   c) key bouncing   d) key decoding

7. Which of the following chip is used to interface keyboard and display with microprocessor?

   a) 8255   b) 8254   c) 8259   d) 8279

8. Which of the following method consider only one key pressed and rejects all other simultaneous keypress in keyboard?

   a) two-key lockout   b) N-key lockout   c) two-key roll over   d) N-key rollover

9. Which of the following mode of 8279 is used when 4-digit 7 segment LEDs are required?

   a) encoded scan keyboard   b) Decoded scan keyboard

   c) Encoded scan sensor matrix   d) none of the above

10. The _____ Intel chip is used to provide timings and to interrupt the processor at periodic interval

    a) 8254   b) 8259   c) 8279   d) none of the above

11. Which of the following operating mode of 8254 is used to generate square wave?

    a) mode-0   b) mode-1   c) mode-2   d) mode-3

12. The minimum analog value that can be represented by the digital data is

    a) average   b) resolution   c) percentage   d) none of the above

13. Which of the following IC is programmable interrupt controller?

    a) 8255   b) 8253   c) 8279   d) 8259

14. How many interrupts can be accepted in cascaded mode in 8259?

    a) 8   b) 16   c) 32   d) 64

15. Which of the following register of 8259 is used to store the interrupt requests?

    a) ISR   b) IMR   c) IRR   d) PR

16. Which of the following command word is used to mask/unmask individual interrupt?

    a) ICW1   b) ICW2   c) OCW1   d) none of the three

17. Which interrupt has the highest priority by default in 8259?

    a) IR7   b) IR0   c) IR1   d) none of the three

18. The INTEL 8279 can be used to interface _____.

    a) hex-keyboard   b) 7 segment LED display   c) both a& b   d) neither a nor b

Answers

1. b	4. d	7. d	10. a	13. d	16. c
2. a	5. d	8. a	11. d	14. d	17. b
3. d	6. b	9. b	12. b	15. c	18. c

IV. Answer the following questions.

E4.1 Write an 8085 program to read the status of 8 switches connected to 8255 port-A and switch status to be displayed on 8 LEDs connected to port-B. Let the address of port-A and port-B and control register are 80_H, 82_H and 86_H respectively.

E4.2 Write an 8085 program to blink an LED connected to pin PC.0 of port-C of 8255 interfaced with 8086.

E4.3 Write an 8085 program to switch ON all the eight LEDs one by one which are connected to Port-C of 8255.

E4.4 Interface 8254 with 8085 and write a program to load counter-0 with 1234_H and enable its output OUT 0 upon termination of count.

E4.5 Repeat exercise E9.4 to count in BCD mode.

E4.6 Write a program to generate a square wave of period 10ms using counter 2 of 8254 operating at 2 MHz clock frequency.

E4.7 Write a program to generate a 'strobe' pulse at the output of counter 1 of 8255 after 1 ms delay. Take 8254 clock frequency = 2 MHz.

E4.8 Find the command word to initialize 8279 in 2 key lockout encoded scan keyboard mode and to display sixteen digits in left entry mode and write an ALP in 8086.

E.4.9 Write a program to display all 'ones' only in left most four LEDs. Use N-key rollover mode.

E4.10 If the clock signal to 8279 is 1.5 MHz, calculate the prescalar value to obtain the operating frequency of 100 kHZ.

E4.11 Identify the RST instruction generated by the following 8 to 3 priority encoder.

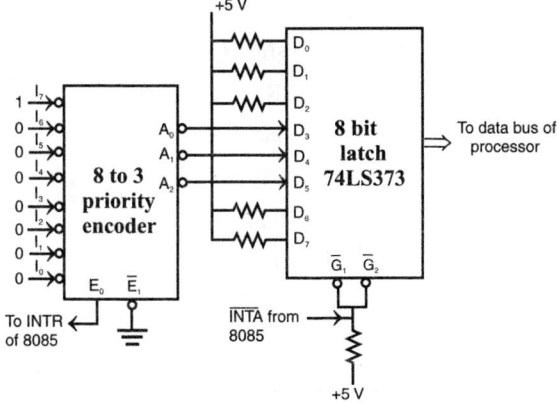

E4.12 Write an operational command word to mask the interrupt on line 3 in 8259.

E4.13 Write an intialization command word to interface 8259A with 8085 in cascaded level triggered mode with call address interval of 8. Assume interrupt vector address 8800H.

E4.14 Write an operational command word to set IR5 as bottom priority level, with rotate on specific EOI command mode.

CHAPTER 5

MICROCONTROLLER PROGRAMMING
AND APPLICATIONS

5.1 SIMPLE PROGRAMMING EXERCISES

EXAMPLE PROGRAM 1: 8-Bit Addition

Write an assembly language program to add two numbers of 8-bit data stored in memory 2400_H and 2401_H and store the result in 2402_H and 2403_H.

Problem Analysis

In order to perform addition in 8051, one of the data should be in the accumulator and the other data can be in any SFR/internal RAM or can be an immediate data. After addition, the sum is stored in the accumulator. The sum of a two 8-bit data can be either 8 bits (sum only) or 9 bits (sum and carry). The accumulator can accommodate only the sum and if there is carry, the 8051 will indicate by setting the carry flag. Hence, one of the internal registers/RAM location is used to account for carry.

Algorithm

1. Set DPTR as pointer for data (load address of data in DPTR).
2. Move first data from external memory to accumulator and save it in R1-register.
3. Increment DPTR.
4. Move second data from external memory to accumulator.
5. Clear R0-register to account for carry.
6. Add the content of R1-register to accumulator.
7. Check for carry. If carry is not set go to step 8, otherwise go to next step.
8. Increment R0-register.
9. Increment DPTR and save the sum(accumulator) in external memory.
10. Increment DPTR, move carry to accumulator and save it in external memory.
11. Stop.

Flowchart

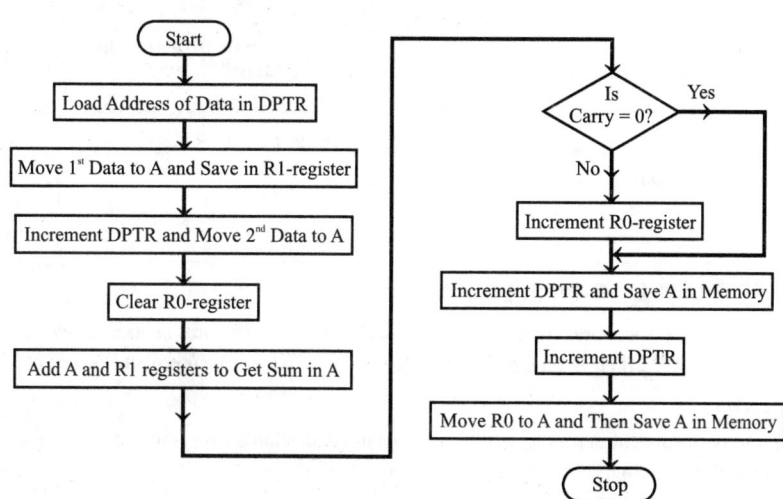

Assembly language program

```
;PROGRAM TO ADD TWO 8-BIT DATA
          ORG   2100H              ;specify program starting address.
          MOV   DPTR,#2400H        ;Load address of 1st data in DPTR.
          MOVX  A,@DPTR            ;Move the 1st data to A.
          MOV   R1,A               ;Save the first data in R1.
          INC   DPTR               ;Increment DPTR to point 2nd data.
          MOVX  A,@DPTR            ;Load 2nd data in A.
          MOV   R0,#00H            ;Clear R0 to account for carry.
          ADD   A,R1               ;Get sum of data in A.

          JNC   AHEAD              ;Check carry flag.
          INC   R0                 ;If carry is set increment R0.
AHEAD:    INC   DPTR               ;Increment DPTR.
          MOVX  @DPTR,A            ;Save sum in external memory.
          INC   DPTR               ;Increment DPTR.
          MOV   A,R0               ;Move carry to A.
          MOVX  @DPTR,A            ;Save carry in external memory.
HALT:     SJMP  HALT               ;Remain idle in infinite loop. Program end.
          END                      ;Assembly end.
```

Assembler Listing for Example Program 1

```
1                                                  ;PROGRAM TO ADD TWO 8-BIT DATA
2
3    2100                   ORG   2100H            ;specify program starting address.

4    2100   90 24 00        MOV   DPTR,#2400H      ;Load address of 1st data in DPTR.
5    2103   E0              MOVX  A,@DPTR          ;Move the 1st data to A.
6    2104   F9              MOV   R1,A             ;Save the first data in R1.
7    2105   A3              INC   DPTR             ;Increment DPTR to point 2nd data.
8    2106   E0              MOVX  A,@DPTR          ;Load 2nd data in A.
9    2107   78 00           MOV   R0,#00H          ;Clear R0 to account for carry.
10   2109   29              ADD   A,R1             ;Get sum of data in A.
11
12   210A   50 01           JNC   AHEAD            ;Check carry flag.
13   210C   08              INC   R0               ;If carry is set increment R0.
14
15   210D   A3    AHEAD:    INC   DPTR             ;Increment DPTR.
16   210E   F0              MOVX  @DPTR,A          ;Save sum in external memory.
17   210F   A3              INC   DPTR             ;Increment DPTR.
18   2110   E8              MOV   A,R0             ;Move carry to A.
19   2111   F0              MOVX  @DPTR,A          ;Save carry in external memory.
20
21   2112   80 FE HALT:     SJMP        HALT       ;Remain idle in infinite loop. Program end.
22   2114         END                              ;Assembly end.
```

Sample Data

Input Data :   Data-1 = F2$_H$

Data-2 = 34$_H$

Output Data: Sum = 26$_H$

Carry = 01$_H$

Memory address	Content	
2400	F2	← Data 1
2401	34	← Data 2
2402	26	← Sum
2403	01	← Carry

EXAMPLE PROGRAM 2:   8-bit Subtraction

Write an assembly language program to subtract two numbers of 8-bit data stored in memory 2400$_H$ and 2401$_H$. Store the magnitude of the result in 2402$_H$. If the result is positive store 00 in 2403$_H$ or if the result is negative store 01 in 2403$_H$.

Problem Analysis

In order to perform subtraction in an 8051, one of the data should be in the accumulator and the other data can be in any one of the internal memory/registers or can be an immediate data. The controller stores the result in the accumulator after subtraction. The 8051 perform 2's complement subtraction and then complements the carry. Therefore, if the result is negative then the carry flag is set and the accumulator will have 2's complement of the result. In order to get the magnitude of the result, again take 2's complement of the result. One of the registers is used to account for sign of the result. The 8051 will consider previous carry while performing subtraction and so the carry should be cleared before performing the subtraction.

Flowchart

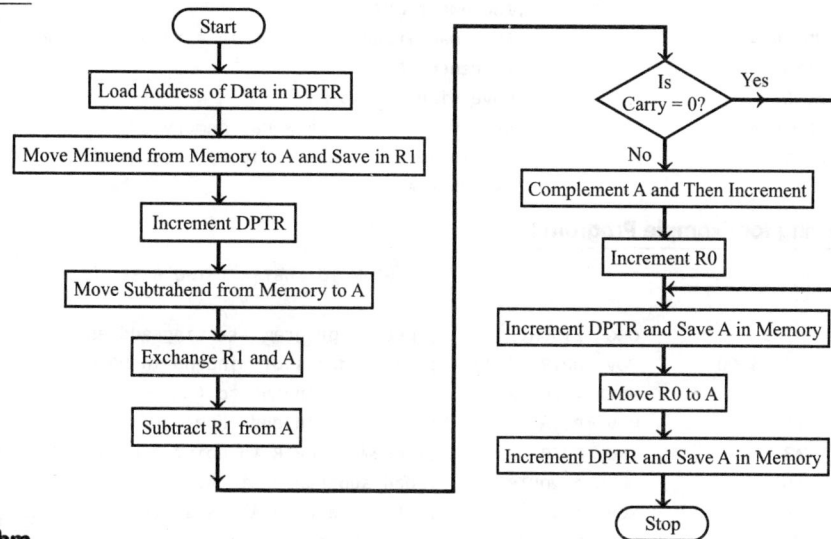

Algorithm

1. Set DPTR as pointer for data (Load address of data in DPTR).
2. Move the minuend from external memory to accumulator (A) and save in R1-register.
3. Increment DPTR and move the subtrahend from external memory to accumulator (A).
4. Exchange the contents of R1 and A, such that minuend is in A and subtrahend is in B.
5. Clear R0-register to account for sign.
6. Clear carry flag.
7. Subtract the content of R1 from A.
8. Check carry flag. If carry flag is not set go to step 10, otherwise go to next step.
9. Complement the content of A and increment by one to get 2's complement of result in A.Also increment R0 by one to indicate negative result.
10. Increment DPTR and save the content of A (which is magnitude of result) in external memory.
11. Increment DPTR, move R0 (sign bit) to A and then save sign bit in external memory.
12. Stop.

Assembly language program

```
;PROGRAM TO SUBTRACT TWO 8-BIT DATA
        ORG   2100H               ;specify program starting address.
        MOV   DPTR,#2400H         ;Load address of minuend in DPTR.
        MOVX  A,@DPTR             ;Move the minuend to A.
        MOV   R1,A                ;Save the minuend in R1.
        INC   DPTR               ;Increment DPTR to point subtrahend.
        MOVX  A,@DPTR             ;Load subtrahend in A.
        XCH   A,R1                ;Get minuend in A and subtrahend in R1.
        MOV   R0,#00H             ;Clear R0 to account for sign.
        CLR   C                   ;Clear carry.
```

```
        SUBB A,R1              ;Subtract R1 from A.
        JNC  AHEAD             ;Check carry flag,If carry is set then,
        CPL  A                 ;get 2's complement of result in A.
        INC  A

        INC  R0               ;Set R0 as one to indicate negative result.

AHEAD:  INC  DPTR            ;Increment DPTR.
        MOVX @DPTR,A          ;Save magnitude of result in external memory.
        INC  DPTR            ;Increment DPTR.
        MOV  A,R0            ;Move sign bit to A.
        MOVX @DPTR,A          ;Save sign bit in external memory.
HALT:   SJMP HALT            ;Remain idle in infinite loop. program end.
        END                  ;Assembly end.
```

Assembler Listing for Example Program 2

```
1                                                    ;PROGRAM TO SUBTRACT TWO 8-BIT DATA
2
3    2100                        ORG   2100H        ;specify program starting address.
4    2100    90 24 00           MOV   DPTR,#2400H   ;Load address of minuend in DPTR.
5    2103    E0                 MOVX  A,@DPTR       ;Move the minuend to A.
6    2104    F9                 MOV   R1,A          ;Save the minuend in R1.
7    2105    A3                 INC   DPTR          ;Increment DPTR to point subtrahend.
8    2106    E0                 MOVX  A,@DPTR       ;Load subtrahend in A.
9    2107    C9                 XCH   A,R1          ;Get minuend in A and subtrahend in R1.
10   2108    78 00              MOV   R0,#00H       ;Clear R0 to account for sign.
11   210A    C3                 CLR   C             ;Clear carry.
12   210B    99                 SUBB  A,R1          ;Subtract R1 from A.
13
14   210C    50 03              JNC   AHEAD         ;Check carry flag,If carry is set then,.
15   210E    F4                 CPL   A             ;get 2's complement of result in A.
16   210F    04                 INC   A
17   2110    08                 INC   R0            ;Set R0 as one to indicate negative result.
18   2111
19   2111    A3            AHEAD:INC   DPTR         ;Increment DPTR.
20   2112    F0                 MOVX  @DPTR,A       ;Save magnitude of result in external memory.
21   2113    A3                 INC   DPTR          ;Increment DPTR.
22   2114    E8                 MOV   A,R0          ;Move sign bit to A.
23   2115    F0                 MOVX  @DPTR,A       ;Save sign bit in external memory.
24
25   2116    80 FE         HALT: SJMP HALT          ;Remain idle in infinite loop. program end.
26   2118                       END                 ;Assembly end.
```

Sample Data

Input Data   : Minuend   = 4C$_H$
                Subtrahend = F7$_H$

Output Data : Difference = AB$_H$
                Sign bit   = 01$_H$

Memory address	Content
2400	4C
2401	F7
2402	AB
2403	01

EXAMPLE PROGRAM 3: 8-bit Multiplication

Write an assembly language program to multiply two numbers of 8-bit data stored in the memory 2400_H and 2401_H and store the product in 2402_H and 2403_H.

Problem Analysis

In order to perform multiplication in 8051, the two 8-bit data should be stored in A and B registers, then multiplication can be performed by using "MUL AB" instruction. After multiplication, the 16-bit product will be in A and B-register such that the low byte is in A and the high byte is in B.

Algorithm

1. Load address of data in DPTR.
2. Move first data from external memory to A and save in B.
3. Increment DPTR and move second data from external memory to A.
4. Perform multiplication to get the product in A and B.
5. Increment DPTR and save A (which is low byte of product) in memory.
6. Increment DPTR, move B (which is high byte of product) to A and save it in memory.
7. Stop.

Flowchart

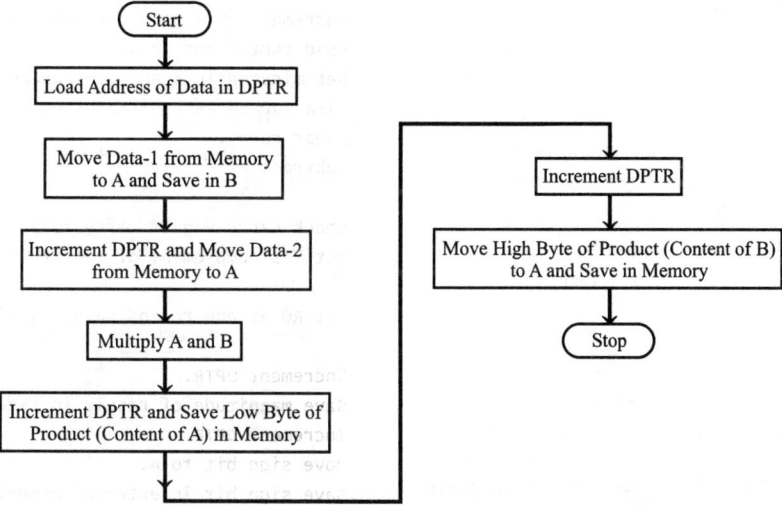

Assembly language program

```
;PROGRAM TO MULTIPLY TWO 8-BIT DATA
       ORG2100H          ;specify program starting address.
       MOV  DPTR,#2400H  ;Load address of 1st data in DPTR.
       MOVX A,@DPTR       ;Move the 1st data to A.
       MOV  B,A           ;Save the first data in B.
       INC  DPTR          ;Increment DPTR to point 2nd data.
       MOVX A,@DPTR       ;Load 2nd data in A.
       MUL  AB            ;Get the product in A and B.
```

```
      INC  DPTR          ;Increment DPTR.
      MOVX @DPTR,A        ;Save low byte of product in external memory.
      INC  DPTR          ;Increment DPTR.
      MOV  A,B           ;Move high byte of product to A,
      MOVX @DPTR,A        ;and save in external memory.

HALT: SJMP HALT          ;Remain idle in infinite loop. program end.
      END                ;Assembly end.
```

Assembler Listing for Example Program 3

```
1                                    ;PROGRAM TO MULTIPLY TWO 8-BIT DATA

2

3   2100                 ORG   2100H   ;specify program starting address.
4   2100   90 24 00      MOV   DPTR,#2400H;Load address of 1st data in DPTR.
5   2103   E0            MOVX  A,@DPTR  ;Move the 1st data to A.
6   2104   F5 F0         MOV   B,A     ;Save the first data in B.
7   2106   A3            INC   DPTR    ;Increment DPTR to point 2nd data.
8   2107   E0            MOVX  A,@DPTR  ;Load 2nd data in A.
9   2108   A4            MUL   AB      ;Get the product in A and B.

10

11  2109   A3            INC   DPTR    ;Increment DPTR.
12  210A   F0            MOVX  @DPTR,A  ;Save low byte of product in external memory.
13  210B   A3            INC   DPTR    ;Increment DPTR.
14  210C   E5 F0         MOV   A,B     ;Move high byte of product to A,
15  210E   F0            MOVX  @DPTR,A  ;and save in external memory.

16

17  210F   80 FE   HALT: SJMP HALT    ;Remain idle in infinite loop. program end.
18  211                  END          ;Assembly end.
```

Sample Data

Input Data   : Data $-1 = C7_H$
              Data $-2 = 4A_H$

Output Data : Product = 3986_H

Memory address	Content	
2400	C7	← Data 1
2401	4A	← Data 2
2402	86	} Product
2403	39	

EXAMPLE PROGRAM 4:   8-bit Division

Write an assembly language program to divide the 8-bit data stored in the memory location 2400ₕ by the 8-bit data in 2401ₕ *Store the quotient in 2402ₕ and remainder in 2403ₕ.*

Problem Analysis

In order to perform division in 8051, the dividend should be stored in A and the divisor should be stored in B. Then the content of A can be divided by B using the instruction "DIV AB". After division, the quotient will be in A and the remainder will be in B.

Algorithm

1. **Load address of data in DPTR.**
2. **Move the dividend from external memory to A and save it in R0-register.**
3. **Increment DPTR and move the divisor from external memory to A and save it in B-register.**
4. **Move the dividend from R0 to A.**

5.   **Perform division to get quotient in A and remainder in B.**
6.   **Increment DPTR and save quotient (content of A)in memory.**
7.   **Increment DPTR.**
8.   **Move the remainder (content of B) to A and save in memory.**
9.   **Stop.**

Flowchart

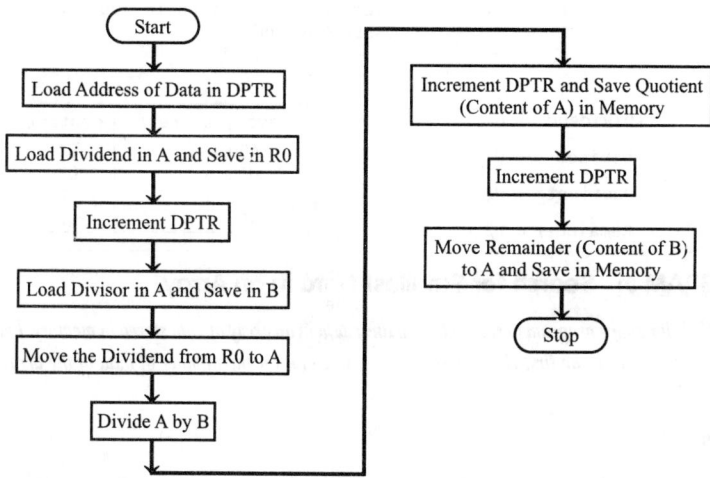

Assembly language program

```
;PROGRAM TO DIVIDE TWO 8-BIT DATA
        ORG     2100H           ;specify program starting address.

        MOV     DPTR,#2400H     ;Load address of dividend in DPTR.
        MOVX    A,@DPTR         ;Load the dividend in A.
        MOV     R0,A            ;Save the dividend in R0.
        INC     DPTR            ;Let DPTR point to divisor.
        MOVX    A,@DPTR         ;Load the divisor in A.
        MOV     B,A             ;Move the divisor to B.
        MOV     A,R0            ;Move the dividend to A.
        DIV     AB              ;Divide the content of A by B.

        INC     DPTR            ;Increment DPTR.
        MOVX    @DPTR,A         ;Save quotient in external memory.
        INC     DPTR            ;Increment DPTR.
        MOV     A,B             ;Move remainder to A,
        MOVX    @DPTR,A         ;and save in external memory.

HALT:   SJMP    HALT            ;Remain idle in infinite loop. program end.
        END                     ;Assembly end.
```

Assembler Listing for Example Program 4

```
 1                              ;PROGRAM TO DIVIDE TWO 8-BIT DATA
 2
 3 2100                 ORG 2100H     ;specify program starting address.

 4 2100    90 24 00     MOV  DPTR,#2400H ;Load address of dividend in DPTR.
 5 2103    E0           MOVX A,@DPTR   ;Load the dividend in A.
 6 2104    F8           MOV  R0,A      ;Save the dividend in R0.
 7 2105    A3           INC  DPTR      ;Let DPTR point to divisor.

 8 2106    E0           MOVX A,@DPTR   ;Load the divisor in A.
 9 2107    F5 F0        MOV  B,A       ;Move the divisor to B.
10 2109    E8           MOV  A,R0      ;Move the dividend to A.
```

11	210A	84		DIV	AB	;Divide the content of A by B.
12						
13	210B	A3		INC	DPTR	;Increment DPTR.
14	210C	F0		MOVX	@DPTR,A	;Save quotient in external memory.
15	210D	A3		INC	DPTR	;Increment DPTR.
16	210E	E5 F0		MOV	A,B	;Move remainder to A,
17	2110	F0		MOVX	@DPTR,A	;and save in external memory.
18						
19	2111	80 FE	HALT:	SJMP	HALT	;Remain idle in infinite loop. program end.
20	2113			END		;Assembly end.

Sample Data

Input Data	:	Dividend	=	64_H
		Divisor	=	07_H
Output Data	:	Quotient	=	$0E_H$
		Remainder	=	02_H

Memory address	Content	
2400	64	← Divider
2401	07	← Divisor
2402	0E	← quotient
2403	02	← Remainder

EXAMPLE PROGRAM 5: Search for Smallest Data in an Array

Write an assembly language program to search the smallest data in an array of data stored in memory. Let the array be stored in memory starting from 2400_H, with the first element of the array as count for the number of data in the array. Store the smallest data in memory location 2500_H.

Problem Analysis

The DPTR is used as the pointer for the array. One of the register of the registers bank is used as counter and another register is used to store the current smallest data. Initially, the first data of the array is considered as the current smallest. The smallest data is searched by performing subtraction of a data of the array with the current smallest. The condition of the carry flag after subtraction is used to determine the smaller among the two and the smallest among the two is moved to the register reserved to store the current smallest data. The comparison by subtraction is performed N – 1 times (where N is the count for the number of data in the array). After N – 1 comparisons, the smallest data in the array will be in the register reserved for current smallest data, which can be stored in the memory.

Algorithm

1. Set DPTR as pointer for data array.
2. Load the count value, N in A and save in R0.
3. Decrement R0 to set count for N–1 comparisons.
4. Increment DPTR.
5. Load the first data of array in A and save it as current smallest in R4-register.
6. Increment DPTR.
7. Get a data of the array in A-register and save it in R2-register.
8. Clear carry flag and subtract the current smallest in R4 from A.
9. Check carry flag. If carry is not set then go to step 11, otherwise go to next step.
10. If carry is set, then the content of R2 is smaller than R4 and so, move R2 to R4 via A.
11. Decrement R0 and check whether it is zero. If R0 is not zero then go to step 6, otherwise go to next step.
12. Load the address of the memory where smallest data to be stored in DPTR.
13. Move the smallest data from R4 to A and save in memory pointed by DPTR.
14. Stop.

Flowchart

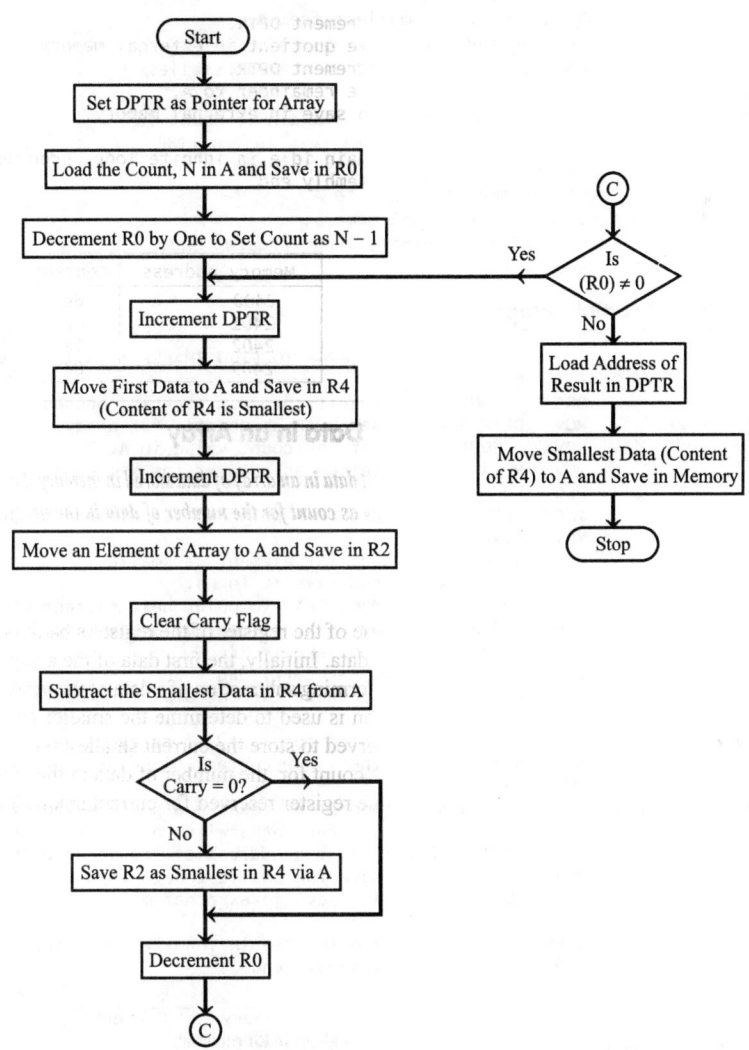

Assembly language program

```
;PROGRAM TO FIND SMALLEST DATA IN AN ARRAY
        ORG   2100H        ;specify program starting address.
        MOV   DPTR,#2400H  ;Set DPTR as pointer for array.
        MOVX  A,@DPTR      ;Get the count value in A.
        MOV   R0,A         ;Set R0 as counter for N-1 comparisons.
        DEC   R0
        INC   DPTR         ;Let DPTR point to 1st element of array.
        MOVX  A,@DPTR
        MOV   R4,A         ;Let 1st element be smallest,
                           ;and save it in R4.
AGAIN:  INC   DPTR         ;Make DPTR to point next element of array.
        MOVX  A,@DPTR      ;Get next element of array in A,
```

```
          MOV   R2,A         ;and save in R2.
          CLR   C            ;Clear carry flag.
          SUBB  A,R4         ;Subtract current smallest from A.
          JNC   AHEAD        ;Check for carry,If carry is set.
          MOV   A,R2         ;then save content of R2 as current smallest.
          MOV   R4,A
AHEAD:    DJNZ  R0,AGAIN     ;Decrement count and go to again if count is
                             ;not zero,otherwise go to next instruction.
          MOV   DPTR,#2500H  ;Load the address of result in DPTR.
          MOV   A,R4         ;Move the smallest data to A,
          MOVX  @DPTR,A      ;and save in external memory.
HALT:     SJMP  HALT         ;Remain idle in infinite loop. program end.
          END                ;Assembly end.
```

Assembler Listing for Example Program 5

```
 1                                      ;PROGRAM TO FIND SMALLEST DATA IN AN ARRAY
 2
 3   2100                   ORG  2100H        ;specify program starting address.
 4   2100   90 24 00        MOV  DPTR,#2400H  ;Set DPTR as pointer for array.
 5   2103   E0              MOVX A,@DPTR      ;Get the count value in A.
 6   2104   F8              MOV  R0,A         ;Set R0 as counter for N-1 comparisons.
 7   2105   18              DEC  R0
 8   2106   A3              INC  DPTR         ;Let DPTR point to 1st element of array.
 9   2107   E0              MOVX A,@DPTR
10   2108   FC              MOV  R4,A         ;Let 1st element be smallest,
11                                            ;and save it in R4.
12   2109   A3       AGAIN: INC  DPTR         ;Make DPTR to point next element of array.
13   210A   E0              MOVX A,@DPTR      ;Get next element of array in A,
14   210B   FA              MOV  R2,A         ;and save in R2.
15   210C   C3              CLR  C            ;Clear carry flag.
16   210D   9C              SUBB A,R4         ;Subtract current smallest from A.
17
18   210E   50 02           JNC  AHEAD        ;Check for carry,If carry is set.
19   2110   EA              MOV  A,R2         ;then save content of R2 as current smallest.
20   2111   FC              MOV  R4,A
21   2112   D8 F5    AHEAD: DJNZ R0,AGAIN     ;Decrement count and go to again if count is
22                                            ;not zero,otherwise go to next instruction.
23   2114   90 25 00        MOV  DPTR,#2500H  ;Load the address of result in DPTR.
24   2117   EC              MOV  A,R4         ;Move the smallest data to A,
25   2118   F0              MOVX @DPTR,A      ;and save in external memory.
26
27   2119   80 FE    HALT:  SJMP HALT         ;Remain idle in infinite loop. program end.
28   211B                   END               ;Assembly end.
```

Sample Data

Input Data :   Count = 06_H

Array = $7F_H$
$1C_H$
42_H
57_H
13_H
FE_H

Output Data: 13

Memory address	Content	
2400	06	Count
2401	7F	⎫
2402	1C	⎪
2403	42	⎬ Array
2404	57	⎪
2405	13	⎪
2406	FE	⎭
2500	13	Smallest data

EXAMPLE PROGRAM 6:   Search for Largest Data in an Array

Write an assembly language program to search the largest data in an array of data stored in the memory. Let the array be stored in the memory starting from 2400$_H$ with the first element of the array as count for the number of data in the array. Store the largest data in memory location 2500$_H$.

Problem Analysis

The DPTR is used as a pointer for the array. One of the registers of register bank is used as the counter and another register is used to store current largest data. Initially, the first data of the array is considered as current largest. The largest data is searched by performing subtraction of a data of the array with current largest. The condition of the carry flag after subtraction is used to determine the larger among the two and the largest among the two is moved to the register reserved to store current largest data. The comparison by subtraction is performed N – 1 times (where N is count for the number of data in the array). After N – 1 comparisons, the largest data in the array will be in the register reserved for the current largest data, which can be stored in the memory.

Flowchart

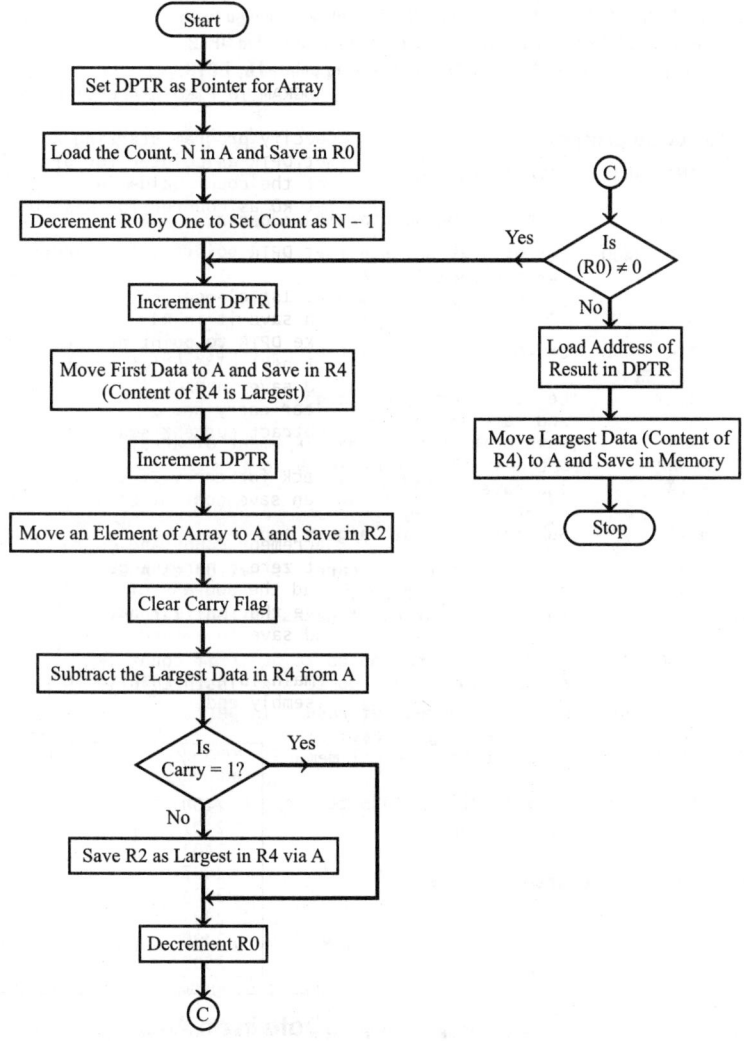

Algorithm

1. Set DPTR as pointer for data array.
2. Load the count value, N in A and save in R0.
3. Decrement R0 to set count for N–1 comparisons.
4. Increment DPTR.
5. Load the first data of array in A and save it as current largest in R4-register.
6. Increment DPTR.
7. Get a data of the array in A-register and save it in R2-register.
8. Clear carry flag and subtract the current largest in R4 from A.
9. Check carry flag. If carry is set then go to step 11, otherwise go to next step.
10. If carry is not set, then the content of R2 is larger than R4 and so, move R2 to R4 via A.
11. Decrement R0 and check whether it is zero. If R0 is not zero then go to step 6, otherwise go to next step.
12. Load the address of the memory where largest data to be stored in DPTR.
13. Move the largest data from R4 to A and save in memory pointed by DPTR.
14. Stop.

Assembly language program

```
;PROGRAM TO FIND LARGEST DATA IN AN ARRAY

        ORG   2100H        ;Specify program starting address.

        MOV   DPTR,#2400H  ;Set DPTR as pointer for array.
        MOVX  A,@DPTR      ;Get the count value in A.
        MOV   R0,A         ;Set R0 as counter for N-1 comparisons.
        DEC   R0

        INC   DPTR         ;Let DPTR point to 1st element of array.
        MOVX  A,@DPTR
        MOV   R4,A         ;Let 1st element be largest,
                           ;and save it in R4.
AGAIN:  INC   DPTR         ;Make DPTR to point next element of array.
        MOVX  A,@DPTR      ;Get next element of array in A,
        MOV   R2,A         ;and save in R2.
        CLR   C            ;Clear carry flag.
        SUBB  A,R4         ;Subtract current largest from A.

        JC    AHEAD        ;Check for carry,If carry is set go to AHEAD.
        MOV   A,R2         ;If carry is not set,
        MOV   R4,A         ;then save content of R2 as current largest.

AHEAD:  DJNZ  R0,AGAIN     ;Decrement count and go to again if count is
                           ;not zero,otherwise go to next instruction.
        MOV   DPTR,#2500H  ;Load the address of result in DPTR.
        MOV   A,R4         ;Move the largest data to A,
        MOVX  @DPTR,A      ;and save in external memory

HALT:   SJMP  HALT         ;Remain idle in infinite loop. Program end.
        END                ;Assembly end.
```

Assembler Listing for Example Program 6

```
1                                    ;PROGRAM TO FIND LARGEST DATA IN AN ARRAY
2
3    2100                   ORG   2100H        ;Specify program starting address.
4
5    2100   90 24 00        MOV   DPTR,#2400H  ;Set DPTR as pointer for array.
6    2103   E0              MOVX  A,@DPTR      ;Get the count value in A
7    2104   F8              MOV   R0,A         ;Set R0 as counter for N-1 comparisons.
8    2105   18              DEC   R0
9
10   2106   A3              INC   DPTR         ;Let DPTR point to 1st element of array.
```

11	2107	E0		MOVX A,@DPTR	
12	2108	FC		MOV R4,A	;Let 1st element be largest,
13					;and save it in R4.
14	2109	A3	AGAIN:	INC DPTR	;Make DPTR to point next element of array.
15	210A	E0		MOVX A,@DPTR	;Get next element of array in A,
16	210B	FA		MOV R2,A	;and save in R2.
17	210C	C3		CLR C	;Clear carry flag.
18	210D	9C		SUBB A,R4	;Subtract current largest from A.
19					
20	210E	40 02		JC AHEAD	;Check for carry,If carry is set go to AHEAD.
21	2110	EA		MOV A,R2	;If carry is not set,
22	2111	FC		MOV R4,A	;then save content of R2 as current largest.
23					
24	2112	D8 F5	AHEAD:	DJNZ R0,AGAIN	;Decrement count and go to again if count is
25					;not zero,otherwise go to next instruction.
26					
27	2114	90 25 00		MOV DPTR,#2500H	;Load the address of result in DPTR.
28	2117	EC		MOV A,R4	;Move the largest data to A,
29	2118	F0		MOVX @DPTR,A	;and save in external memory.
30					
31	2119	80 FE	HALT:	SJMP HALT	;Remain idle in infinite loop. Program end.
32					
33	211B			END	;Assembly end.

Sample Data

			Memory address	Content	
Input Data : Count =	06$_H$		2400	06	Count
Array =	7F$_H$		2401	7F	⎫
	1C$_H$		2402	1C	⎪
	42$_H$		2403	42	⎬ Array
	57$_H$		2404	57	⎪
	13$_H$		2405	13	⎪
	FE$_H$		2406	FE	⎭
Output Data : FE$_H$			2500	FE	Largest data

EXAMPLE PROGRAM 7: Sorting an Array in Ascending Order

Write an assembly language program to sort an array of data in ascending order. The array is stored in memory starting from 2400$_H$. The first element of the array gives the count value for the number of elements in the array.

Problem Analysis

The algorithm for bubble sorting is given below. In bubble sorting of N-data, N–1 comparisons are carried by taking two consecutive data at a time. After each comparison, the data are rearranged such that the smallest among the two is in the first memory location and the largest in the next memory location. (Here the data are rearranged within the two memory locations whose contents are compared.) When we perform N–1 comparisons as mentioned above for N–1, times then the array consisting of N-data will be sorted in the ascending order.

Algorithm

1. Load address of data array in DPTR, and using DPTR load the count value in R2 via A.
2. Decrement R2-register (R2 is counter for N–1 repetitions).
3. Set DPTR as data array address pointer.
4. Set R1-register as counter for N–1 comparisons.
5. Increment DPTR, and using DPTR load two consecutive data of the array in R3 and A.
6. Compare the content of A and R3 by performing subtraction.
7. If carry flag is not set (If the content of A is greater than R3) then go to step 9, otherwise go to next step.
8. If carry flag is set (If the content of A is less than R3), then exchange the content of memory pointed by DPTR and previous memory location.

9. Decrement R1-register. If zero flag =0, go to step 5 otherwise go to next step.
10. Decrement R2-register. If zero flag =0, go to step 3 otherwise go to next step.
11. Stop.

Flowchart

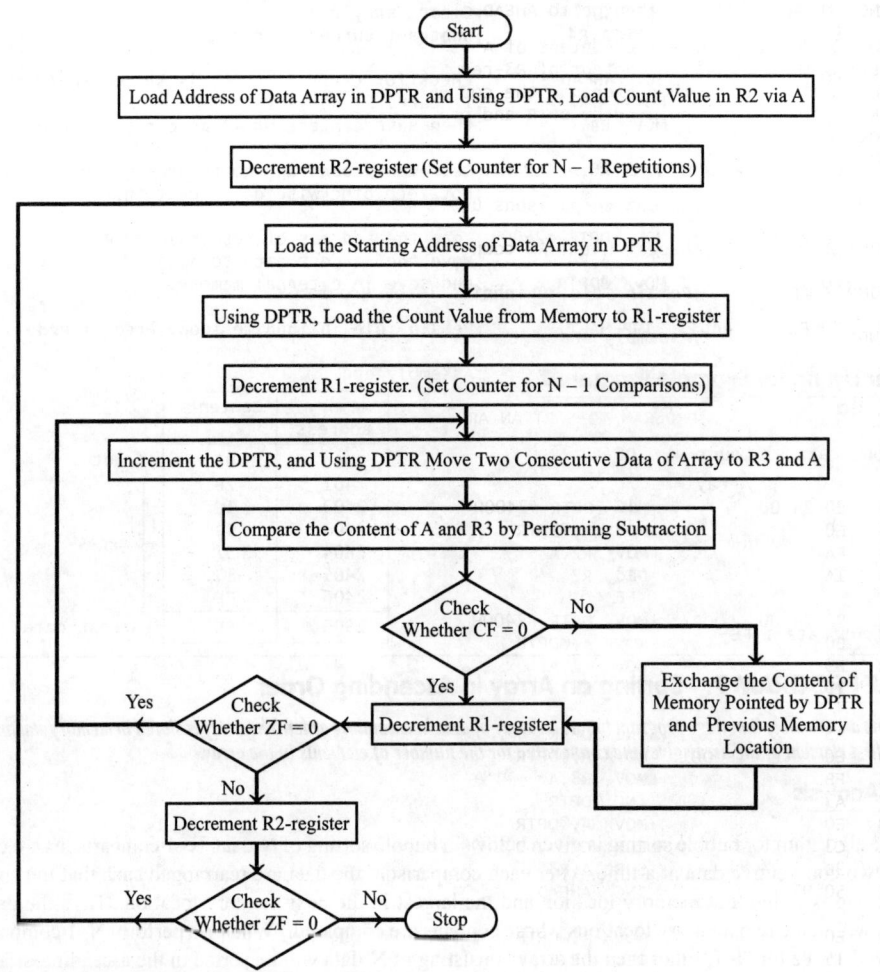

Assembly language program

```
;PROGRAM TO SORT AN ARRAY OF DATA IN ASCENDING ORDER

        ORG  2100H       ;Specify program starting address.

        MOV  DPTR,#2400H ;Load the count value in A-register.
        MOVX A,@DPTR
        MOV  R2,A        ;Set counter for (N-1) repetitions
        DEC  R2          ;of (N-1) comparisons.
LOOP2:  MOV  DPTR,#2400H ;Set pointer for array.
        MOVX A,@DPTR     ;Set count for (N-1) comparisons.
        MOV  R1,A
        DEC  R1

        INC  DPTR        ;Increment pointer.
```

```
LOOP1:  MOVX  A,@DPTR      ;Get two consecutive data of array in
        MOV   R3,A         ;R3 and A-register.
        INC   DPTR
        MOVX  A,@DPTR
        CLR   C            ;Compare A and R3-register.
        SUBB  A,R3         ;If content of A is greater than
        JNC   AHEAD        ;R3 then go to AHEAD.

        MOVX  A,@DPTR      ;If the content of A is less than
        DEC   DPTR         ;the content of R3-register.
        MOVX  @DPTR,A      ;then exchange the content of memory
        INC   DPTR         ;pointed by DPTR and previous location.
        MOV   A,R3
        MOVX  @DPTR,A

AHEAD:  DJNZ  R1,LOOP1     ;Repeat comparisons until R1 count is zero.

        DJNZ  R2,LOOP2     ;Repeat until R2 count is zero.

HALT:   SJMP  HALT         ;Remain idle in infinite loop. Program end.

        END                ;Assembly end.
```

Assembler Listing for Example Program 7

```
1                                ;PROGRAM TO SORT AN ARRAY OF DATA IN ASCENDING ORDER
2
3    2100                        ORG   2100H        ;Specify program starting address.
4
5    2100   90 24 00             MOV   DPTR,#2400H  ;Load the count value in A-register.
6    2103   E0                   MOVX  A,@DPTR
7    2104   FA                   MOV   R2,A         ;Set counter for (N-1) repetitions
8    2105   1A                   DEC   R2           ;of (N-1) comparisons.
9
10   2106   90 24 00  LOOP2:     MOV   DPTR,#2400H  ;Set pointer for array.
11   2109   E0                   MOVX  A,@DPTR      ;Set count for (N-1) comparisons.
12   210A   F9                   MOV   R1,A
13   210B   19                   DEC   R1
14
15   210C   A3                   INC   DPTR         ;Increment pointer.
16   210D   E0        LOOP1:     MOVX  A,@DPTR      ;Get two consecutive data of array in
17   210E   FB                   MOV   R3,A         ;R3 and A-register.
18   210F   A3                   INC   DPTR
19   2110   E0                   MOVX  A,@DPTR
20   2111   C3                   CLR   C            ;Compare A and R3-register.
21   2112   9B                   SUBB  A,R3         ;If content of A is greater than
22   2113   50 07                JNC   AHEAD        ;R3 then go to AHEAD.
23
24   2115   E0                   MOVX  A,@DPTR      ;If the content of A is less than
25   2116   15 82                DEC   DPTR         ;the content of R3-register.
26   2118   F0                   MOVX  @DPTR,A      ;then exchange the content of memory
27   2119   A3                   INC   DPTR         ;pointed by DPTR and previous location.
28   211A   EB                   MOV   A,R3
29   211B   F0                   MOVX  @DPTR,A
30
31   211C   D9 EF     AHEAD:     DJNZ  R1,LOOP1     ;Repeat comparisons until R1 count is zero.
32   211E
33   211E   DA E6                DJNZ  R2,LOOP2     ;Repeat until R2 count is zero.
34
35   2120   80 FE     HALT:      SJMP  HALT         ;Remain idle in infinite loop. Program end.
36
37   2122                        END                ;Assembly end.
```

Sample Data

	Memory address	Content			Memory address	Content	
Input Data: 07	2400	07	Output data : 07		2400	07	← Count
AB	2401	AB	34		2401	34	⎫
92	2402	92	4F		2402	4F	⎪
84	2403	84	69		2403	69	Array in
4F	2404	4F	84		2404	84	ascending
69	2405	69	92		2405	92	order
F2	2406	F2	AB		2406	AB	⎪
34	2407	34	F2		2407	F2	⎭

(Before sorting)   (After sorting)

EXAMPLE PROGRAM 8: Sorting an Array in Descending Order

Write an assembly language program to sort an array of data in descending order. The array is stored in memory starting from 2400$_H$. The first element of the array gives the count value for the number of elements in the array.

Problem Analysis

The algorithm for bubble sorting is given below. In bubble sorting of N-data, N–1 comparisons are carried by taking two consecutive data at a time. After each comparison, the data are rearranged such that the largest among the two is in the first memory location and the smallest in the next memory location. (Here the data are rearranged within the two memory locations whose contents are compared.) When we perform N –1 comparisons as mentioned above for N–1 times, then the array consisting of N-data will be sorted in the descending order.

Algorithm

1. Load address of data array in DPTR, and using DPTR load the count value in R2 via A.

2. Decrement R2-register (R2 is counter for N–1 repetitions).

3. Set DPTR as data array address pointer.

4. Set R1-register as counter for N–1 comparisons.

5. Increment DPTR, and using DPTR load two consecutive data of the array in R3 and A.

6. Compare the content of A and R3 by performing subtraction.

7. If carry flag is set (If the content of A is smaller than R3) then go to step 9, otherwise go to next step.

8. If carry flag is not set (If the content of A is larger than R3), then exchange the content of memory pointed by DPTR and next memory location.

9. Decrement R1-register. If zero flag =0, go to step 5 otherwise go to next step.

10. Decrement R2-register. If zero flag =0, go the step 3 otherwise go to next step.

11. Stop.

Flowchart

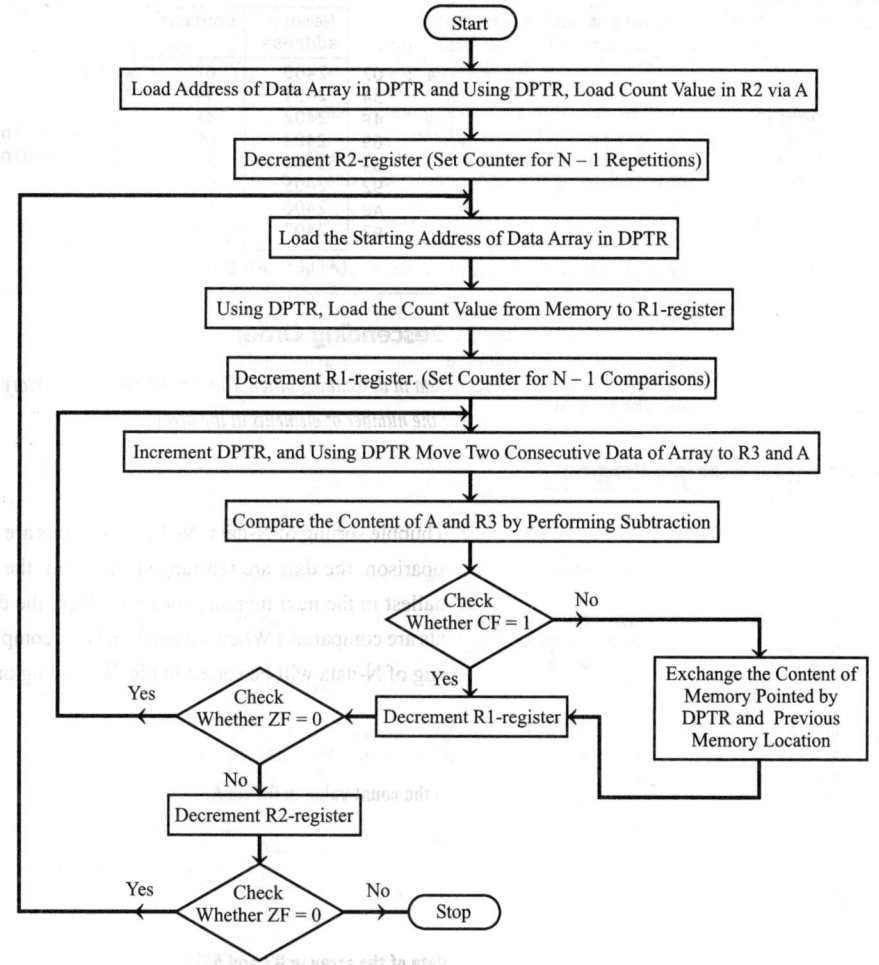

Assembly language program

```
;PROGRAM TO SORT AN ARRAY OF DATA IN DESCENDING ORDER

        ORG   2100H       ;Specify program starting address.

        MOV   DPTR,#2400H ;Load the count value in A-register.
        MOVX  A,@DPTR
        MOV   R2,A        ;Set counter for (N-1) repetitions
        DEC   R2          ;of (N-1) comparisons.

LOOP2:  MOV   DPTR,#2400H ;Set pointer for array.
        MOVX  A,@DPTR     ;Set count for (N-1) comparisons.
        MOV   R1,A
        DEC   R1

        INC   DPTR        ;Increment pointer.

LOOP1:  MOVX  A,@DPTR     ;Get two consecutive data of array in
        MOV   R3,A        ;R3 and A-register.
```

```
          INC   DPTR
          MOVX  A,@DPTR
          CLR   C            ;Compare A and R3-register.
          SUBB  A,R3         ;If content of A is less than
          JC    AHEAD        ;R3 then go to AHEAD.

          MOVX  A,@DPTR      ;If the content of A is greater than
          DEC   DPTR         ;the content of R3-register
          MOVX  @DPTR,A      ;then exchange the content of memory
          INC   DPTR         ;pointed by DPTR and previous location.
          MOV   A,R3
          MOVX  @DPTR,A

AHEAD:    DJNZ  R1,LOOP1     ;Repeat comparisons until R1 count is zero.

          DJNZ  R2,LOOP2     ;Repeat until R2 count is zero.

HALT:     SJMP  HALT         ;Remain idle in infinite loop. Program end.

          END                ;Assembly end.
```

Assembler Listing for Example Program 8

```
1                              ;PROGRAM TO SORT AN ARRAY OF DATA IN DECENDING ORDER
2
3    2100                      ORG   2100H        ;Specify program starting address.
4
5    2100   90 24 00           MOV   DPTR,#2400H  ;Load the count value in A-register.
6    2103   E0                 MOVX  A,@DPTR
7    2104   FA                 MOV   R2,A         ;Set counter for (N-1) repetitions
8    2105   1A                 DEC   R2           ;of (N-1) comparisons.
9
10   2106   90 24 00  LOOP2:   MOV   DPTR,#2400H  ;Set pointer for array.
11   2109   E0                 MOVX  A,@DPTR      ;Set count for (N-1) comparisons.
12   210A   F9                 MOV   R1,A
13   210B   19                 DEC   R1
14
15   210C   A3                 INC   DPTR         ;Increment pointer.
16   210D   E0        LOOP1:   MOVX  A,@DPTR      ;Get two consecutive data of array in
17   210E   FB                 MOV   R3,A         ;R3 and A-register.
18   210F   A3                 INC   DPTR
19   2110   E0                 MOVX  A,@DPTR
20   2111   C3                 CLR   C            ;Compare A and R3-register.
21   2112   9B                 SUBB  A,R3         ;If content of A is less than
22   2113   40 07              JC    AHEAD        ;R3 then go to AHEAD.
23
24   2115   E0                 MOVX  A,@DPTR      ;If the content of A is greater than
25   2116   15 82              DEC   DPTR         ;the content of R3-register.
26   2118   F0                 MOVX  @DPTR,A      ;then exchange the content of memory
27   2119   A3                 INC   DPTR         ;pointed by DPTR and previous location.
28   211A   EB                 MOV   A,R3
29   211B   F0                 MOVX  @DPTR,A
30
31   211C   D9 EF     AHEAD:   DJNZ  R1,LOOP1     ;Repeat comparisons until R1 count is zero.
32
33   211E   DA E6              DJNZ  R2,LOOP2     ;Repeat until R2 count is zero.
34
35   2120   80 FE     HALT:    SJMP  HALT         ;Remain idle in infinite loop. Program end.
36
37   2122                      END               ;Assembly end.
```

Sample Data

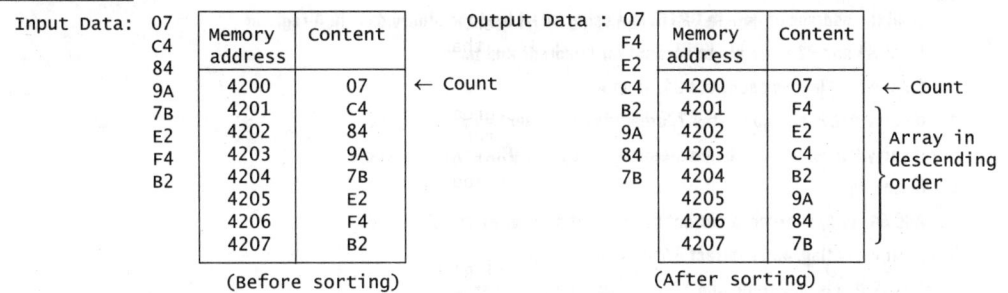

Input Data:	07
	C4
	84
	9A
	7B
	E2
	F4
	B2

Memory address	Content
4200	07 ← Count
4201	C4
4202	84
4203	9A
4204	7B
4205	E2
4206	F4
4207	B2

(Before sorting)

Output Data :	07
	F4
	E2
	C4
	B2
	9A
	84
	7B

Memory address	Content
4200	07 ← Count
4201	F4
4202	E2
4203	C4
4204	B2
4205	9A
4206	84
4207	7B

Array in descending order

(After sorting)

EXAMPLE PROGRAM 9:   Binary to BCD Conversion (or) Hexadecimal to BCD conversion

Write an assembly language program to convert an 8-bit binary data (2-digit hexa) to BCD. The binary data is stored in 2400_H. Store the ten's and unit's digits in 2401_H. Store the hundred's digit in 2402_H.

Problem Analysis

The maximum value of 8-bit binary is $FF_H = 256_{10}$. Hence, the maximum size of the data will have hundreds, tens and units. The algorithm given below uses two counters to count hundreds and tens. Initially the counters are cleared. First, let us subtract all hundreds from the binary data. For each subtraction, the hundred's register is incremented by one. Then let us subtract all tens. For each subtraction ten's register is incremented by one. The remaining will be units. The tens and units are combined to form 2-digit BCD (8-bit binary representation of BCD).

Flowchart for Example Program 9

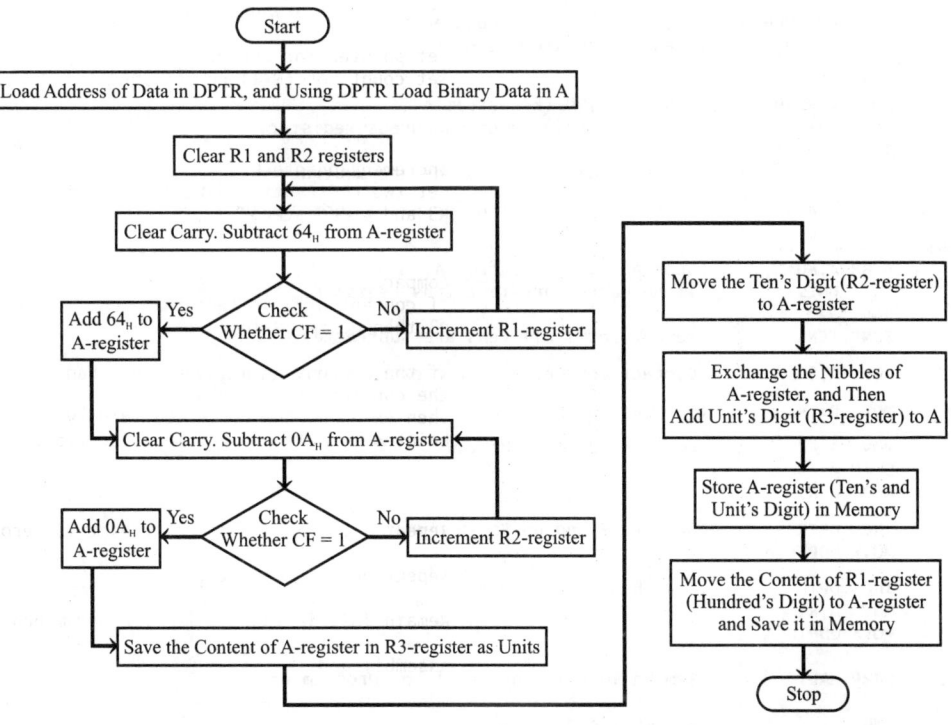

Algorithm

1. Load the address of data in DPTR, and using DPTR load the binary data in A-register.
2. Clear R1 and R2 registers to account for hundreds and tens.
3. Clear carry flag, and subtract 64_H from A-register.
4. If carry flag is set, go to step 7, otherwise go to next step.
5. If carry flag is not set, then increment R1-register (Hundred's register).
6. Go to step 3.
7. Add 64_H to A, in order to correct the content of A for extra subtraction.
8. Clear carry flag, and subtract $0A_H$ from A-register.
9. If carry flag is set, go to step 12, otherwise go to next step.
10. If carry flag is not set, then increment R2-register (Ten's register).
11. Go to step 8.
12. Add $0A_H$ to A, in order to correct the content of A for extra subtraction.
13. Save the content of A as units in R3-register.
14. Combine the units and tens to form 8-bit result.
15. Save the units, tens and hundreds in memory.

Assembly language program

```
;PROGRAM TO CONVERT A BINARY DATA TO BCD DATA

        ORG   2100H       ;Specify program starting address.

        MOV   DPTR,#2400H  ;Load the data address in DPTR.
        MOVX  A,@DPTR      ;Get the binary data in A-register.

        MOV   R1,#00H      ;Clear R1 to store hundreds.
        MOV   R2,#00H      ;Clear R2 to store tens.

HUND:   CLR   C
        SUBB  A,#64H       ;Subtract 100 (64h) from A.
        JC    AHEAD        ;If no carry increment hundreds register.
        INC   R1
        SJMP  HUND         ;Repeat until all hundreds are subtracted.

AHEAD:  ADD   A,#64H       ;Correct content of A for extra subtraction.

TENS:   CLR   C
        SUBB  A,#0AH       ;Subtract 10 (0Ah) from A.
        JC    UNIT         ;If no carry increment tens register.
        INC   R2
        SJMP  TENS         ;Repeat until all tens are subtracted.

UNIT:   ADD   A,#0AH       ;Correct content of A for extra subtraction.

        MOV   R3,A         ;Save units in R3.

        MOV   A,R2         ;Combine tens and units.
        SWAP  A
        ADD   A,R3

        INC   DPTR         ;Save tens and units in memory.
        MOVX  @DPTR,A

        INC   DPTR         ;Save hundreds in memory.
        MOV   A,R1
        MOVX  @DPTR,A

HALT:   SJMP  HALT         ;Remain idle in infinite loop. Program end.

        END                ;Assembly end.
```

Assembler Listing for Example Program 9

```
1                              ;PROGRAM TO CONVERT A BINARY DATA TO BCD DATA
2
3    2100                      ORG   2100H      ;Specify program starting address.
4
5    2100   90 24 00           MOV   DPTR,#2400H ;Load the data address in DPTR.
6    2103   E0                 MOVX  A,@DPTR     ;Get the binary data in A-register.
7
8    2104   79 00              MOV   R1,#00H     ;Clear R1 to store hundreds.
9    2106   7A 00              MOV   R2,#00H     ;Clear R2 to store tens.
10
11   210A   C3         HUND:   CLR   C
12   210B   94 64              SUBB  A,#64H      ;Subtract 100 (64h) from A.
13   210D   40 03              JC    AHEAD       ;If no carry, increment hundreds register.
14   210F   09                 INC   R1
15   2110   80 F8              SJMP  HUND        ;Repeat until all hundreds are subtracted.
16
17   2112   24 64      AHEAD:  ADD   A,#64H      ;Correct content of A for extra subtraction.
18
19   2114   C3         TENS:   CLR   C
20   2115   94 0A              SUBB  A,#0AH      ;Subtract 10 (0Ah) from A.
21   2117   40 03              JC    UNIT        ;If no carry, increment tens register.
22   2119   0A                 INC   R2
23   211A   80 F8              SJMP  TENS        ;Repeat until all tens are subtracted.
24
25   211C   24 0A      UNIT:   ADD   A,#0AH      ;Correct content of A for extra subtraction.
26
27   211E   FB                 MOV   R3,A        ;Save units in R3.
28
29   211F   EA                 MOV   A,R2        ;Combine tens and units.
30   2120   C4                 SWAP  A
31   2121   2B                 ADD   A,R3
32
33   2122   A3                 INC   DPTR        ;Save tens and units in memory.
34   2123   F0                 MOVX  @DPTR,A
35
36   2124   A3                 INC   DPTR        ;Save hundreds in memory.
37   2125   E9                 MOV   A,R1
38   2126   F0                 MOVX  @DPTR,A
39
40   2127   80 FE      HALT:   SJMP  HALT        ;Remain idle in infinite loop. Program end.
41
42   2129                      END               ;Assembly end.
```

Sample Data

Input Data : $B9_H$

Output Data : 0185_{10}

Memory address	Content	
2400	B9	Binary data
2401	85	} BCD data
2402	01	

EXAMPLE PROGRAM 10: 2-Digit Hexa to ASCII Conversion or b-bit Binary to ASCII Conversion

Write an assembly language program to convert a 2-digit hexa (8-bit binary) to ASCII code. The binary data is stored in 2400_H *and store the ASCII code in* 2401_H *and* 2402_H.

Problem Analysis

Each hexa digit (4-bit binary) is represented by an 8-bit ASCII. The hexa digit **0** through **9** are represented by 30_H to 39_H in ASCII. Hence for hexa **0** to **9**, if we add 30_H, we will get the corresponding ASCII. The hexa digit **A** through **F** are represented by 41_H to 46_H in ASCII. Hence for hexa digit **A** to **F** if we add 37_H we will get the corresponding ASCII.

In the following algorithm, the given 8-bit data is split into two nibbles. The ASCII code for each nibble is found by calling a subroutine, which takes care of adding 30_H to the nibble if it is less than $0A_H$, or adding 37_H if the nibble is greater than 09_H.

Flowchart for Example Program 10

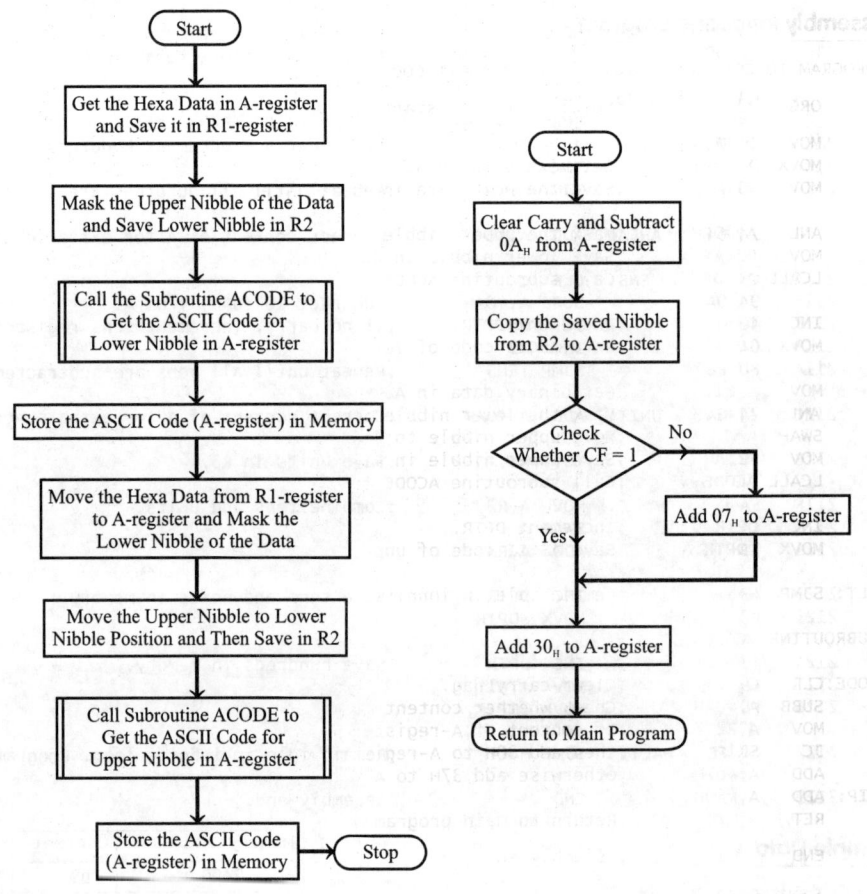

Algorithm

1. **Load the given data in A-register and move to R1-register.**
2. **Mask the upper nibble of the binary (hexa) data in A-register and save the lower nibble in R2-register.**
3. **Call subroutine acode to get ASCII code of the lower nibble and store in memory.**
4. **Move the content of R1-register to A-register and mask the lower nibble.**
5. **Move the upper nibble to lower nibble position and save in R2-register.**
6. **Call subroutine acode to get the ASCII code of upper nibble and store in memory.**
7. **Stop.**

Algorithm for Subroutine ACODE

1. Clear carry flag, and subtract $0A_H$ from the content of A-register. Move R2 to A.
2. If CF = 1, go to step 4. If CF = 0, go to next step.
3. Add 07_H to A-register.
4. Add 30_H to A-register.
5. Return to main program.

Assembly language program

```
;PROGRAM TO CONVERT 2-DIGIT HEXA TO ASCII CODE

        ORG    2100H        ;Specify program starting address.

        MOV    DPTR,#2400H
        MOVX   A,@DPTR      ;Get hexa data in A.
        MOV    R1,A         ;Save the hexa data in R1-register.

        ANL    A,#0FH       ;Mask the upper nibble.
        MOV    R2,A         ;Save lower nibble in R2.
        LCALL  ACODE        ;Call subroutine ACODE to get ASCII code.

        INC    DPTR         ;Increment DPTR.
        MOVX   @DPTR,A      ;save ASCII code of lower nibble in memory.

        MOV    A,R1         ;Get binary data in A.
        ANL    A,#F0H       ;Mask the lower nibble.
        SWAP   A            ;Move upper nibble to lower nibble position.
        MOV    R2,A         ;Save upper nibble in R2.
        LCALL  ACODE        ;Call subroutine ACODE to get ASCII code.

        INC    DPTR         ;Increment DPTR.
        MOVX   @DPTR,A      ;Save ASCII code of upper nibble in memory.

HALT:   SJMP   HALT         ;Remain idle in infinite loop. Program end.

;SUBROUTINE ACODE

ACODE:CLR     C            ;Clear carry flag.
        SUBB   A,#0AH       ;Check whether content A is less than 0AH.
        MOV    A,R2         ;If content of A-register is less than 0AH,
        JC     SKIP         ;then add 30H to A-register,
        ADD    A,#07H       ;otherwise add 37H to A-register.
SKIP:   ADD    A,#30H
        RET                 ;Return to main program.

        END                 ;Assembly end.
```

Assembler Listing for Example Program 10

```
 1                           ;PROGRAM TO CONVERT 2-DIGIT HEXA TO ASCII CODE
 2
 3    2100                    ORG    2100H     ;Specify program starting address.
 4
 5    2100    90 24 00        MOV    DPTR,#2400H
 6    2103    E0              MOVX   A,@DPTR   ;Get hexa data in A.
 7    2104    F9              MOV    R1,A      ;Save the hexa data in R1-register.
 8
 9    2105    54 0F           ANL    A,#0FH    ;Mask the upper nibble.
10    2107    FA              MOV    R2,A      ;Save lower nibble in R2.
11    2108    12 21 19        LCALL  ACODE     ;Call subroutine ACODE to get ASCII code.
12
13    210B    A3              INC    DPTR      ;Increment DPTR.
14    210C    F0              MOVX   @DPTR,A   ;Save ASCII code of lower nibble in memory.
15
```

```
16    210D    E9              MOV    A,R1          ;Get binary data in A.
17    210E    54 F0           ANL    A,#F0H        ;Mask the lower nibble.
18    2100    C4              SWAP   A             ;Move upper nibble to lower nibble position.
19    2111    FA              MOV    R2,A          ;Save upper nibble in R2.
20    2112    12 21 19        LCALL  ACODE         ;Call subroutine ACODE to get ASCII code.
21
22    2115    A3              INC    DPTR          ;Increment DPTR.
23    2116    F0              MOVX   @DPTR,A       ;Save ASCII code of upper nibble in memory.
24
25    2117    80 FE    HALT:  SJMP   HALT          ;Remain idle in infinite loop. Program end.
26
27                    ;SUBROUTINE ACODE
28
29    2119    C3       ACODE:CLR    C             ;Clear carry flag.
30    211A    94 0A           SUBB   A,#0AH        ;Check whether content A is less than 0AH.
31    211C    EA              MOV    A,R2          ;If content of A-register is less than 0AH,
32    211D    40 02           JC     SKIP          ;then add 30H to A-register,
33    211F    24 07           ADD    A,#07H        ;otherwise add 37H to A-register.
34    2121    24 30    SKIP:  ADD    A,#30H
35    2123    22              RET                  ;Return to main program.
36
37    2124                    END                  ;Assembly end.
```

Sample Data

Input Data :	E4$_H$
Output Data :	34 (ASCII code for 4)
	45 (ASCII code for E)

Memory address	Content
2400	E4
2401	34
2402	45

←Hexadecimal

} ASCII

Example Program 11: GCD of Two 8-Bit Data

Write an assembly language program to determine the GCD of two 8-bit data.

Problem Analysis

First divide the smaller data by the larger data and check for remainder. If remainder is zero then the smaller data is the GCD.

If the remainder is not zero then take the remainder as the divisor and the previous divisor as the dividend and repeat the division until the remainder is zero. When the remainder is zero, we can store the divisor as GCD.

Before performing division we can even check whether the dividend and divisor are equal. If they are equal then we can directly store the divisor as GCD without performing division.

Algorithm

1. Set DPTR as pointer for input data.
2. Get one data in A-register and save in R1-register.
3. Increment DPTR and get another data in A-register.
4. Check whether the content of A and R1 are equal using compare instruction.
5. If zero flag is set then go to step 13, otherwise go to next step.
6. Save A in R2. Check whether A is greater than R1 using subtract instruction. Then restore the content of A from R2.
7. If carry flag is not set then go to step 10, otherwise go to next step.
8. Exchange the content of A and R1, so that the larger among the two is dividend and smaller is the divisor.
9. Copy R1 to B, and divide A-register by B-register.
10. Move remainder in B to A. Check whether remainder is zero using compare instruction.
11. If zero flag is set then go to next step, otherwise go to step 4.
12. Move the content of R1 to A.
13. Increment DPTR and store the content of A-register as GCD in memory.
14. Stop.

Flowchart

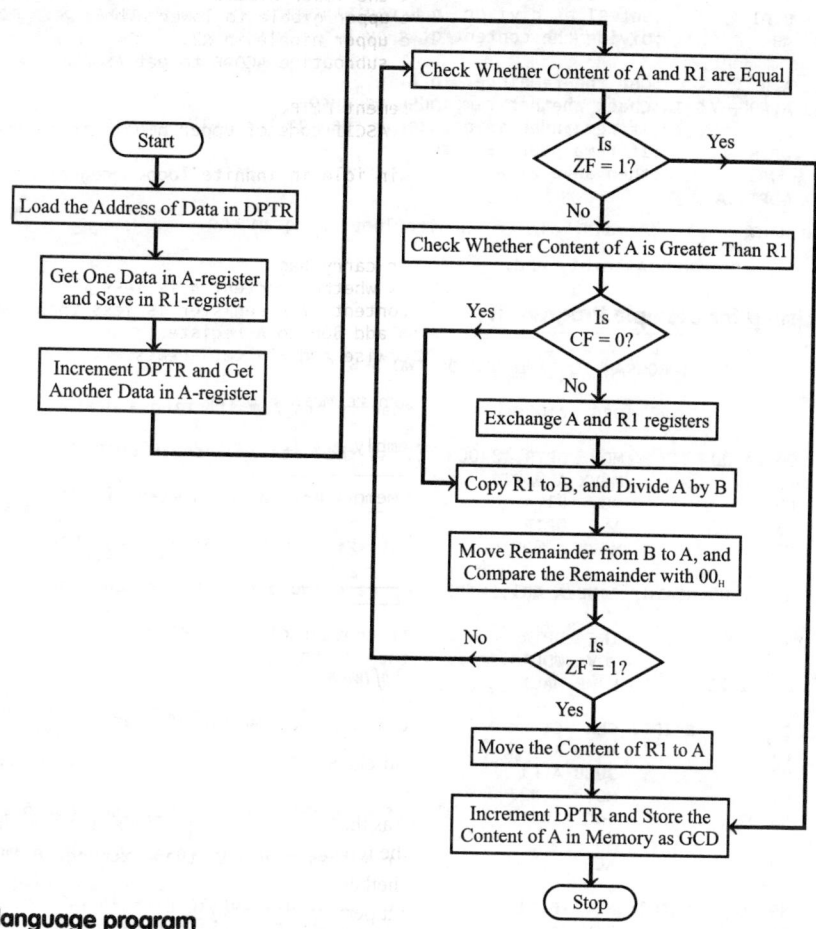

Assembly language program

```
;PROGRAM TO FIND GCD OF TWO 8-BIT DATA

        ORG   2100H        ;Specify program starting address.

        MOV   DPTR,#2400H  ;Load the data address in DPTR.
        MOVX  A,@DPTR
        MOV   R1,A          ;Get the first data in R1-register.
        INC   DPTR
        MOVX  A,@DPTR        ;Get the second data in A-register.

AGAIN:  CJNE  A,#R1,SKIP1   ;Compare two data, if not equal go to SKIP1.

        INC   DPTR          ;If content of A and R1 are equal,
        MOVX  @DPTR,A        ;then store A as GCD.
        LJMP  HALT

SKIP1:  CLR   C             ;Compare the content of A and R1.
        MOV   R2,A
        SUBB  A,R1
        MOV   A,R2
        JNC   SKIP2         ;If A is greater than R1, then go to SKIP2.
```

```
        XCH  A,R1          ;If A less than R1, then exchange A and R1.
SKIP2:  MOV  B,R1          ;Set R1 as divisor in B-register.
        DIV  AB            ;Divide the content of A by B.

        MOV  A,B           ;Get the remainder in A.
        CJNE A,#00H,AGAIN  ;Check whether remainder is zero,
                           ;if remainder is not zero, then go to AGAIN.
        INC  DPTR          ;If remainder is zero,
        MOV  A,R1          ;then save R1 as GCD.
        MOVX @DPTR,A
HALT:   SJMP HALT          ;Remain idle in infinite loop. Program end.

        END                ;Assembly end.
```

Assembler Listing for Example Program 11

```
1                        ;PROGRAM TO FIND GCD OF TWO 8-BIT DATA
2
3    2100                      ORG  2100H            ;Specify program starting address.
4
5    2100  90 24 00            MOV  DPTR,#2400H      ;Load the data address in DPTR.
7    2103  E0                  MOVX A,@DPTR
8    2104  F9                  MOV  R1,A             ;Get the first data in R1-register.
9    2105  A3                  INC  DPTR
10   2106  E0                  MOVX A,@DPTR          ;Get the second data in A-register.
11
12   2107  B4 01 05  AGAIN: CJNE A,#R1,SKIP1         ;Compare two data, if not equal go to SKIP1.
13
14   210A  A3                  INC  DPTR             ;If content of A and R1 are equal,
15   210B  F0                  MOVX @DPTR,A          ;then store A as GCD.
16   210C  02 21 21            LJMP HALT
17
18   210F  C3        SKIP1: CLR  C                   ;Compare the content of A and R1.
19   2110  FA                  MOV  R2,A
20   2111  99                  SUBB A,R1
21   2112  EA                  MOV  A,R2
22   2113  50 01               JNC  SKIP2            ;If A is greater than R1, then go to SKIP2.
23
24   2115  C9                  XCH  A,R1             ;If A less than R1, then exchange A and R1.
25
26   2116  89 F0     SKIP2: MOV  B,R1                ;Set R1 as divisor in B-register.
27   2118  84                  DIV  AB               ;Divide the content of A by B.
28
29   2119  E5 F0               MOV  A,B              ;Get the remainder in A.
30   211B  B4 00 E9            CJNE A,#00H,AGAIN     ;Check whether remainder is zero,
31                                                    ;if remainder is not zero, then go to AGAIN.
32   211E  A3                  INC  DPTR             ;If remainder is zero,
33   211F  E9                  MOV  A,R1             ;then save R1 as GCD.
34   2120  F0                  MOVX @DPTR,A
35
36   2121  80 FE     HALT:  SJMP HALT                ;Remain idle in infinite loop. Program end.
37
38   2123                      END                   ;Assembly end.
```

Sample Data

Input Data : Data1 = 0C_H   Output Data : 04_H

Data2 = 04_H

Memory address	Content	
2400	0C	← Data 1
2401	04	← Data 2
2402	04	← GCD

Example Program 12: LCM of Two 8-Bit Data

Write an assembly language program to determine the LCM of the two 8-bit data.

Problem Analysis

First determine the GCD of two data. Then determine the product of the two data. Here, it is assumed that the product does not exceed 8 bits. When the product is divided by GCD, the quotient will be the LCM of the two data. (For the GCD of two data please refer to example program 11.)

Algorithm

1. Set DPTR as pointer for input data.
2. Get one data in A-register and save in R1 and R4 registers.
3. Increment DPTR and get another data in A-register and save in R5-register.
4. Call subroutine GCD to get the GCD in A-register.
5. Save GCD in R3-register.
6. Copy two input data from R4 and R5 into A and B registers, and get the product of two data in A-register.
7. Copy GCD from R3 to B. Divide the product in A by GCD in B. The quotient is LCM.
8. Increment DPTR, and save the LCM in memory.
9. Stop.

> *Note: The algorithm for subroutine GCD can be obtained from example program 11.*

Flowchart

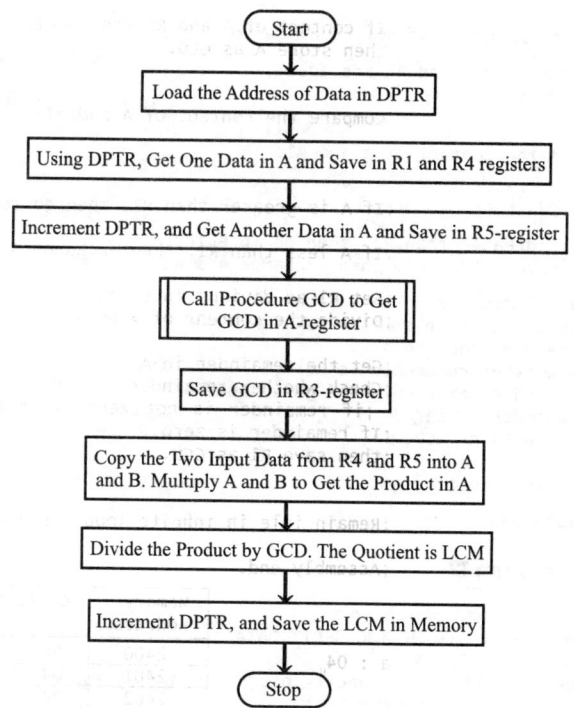

> *Note: The flowchart for procedure GCD can be obtained from example program 11.*

Assembly language program

```
;PROGRAM TO FIND LCM OF TWO 8-BIT DATA

        ORG    2100H      ;Specify program starting address.

        MOV    DPTR,#2400H ;Load the data address in DPTR.
        MOVX   A,@DPTR
        MOV    R1,A       ;Get the first data in R1-register.
        MOV    R4,A       ;Save the first data in R4-register.
        INC    DPTR
        MOVX   A,@DPTR    ;Get the second data in A-register.
        MOV    R5,A       ;Save the second data in R5.

        LCALLGCD
        MOV    R3,A

        MOV    B,R4       ;Get the product of two data in A.
        MOV    A,R5
        MUL    AB
        MOV    B,R3       ;Set R3 (GCD) as divisor in B-register.
        DIV    AB         ;Divide product by GCD, the quotient is LCM.

        INC    DPTR       ;Store LCM in memory.
        MOVX   @DPTR,A
HALT:   SJMP   HALT       ;Remain idle in infinite loop. Program end.

;SUBROUTINE TO FIND GCD OF TWO 8-BIT DATA

GCD:    CJNE   A,#R1,SKIP1 ;Compare two data, if not equal go to SKIP1.

        INC    DPTR       ;If content of A and R1 are equal,
        MOVX   @DPTR,A    ;then store A as GCD.
        LJMP   RET1
SKIP1:  CLR    C          ;Compare the content of A and R1.
        MOV    R2,A
        SUBB   A,R1
        MOV    A,R2
        JNC    SKIP2      ;If A is greater than R1, then go to SKIP2.
        XCH    A,R1       ;If A is less than R1, then exchange A & R1.
SKIP2:  MOV    B,R1       ;Set R1 as divisor in B-register.
        DIV    AB         ;Divide the content of A by B.
        MOV    A,B        ;Get the remainder in A.
        CJNE   A,#00H,GCD ;Check whether remainder is zero,
                          ;if remainder not zero, then go to AGAIN.
        INC    DPTR       ;If remainder is zero,
        MOV    A,R1       ;then save R1 as GCD.
        MOVX   @DPTR,A
RET1:   RET               ;Return to main program.

        END     ,         ;Assembly end.
```

Assembler Listing for Example Program 12

```
1                         ;PROGRAM TO FIND LCM OF TWO 8-BIT DATA
2
3    2100                 ORG  2100H      ;Specify program starting address.
4
5    2100  90 24 00       MOV  DPTR,#2400H ;Load the data address in DPTR.
6    2103  E0             MOVX A,@DPTR
7    2104  F9             MOV  R1,A        ;Get the first data in R1-register.
```

8	2105	FC		MOV R4,A	;Save the first data in R4-register.
9	2106	A3		INC DPTR	
10	2107	E0		MOVX A,@DPTR	;Get the second data in A-register.
11	2108	FD		MOV R5,A	;Save the second data in R5.
12					
13	2109	12 21 18		LCALL GCD	
14	210C	FB		MOV R3,A	
15					
16	210D	8C F0		MOV B,R4	;Get the product of two data in A.
17	210F	ED		MOV A,R5	
18	2110	A4		MUL AB	
19	2111	8B F0		MOV B,R3	;Set R3 (GCD) as divisor in B-register.
20	2113	84		DIV AB	;Divide product by GCD, the quotient is LCM.
21					
22	2114	A3		INC DPTR	;Store LCM in memory.
23	2115	F0		MOVX @DPTR,A	
24					
25	2116	80 FE	HALT:	SJMP HALT	;Remain idle in infinite loop. Program end.
26					
27				;SUBROUTINE TO FIND GCD OF TWO 8-BIT DATA	
28					
29	2118	B4 01 05	GCD:	CJNE A,#R1,SKIP1	;Compare two data, if not equal go to SKIP1.
30					
31	211B	A3		INC DPTR	;If content of A and R1 are equal,
32	211C	F0		MOVX @DPTR,A	;then store A as GCD.
33	211D	02 21 32		LJMP RET1	
34					
35	2120	C3	SKIP1:	CLR C	;Compare the content of A and R1.
36	2121	FA		MOV R2,A	
37	2122	99		SUBB A,R1	
38	2123	EA		MOV A,R2	
39	2124	50 01		JNC SKIP2	;If A is greater than R1, then go to SKIP2.
40	2126	C9		XCH A,R1	;If A is less than R1, then exchange A & R1.
41					
42	2127	89 F0	SKIP2:	MOV B,R1	;Set R1 as divisor in B-register.
43	2129	84		DIV AB	;Divide the content of A by B.
44	212A	E5 F0		MOV A,B	;Get the remainder in A.
45	212C	B4 00 E9		CJNE A,#00H,GCD	;Check whether remainder is zero,
46					;if remainder not zero, then go to AGAIN.
47	212F	A3		INC DPTR	;If remainder is zero,
48	2130	E9		MOV A,R1	;then save R1 as GCD.
49	2131	F0		MOVX @DPTR,A	
50	2132	22	RET1:	RET	;Return to main program.
51					
52	2133			END	;Assembly end.

Sample Data

Input Data :

    Data1 = 0C$_H$

    Data2 = 04$_H$

Output Data :

    GCD = 04$_H$

    LCM = 0C$_H$

Memory address	Content	
2400	0C	} Input data
2401	04	
2402	04	← GCD
2403	0C	← LCM

5.2 PROGRAMMING 8051 TIMERS

The 8051 has two internal 16-bit timers/counters that can be programmed to work independently. They are called timer-0/counter-0 and timer-1/counter-1. In the counter mode of operation, the timer will count the high-to-low transition of the signal applied at the corresponding timer pin (port-3 pin, P3.2 for timer-0 and P3.3 for timer-1), by incrementing the content of the timer register associated with the timer by one for every high-to-low transition. (The signal applied at the port pin will act as a clock for incrementing the content of the timer register.)

In the timer mode of operation, the internal timer clock will increment the content of the associated timer register for every clock pulse. The timer clock is internally derived by dividing the crystal frequency by 12. Therefore, the timer clock is an independent clock, and the frequency of the timer clock will be 1/12 of the system clock frequency.

The various special function registers associated with internal timers/counters of 8051 are,

TMOD : Timer/Counter mode control register

TCON : Timer/Counter control register

TL0 : Timer-0 low order register

TH0 : Timer-0 high order register

TL1 : Timer-1 low order register

TH1 : Timer-1 high order register

The TMOD register is programmed to select various operating modes of the timer and the TCON register is programmed to control the timer operation. The TH0 and TL0 together form the 16-bit count register of the timer-0. The TH1 and TL1 together form the 16-bit count register of the timer-1.

A timer can be programmed to initiate a task after a specified time delay. Alternatively, a timer can be programmed to initiate a repetitive task again and again after a specified time delay. For both these applications, first the register TMOD has to be programmed for the desired mode of operation, then an initial count value calculated for the specified time delay should be loaded in the timer count register. Then the timer is started by programming the TCON register, and at the end of time delay, an interrupt is generated which can be used to initiate the specified task.

5.2.1 TIMER MODE CONTROL (TMOD) REGISTER

The TMOD register is used to select the operating mode and the timer/counter operation of the timers. The format of a TMOD register is shown in Fig. 5.1. The lower four bits of the TMOD register are used to control timer-0 and the upper four bits are used to control timer-1.

The two timers can be independently programmed to operate in various modes. The TMOD register has two separate two-bit fields, M0 and M1, to program the operating mode of the timers. The operating modes of the timers are mode-0, mode-1, mode-2 and mode-3. In all these operating modes, the oscillator clock is divided by 12 and applied as the input clock to the timer.

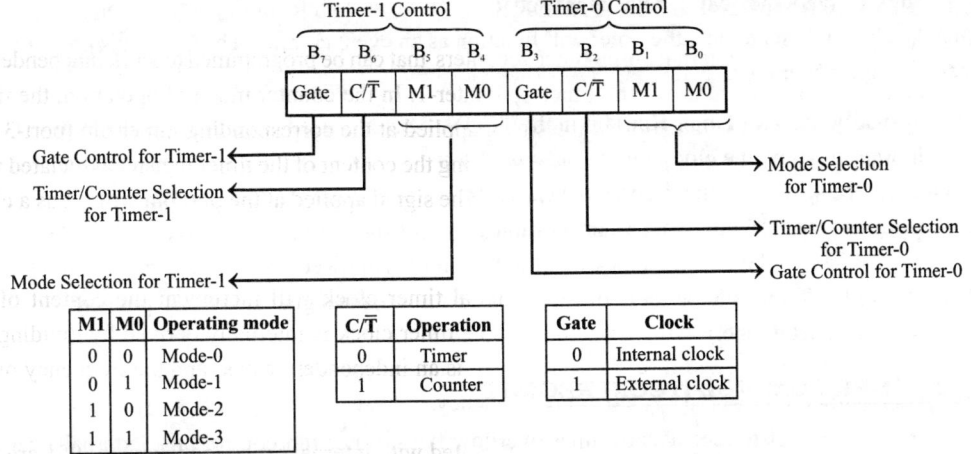

Fig. 5.1: Format of the TMOD register of the 8051 family of microcontrollers.

Timer Port Mode-0

In mode-0, the timer register is configured as a 13-bit register. For timer-1, the 8 bits of TH1 and lower 5 bits of TL1 are used to form the 13-bit register. For timer-0, the 8-bit of TH0 and lower 5 bits of TL0 are used to form the 13-bit register. (The upper three bits of the TL registers are ignored). For every clock input to the timer, the 13-bit timer register is incremented by one. When the timer count rolls over from all 1s to all 0s (i.e., 11111 1111 1111 to 00000 0000 0000), the timer interrupt flag in the TCON register is set to one.

Timer Port Mode-1

Mode-1 is same as mode-0 except the size of the timer register. In mode-1, the TH and TL registers are cascaded to form a 16-bit timer register.

Timer Port Mode-2

In mode-2, the timers function as 8-bit timers with automatic reload feature. The TL register will function as an 8-bit timer count register and the TH-register will hold an initial count value. When the timer is started, the initial value in TH is loaded to TL and for each clock input to the timer, the 8-bit timer count register is incremented by one. When the timer count rolls over from all 1s to all 0s (i.e., 1111 1111 to 0000 0000), the timer interrupt flag in the TCON register is set to one and the content of the TH register is reloaded in the TL register and the count process starts again from this initial value.

Timer Port Mode-3

In mode-3, the timer-0 is configured as two separate 8-bit timers and timer-1 is stopped. In mode-3, the TL0 will function as an 8-bit timer controlled by standard timer-0 control bits and the TH0 will function as an 8-bit timer controlled by timer-1 control bits. While timer-0 is programmed in mode-3, timer-1 can be programmed in mode-0, 1 or 2 and can be used for applications that do not require an interrupt.

The C/$\overline{\text{T}}$ bit of the TMOD register is used to program the counter or timer operation of the timer. When the C/$\overline{\text{T}}$ bit is set to one, the timer will function as an event counter. The C/$\overline{\text{T}}$ bit is programmed to zero for timer operation.

Normally, the TR (**T**imer **R**un) bit in the TCON register is used to control the clock input to the timer. In order to allow the clock input (*Note: the timer will run only if the clock input is allowed*), the TR bit should be set to one. In addition, the GATE bit in the TMOD register will facilitate the external signal applied to the $\overline{\text{INT}}$ pin to act as an additional control signal to allow or disallow the clock input to the timer. When GATE = 1, the clock input to the timer is allowed only if the signal at the $\overline{\text{INT}}$ pin is **high** (as well as TR should be set to one.) When GATE = 0, the signal at the $\overline{\text{INT}}$ pin is ignored (but the TR should be set to one.)

5.2.2 TIMER CONTROL (TCON) REGISTER

The TCON register consists of timer overflow flags, timer run control bits, external interrupt flags and external interrupt-type control bits. The format of a TCON register is shown in Fig. 5.2.

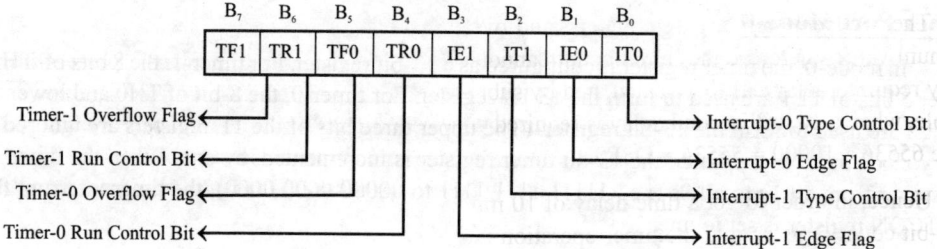

Fig. 5.2: Format of a TCON register of the 8051 family of microcontrollers.

The timers in an 8051 microcontroller are upcounters and keep on incrementing as long as the clock is applied. Therefore, when the clock is applied after reaching the maximum value (i.e., the content of the counter is all 1s), the content of the counter will become zero (i.e., all 0s). This condition is called timer overflow and it is also the end of timing which a program wants to maintain by using the timer. The TCON register has a 1-bit flag, TF for each timer to indicate the timer overflow or end of timing. Whenever the timer/counter overflows, the TF flag is set to one. The TF flag is also used as an interrupt signal to initiate the execution of a subroutine. When the controller vectors to subroutine, the TF flag is cleared.

The TR bit is used to start/stop the timer/counter. When the TR bit is set to one, the timer/counter will start counting and continue the counting as long as the TR bit is one. The timer/counter will stop counting when the TR bit is cleared to zero.

When a valid external interrupt signal is detected, the IE flag is set to one. When the controller accepts the external interrupt and starts processing it, the IE flag is cleared to zero. The IT bit is used to program the type of external interrupt signal to be recognized by the controller. The IT bit is programmed as one to recognize the falling edge triggered external interrupt and it is programmed as zero to recognize logic **low**-level external interrupt.

5.2.3 CALCULATING THE TIME DELAY

Case-1: Crystal Frequency is 12 MHz

When the crystal frequency is 12 MHz, the timer clock frequency will be 1 MHz. Therefore, the time period of 1 clock is 1/1 MHz = 1 microsecond. Therefore, when the timer clock is a 1 MHz clock, the timer count gets incremented once in every 1 microsecond. Hence, the timer count value for a specified delay in microseconds will be same as the value of delay in microsecond itself. For example, for 100 microseconds delay, the count value is 100. In 8051 timers, there are three choices for the size of count value, and they are 8-bit, 13-bit and 16-bit.

When the 8-bit count register is selected, the maximum count value is $2^8 = 256_{10}$ = FFH, and maximum possible time delay is 256 microseconds.

When the 13-bit count register is selected, the maximum count value is $2^{13} = 8192_{10}$ = 2000H, and maximum possible time delay is 8192 microseconds = 8.192 milliseconds.

When the 16-bit count register is selected, the maximum count value is $2^{16} = 65536_{10}$ = FFFFH, and maximum possible time delay is 65536 microseconds = 65.536 milliseconds.

The 8051 counters are up-counters, and overflow or end of count occurs only after reaching maximum value, therefore the initial count cannot be 0 for any desired time delay. The initial count for any required delay has to be calculated by subtracting the delay count from maximum value. For example, when a 10 millisecond delay is required while using a 16-bit count register, the initial count will be 65536 – 10000 = 55536 = D8F0H.

Hence, in order to get a time delay of 10 milliseconds, if the initial value D8F0H is loaded in the 16-bit count register, and the timer operation is started, then the overflow occurs when the count reaches FFFFH and the time delay achieved will be 10 milliseconds.

Case-2: Crystal Frequency is 11.0592 MHz

The crystal frequency of 11.0592 MHz is used when serial communication with standard PC (Personal Computer) is employed in order to generate the clock for right baud rate for serial communication with standard PC. When the crystal frequency is 11.0592 MHz, the timer clock frequency will be 11.0592/12 = 0.9216 MHz. Therefore, the time period of 1 clock is 1/0.9216 MHz = 1.085 microseconds. Now, the timer can increment the count by one in 1.085 microseconds. The count value for a specified delay in microseconds will be microseconds delay value divided by 1.085. For example, for 100 microseconds delay, the count value is 100/1.085 = 92. In 8051 timers, there are three choice for the size of count value, and they are 8-bit, 13-bit and 16-bit.

When an 8-bit count register is selected, the maximum count value is $2^8 = 256_{10}$ = FFH, and maximum possible time delay is 256 × 1.085 = 277 microseconds.

When a 13-bit count register is selected, the maximum count value is $2^{13} = 8192_{10}$ = 2000H, and maximum possible time delay is 8192 × 1.085 = 8888 microseconds = 8.888 milliseconds.

When the 16-bit count register is selected, the maximum count value is, $2^{16} = 65536_{10}$ = FFFFH, and maximum possible time delay is 65536 × 1.085 = 71106 microseconds = 71.106 milliseconds.

5.2.4 EXAMPLES OF TIMER PROGRAMMING IN THE 8051 CONTROLLER

EXAMPLE PROGRAM 13

This example program is developed for generation of unipolar square waveform of 1 kHz frequency using the timer-0 of 8051 in mode-1. Assume that the crystal frequency of the controller is 12 MHz.

Problem Analysis

In order to generate a square wave, a port pin can be set to high first and then after a time delay, the port pin is reset to zero; then the process of set and reset are repeated continuously with a uniform time delay. The desired time delay can be achieved using the timer of 8051.

The time period of 1 kHz square wave is $1/1 \times 10^3 = 1$ millisecond = 1000 microseconds. But in a square wave during half the period, the square wave will be high and during the next half of the period, the square wave will be low. So the time delay required is $1000/2 = 500$ micro-seconds.

Since the crystal frequency is 12 MHz, the timer clock will $1/12 = 1$ MHz, and so the time period of the timer clock will be 1 microsecond. Therefore, the count for 500 microseconds time delay is 500. The count register in mode-1 is 16-bit and so the maximum count is $2^{16} = 65536_{10}$. The initial count for any required delay has to be calculated by subtracting the delay count from the maximum count value.

Therefore, the initial count = $65536_{10} - 500_{10} = 65036_{10} = \text{FE0CH}$

The byte to be loaded in the TMOD register to select mode-1 operation of timer-0 is framed as shown below:

$$\text{TMOD} = \boxed{\text{Gate} \mid \text{C/}\overline{\text{T}} \mid \text{M1} \mid \text{M0} \mid \text{Gate} \mid \text{C/}\overline{\text{T}} \mid \text{M1} \mid \text{M0}} = \text{X X X X 0 0 0 1} = 01_{\text{H}}$$

Flowchart

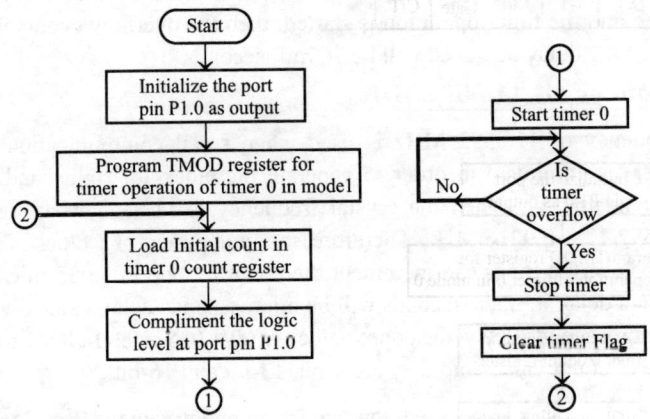

Assembly language program

```
;Program to generate square wave using 8051 timer in mode-1

        CLR   P1.0          ;Initialize port pin P1.0 as output
        MOV   TMOD,#01H      ;Program TMOD register for mode-1 operation of timer-0
AGAIN:  MOV   TL0,#0CH       ;Load low byte of count in timer-0 low order count register
        MOV   TH0,#FEH       ;Load high byte of count in timer-0 high order count register
        CPL   P1.0           ;Compliment the port pin P1.0
        SETB  TR0            ;Set timer run flag, to start timer
WAIT:   JNB   TF0,WAIT       ;Wait for timer overflow
```

```
        CLR  TR0          ;Clear timer run flag, to stop timer
        CLR  TF0          ;Clear timer flag

        SJMP AGAIN        ;Repeat generation of next cycle

        END               ;Assemby end
```

EXAMPLE PROGRAM 14

This example program is developed for generation of unipolar square waveform of 1 kHz frequency using the timer-0 of 8051 in mode-0. Assume that the crystal frequency of the controller is 12 MHz.

Problem Analysis

The program logic is same as Example Program 1.

The count register in mode-0 is 13-bit and so the maximum count is $2^{13} = 8192_{10}$. The initial count for any required delay has to be calculated by subtracting the delay count from the maximum count value.

Therefore, the initial count = $8192_{10} - 500_{10} = 7692_{10} = 1E0CH$

The 13-bit initial count has to be divided into upper 8 bits and lower 5 bits as shown below and the upper 8 bits are loaded in timer high order register and the lower 5 bits are loaded in timer low-order register.

1E0CH = 1 1110 0000 1100 = 11110000 01100

11110000 = 1111 0000 = F0H (Initial count for high order timer register)

01100 = 0 1100 = 0000 1100 = 0CH (Initial count for low order timer register)

The byte to be loaded in the TMOD register to select mode-0 operation of timer-0 is framed as shown below:

$$\text{TMOD} = \boxed{\text{Gate} \mid \text{C/}\overline{\text{T}} \mid \text{M1} \mid \text{M0} \mid \text{Gate} \mid \text{C/}\overline{\text{T}} \mid \text{M1} \mid \text{M0}} = \text{X X X X 0 0 0 0} = 00_\text{H}$$

Flowchart

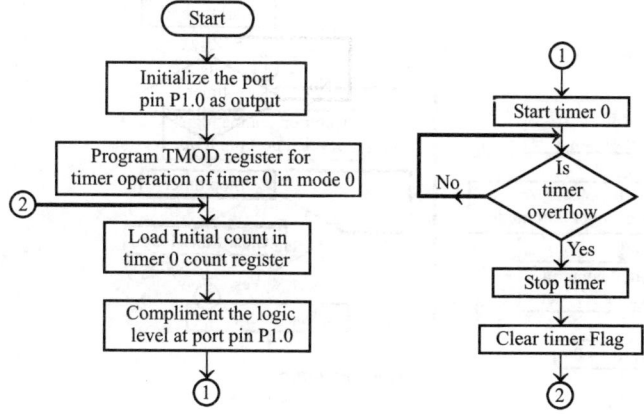

Assembly language program

```
;Program to generate square wave using 8051 timer in mode-0

        CLR  P1.0           ;Initialize port pin P1.0 as output
        MOV  TMOD,#00H      ;Program TMOD register for mode-0 operation of timer-0
AGAIN:  MOV  TL0,#0CH       ;Load low byte of count in timer-0 low order count register
        MOV  TH0,#F0H       ;Load high byte of count in timer-0 high order count register
```

```
          CPL   P1.0              ;Compliment the port pin P1.0
          SETB  TR0               ;Set timer run flag, to start timer
WAIT:     JNB   TF0,WAIT          ;Wait for timer overflow
          CLR   TR0               ;Clear timer run flag, to Stop timer
          CLR   TF0               ;Clear timer flag
          SJMP  AGAIN             ;Repeat generation of next cycle
          END                     ;Assemby end
```

EXAMPLE PROGRAM 15

This example program is developed for generation of unipolar square waveform of 2 kHz frequency using the timer-0 of 8051 in mode-2. Assume that the crystal frequency of the controller is 12 MHz.

Problem Analysis

The program logic is same as Example Program 1. Here we have chosen the frequency as 2 kHz to reduce the time delay count.

Now, time delay count = $(1/2 \times 10^3)/2 = 250$

The count register in mode-2 is 8-bit and so the maximum count is $2^8 = 256_{10}$. The initial count for any required delay has to be calculated by subtracting the delay count from maximum count value.

Therefore, the initial count = $256_{10} - 250_{10} = 6_{10} = 06H$

In mode-2, the 8-bit count value is loaded in high order count register only once. The controller will copy the value of count in low order register and start the count operation. At the end of timer operation, the count value is reloaded from high order register to low order register and again restart the count operation. Therefore, the count register need not be loaded with the count value for every cycle of wave generation.

The byte to be loaded in the TMOD register to select mode-2 operation of timer-0 is framed as shown below:

$$\text{TMOD} = \boxed{\text{Gate} \mid \text{C/}\overline{\text{T}} \mid \text{M1} \mid \text{M0} \mid \text{Gate} \mid \text{C/}\overline{\text{T}} \mid \text{M1} \mid \text{M0}} = \text{X X X X 0 0 1 0} = 02_{\text{H}}$$

Flowchart

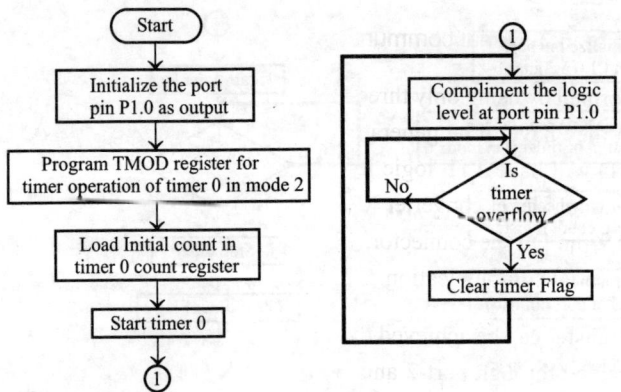

Assembly language program

```
;Program to generate square wave using 8051 timer in mode-2

          CLR   P1.0              ;Initialize port pin P1.0 as output
          MOV   TMOD,#02H         ;Program TMOD register for mode-2 operation of timer-0

          MOV   TH0,#06H          ;Load initial count in timer-0 high order count register
          SETB  TR0               ;Set timer run flag, to start timer
```

```
AGAIN:   CPL   P1.0        ;Compliment the port pin P1.0

WAIT:    JNB   TF0,WAIT    ;Wait for timer overflow
         CLR   TF0         ;Clear timer flag

         SJMP  AGAIN       ;Repeat generation of next cycle

         END               ;Assemby end
```

5.3   SERIAL PORT PROGRAMMING

The 8051 microcontroller has an internal serial port which can be operated in four modes. The baud rates for serial communication are programmable using internal timer-1 of the 8051 controller.

The 8051 controller can be used as a full-duplex serial communication device. A simple schematic for a serial communication using 8051 is shown in Fig. 5.3. The system requires a 8051 microcontroller with internal program memory and RS232 level converter like MAX 232. A quartz crystal and a reset circuit should be connected to the controller. The program for serial communication can be stored permanently in the internal ROM and so there is no need for external memory.

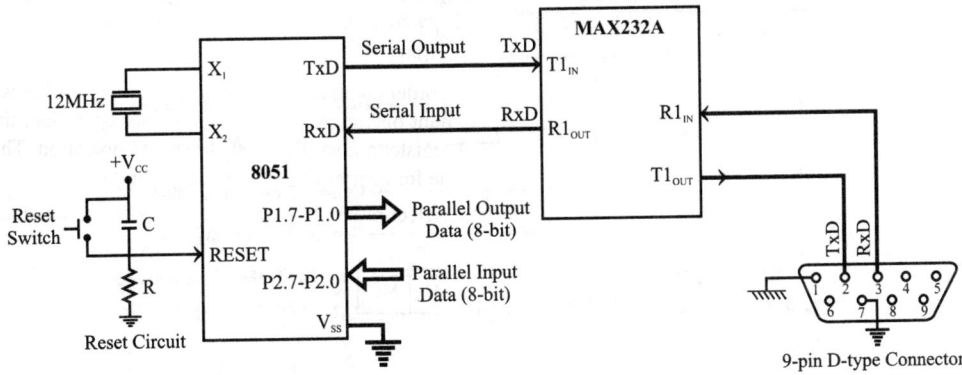

Fig. 5.3: Serial communication using 8051 microcontroller.

The serial bus is formed by using only three lines TxD, RxD and V_{ss}(Ground). If any additional signals are required then they have to be generated by software and output/input through port pins. The signals TxD and RxD will have TTL logic levels and they can be converted to standard RS232 logic levels using bi-directional level converter MAX 232. The RS232 level serial port signal can be terminated on a standard 9-pin D-type connector, so that any standard serial device can be connected to the 8051 controller for serial communication.

The parallel data transfer can be achieved through ports. In the schematic shown in Fig. 5.3, the parallel input data is received through port-2 and the parallel output data is sent through port-1. For example, the controller can receive parallel data from an ADC and convert it to serial and transmit via the serial port to another serial device. Also, the controller can receive a serial data from another serial device via the serial port and convert to parallel, and then outputs through port-1 to a parallel device like DAC. The 8051 controller supports full-duplex communication and so the transmission and reception can be performed simultaneously.

5.3.1 SERIAL DATA BUFFER (SBUF) REGISTER

The SBUF register is used to hold the parallel data during transmission and reception. During serial reception, the serial data is received via RxD pin and converted to parallel data and stored in the receive buffer. During serial transmission, the parallel data is stored in the transmit buffer and then converted to serial data to transmit via TxD pin.

The transmit and receive buffers are assigned the same internal address 99_H but the transmit buffer can be accessed only for write operation and the receive buffer can be accessed only for read operation. When data is written to SBUF, it goes to the transmit buffer and when data is read from SBUF it comes from the receive buffer.

5.3.2 POWER CONTROL REGISTER (PCON)

The PCON register is used for power control and baud rate selection. The format of the PCON register is shown in Fig. 5.4.

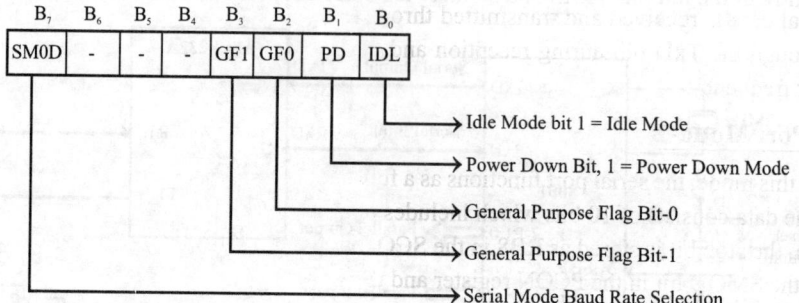

Fig. 5.4: Format of a PCON register of 8051 family of microcontrollers.

The SMOD bit is used to decide the baud rate in serial port operating modes 1, 2 or 3. In mode-2, if SMOD = 0 then the baud rate is 1/64 of the oscillator frequency and if SMOD = 1 then the baud rate is 1/32 of oscillator frequency. (In 8051, the oscillator frequency and microcontroller internal frequency are same). In modes 1 and 3, the baud rate depends on the SMOD and timer-1 overflow rate.

The baud rate in mode 1 or $3 = \dfrac{2^{SMOD}}{32} \times$ (Timer-1 overflow rate).

When timer-1 is configured for auto reload mode then,

$$\text{Timer-1 overflow rate} = \frac{\text{Oscillator frequency}}{12 \times [256 - (TH1)^2]}$$

where, TH1 = Reload count value (8-bit) in higher order timer-1 count register.

5.3.3 SERIAL PORT CONTROL REGISTER (SCON)

The format of a SCON register is shown in Fig. 5.5. The SCON register consists of mode selection bits, the 9th data bit (bit-B_8) for transmit and receive, and the serial port interrupt bits TI and RI. The bits SM0 and SM1 are used to select any one of the four operating modes for serial transmission and reception. The four modes of a serial port are mode-0, mode-1, mode-2 and mode-3.

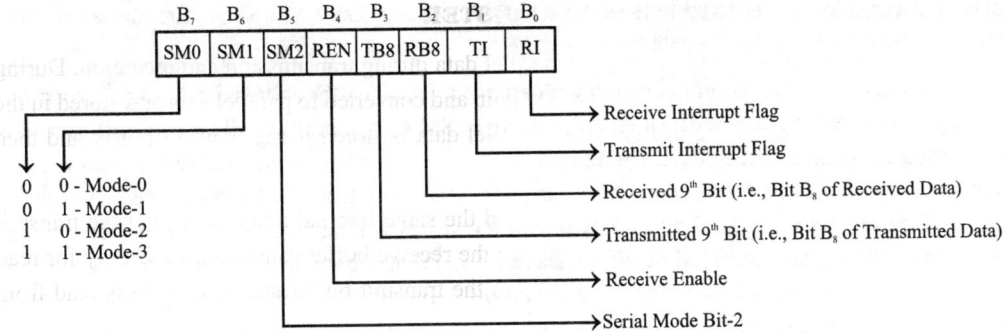

Fig. 5.5: Format of an SCON register of the 8051 family of microcontrollers.

Serial Port Mode-0

In this mode, the serial port functions as a half-duplex serial port with a fixed baud rate. The 8-bit serial data is received and transmitted through the RxD pin and the controller outputs the shift clock through the TxD pin during reception and transmission. The baud rate is fixed at 1/12 of the oscillator frequency.

Serial Port Mode-1

In this mode, the serial port functions as a full-duplex serial port with variable baud rates. In this mode, one data consists of 10 bits which includes one start bit, eight data bit and one stop bit. During reception, the stop bit is stored as RB8 in the SCON register. The baud rate in mode-1 depends on the value of the SMOD bit in the PCON register and the timer-1 overflow rate.

Serial Port Mode-2

In this mode, the serial port functions as a full duplex serial port with a baud rate of either 1/32 or 1/64 of the oscillator frequency. In this mode, one data consists of 11 bits which includes one start bit, eight data bits a programmable 9th data bit and one stop bit. During transmission, the TB8 of the SCON register is added as the 9th data bit and during reception, the 9th data bit is stored as RB8 in the SCON register. (This 9th data bit can be used as the parity bit). The baud rate depends on the value of the SMOD bit in the PCON register.

Serial Port Mode-3

Mode-3 is same as mode-2, except the baud rate. In mode-3, the baud rate is variable. The baud rate depends on the value of the SMOD bit in the PCON register and the timer-1 overflow rate.

The serial mode bit-2 (SM2) has no effect in mode-0 and when programmed for mode-0, the SM2 should be equal to zero. In mode-1, SM2 is used to check a valid stop bit during reception. In mode-1, if SM2 = 0 then Receive Interrupt (RI) is activated only when a valid stop bit is received.

In mode-2 and mode-3, the SM2 bit is used to enable multiprocessor communication. In multiprocessor communication, the serial port of a number of microcontrollers can be connected to a common serial bus. One controller will act as a master and all other controllers will act as slaves. A unique 8-bit address is assigned to each slave and the SM2 bit in all the slaves is set to 1. When the SM2 bit is one, the slaves will consider the received byte as address and when the SM2 bit is zero, the slaves will consider the received byte as data. For communication with a slave, the master will first send an address byte and then a data byte.

The master initiates communication with a slave by sending the address of the slave on the bus. The address byte will be received by all the slaves. Since SM2 = 1, initially in all the slaves, the received byte will be considered as the address and the slaves will verify whether the received address matches with the assigned address. The slave whose assigned address matches with the received address will clear its SM2 bit. Now, the SM2 bit of only one of the slaves will be zero. Next, the master will send a data byte which is also received by all the slaves, but the data byte is accepted by the slave whose SM2 = 0 and so the receive interrupt is activated only in one of the slaves whose SM2 = 0. After reading the received data from the SBUF register, the SM2 bit of the slave should be set to one again to receive the next data.

The REN bit of the SCON register can be used to enable or disable the serial reception. When REN is set to one, the serial reception is enabled and when REN is cleared to zero, the serial reception is disabled.

The bits TI and RI of the SCON register are a transmit interrupt flag and a receive interrupt flag respectively. They are also called serial data interrupt flags. The controller will set the TI bit during the transmission of stop bit of a data character in modes 1 to 3 and during the transmission of the 8th bit of a data character in mode-0. Similarly, the controller will set the RI bit during the reception of stop bit of a data character in modes 1 to 3 and during the reception of the 8th bit of a data character is mode-0. These two flags are logically ORed internally and used as an internal interrupt signal (called serial port interrupt) to interrupt the current program being executed by the controller. On receiving the serial port interrupt, the controller has to suspend the current program execution and begins to execute a subroutine program to check the value of bits TI and RI of the SCON register. If TI is one then the controller can understand that the previous character has been transmitted and so the controller can execute another subroutine to clear TI flag and load the next data in the SBUF register. If RI is one, the controller can understand that a character has been received and so the controller can execute another subroutine to clear the RI flag and read the data from the SBUF register.

5.3.4 BAUD RATE IN 8051

In the 8051 microcontroller, the baud rate is programmable in serial communication modes 1 and 3, and the baud rate is fixed in modes 0 and 2. In the 8051 microcontroller, the timer-1 is dedicated to generate the required baud rate clock for serial communication modes 1 and 3. The timer-1 overflow rate or output frequency is divided by 32 or 64 and used as baud rate clock (or transmit and receive clock) in serial communication modes 1 and 3.

The crystal frequency of 11.0592 MHz is used for serial communication with standard PC (**P**ersonal **C**omputer) in order to generate the clock for right baud rate for serial communication. The standard baud rates for serial communication with PC are 110, 150, 300, 600, 1200, 2400, 4800, 9600 and 19200. When the crystal frequency is 11.0592 MHz, the timer input clock frequency will be 11.0592/12 = 0.9216 MHz = 921.6 kHz.

The SMOD bit in the PCON register can be programmed to divide the timer overflow rate or clock frequency by either 64 or 32, to generate a timer clock frequency for right baud rate. The various timer clock frequency, and the initial count for timer overflow to achieve right baud rate clock for serial transmission and reception are listed in Table 5.1.

Table 5.1: Baud Rate Frequency and Timer Initial Count For Crystal Frequecy of 11.059 MHz

Timer overflow rate	Initial count	SMOD	Baud rate frequency
$\dfrac{921.6 \times 10^3}{3}$	$256 - 3 = 253 = FD_H$	0	$\dfrac{921.6 \times 10^3}{3 \times 64} \approx 4800\,Hz$
		1	$\dfrac{921.6 \times 10^3}{3 \times 32} = 9600\,Hz$
$\dfrac{921.6 \times 10^3}{6}$	$256 - 6 = 250 = FA_H$	0	$\dfrac{921.6 \times 10^3}{6 \times 64} = 2400\,Hz$
		1	$\dfrac{921.6 \times 10^3}{6 \times 32} = 4800\,Hz$
$\dfrac{921.6 \times 10^3}{12}$	$256 - 12 = 244 = F4_H$	0	$\dfrac{921.6 \times 10^3}{12 \times 64} = 1200\,Hz$
		1	$\dfrac{921.6 \times 10^3}{12 \times 32} = 2400\,Hz$
$\dfrac{921.6 \times 10^3}{24}$	$256 - 24 = 232 = E8_H$	0	$\dfrac{921.6 \times 10^3}{24 \times 64} = 600\,Hz$
		1	$\dfrac{921.6 \times 10^3}{24 \times 32} = 1200\,Hz$

5.3.5 EXAMPLES OF SERIAL PORT PROGRAMMING IN 8051 CONTROLLER

EXAMPLE PROGRAM 16

This example program is developed to receive serial data from a standard PC at 9600 baud rate and store as parallel data in internal RAM of microcontroller.

Problem Analysis

In order to receive serial data at 9600 baud rate, the timer-1 should be programmed for mode-1, and an initial count for 9600 baud rate should be loaded in the timer count register. The SCON register is programmed for mode-1 serial communication. Then the timer-1 is started by setting the timer 1 run flag and the controller has to wait for serial receive interrupt flag to go high. When a serial data is received and converted to parallel data and loaded in the SBUF register, the receive interrupt flag will go high. Now the controller has to read the SBUF data and load in memory and clear the receive interrupt flag, in order to enable serial port to receive next data. Here, the data is stored in the memory location 7FH, and memory will have the last received data.

The byte to be loaded in the TMOD register to select mode-2 operation of timer-1 is framed as shown below:

$$\text{TMOD} = \boxed{\text{Gate} \mid \text{C/}\overline{\text{T}} \mid \text{M1} \mid \text{M0} \mid \text{Gate} \mid \text{C/}\overline{\text{T}} \mid \text{M1} \mid \text{M0}} = 0\ 0\ 1\ 0\ \text{X X X X} = 20_H$$

The byte to be loaded in the SCON register to select mode-1 serial communication is framed as shown below:

$$\text{SCON} = \boxed{\text{SM0} \mid \text{SM1} \mid \text{SM2} \mid \text{REN} \mid \text{TB8} \mid \text{RB8} \mid \text{TI} \mid \text{RI}} = 0\ 1\ 0\ 1\ 0\ 0\ 0\ 0 = 50_H$$

Flowchart

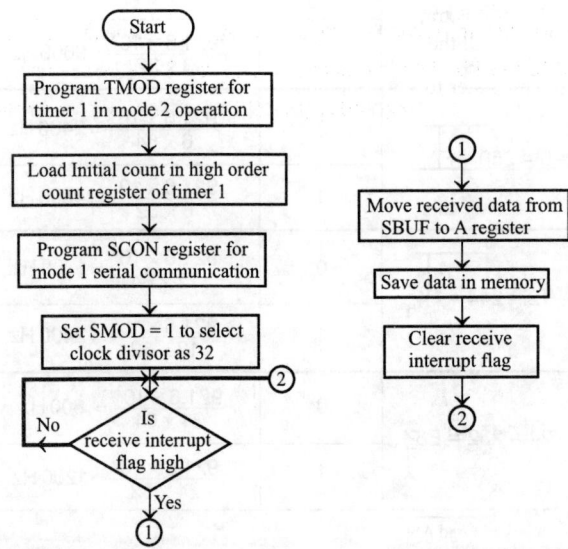

Assembly language program

```
;Program to receive serial data in 8051

        MOV   TMOD,#20H    ;Program timer-1 for mode-2 operation
        MOV   TH1,#FDH     ;Load initial count for 9600 baud rate in timer-1 high order register
        MOV   SCON,#50H    ;Program SCON reister for mode-1 serial communication
        MOV   A,PCON       ;Set SMOD = 1, via A register
        SETB  ACC.7        ;to choose clock divisor as 32
        MOV   PCON, A
        SETB  TR1          ;Start timer-1

WAIT:   JNB   RI,WAIT      ;Wait for receive interrupt
        MOV   A,SBUF       ;Save serial data in A register
        MOV   7FH,A        ;Save data in internal RAM
        CLR   RI           ;Clear receive interrupt flag to receive next data
        SJMP  WAIT         ;Wait for next receive interrupt

        END                ;Assemby end
```

EXAMPLE PROGRAM 17

This example program is developed to transmit the word 'HELO' by serial communication to a standard PC at 9600 baud rate.

Problem Analysis

In order to transmit serial data at 9600 baud rate, the timer-1 should be programmed for mode-1, and an initial count suitable for 9600 baud rate should be loaded in timer count register. The SCON register is programmed for mode-1 serial communication. Then the timer-1 is started by setting the timer-1 run flag. The ASCII value of first character to be transmitted should be loaded in SBUF register. At the end of serial transmission, the transmit interrupt flag is set. Now the controller can clear this transmit interrupt flag and load the next character in SBUF, and this process is repeated until transmission of all the characters one by one.

The byte to be loaded in the TMOD register to select mode-2 operation of timer-1 is framed as shown below:

TMOD = | Gate | C/$\overline{T}$ | M1 | M0 | Gate | C/$\overline{T}$ | M1 | M0 | = 0 0 1 0 X X X X = 20_H

The byte to be loaded in the SCON register to select mode-1 serial communication is framed as shown below:

SCON = | SM0 | SM1 | SM2 | REN | TB8 | RB8 | TI | RI | = 0 1 0 1 0 0 0 0 = 50_H

Flowchart

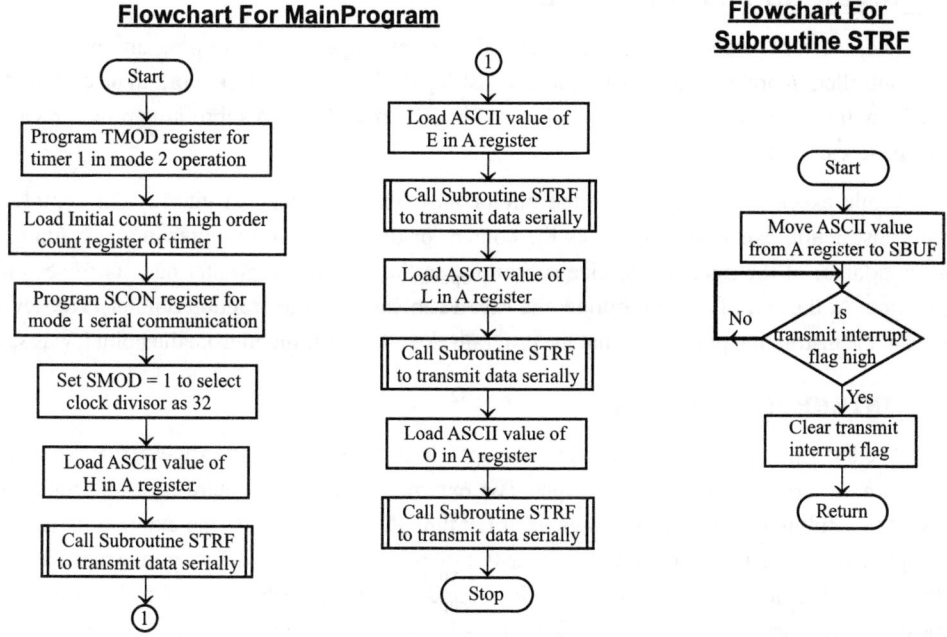

Flowchart For MainProgram

Start → Program TMOD register for timer 1 in mode 2 operation → Load Initial count in high order count register of timer 1 → Program SCON register for mode 1 serial communication → Set SMOD = 1 to select clock divisor as 32 → Load ASCII value of H in A register → Call Subroutine STRF to transmit data serially → ①

① → Load ASCII value of E in A register → Call Subroutine STRF to transmit data serially → Load ASCII value of L in A register → Call Subroutine STRF to transmit data serially → Load ASCII value of O in A register → Call Subroutine STRF to transmit data serially → Stop

Flowchart For Subroutine STRF

Start → Move ASCII value from A register to SBUF → Is transmit interrupt flag high (No → loop back) / Yes → Clear transmit interrupt flag → Return

Assembly language program

```
;Program for serial data transmission in 8051
        MOV    TMOD,#20H       ;Program timer-1 for mode-2 operation
        MOV    TH1,#FDH        ;Load initial count for 9600 baud rate in timer-1 high order register
        MOV    SCON,#50H       ;Program SCON reister for mode 1 serial communication
        MOV    A,PCON          ;Set SMOD = 1, via A register
        SETB   ACC.7           ;to choose clock divisor as 32
        MOV    PCON, A
        SETB   TR1             ;Start timer-1

        MOV    A,#'H'          ;Load ASCII value of H in A register
        ACALL  STRF            ;Transfer the content of A register serially

        MOV    A,#'E'          ;Load ASCII value of E in A register
        ACALL  STRF            ;Transfer the content of A register serially

        MOV    A,#'L'          ;Load ASCII value of L in A register
        ACALL  STRF            ;Transfer the content of A register serially

        MOV    A,#'O'          ;Load ASCII value of O in A register
        ACALL  STRF            ;Transfer the content of A register serially

NEXT:   SJMP   NEXT            ;Wait in infinite loop for next task

STRF:   MOV    SBUF,A          ;Load data to be transmitted serially in SBUF
WAIT:   JNB    TI,WAIT         ;Wait for end of serial transmission
        CLR    TI              ;Clear transmit interrupt flag
        RET

        END                    ;Assembly end
```

5.4   INTERRUPT PROGRAMMING

The interrupts are signals that are generated to interrupt the normal program execution of the microcontroller, in order to carry out a specific task/work. The specific task/work to be executed by an interrupt will be written as a program and stored in code memory as a subroutine program called the interrupt service subroutine.

While executing a program if the microcontroller encounters an interrupt, it completes the current instruction execution and saves the content of the Instruction pointer in stack and load the starting address of the interrupt service subroutine in the instruction pointer and start executing the subroutine. At the end of the subroutine, the saved content of the instruction pointer is retrieved and loaded in the instruction pointer, so that the program execution continued from the point it was stopped.

5.4.1   INTERRUPTS OF 8051

The 8051 microcontroller has five interrupts. In this, two interrupts are external interrupts and the remaining three are internal interrupts. The external interrupts are hardware interrupts that are initiated by applying an appropriate signals at the pins INT0 or INT1 by the external hardware. The three internal interrupts are initiated by timer-0, timer-1 and serial port. Every interrupt has a vector address to which program control is transferred and a flag. The interrupts of 8051 in the order of highest to lowest priority along with their vector address are listed in Table 5.2.

The priority of 8051 interrupts is alterable by programming the interrupt priority register. The interrupts of 8051 can be also disabled and enabled by programming the interrupt enable register. Sometimes, the reset is also considered as an interrupt with vector address 0000H.

Table 5.2: Interrupts of 8051 and their Vector Address

Interrupt	Vector address	Normal priority
External interrupt-0	0003_H	highest
Timer-0 interrupt	$000B_H$	
External interrupt-1	0013_H	
Timer-1 interrupt	$001B_H$	
Serial port interrupt	0023_H	lowest

5.4.2 INTERRUPT ENABLE (IE) REGISTER

The IE-register is used to enable/disable the interrupts of an 8051. The interrupts are recognized (or accepted) by the controller only if they are enabled. The IE-register can be programmed to enable/disable all the five interrupts of an 8051 totally or individually. The format of an IE-register is shown in Fig. 5.6. The EA bit of the IE-register can be programmed as zero, to disable all the five interrupts of 8051. When the EA bit is programmed as one, the interrupts are enabled provided their individual enable bits are programmed as one. (The EA bit is also called global enable.)

Each interrupt has a one-bit field to enable or disable it individually. When EA = 1, if the enable bit of a particular interrupt is programmed as one then it is enabled, and if the enable bit is programmed as zero then it is disabled.

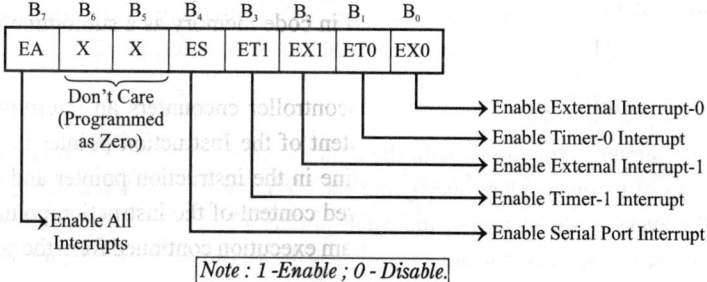

Fig. 5.6: Format of the IE register of the 8051 family of microcontrollers.

5.4.3 INTERRUPT PRIORITY (IP) REGISTER

The 8051 has five interrupts and the normal priority of these interrupts from highest to lowest are external interrupt-0, Timer-0 interrupt, External interrupt-1, Timer-1 interrupt and serial port interrupt.

The IP-register can be programmed to make the priority of any of the interrupt as highest. The format of an IP-register is shown in Fig. 5.7. The IP-register has one-bit field for the priority of each interrupt. When the priority bit of a particular bit is programmed as one then its priority will be highest. In 8051, while servicing a lower priority interrupt a higher priority interrupt will be recognized but another lower priority interrupt will not be recognized.

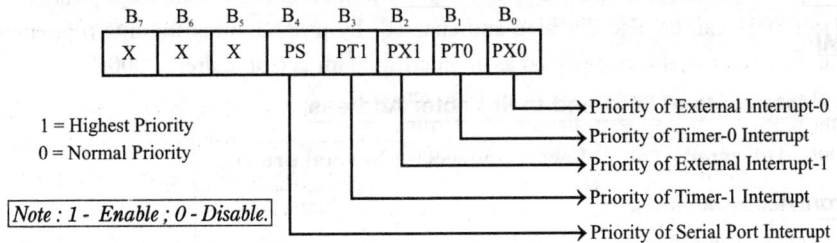

Fig. 5.7: Format of the IP-register of an 8051 family of microcontrollers.

5.4.4 HANDLING INTERRUPTS OF 8051

Enabling and Disabling of Interrupts

The 8051 interrupts are disabled upon power ON or after a reset. The desired interrupts has to be enabled by programming the appropriate bits in the interrupt enable register. The interrupts has a global enable which can be programmed to enable/disable all the interrupts. Also, every interrupt has an individual enable that can be programmed to enable/disable a particular interrupt. In order to enable a particular interrupt, both global enable and particular enable bit has to be set to high. For disabling an interrupt, it is enough if the particular enable bit of an interrupt is reset to low.

Storing the Interrupt Subroutine Program in Memory

For every interrupt, a subroutine program has to be developed and stored in code memory starting from the vector address of that interrupt. Normally, very few memory locations are reserved for storing the interrupt subroutine. If the interrupt subroutine requires larger memory space then it is stored in any available code memory space. A branch instruction with starting address of interrupt subroutine as branch address is stored in vector address of the interrupt.

Initializing the Stack Pointer

The stack pointer is initialized with the data memory address 07H upon power ON or after a reset. The content of memory addressed by the stack pointer is considered occupied stack (or top of stack). Then for every push operation, the stack pointer is incremented and using the incremented address the content of memory/register is stored in the stack. Now, the default data memory space available for the stack will be 08H to 1FH, with a memory size of 24 bytes. Also, this memory space is common for register banks 1,2 and 3. If the programmer needs register banks 1,2 and 3, or a larger memory space for stack then the stack pointer has to be initialized with an address of available data memory space.

> **Note:** The data memory address 00H to 07H is reserved for Bank 0 registers R0 to R7. The data memory spaces 20H to 2FH are reserved for bit addressable RAM.

5.4.5 EXAMPLES OF INTERRUPT PROGRAMMING IN 8051 CONTROLLER

EXAMPLE PROGRAM 18

This example program is developed for generation of unipolar square waveform of 1 kHz frequency using the timer-0 in mode-1 by using timer-0 interrupt, while the controller is performing continuous data transfer from port-0 to port-2. Assume that the crystal frequency of the controller is 12 MHz.

Program Analysis

In order to generate a square wave, a port pin can be set to high first and then after a time delay the port pin is reset to zero, then the process of set and reset are repeated continuously with a uniform time delay. The desired time delay can be achieved using the timer of 8051. This is achieved by the timer-0 interrupt service subroutine program.

The time period of 1 kHz square wave is $1/1 \times 10^3 = 1$ millisecond $= 1000$ microseconds. But in a square wave during half the period, the square wave will be high and during the next half of the period, the square wave will be low. So the time delay requires is $1000/2 = 500$ microseconds.

Since the crystal frequency is 12 MHz, the timer clock will $1/12 = 1$ MHz, and so the time period of the timer clock will be 1 microsecond. Therefore, the count for 500 microseconds time delay is 500. The count register in mode-1 is 16-bit and so the maximum count is, $2^{16} = 65536_{10}$. The initial count for any required delay has to be calculated by subtracting the delay count from the maximum count value.

Therefore, the initial count $= 65536_{10} - 500_{10} = 65036_{10} = $ FE0CH

The byte to be loaded in TMOD register to select mode-1 operation of timer-0 is framed as shown below:

$$\text{TMOD} = \boxed{\text{Gate} \mid \text{C/}\overline{\text{T}} \mid \text{M1} \mid \text{M0} \mid \text{Gate} \mid \text{C/}\overline{\text{T}} \mid \text{M1} \mid \text{M0}} = X \ X \ X \ X \ 0 \ 0 \ 0 \ 1 = 01_{\text{H}}$$

The byte to be loaded in the IE register to enable timer-0 interrupt is framed as shown below:

$$\text{IE} = \boxed{\text{EA} \mid X \mid X \mid \text{ES} \mid \text{ET1} \mid \text{EX1} \mid \text{ET0} \mid \text{EX0}} = 1 \ 0 \ 0 \ 0 \ 0 \ 1 \ 0 = 82_{\text{H}}$$

Flowchart

Flowchart for Main Program

Flowchart for Interrupt Service Subroutine

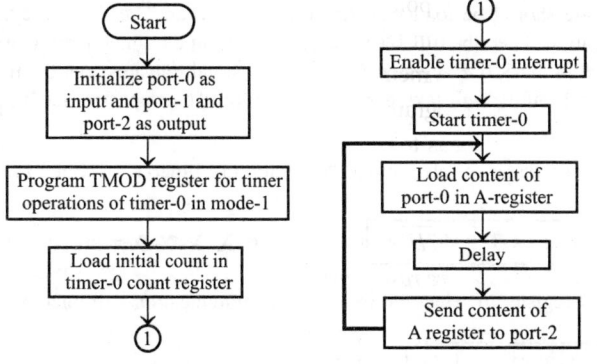

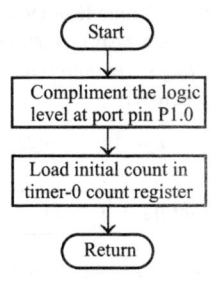

Assembly language program

```
;Program to generate square wave using 8051 timer-0 interrupt in mode-1

        ORG    0000H          ;Skip interrupt vector address and jump to main program
        SJMP   MAIN

;Inerrupt subroutine for timer-0

        ORG    000BH          ;Start from vector address of timer-0 interrupt
        CPL    P1.0           ;Compliment the port pin P1.0
        MOV    TL0,#0CH       ;Load low byte of count in timer-0 low order count register
        MOV    TH0,#FEH       ;Load high byte of count in timer-0 high order count register
        RETI                  ;Return from interrupt to main program

;Main Program

        ORG    0040H          ;Starting address to store main program

MAIN:   MOV    P0,#FFH        ;Initialize port-0 as input port
        MOV    P1,#00H        ;Initialize port-1 as output port
        MOV    p2,#00H        ;Initialize port-2 as output port

        MOV    TMOD,#01H      ;Program TMOD register for mode-1 operation of timer-0
        MOV    TL0,#0CH       ;Load low byte of count in timer-0 low order count register
        MOV    TH0,#FEH       ;Load high byte of count in timer-0 high order count register
        MOV    IE,#82H        ;Enable timer-0 interrupt
        SETB   TR0            ;Set timer run flag, to start timer
AGAIN:  MOV    A,P0           ;Get data from port-0
        NOP                   ;Wait for sometime
        NOP
        NOP
        MOV    P2,A           ;Send data to port-2
        SJMP   AGAIN          ;Repeat data transfer from port-0 to port-2

        END                   ;Assemby end
```

EXAMPLE PROGRAM 19

This example program is developed to transmit the parallel data in port-0, by serial communication to a standard PC at 9600 baud rate, by using serial port interrupt, while the controller is performing continuous data transfer from port-0 to port-2.

Program Analysis

In order to transmit serial data at 9600 baud rate, the timer-1 should be programmed for mode-2, and an initial count suitable for 9600 baud rate should be loaded in timer count register. The SCON register is programmed for mode-1 serial communication. Then the timer-1 is started by setting the timer-1 run flag. Then the data to be transmitted should be loaded in the SBUF register. At the end of serial transmission, the transmit interrupt flag is set, and interrupt service subroutine will take care of clearing this interrupt flag and allow the next data transfer.

The byte to be loaded in the TMOD register to select mode-2 operation of timer-1 is framed as shown below:

$$\text{TMOD} = \boxed{\text{Gate} \mid \text{C/}\overline{\text{T}} \mid \text{M1} \mid \text{M0} \mid \text{Gate} \mid \text{C/}\overline{\text{T}} \mid \text{M1} \mid \text{M0}} = 0\ 0\ 1\ 0\ \text{X}\ \text{X}\ \text{X}\ \text{X} = 20_\text{H}$$

The byte to be loaded in the SCON register to select mode-1 serial communication is framed as shown below:

$$\text{SCON} = \boxed{\text{SM0} \mid \text{SM1} \mid \text{SM2} \mid \text{REN} \mid \text{TB8} \mid \text{RB8} \mid \text{TI} \mid \text{RI}} = 0\ 1\ 0\ 1\ 0\ 0\ 0\ 0 = 50_\text{H}$$

The byte to be loaded in the IE register to enable serial port interrupt is shown below:

IE	=	EA	X	X	ES	ET1	EX1	ET0	EX0	$= 1\ 0\ 0\ 1\ 0\ 0\ 0\ 0 = 90_H$

Flowchart

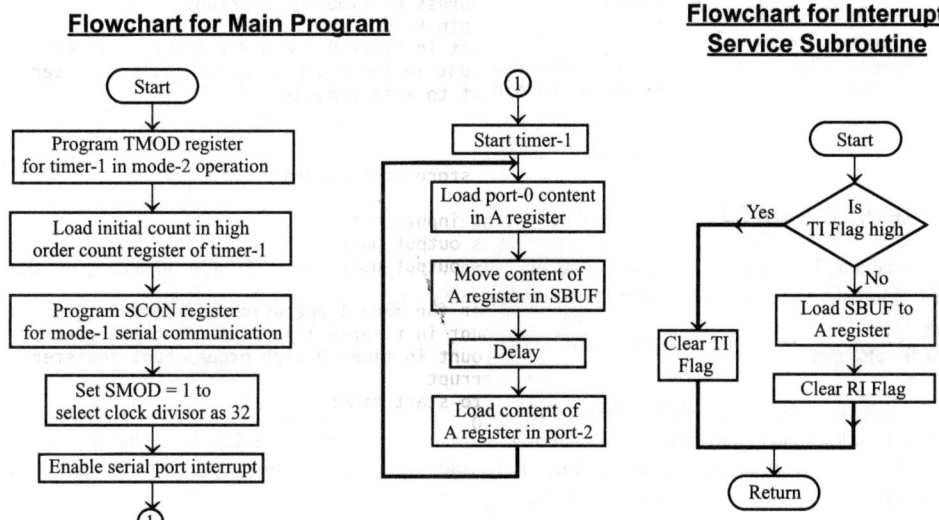

Flowchart for Main Program

Flowchart for Interrupt Service Subroutine

Assembly language program

```
;Program for serial data transmission in 8051 using serial port interrupt

        ORG  0000H        ;Skip interrupt vector address and jump to main program
        SJMP MAIN

;Inerrupt subroutine for serial data transmission

        ORG  0023H        ;Start from vector address of serial port interrupt
        JB   TI,CLRTI     ;If TI flag is high then clear it to allow next data transfer
        MOV  A,SBUF       ;If TI flag not high then clear receive buffer to Set RI flag
        CLR  RI           ;and then clear RI flag to return from interrupt
        RETI              ;Return from interrupt to main program
CLRTI:  CLR  TI
        RETI

;Main Program

        ORG  0040H        ;Starting address to store main program

MAIN:   MOV  P0,#FFH      ;Initialize port-0 as input port
        MOV  p2,#00H      ;Initialize port-2 as output port

        MOV  TMOD,#20H    ;Program timer-1 for mode-2 operation
        MOV  TH1,#FDH     ;Load the initial count for 9600 baud rate in timer-1 high
                          ; order register
        MOV  SCON,#50H    ;Program SCON reister for mode-1 serial communication
        MOV  A,PCON       ;Set SMOD=1, via A register
```

```
           SETB  ACC.7        ;to choose clock dividor as 32
           MOV   PCON,A
           MOV   IE,90H       ;Enable serial port interrupt
           SETB  TR1          ;Start timer-1

AGAIN:     MOV   A,P0         ;Get data from port-0
           MOV   SBUF,A       ;Transmit data serially
           NOP                ;Wait for sometime
           NOP
           NOP
           MOV   P2,A         ;Send data to port-2
           SJMP  AGAIN        ;Repeat data transfer from port-0 to port-2

           END                ;Assemby end
```

EXAMPLE PROGRAM 20

This example program is developed to receive serial data from a standard PC at 9600 baud rate and output to port-1, while the controller is performing continuous data transfer from port-0 to port-2.

Program Analysis

In order to receive serial data at 9600 baud rate, the timer-1 should be programmed for mode-1, and an initial count for 9600 baud rate should be loaded in the timer count register. The SCON register is programmed for mode-1 serial communication. Then the timer-1 is started by setting the timer-1 run flag and the controller has to wait for serial receive interrupt flag to go high.

When the receive interrupt flag goes high, the interrupt service routine is called to read the SBUF register and transmit data to port-1 and then the receive interrupt flag is cleared, in order to enable serial port to receive next data.

The byte to be loaded in the TMOD register to select mode-2 operation of timer-1 is framed as shown below:

TMOD = | Gate | C/$\overline{T}$ | M1 | M0 | Gate | C/$\overline{T}$ | M1 | M0 | = 0 0 1 0 X X X X = 20_H

The byte to be loaded in the SCON register to select mode-1 serial communication is framed as shown below:

SCON = | SM0 | SM1 | SM2 | REN | TB8 | RB8 | TI | RI | = 0 1 0 1 0 0 0 0 = 50_H

The byte to be loaded in the IE register to enable serial port interrupt is shown below:

IE   = | EA | X | X | ES | ET1 | EX1 | ET0 | EX0 | = 1 0 0 1 0 0 0 0 = 90_H

Flowchart

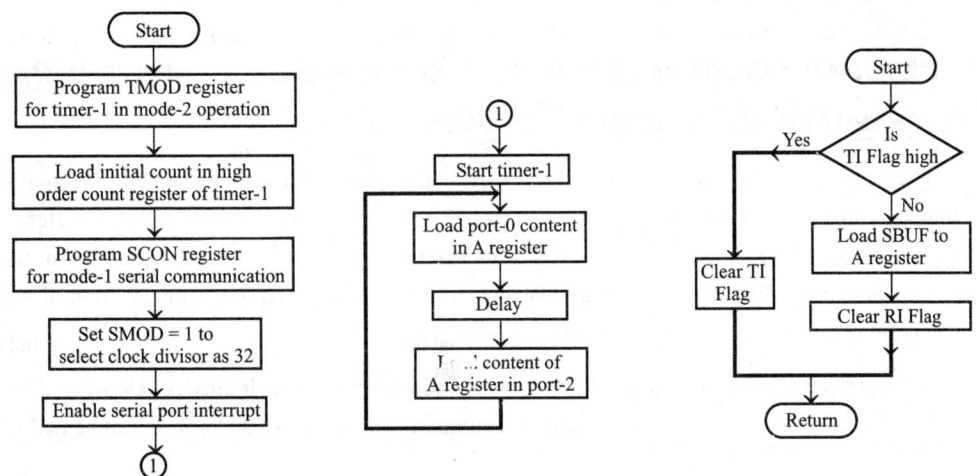

Flowchart for Main Program

Flowchart for Interrupt Service Subroutine

Assembly language program

```
;Program to receive serial data in 8051 using serial port interrupt
          ORG  0000H         ;Skip interrupt vector address and jump to main program
          SJMP MAIN
;Inerrupt subroutine to receive serial data
          ORG  0023H         ;Start from vector address of serial port interrupt
          JB   TI,CLRTI      ;If TI flag is high then clear TI flag to allow next data
                              transfer
          MOV  A,SBUF        ;Load received data in A register
          MOV  P1,A          ;Send data to port-1
          CLR  RI            ;and then clear RI flag to return from interrupt
          RETI              ;Return from interrupt to main program
CLRTI:    CLR  TI
          RETI
;Main Program
          ORG  0040H         ;Starting address to store main program
MAIN:     MOV  P0,#FFH       ;Initialize port-0 as input port
          MOV  p2,#00H       ;Initialize port-2 as output port

          MOV  TMOD,#20H     ;Program timer-1 for mode-2 operation
          MOV  TH1,#FDH      ;Load the initial count for 9600 baud rate in timer-1 high
                              order register
          MOV  SCON,#50H     ;Program SCON reister for mode-1 serial communication
          MOV  A,PCON        ;Set SMOD=1, via A register
          SETB ACC.7         ;to choose clock dividor as 32
          MOV  PCON,A
          MOV  IE,90H        ;Enable serial port interrupt
          SETB TR1           ;Start timer-1

AGAIN:    MOV  A,P0          ;Get data from port-0
          NOP               ;Wait for sometime
```

```
NOP
NOP
MOV   P2,A          ;Send data to port-2
SJMP  AGAIN         ;Repeat data transfer from port-0 to port-2
END                ;Assemby end
```

5.5 KEYBOARD INTERFACING IN 8051 CONTROLLER   *(AU, Nov/Dec' 19, 15 Marks)*

5.5.1 KEYBOARD INTERFACING USING PORTS OF 8051

The general concepts of keyboard interface using ports are discussed in Chapter 4, Section-4.5. The interfacing of hex keyboard with the ports of a 8051 microcontroller is shown in Fig. 5.8. Here, the 4×4 matrix keyboard is formed using the lower 4 lines of port-0 and upper 4 lines of port-2. Normally, all the port pins are pulled high by pullup resistors. A hex key is placed at the intersection of a row and a column.

A keypress is sensed by sending a low to a row and reading the columns. A key pressed in the row that made low, will make the corresponding column also low, and so a keypress can be deducted from the data read from columns.

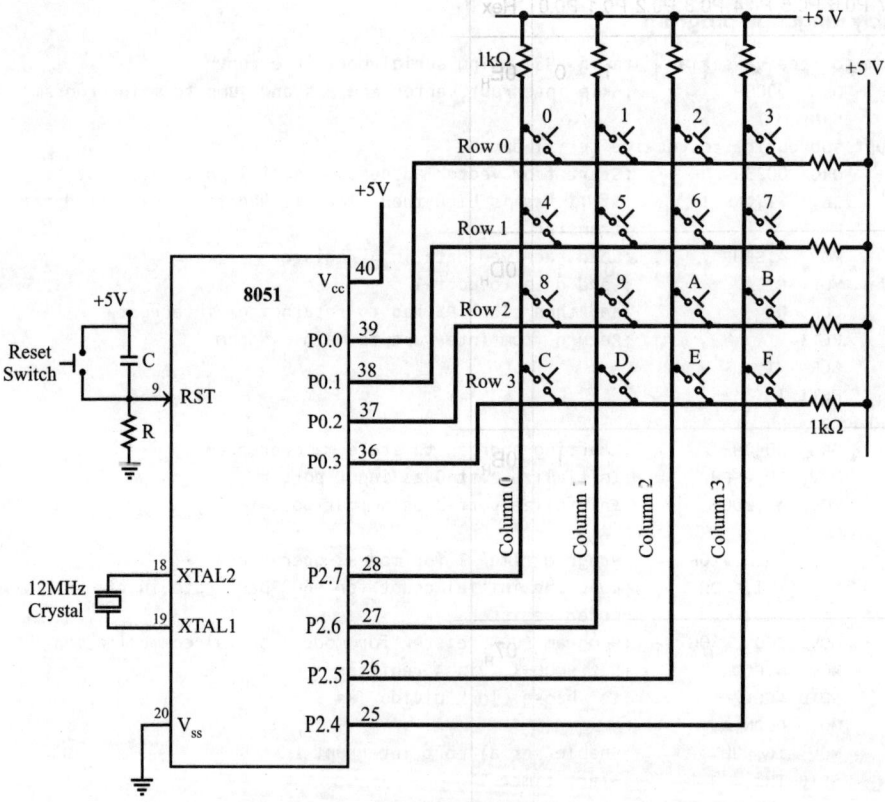

Fig. 5.8: Hex keyboard interfacing to 8051 microcontroller.

EXAMPLE PROGRAM 21

This example program is developed for the hex keyboard interface shown in Fig. 5.8. The scan codes and the corresponding hex keys are listed in the following table.The row scan codes are stored in 8051 internal RAM starting from address 60H and the column scan codes starting from address 70H.

The program has been developed to wait for a keypress. When a key is pressed, the program starts scanning. The R0 register is initialized with the hex code 00H. The scanning is performed by sending a row scan code to port-0 and reading column data through port-2 and check with column scan code. After every column scan, the hex value in the R0 register is incremented. When a column data matches with the column scan code, the scanning process stopped, and the hex key value in R0 is stored in the 8051 microcontroller internal memory at the address 7FH. Then the program waits for key debouncing and start scanning next key press. The content of memory at 7FH will be the hex value of the last pressed key.

TABLE 1: SCAN CODE AND CORRESPONDING HEX KEY

Row Scan Code									Column Scan Code									Hex Key
P0.7	P0.6	P0.5	P0.4	P0.3	P0.2	P0.1	P0.0	Hex	P2.7	P2.6	P2.5	P2.4	P2.3	P2.2	P2.1	P2.0	Hex	
*	*	*	*	1	1	1	0	$0E_H$	0	1	1	1	*	*	*	*	70_H	0
									1	0	1	1	*	*	*	*	$B0_H$	1
									1	1	0	1	*	*	*	*	$D0_H$	2
									1	1	1	0	*	*	*	*	$E0_H$	3
*	*	*	*	1	1	0	1	$0D_H$	0	1	1	1	*	*	*	*	70_H	4
									1	0	1	1	*	*	*	*	$B0_H$	5
									1	1	0	1	*	*	*	*	$D0_H$	6
									1	1	1	0	*	*	*	*	$E0_H$	7
*	*	*	*	1	0	1	1	$0B_H$	0	1	1	1	*	*	*	*	70_H	8
									1	0	1	1	*	*	*	*	$B0_H$	9
									1	1	0	1	*	*	*	*	$D0_H$	A
									1	1	1	0	*	*	*	*	$E0_H$	B
*	*	*	*	0	1	1	1	07_H	0	1	1	1	*	*	*	*	70_H	C
									1	0	1	1	*	*	*	*	$B0_H$	D
									1	1	0	1	*	*	*	*	$D0_H$	E
									1	1	1	0	*	*	*	*	$E0_H$	F

Flowchart

Flowchart for Main Program

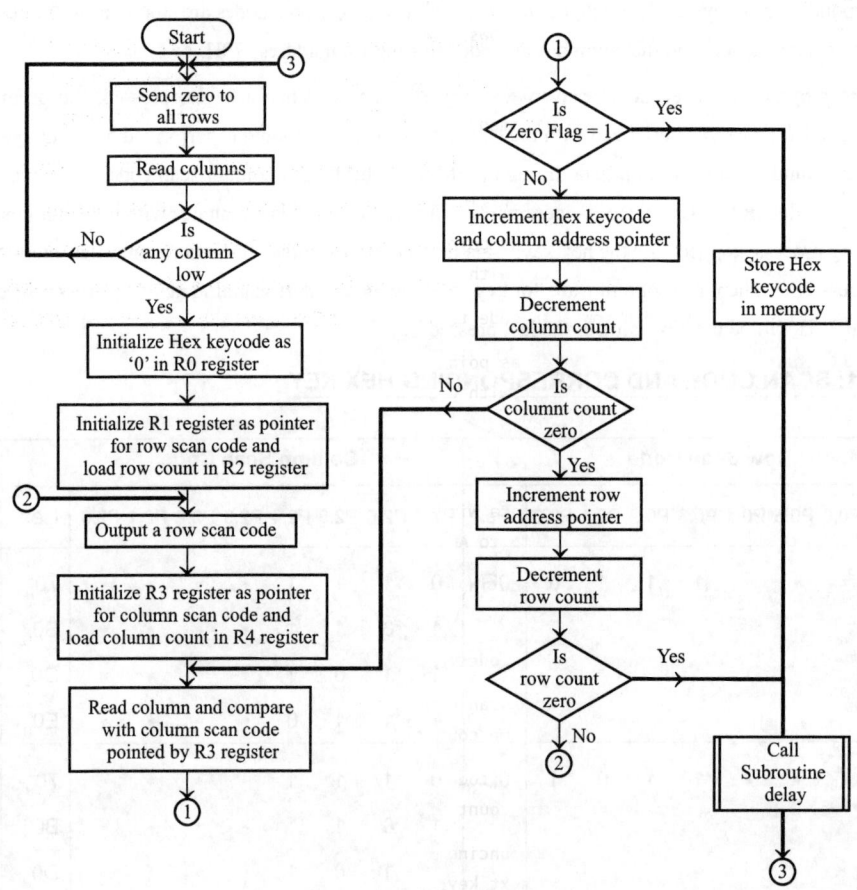

Flowchart for Subroutine Delay

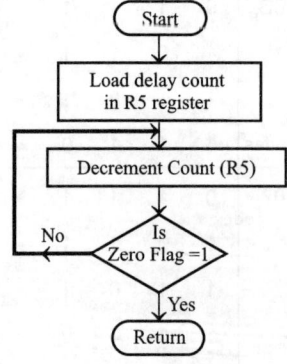

Assembly language program

```
;Program to scan hex keyboard and store the hex value of key pressed in memory


WAIT:   MOV  P0,#00H        ;Send zero to all rows
        MOV  A,P2           ;Read columns
        ANL  A,#F0H         ;Mask lower nibble
        CJNE A,#F0H,SCAN    ;If keypress deducted, start scanning
        SJMP WAIT           ;Wait for keypress

SCAN    MOV  R0,#00H        ;Set initial keycode in R0 as zero

        MOV  R1,#60H        ;Set R1 register as pointer for row scan code
        MOV  R2,#04H        ;Load R2 register with row count

L2:     MOV  P0,@R1         ;Send a row scan code to port-0

        MOV  R3,#70H        ;Set R3 register as pointer for column scan code
        MOV  R4,#04H        ;Load R4 register with column count

L1:     MOV  A,P2           ;Read column data
        ANL  A,#F0H         ;Mask lower nibble

        CJNE A,03H,AHEAD    ;Compare column data with column scan code
                            ;if not equal jump to AHEAD(Here 03H is address of R3 register)

STORE:  MOV  7FH, R0        ;Store hex key code in internal RAM
        SJMP SKIP

AHEAD:  INC  R0             ;Get next hex key code in R0

        INC  R3             ;Increment column scan code pointer
        DJNZ R4,L1          ;Repeat until column count is zero

        INC  R1             ;Increment row scan code pointer
        DJNZ R2,L2          ;Repeat until row count is zero

SKIP:   ACALL DELAY         ;Wait for key debouncing
        SJMP WAIT           ;Start scanning next keypress


;Delay Subroutine

DELAY: MOV R5,#FFH
AGAIN: DJNZ R5,AGAIN
       RET

       ORG 60H             ;Store row scan codes in memory
RSCAN: DB  0EH,0DH,0BH,07H ;Row scan code

       ORG 70H             ;Store column scan codes in memory
CSCAN: DB  70H,B0H,D0H,E0H ;Column scan code

       END                 ;Assemby end
```

5.6 DISPLAY INTERFACING WITH THE 8051 MICROCONTROLLER

The LCD module has eight data pins and three control pins for interface with the microcontroller. The LCD module can be interfaced with the microcontroller for either 4-bit or 8-bit data. When interfaced for 4-bit data, the number of IO lines required for interfacing will be very less.

The interfacing of the LCD module with the 8051 microcontroller is shown in Fig. 5.9. In this interface, the LCD data bus lines DB0-DB7 are connected to Port-0 of 8051. Three pins of Port-2 (P2.7, P2.6 and P2.5) are used to provide control signals E, R/$\overline{W}$ and RS for the LCD module.

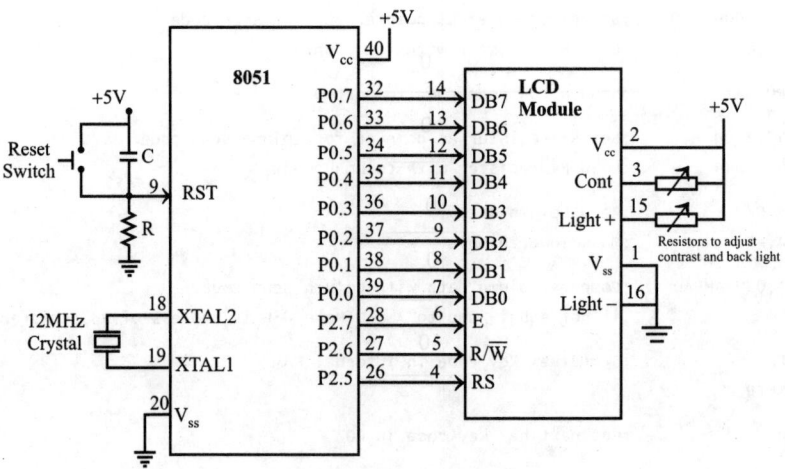

Fig. 5.9: Interfacing LCD module to 8051 Microcontroller.

The instruction set of the LCD module are presented in Chapter 3, Table 3.38. The brief discussion presented in Chapter 3, Section 3.14.3, about instruction set of LCD module and programming the LCD module are also applicable for programming LCD modules interfaced with the 8051 microcontroller.

5.6.1 EXAMPLE PROGRAM FOR LCD DISPLAY

EXAMPLE PROGRAM 22

This example program is developed to display 51 in the LCD interface shown in Fig. 5.10. The LCD pins are directly connected to port pins of the 8051 microcontroller as shown in the following table.

DETAILS OF LCD PINS CONNECTED TO 8051 PORT PINS

Port-0	P0.7	P0.6	P0.5	P0.4	P0.3	P0.2	P0.1	P0.0
LCD	DB7	DB6	DB5	DB4	DB3	DB2	DB1	DB0
Port-2	P2.7	P2.6	P2.5	P2.4	P2.3	P2.2	P2.1	P2.0
LCD	E	R/$\overline{W}$	RS	*	*	*	*	*

The LCD is intialized for 5×8 dot matrix pattern, 8-bit interface and address increment. The formation of command words are listed in the following table.

COMMAND WORD/CONTROL SIGNALS FOR LCD INTERFACE PROGRAM

Command/ Port-2	Feature Selected	Binary code								Hexa code
Function set	8-bit interface 2-line display and 5×8 dot pattern	0	0	1	DL	N	F	*	*	
	DL = 1, N = 1, F = 0	0	0	1	1	1	0	0	0	38_H
Display ON/OFF control	Display ON, cursor OFF and No blinking	0	0	0	0	1	D	C	B	
	D = 1, C = 0, B = 0	0	0	0	0	1	1	0	0	$0C_H$
Entry mode set	Address increment with no shift	0	0	0	0	0	1	I/D	S	
	I/D = 1, S = 0	0	0	0	0	0	1	1	0	06_H
Display clear	Clear display RAM and bring cursor home	0	0	0	0	0	0	0	1	01_H
Port-2	Write IR of LCD with high enable	E	$R/\overline{W}$	RS	*	*	*	*	*	
	RS = 0, R/$\overline{W}$ = 0, E = 1	1	0	0	0	0	0	0	1	80_H
Port-2	Write IR of LCD with low enable	E	$R/\overline{W}$	RS	*	*	*	*	*	
	RS = 0, R/$\overline{W}$ = 0, E = 0	0	0	0	0	0	0	0	0	00_H
Port-2	Write DR of LCD with high enable	E	$R/\overline{W}$	RS	*	*	*	*	*	
	RS = 1, R/$\overline{W}$ = 0, E = 1	1	0	1	0	0	0	0	0	$A0_H$
Port-2	Write DR of LCD with low enable	E	$R/\overline{W}$	RS	*	*	*	*	*	
	RS = 1, R/W = 0, E = 0	0	0	1	0	0	0	0	0	20_H

Flowchart

Flowchart for Main Program

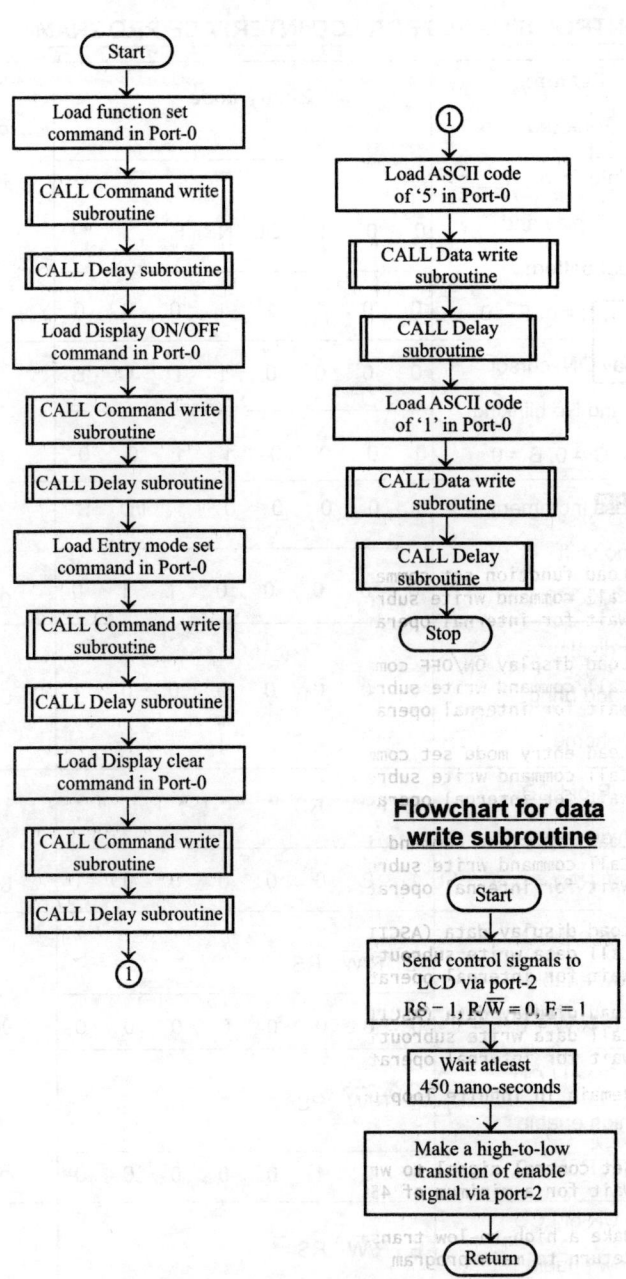

Flowchart for data write subroutine

Flowchart for command write subroutine

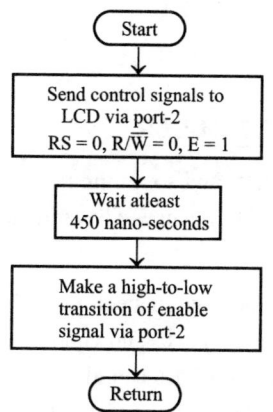

Flowchart for delay subroutine

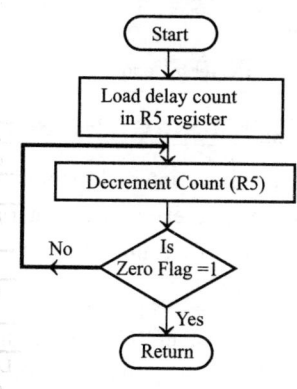

Assembly language program

```
;Program for LCD display

        MOV  P0,#38H    ;Load function set command in LCD data bus
        ACALL CMDWRITE  ;Call command write subroutine
        ACALL DELAY     ;Wait for internal operations to complete

        MOV  P0,#0CH    ;Load display ON/OFF command in LCD data bus
        ACALL CMDWRIT   ;Call command write subroutine
        ACALL DELAY     ;Wait for internal operations to complete

        MOV  P0,#06H    ;Load entry mode set command in LCD data bus
        ACALL CMDWRIT   ;Call command write subroutine
        ACALL DELAY     ;Wait for internal operations to complete

        MOV  P0,#01H    ;Load LCD clear command in LCD data bus
        ACALL CMDWRIT   ;Call command write subroutine
        ACALL DELAY     ;Wait for internal operations to complete

        MOV  P0,'5'     ;Load display data (ASCII value of 5) in LCD data bus
        ACALL DATWRIT   ;Call data write subroutine
        ACALL DELAY     ;Wait for internal operations to complete

        MOV  P0,'1'     ;Load display data (ASCII value of 1) in LCD data bus
        ACALL DATWRIT   ;Call data write subroutine
        ACALL DELAY     ;Wait for internal operations to complete

HALT:   SJMP HALT       ;Remain in infinite loop until reset, program end

;Command write subroutine

CMDWRIT: MOV P2,#80H    ;Set control signal to write IR of LCD
        NOP             ;Wait for a minimum of 450 nano-seconds
        NOP
        MOV  P2,#00H    ;Make a high-to-low transition of enable
        RET             ;Return to main program

;Command data subroutine

DATWRIT: MOV P2,#A0H    ;Set control signal to write DR of LCD
        NOP             ;Wait for a minimum of 450 nano-seconds
        NOP
        MOV  P2,#20H     ;Make a high-to-low transition of enable
```

```
        RET              ;Return to main program

;Delay Subroutine
DELAY:  MOV R5,#FFH
AGAIN:  DJNZ R5,AGAIN
        RET
        END              ;Assemby end
```

5.7 SYSTEM DESIGN USING THE 8051 MICROCONTROLLER

The microcontrollers have internal memory and internal IO ports, and so a minimum system can be formed by connecting a reset circuit and quartz crystal of required frequency. The input devices and output devices can be directly connected to IO ports.

The 8051 microcontroller-based minimum system can be formed by connecting a 12 MHz quartz crystal to X1 and X2 pins, an RC reset circuit and a 5 volt power supply as shown in Fig. 5.10. The system will have 32 IO lines organized as four 8-bit IO ports. The IO lines can also be used as individual IO lines. The devices like push-button keys, 7-segment LED display, LCD, ADC, DAC, etc., can be directly interfaced to IO lines.

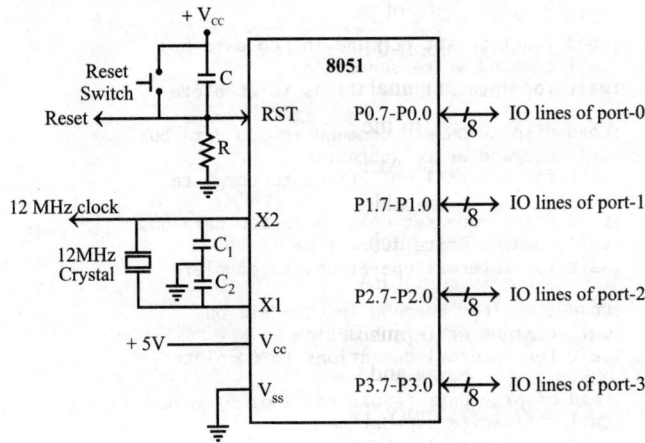

Fig. 5.10: The 8051 microcontroller-based minimum system.

The 8051 microcontroller has 4 kB internal ROM for program storage and 128 bytes internal RAM for user data storage. The 8051 supports optional external program memory and data memory. Usually, if the controller internal memory is not sufficient then external memory of required capacity is interfaced to the system. For interfacing external memory, an external system bus has to be formed as shown in Fig. 5.11. The address and data bus is formed using port-0 and port-2. The address bus is 16-bit wide and the data bus is 8-bit wide. The port-0 lines are multiplexed low-byte address and data lines, and the port-2 lines are high-byte address lines.

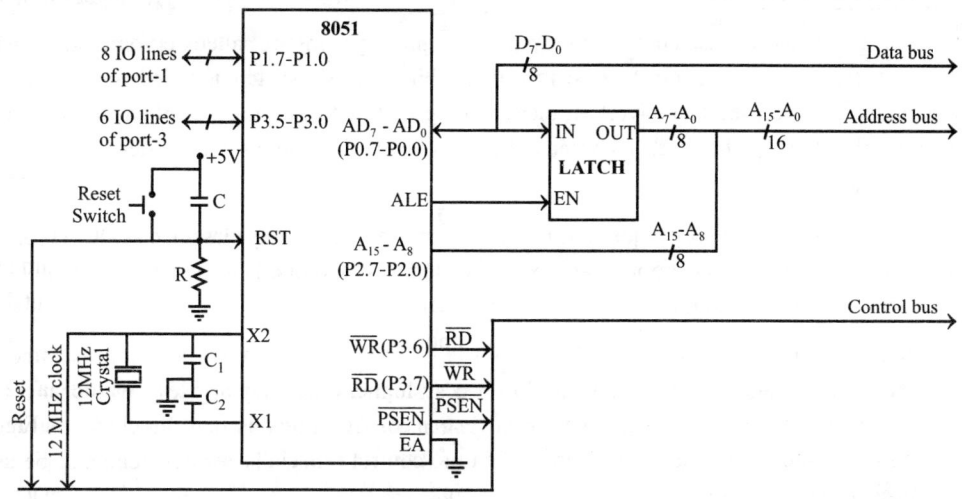

Fig. 5.11: External bus structure of the 8051 microcontroller-based system.

The multiplexed address and data lines of port-0 are demultiplexed by using an 8-bit external latch. The input of the latch is connected to port-0 lines and the output of the latch forms the low-byte address bus. The 8051 controller provides a signal called ALE (Address Latch Enable) which is used as enable/clock signal for the external latch. In the beginning of every machine cycle, the address is output on port-0 and port-2 lines, and then the ALE is asserted high and then low. The ALE acts as a clock/enable for the external latch, and so the address at the input of the latch (which is connected to port-0), will be transferred to the output of the latch, and so the port-0 lines are free to carry data. When external memory access is employed, the signal $\overline{EA}$ is tied low.

The control bus is formed using port-3 pins P3.6 and P3.7 (through which the controller sends $\overline{RD}$ and $\overline{WR}$), and other signals $\overline{PSEN}$, Reset and Clock signals. The $\overline{RD}$ and $\overline{WR}$ are used for read and write operations of the external data memory. The $\overline{PSEN}$ is used for read operation of the external program memory. When external bus is formed for external memory access, the port-0, port-2 and port-3 pins P3.6 and P3.7 cannot be used as IO lines. Therefore, in the thirty-two IO lines of 8051, eighteen IO lines are used to form the system bus, and the remaining fourteen lines can be used as IO lines.

Besides accessing external memory, the external bus can be used to interface the peripheral devices like ADC, DAC, etc., as memory-mapped IO, mapped in the data memory address space. The devices that have matched read and write timing as that of the microcontroller can be directly interfaced to the 8051 system bus. The slow devices have to be interfaced to system bus via buffers/latches/8255 ports. When other devices are interfaced in the system bus, some of the memory addresses will be used to select peripheral devices, and so full data memory space cannot be used for memory.

5.8 SERVO MOTOR   *(AU, Nov/Dec' 19, 13 Marks)*

The motors that are used in automatic position control systems are called Servomotors. When the objective of the system is to control the position of an object then the system is called Servomechanism. The servomotors are used to convert an electrical signal (control voltage) applied to them into an angular displacement of the shaft. They can either operate in a continuous duty or step duty depending on construction.

Depending on the supply required to run the motor, they are broadly classified as DC servomotors and AC servomotors. The DC motors are expensive than AC motors. But the DC servomotors have linear characteristics and so it is easier to control.

The DC servomotors are generally used for large power applications such as in machine tools and robotics. Advantages of AC motors are lower cost, higher efficiency and less maintenance since there is no commutator and brushes. The disadvantage of AC motor is that the characteristics are quite non-linear and these motors are more difficult to control especially for positioning applications (servomechanisms).

The AC motor are best suited for low power applications such as instrument servo (Example: Control of pen in X-Y recorders) and computer related equipment (Example: Disk drives, Tape drives, Printers, etc.)

5.8.1 CONTROL OF SERVO MOTOR

The DC servo motors are specially designed DC motors to satisfy the characteristics of servo applications. A normal DC motor can be converted to servo motor by adding speed reduction gear system to alter the speed torque ratio. Power output a motor is directly proportional to product of speed and torque. Therefore, at lesser speeds it is possible to achieve very high torque. Most of servo motors are used to provide small incremental motions proportional to input voltage and so they can offer very high driving torque.

The speed and hence angular rotation of servo motor is directly proportional to armature voltage when field excitation is constant. Hence in servo applications the field of DC motor is excited by a constant DC voltage. Alternatively, field is constructed with high retentivity permanent magnet.

For continuous rotation the speed is controlled by varying armature voltage by PWM (Pulse Width Modulation) technique. In PWM technique a pulsating DC voltage is applied to armature so that the average DC voltage is varied by varying the ON time.

If the ON time is 10% then average DC voltage is 10%.

If the ON time is 20% then average DC voltage is 20% and so on.

For incremental angular motion, the armature is supplied with time-based pulses. The angular motion is limited to fixed number of revolutions. The resolution of angular motion and the duration of pulses required to achieve angular motions depends on the type and construction of the servo motor. For example, in a DC servo motor, if 0.05 millisecond pulse can produce an angular motion of 2°, then for 180° motion (180°/2 = 90°), 90 pulses of 0.05 millisecond duration is required.

In DC servo motor the reversal of direction of rotation can be achieved by reversing the polarity or sign of DC voltage applied to armature. For accurate speed or motion control, closed loop control is required. In continuous type DC servo motor, speed sensor is employed for sensing the speed and compare with desired set speed and error is generated to correct the speed if there is a difference between actual and set speed. In incremental motion DC servo motor the angular position is sensed by using potentiometer and feedback signal from potentiometer is used to correct error in desired angular position.

Servo Motors with Inbuilt Control Circuit

Servo motors with inbuilt control circuit are popularly used in wireless control toys, robots, etc, are they are also called RC Servo motors. A RC servo motor is permanent magnet type DC servo motor fitted with a train of reduction gears, a feedback potentiometer to sense angular position and driver circuit for forward and reverse angular motion control. The angular motion is limited to 0 to 180° (or – 90° to +90°) The RC servo motors are popularly used in microcontroller-based applications due to simplicity in controlling the angular motion using time-based pulses. A periodic pulse train of fixed ON time duration can turn the motor shaft in a fixed angular motion.

In RC servo motors, the field is constructed using permanent magnet and so field flux is constant and speed or angular motion is directly proportional to armature voltage. The armature of a typical servo motor is excited by a periodic pulse train with time period of 20 milliseconds. The ON time of the periodic pulse train decides the angular motion of the shaft.

Microcontroller based closed loop position control system of a typical DC servo motor is shown in Fig. 5.12. The system consists of a DC motor fitted with gear wheels specially designed for servo applications, a potentiometer to generate a voltage called feedback voltage proportional to shaft position and driver transistors to supply the required volage to armature of DC motor.

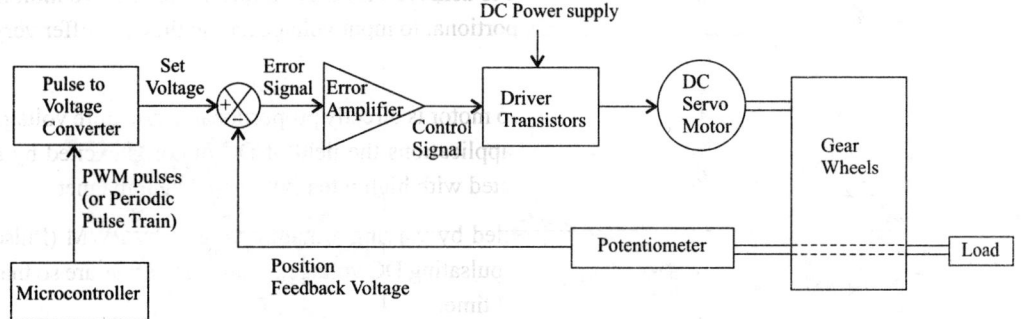

Fig. 5.12: Microcontroller Based Position Control System using DC Servo Motor.

The microcontroller will generate the required PWM pulses which is converted to a voltage called set voltage for the desired position of motor shaft. This set voltage is compared with feedback voltage from potentiometer and an error signal is generated. The amplified error signal is used as control signal for driver transistors so that motor will be ON for the required time duration and move the shaft to desired position. Most of the commercial motors require a response time of 15 to 20 milliseconds for the motor shaft to move from one position to another when there is a change in pulse width and so the time period of PWM signal or the periodic pulse train is normally selected as 20 milliseconds.

0° to 180° Motion Control

For 0 to 180° angular motion, a periodic pulse train with 1 millisecond ON time will move the motor shaft to one extreme end referred to as 0° position and a periodic pulse train with 2 millisecond ON time will move the motor shaft to the other extreme end referred to as 180° position.

Hence, in a time period of 20 milliseconds, the ON pulse time duration is 1 millisecond to 2 milliseconds, with a total time difference of 1 millisecond (1000 microseconds) for 180° rotation. Most of the servo motors are available with a resolution of 1° motion.

ON time for 1° motion = 1000 + 1000/180 = 1000 + 5.5556 microseconds.

ON time for 2° motion = 1000 + (2 × 1000/180) = 1000 + (2 × 5.5556) microseconds.

ON time for N° motion = 1000 + (N × 1000/180) = 1000 + (N × 5.5556) microseconds.

Table 5.3: 0° to 180° Rotation in Forward Direction (45° Steps)

Initial position	PWM/Periodic pulse train	Final position
Any position	1 ms pulse, 0 to 20 ms	90° / 45° / 0° / 180° / 135° (0°)
90° / 45° / 0° / 180° / 135° (0°)	1.25 ms pulse, 0 to 20 ms	(45°)
(45°)	1.5 ms pulse, 0 to 20 ms	(90°)
(90°)	1.75 ms pulse, 0 to 20 ms	(135°)
(135°)	2 ms pulse, 0 to 20 ms	(180°)

An example of PWM pulses and motor shaft position of servo motor in 0 to 180° position control for forward rotation with 45° incremental motion is shown in Table 5.3.

The servo motor can be operated in forward and reverse direction. For forward rotation the ON time of periodic pulses are can be estimated as shown above. But for reverse rotation, the ON time should be decreased from current value (Current Ton).

ON time for N° reverse motion = Current Ton − (N × 1000/180) microseconds.

An example of PWM pulses and motor shaft position of servo motor in 0 to 180° position control for reverse rotation with 45° incremental motion is shown in Table 5.4.

Table 5.4: 0° to 180° Rotation in Reverse Direction (45° Steps)

Initial position	PWM/Periodic pulse train	Final position
Any position	2 ms, 0 to 20 ms	90°, 135°, 45°, 180°, 0°
90°, 135°, 45°, 180°, 0°	1.75 ms, 0 to 20 ms	90°, 135°, 45°, 180°, 0°
90°, 135°, 45°, 180°, 0°	1.5 ms, 0 to 20 ms	90°, 135°, 45°, 180°, 0°
90°, 135°, 45°, 180°, 0°	1.25 ms, 0 to 20 ms	90°, 135°, 45°, 180°, 0°
90°, 135°, 45°, 180°, 0°	1 ms, 0 to 20 ms	90°, 135°, 45°, 180°, 0°

– 90° to + 90° Motion Control

For – 90° to + 90° angular motion, 1.5 millisecond pulse will move the motor shaft to midpoint referred to as neutral or 0° position, 1 millisecond pulse will move the motor shaft to one extreme end referred to as – 90° position and 2 millisecond pulse will move the motor shaft to the other extreme end referred to as + 90° position.

0° to + 90° Motion Control

In a time period of 20 milliseconds, for 0° to + 90° angular motion, the ON pulse time duration is 1.5 millisecond to 2 milliseconds, with a total time difference of 0.5 millisecond (500 microseconds) for 90° rotation. Therefore, for a servo motor with 1° resolution,

Timing pulse for 1° motion = 1500 + 500/90 = 1500 + 5.5556 microseconds.

Timing pulse for 2° motion = 1500 + (2 × 500/90) = 1500 + (2 × 5.5556) microseconds.

Timing pulse for N° motion = 1500 + (N × 500/90) = 1500 + (N × 5.5556) microseconds.

For forward rotation the ON time of periodic pulses are can be estimated as shown above. But for reverse rotation, the ON time should be decreased from current value (Current Ton).

ON time for N° reverse motion = Current Ton – (N × 500/90) microseconds.

An example of PWM pulses and motor shaft position of servo motor in 0° to + 90° position control for forward and reverse rotation with 45° incremental motion is shown in Table 5.5.

Table 5.5: 0° to +90° Rotation in Forward and Reverse Direction (45° Steps)

Initial position	PWM/Periodic pulse train	Final position
Any position	1.5 ms pulse, 0 to 20 ms	90°, 45°, 0°, –45°, –90° (shaft at 0°)
90°, 45°, 0°, –45°, –90° (shaft at 0°)	1.75 ms pulse, 0 to 20 ms	90°, 45°, 0°, –45°, –90° (shaft at 45°)

Table 5.5 continued....

Initial position	PWM/Periodic pulse train	Final position
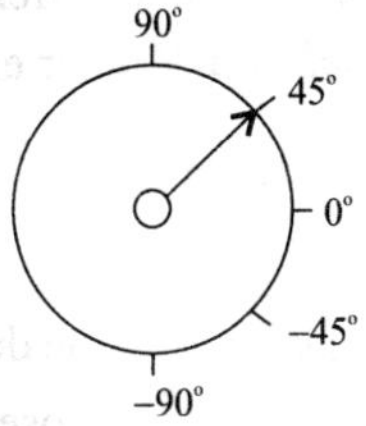 90°, 45°, 0°, −45°, −90° (pointer at 45°)	 2 ms pulse; 0 to 20 ms	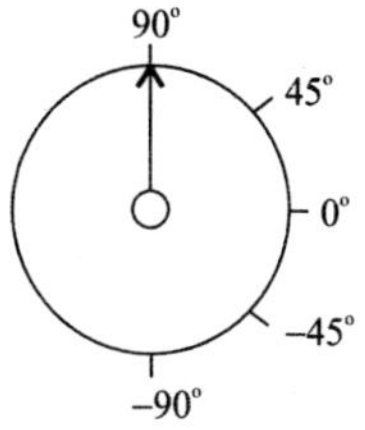 90°, 45°, 0°, −45°, −90° (pointer at 90°)
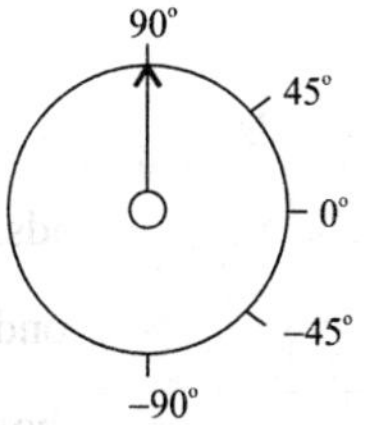 90°, 45°, 0°, −45°, −90° (pointer at 90°)	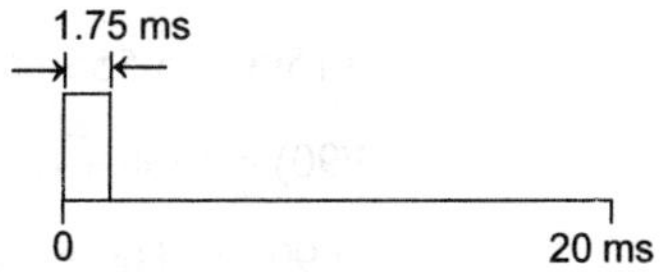 1.75 ms pulse; 0 to 20 ms	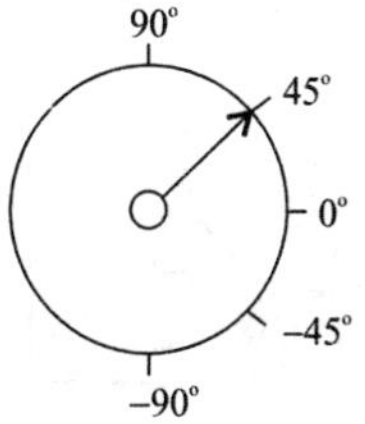 90°, 45°, 0°, −45°, −90° (pointer at 45°)
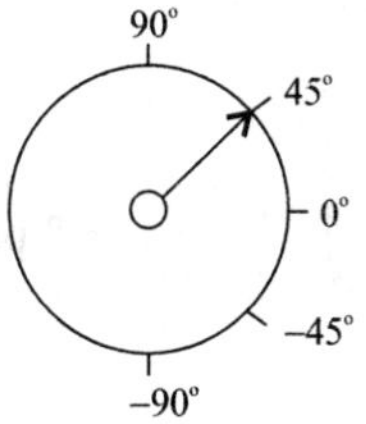 90°, 45°, 0°, −45°, −90° (pointer at 45°)	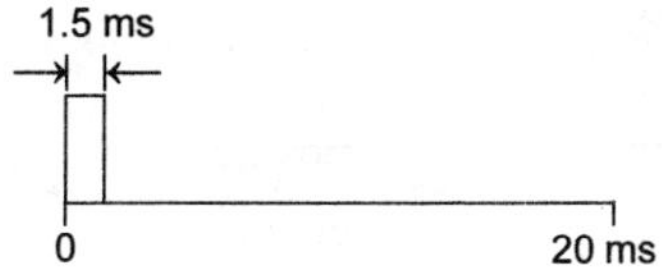 1.5 ms pulse; 0 to 20 ms	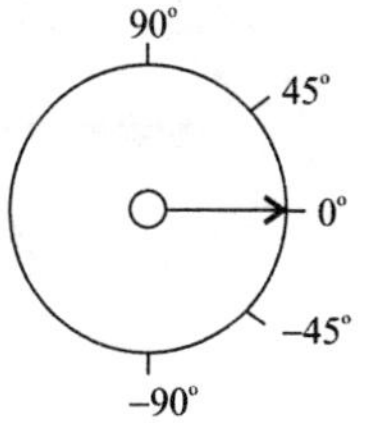 90°, 45°, 0°, −45°, −90° (pointer at 0°)

0° to − 90° Motion Control

In a time period of 20 milliseconds, for 0° to -90° angular motion, the ON pulse time duration is 1.5 millisecond to 1 milliseconds, with a total time difference of 0.5 millisecond (500 microseconds) for 90° rotation. Therefore, for a servo motor with 1° resolution,

Timing pulse for 1° motion = 1500 − 500/90 = 1500 − 5.5556 microseconds.

Timing pulse for 2° motion = 1500 − (2 × 500/90) = 1500 − (2 × 5.5556) microseconds.

Timing pulse for N° motion = 1500 − (N × 500/90) = 1500 − (N × 5.5556) microseconds.

For reverse rotation the ON time of periodic pulses are can be estimated as shown above. But for forward rotation, the ON time should be increased from current value (Current Ton).

ON time for N° forward motion = Current Ton + (N × 500/90) microseconds.

An example of PWM pulses and motor shaft position of servo motor in 0° to − 90° position control for forward and reverse rotation with 45° incremental motion is shown in Table 5.6.

Table 5.6: 0° to − 90° Rotation in Forward and Reverse Direction (45° Steps)

Initial position	PWM/Periodic pulse train	Final position
Any position	1.5 ms 0 20 ms	90° / 45° / 0° / −45° / −90°
90° / 45° / 0° / −45° / −90°	1.25 ms 0 20 ms	90° / 45° / 0° / −45° / −90°
90° / 45° / 0° / −45° / −90°	1 ms 0 20 ms	90° / 45° / 0° / −45° / −90°
90° / 45° / 0° / −45° / −90°	1.25 ms 0 20 ms	90° / 45° / 0° / −45° / −90°
90° / 45° / 0° / −45° / −90°	1.5 ms 0 20 ms	90° / 45° / 0° / −45° / −90°

5.9 STEPPER MOTOR CONTROL USING THE 8051 MICROCONTROLLER

(AU, Nov/Dec' 19, 13 Marks)

Stepper motors are popularly used in computer peripherals, plotters, robots and machine tools for precise incremental rotation. In a stepper motor, the stator windings are excited by electrical pulses and for each pulse, the motor shaft advances by one angular step. (Since the stepper motor can be driven by digital pulses, it is also called digital motor.) The step size in the motor is determined by the number of poles in the rotor and the number of pairs of stator windings (one pair of stator winding is called one phase). The stator windings are also called control windings.

The motor is controlled by switching ON/OFF the control winding. The popular stepper motor used for demonstration in laboratories has a step size of 1.8° (i.e., 200 steps per revolution). This motor consists of four stator windings and requires four switching sequences as shown Table 5.7. The basic step size of the motor is called full-step. By altering the switching sequence, the motor can be made to run with incremental motion of half the full-step value. The switching sequence for half step rotation is shown in Table 5.8.

The stepper motor interfaced to the 8051 microcontroller is shown in Fig. 5.13. The stepper motor has four windings and a Darlington pair transistor is provided for each winding to control the current through the winding. When the transistor is ON, current flows through the winding and when OFF no current flows.

The windings of the stepper motor are connected to the collector of the Darlington pair transistors. The transistors are switched ON/OFF by the microcontroller through port-0, by applying high/low to base of transistors. A free-wheeling diode is connected across each winding for fast switching. The controller has to output a switching sequence and wait for 1 to 5 milliseconds before sending the next switching sequence. (The delay is necessary to allow the motor transients to die out.)

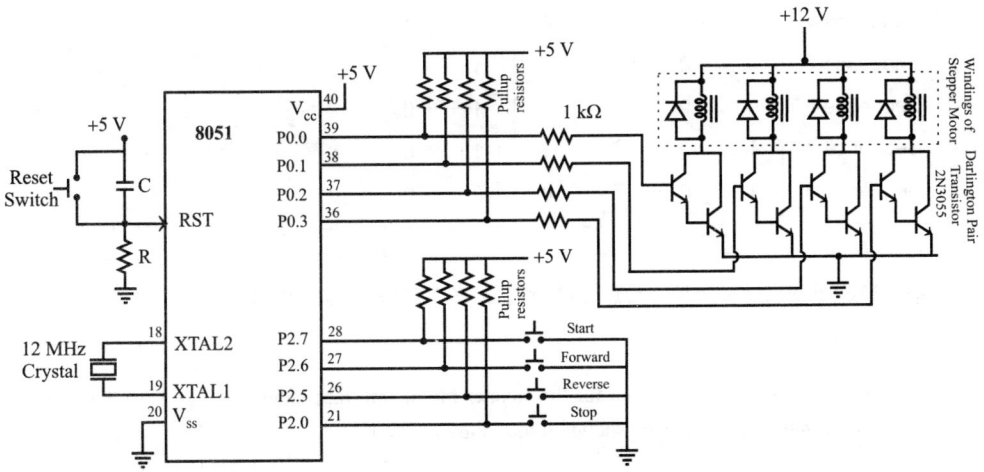

Fig 5.13: Stepper motor interface to 8051.

The stepper motor interface has four keys interfaced through port-2 pins in order to control the motor operation. The keys can be used to start and stop the motor and also used to select forward or reverse rotation of the motor. Normally, the port pins are pulled high and pressing a key will make the port pin low, and this low signal is sensed by the microcontroller to initiate the operation selected by the key press.

Table 5.7: Switching Sequence For Full-step Rotation

Switching sequence	Forward (clockwise) rotation				Reverse (anticlockwise) rotation			
	PA_3	PA_2	PA_1	PA_0	PA_3	PA_2	PA_1	PA_0
Sequence-1	1	1	0	0	0	0	1	1
Sequence-2	0	1	1	0	0	1	1	0
Sequence-3	0	0	1	1	1	1	0	0
Sequence-4	1	0	0	1	1	0	0	1

Table 5.8: Switching Sequence for Half-Step Rotation

Switching sequence	Forward (clockwise) rotation				Reverse (anticlockwise) rotation			
	PA_3	PA_2	PA_1	PA_0	PA_3	PA_2	PA_1	PA_0
Sequence-1	1	1	0	0	0	0	1	1
Sequence-2	0	1	0	0	0	0	1	0
Sequence-3	0	1	1	0	0	1	1	0
Sequence-4	0	0	1	0	0	1	0	0
Sequence-5	0	0	1	1	1	1	0	0
Sequence-6	0	0	0	1	1	0	0	0
Sequence-7	1	0	0	1	1	0	0	1
Sequence-8	1	0	0	0	0	0	0	1

5.9.1 EXAMPLE PROGRAM FOR STEPPER MOTOR CONTROL

EXAMPLE PROGRAM 23

This example program is developed for the stepper motor interface shown in Fig. 5.12. The switching sequence data for forward and reverse are shown in the following table.

Forward/Reverse Sequence Data

Direction	Sequence No	P0.7	P0.6	P0.5	P0.4	P0.3	P0.2	P0.1	P0.0	Hex Value
Forward	Sequence - 1	*	*	*	*	1	1	0	0	$0C_H$
	Sequence - 2	*	*	*	*	0	1	1	0	06_H
	Sequence - 3	*	*	*	*	0	0	1	1	03_H
	Sequence - 4	*	*	*	*	1	0	0	1	09_H
Reverse	Sequence - 1	*	*	*	*	0	0	1	1	03_H
	Sequence - 2	*	*	*	*	0	1	1	0	06_H
	Sequence - 3	*	*	*	*	1	1	0	0	$C0_H$
	Sequence - 4	*	*	*	*	1	0	0	1	09_H

Flowchart

Flowchart for Main Program

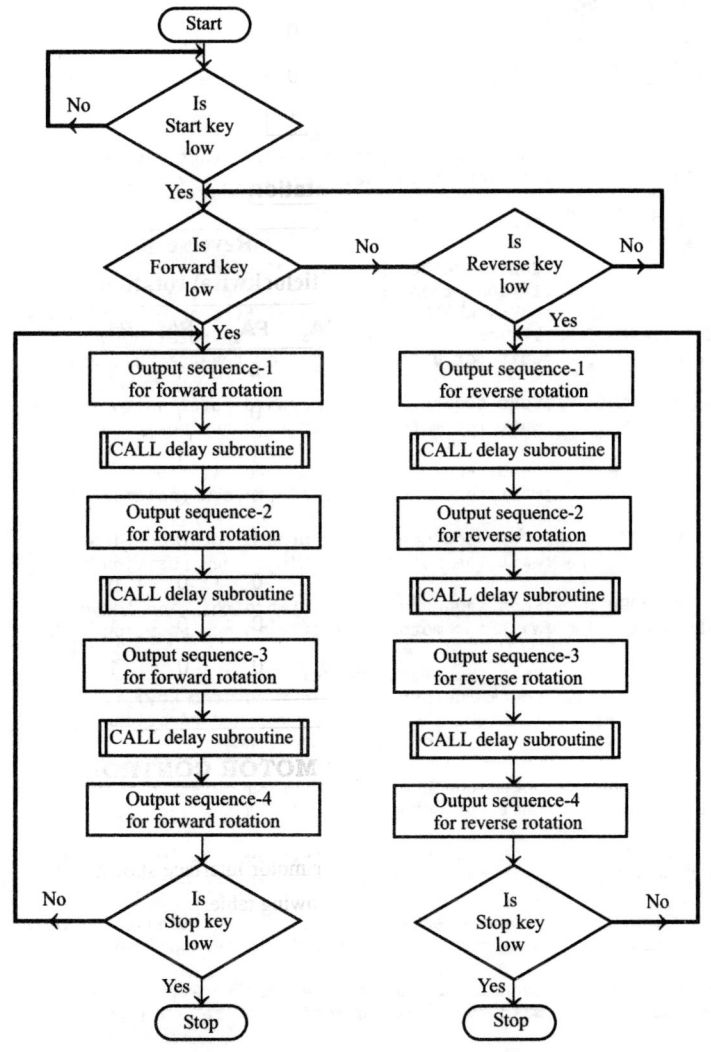

Flowchart for Subroutine Delay

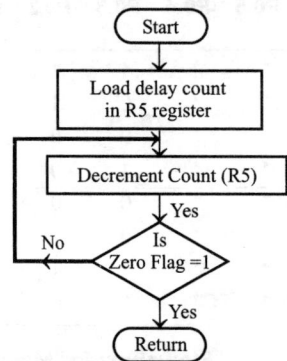

Assembly language program

```
;Program for Stepper Motor
        MOV  P0,#00H      ;Set port-0 as output port
        MOV  P2,#FFH      ;Set port-2 as input port

START:  MOV  A,P2         ;Read the status of keys
        RLC A             ;Check whether port pin P2.7(start key) is low
        JC START          ;If high, then  Wait for start key to go low

WAIT:   MOV  A,P2         ;Read the status of keys
        RLC A
        RLC A             ;Check whether port pin P2.6(forward key) is low
        JNC FORWARD       ;If low, then initiate forward rotation
        RLC A             ;Check whether port pin P2.5(reverse key) is low
        JNC REVERSE       ;If low, then initiate reverse rotation
        SJMP WAIT         ;Wait for either forward or reverse keypress

FORWARD: MOV  P0,#0CH     ;Send sequence-1 for forward rotation to port-0
        ACALL DELAY       ;Wait for motor to settle in new position

        MOV  P0,#06H      ;Send sequence-2 for forward rotation to port-0
        ACALL DELAY       ;Wait for motor to settle in new position

        MOV  P0,#03H      ;Send sequence-3 for forward rotation to port-0
        ACALL DELAY       ;Wait for motor to settle in new position

        MOV  P0,#09H      ;Send sequence-4 for forward rotation to port-0
        ACALL DELAY       ;Wait for motor to settle in new position

        MOV  A,P2         ;Read the status of keys
        RRC A             ;Check whether port pin P2.0(stop key) is low
        JC FORWARD        ;If high, then continue forward rotation
        SJMP STOP         ;If low, stop motor rotation

REVERSE: MOV  P0,#03H     ;Send sequence-1 for reverse rotation to port-0
        ACALL DELAY       ;Wait for motor to settle in new position

        MOV  P0,#06H      ;Send sequence-2 for reverse rotation to port-0
        ACALL DELAY       ;Wait for motor to settle in new position

        MOV  P0,#C0H      ;Send sequence-3 for reverse rotation to port-0
        ACALL DELAY       ;Wait for motor to settle in new position

        MOV  P0,#09H      ;Send sequence-4 for reverse rotation to port-0
        ACALL DELAY       ;Wait for motor to settle in new position
```

```
        MOV  A,P2           ;Read the status of keys
        RRC  A              ;Check whether port pin P2.0(stop key) is low
        JC REVERSE          ;If high, then continue reverse rotation

STOP:   MOV P0,#00H         ;Stop motor rotation
HALT:   SJMP HALT           ;Remain in infinite loop until reset, program end


;Delay Subroutine

DELAY:  MOV R5,#FFH
AGAIN:  DJNZ R5,AGAIN
        RET
        END                 ;Assemby end
```

5.10 APPLICATION TO AUTOMATION SYSTEMS   *(AU, Nov/Dec' 19, 2 Marks)*

Automation systems require one or more system variables to be monitored and maintained at desired values called set values. If there is any deviation or error in desired set values, the system automatically corrects the error and maintain the value of variables at the desired set values.

The modern domestic and industrial automation systems employ microcontrollers for monitoring and controlling the variables of the systems. There is a wide choice of microcontrollers with variety of features and so a proper choice of microcontroller can lead to development simple, compact and single chip microcontroller based automatic control systems.

Examples of automation systems

Temperature Control System: Consider the temperature of a liquid in a tank fitted with electric heater. A microcontroller can sense the temperature by using a temperature sensor and compare with desired set temperature and generate an error signal to control the electric supply to heater in order to maintain the required temperature of the liquid.

Level Control System: Consider the level of a liquid in a tank fitted with inlet and outlet valves. A microcontroller can sense the level by using a level sensor and compare with desired set level and generate an error signal to open/close the inlet/outlet valves of the tank in order to maintain the required level of the liquid.

Motor Speed Control System: Consider the armature-controlled DC motor in which the field is excited by a fixed voltage and armature voltage is varied to maintain the desired set speed. A microcontroller can sense the speed by using a speed sensor and compare with desired set speed and generate an error signal to control the armature voltage in order to maintain the required speed of the motor.

Fire Control Safety System: Fire is associated with smoke, heat and light. In fire safety systems appropriate sensors are interfaced to microcontroller in order to sense normal and abnormal conditions. When a fire is deducted, the microcontroller can automatically turn ON the fire quenching devices.

Traffic Light Control System: In traffic lights at road junctions the traffic lights are connected to ports of microcontroller via drivers. The ON and OFF timings of traffic lights can be automated by using the timers in microcontroller.

5.10.1 SENSOR INTERFACE IN 8051 FOR AUTOMATION SYSTEM

The sensors are basically transducers which are energy-converting devices. The sensors that convert energy in any form to electrical signals can be interfaced to microcontrollers. These sensors generate electrical signals proportional to the quantity being sensed and so can be used for measurement of the physical quantity being sensed. Examples of such applications are measurement of temperature, humidity, pressure, flow, etc.

When a sensor is used for measurement, the magnitude of the electrical signal from the sensor should be calibrated in order to relate the quantity being measured with electrical output. Also, the electrical signal has to be amplified and scaled by using operational amplifiers to a range of voltage acceptable by ADC (Analog-to-Digital-Converter). This process is also known as signal conditioning. The output signal from the signal conditioning unit will be an analog voltage. Using ADC, this analog voltage can be converted to digital data (8-bit/10-bit/12-bit data) and fed to a microcontroller.

In certain applications, the sensors can be used to sense normal and abnormal conditions of a physical quantity or presence and absence of a physical quantity. Examples of such applications are gas-leakage detection, fire detection, etc. In such cases, the analog signal from the sensor can be converted to digital signal using comparator and fed to one of the interrupts of a microcontroller or fed to one of the IO port pins of the microcontroller. The comparator will compare two analog inputs and generate a high or low signal that can be used as a digital signal. One input to the comparator is a preset analog voltage and another input is the signal conditioned voltage from the sensor.

Temperature Sensors

The various types of temperature sensors are thermistors, thermocouples, bimetallic strips, integrated circuits (IC), temperature sensors, etc. Since the output voltage of an IC version of temperature sensors is internally linearized, they may not require calibration and so they are superior versions of temperature sensors for interfacing with microcontrollers. Some examples of the IC version of temperature sensors are AD590, LM34, LM35, etc.

In an IC version of the temperature sensor, the change in Base-Emitter voltage of a transistor Base-Emitter junction is used to sense the temperature. The IC includes circuits for amplification and linearization of this voltage in order to generate an output voltage proportional to temperature. Therefore, no additional hardware is required for interfacing the IC version of the temperature sensor to a microcontroller via an ADC.

Ultrasonic Sensors

The ultrasonic sensors are widely used for distance and speed measurement. The other applications include humidifiers, sonar, medical ultrasonography, burglar alarms, non-destructive testing and wireless charging.

The ultrasonic sensor basically employs a transducer-like piezoelectric crystal which generates high-frequency sound waves. Piezoelectric crystals change size and shape when an AC voltage is applied which creates sound waves whose frequency is same as that of the applied AC voltage. Since piezoelectric materials generate a voltage when force is applied to them, they can also work as ultrasonic detectors.

The ultrasonic sensors for distance measurement are commercially available as a module consisting of an ultrasonic transmitter and receiver. The transmitter is basically a sound-wave generator, that generates high-frequency sound waves, above 20 kHz (typically values are in the range 28 kHz to 40 kHz). When the ultrasonic signals are transmitted in a medium, they get reflected at the interface where there is a change in the medium. The reflected wave can be detected by an ultrasonic receiver. The time difference between the transmitted wave and reflected wave can be used to estimate the distance between the transmitter and reflected object.

Humidity Sensor

Humidity is the presence of water in air. The amount of water vapour in air can affect human comfort as well as many manufacturing processes in industries. Electronic-type humidity sensors are of two types: capacitive type and resistive type.

The capacitive-type humidity sensor is based on the change in dielectric due to change in humidity. Humidity sensors relying on this principle consist of a hygroscopic dielectric material sandwiched between a pair of electrodes forming a small capacitor. The change in capacitance can be converted to a change in analog voltage by an appropriate circuit.

Resistive-type humidity sensors depend on change in the resistance of the sensor element due to change in humidity. The humidity-sensitive resistor is composed of an organic polymer, such as polyvinyl chloride or polyethylene or a metal oxide. The change in resistance can be converted to a change in analog voltage by an appropriate circuit.

Humidity sensors are available as modules that give an electrical voltage output which can be amplified and scaled using signal-conditioning circuit and applied to the input of an ADC interfaced to the microcontroller.

Gas Sensor

Gas sensors can be used to detect combustible, flammable and toxic gases, and non-availability of oxygen. The gas sensors are mostly part of a safety system in industrial plants, refineries, wastewater treatment facilities, vehicles and homes to detect potentially hazardous gas leakage.

The various types of gas sensors are infrared gas sensors, ultrasonic gas sensors, electrochemical gas sensors and semiconductor gas sensors.

Semiconductor sensors employ chemicals such as tin dioxide to detect gases. The resistance of tin dioxide decreases with increase in the concentration of gas surrounding it. This change in resistance is used to estimate the gas concentration. Semiconductor sensors are commonly used to detect hydrogen, oxygen, alcohol vapour, and harmful gases such as carbon monoxide.

Infra-Red(IR) Sensors

Infrared (IR) rays are invisible electromagnetic waves which have a wide frequency range between microwaves and visible light in the electromagnetic spectrum. The frequency range of IR waves will be from 300 GHz to 430 THz (with wavelength in the range 1000 micrometre to 0.75 micrometre). The wavelength region of 0.75 μm to 3 μm is called near infrared, the region from 3 μm to 6 μm is called mid infrared, and the region higher than 6 μm is called far infrared.

The infrared sensors basically employ IR diodes as sources to generate IR waves and photo transistors that work in the IR frequency range to deduct IR waves. The IR diodes can emit infrared waves, when forward biased. The conduction of an IR photo transistor will depend on the strength of an IR wave incident on its base-emitter junction.

One of the major applications of an IR sensor is object sensing. When IR waves are transmitted in a medium, they will get reflected when they encounter an object in their path. The reflected wave can be sensed by the IR receiver to recognize the presence of an object.

5.10.2 TEMPERATURE CONTROL SYSTEM WITH THE 8051 MICROCONTROLLER

The interfacing of the IC version of a temperature sensor, LM35 with the 8051 controller is shown in Fig. 5.14. The LM35 devices are precision integrated-circuit temperature sensors, with an output voltage linearly proportional to the Centigrade temperature and can be used to measure temperature in the range $-55°C$ to $150°C$, with typical accuracies of $\pm \frac{1}{4}$ °C at room temperature and $\pm \frac{3}{4}$ °C over a full temperature range.

The device has only three pins. Two pins for connecting a dc power supply in the range +4 V to +20 V and one pin for output voltage proportional to temperature with respect to power supply ground. The output voltage will be 10 mV/°C. In the interface circuit shown in Fig. 5.14, a capacitor is connected across the output for noise immunity. The sensor output is connected to a high input impedance operational amplifier, to avoid loading of the sensor. Also, the operational amplifier is designed to provide a gain in order to scale the output value suitable for ADC input.

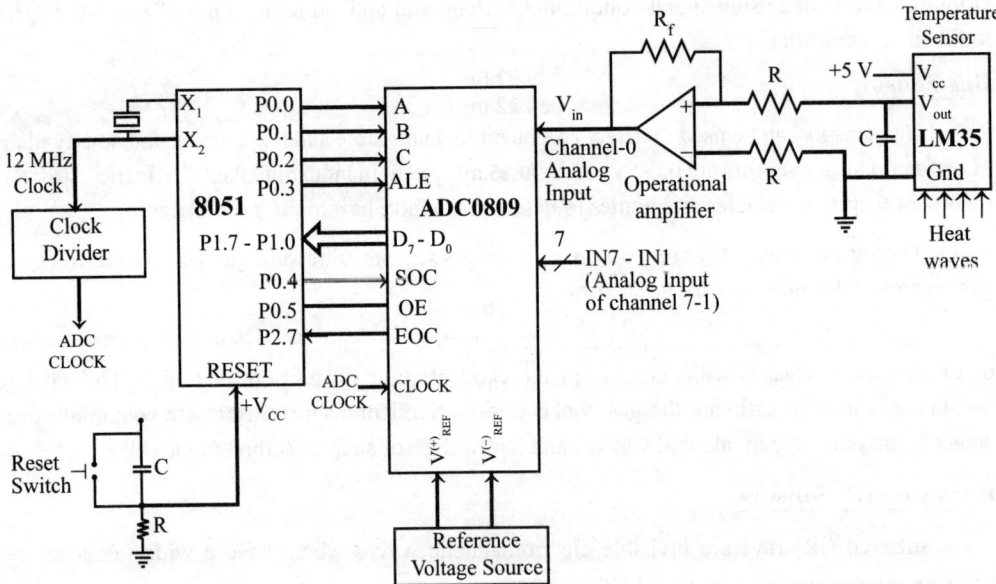

Fig. 5.14: Interfacing temperature sensor LM35 to 8051 microcontroller via ADC.

The ADC program developed in Section 5.6 can be used to read the temperature sensor output through ADC as digital data. The digital data read from ADC should be converted to temperature value by multiplying by an appropriate constant and then converted to ASCII value and send to LCD for display. The LCD program developed in Section 5.8 can be used to display the temperature value on the LCD.

EXAMPLE PROGRAM 24

Consider the temperature sensor interface shown in Fig. 5.13. Let the interface be designed to read temperature in the range 0°C to 100°C. Let the signal conditioning circuit using operational amplifier be designed to give zero voltage at 0°C and an increment of 19.61 mV for every one degree rise in temperature. Let us assume that ASCII conversion program is available as a subroutine called ASCICON and LCD display program is available as a subroutine called LCDDISP.

Sample Calculation of Temperature

The 8-bit ADC data for the input analog voltage, V_{in}, when $+V_{ref}$ is +5 V and $-V_{ref}$ is 0 V is given by,

$$\text{ADC data} = \frac{V_{in}}{+V_{ref}} \times 255$$

The ADC data is calculated for various values of input analog voltage, V_{in} and tabulated in the following table.

Temperature	V_{in}	ADC data
0°C	0 V	$0 = 00_H$
1°C	19.61 mV	$1 = 01_H$
2°C	39.22 mV	$2 = 02_H$
:	:	:
25°C	490.25 mV	$25 = 19_H$
:	:	:
50°C	980.5 mV	$50 = 32_H$
:	:	:
75ᵇC	1470.75 mV	$75 = 4B_H$
:	:	:
100°C	1961 mV	$100 = 64_H$

Flowchart

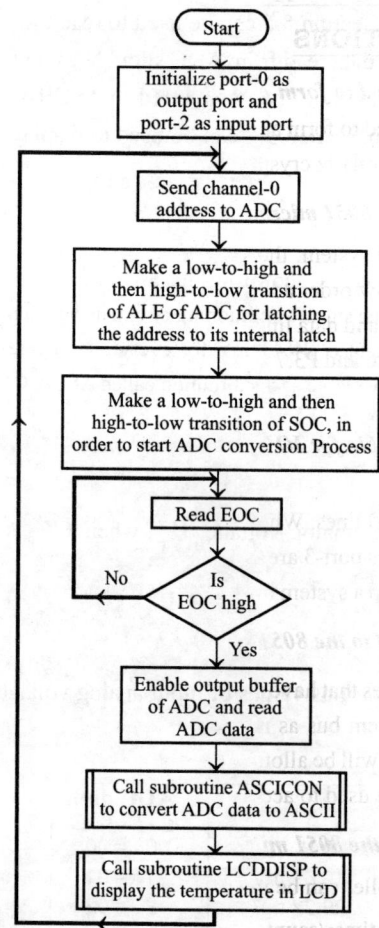

Assembly language program

```
;Program to measure temperature using LM35 interfaced to 8051 via ADC

START:  MOV   P0,#00H      ;Initialize port-0 as output port
        MOV   P2,#FFH      ;Initialize port-2 as input port

AGAIN:  MOV   P0,#00H      ;Send channel-0 address to port-0
        SETB  P0.3         ;Set ALE high
        NOP                ;Delay and make ALE low
        NOP
        CLR   P0.3

        SETB  P0.4         ;Set SOC high
        NOP                ;Delay and make SOC low
        NOP
        CLR   P0.4

WAIT:   JNB   P2.7,WAIT    ;Wait until EOC is high

        SETB  P0.5         ;Set OE high, to enable ADC output buffer
        MOV   A,P1         ;Get ADC data in A register

        ACALL ASCICON      ;Call subroutine ASCICON to convert ADC data to ASCII
        ACALL LCDDISP      ;Call subroutine LCDDISP to display the temperature value in LCD

        SJMP  AGAIN

        END                ;Assemby end
```

5.11 SHORT-ANSWER QUESTIONS

Q5.1 **What are the components required to form a minimum system using the 8051 microcontroller?**

The minimum components required to form an 8051 microcontroller-based minimum system are an RC circuit with switch for reset, a quartz crystal and a +5 V power supply.

Q5.2 **How is a system bus formed in an 8051 microcontroller-based system?**

In the 8051 microcontroller-based system, the system bus is formed using port-0 and port-2 pins. The port-0 pins are multiplexed low order address and data bus, and the port-2 pins are high order data bus. The multiplexed address and data lines are demultiplexed using external latch. The control bus is formed using port-3 pins P3.6 and P3.7 as read and write control signals and PSEN, Reset and Clock signals.

Q5.3 **How many IO pins will be available for IO operations in the 8051-based system when a system bus is formed?**

The 8051 microcontroller has 32 IO lines. When a system bus is formed, the 8 IO lines of port-0, the 8 IO lines of port-2 and two IO lines port-3 are used to form the system bus. Therefore, in 32 IO lines of 8051, 18 IO lines are used to form a system bus, and the remaining 14 lines can be used as IO lines.

Q5.4 **How can IO devices be interfaced to the 8051 system bus?**

The IO devices or peripheral devices that have matched read and write time as that of microcontroller can be interfaced to the 8051 system bus as memory-mapped IO, mapped in the data memory ad dress space. Therefore, IO devices will be allotted 16-bit address and all instructions that can be used for external memory access can be used to access IO devices.

Q5.5 **How can you initialize a timer of the 8051 microcontroller?**

The timer of the 8051 microcontroller can be initialized as follows:

1. **Program the TMOD register for timer/counter operation, and desired timer operating mode.**
2. **Load the timer count register with initial count.**
3. **Set the timer run bit as high to start the timer.**

Q5.6 **What is the auto reload feature in 8051 timer?**

In mode-2 operation of the 8051 timer, an 8-bit initial count is loaded in timer high order register. When the timer is started, the 8-bit initial count in timer high order register is loaded in timer low order register and the timer starts incrementing the content of low order count register by one for every timer clock pulse. When the timer overflow occurs, the initial count in the high order count register is automatically reloaded in the low order count register and so the timer continue to count from the initial value after an overflow. This process of auto reloading the initial count will take place after every timer overflow. This process of reloading the initial count again and again is called auto reload feature.

Q5.7 How can the initial count be loaded in the count register calculated for a timer of the 8051 microcontroller?

The timer clock frequency is obtained by dividing crystal frequency by 12.

Timer clock frequency, $F_{timer} = F_{crystal}/12$

The time delay for one count is given by inverse of timer clock frequency.

Time delay for one count, $t_d = 1/F_{timer}$

The timer count value for a specified time delay, t_{spe} is obtained by dividing specified time delay by time delay for one count, t_d.

Timer count value $= t_{spe}/t_d$

For an n-bit count, the maximum possible count is 2^n. Since, the timer is an up counter, an initial count has to calculated by subtracting the timer count value from the maximum count.

Initial count $= 2^n -$ Timer count value

Based on above discussion, the equation for initial count is,

Initial count $= 2^n - (t_{spe} \times F_{crystal}/12)$

<u>Example:</u>

Let, $F_{crystal} = 11.059$ MHz, $t_{spe} = 1$ millisecond, $n = 13$

Initial count $= = 2^{13} - (1 \times 10^{-3} \times 11.059 \times 10^6 / 12) = 7270_{10} = 1C66_H$

Q5.8 What is an SBUF register?

The SBUF register is a Special Function Register of 8051 dedicated to hold the parallel data during serial transmission and reception. The SBUF has a transmit buffer that can be accessed for write operation and a receive buffer that can be accessed for a read operation. Both the buffers are ac cessed by using the same internal address 99H.

Q5.9 What is the function of the SBUF register in serial communication?

In serial communication, the received serial data is converted to parallel data and loaded in SBUF, so that the programmer can read the parallel data from SBUF. During serial transmission, the parallel data is loaded in SBUF, so that the serial port of the microcontroller will take parallel data from SBUF and convert to serial data and transmit.

Q5.10 How is clock (or baud rate clock) for serial communication generated in the 8051 microcontroller?

In mode-0 serial communication, the baud rate is fixed at 1/12 of crystal frequency.

In mode-2 serial communication, the baud rate is fixed at 1/32 or 1/64 of crystal frequency.

In mode-1 and 3, the baud rate is variable and decided by the timer-1 overflow rate or timer output clock and the SMOD bit in the PCON register. The timer overflow rate is achieved by dividing the timer input clock by a suitable divisor, for which an initial count is loaded in the timer-1 count register. When SMOD = 0, the timer overflow rate is divided by 64, and when SMOD = 1, the timer overflow rate is divided by 32, in order to achieve the required baud rate.

Example:

Let, $F_{crystal}$ = 11.059 MHz, Timer-1 divisor = 3, SMOD = 1

Timer clock frequency, $F_{timer} = F_{crystal}/12 = 0.9216$ MHz

Baud rate clock $= 0.9216 \times 10^6/(3 \times 32) = 9600$ Hz

Q5.11 Mention few differences between a microprocessor and a microcontroller.

1. A microprocessor is concerned with the rapid movement of code and data between the external memory and the processor, whereas a microcontroller is concerned with the rapid movement of code and data within the controller.

2. A microprocessor will have few bit manipulating instructions, whereas a microcontroller will have a large number of bit manipulating instructions.

3. Microprocessors are generally used for designing general purpose systems, whereas microcontrollers are used for designing dedicated application specific systems.

5.12 EXERCISES

I. Fill in the blanks with appropriate words

1. The _____ converts the temperature into proportional analog voltage or current.

2. The _____ voltage of the DC motor is variable while the _____ voltage is kept constant for varying the DC motor speed in armature voltage control method.

3. The stepper motor which has step size of 1.8° can rotate _____ number of steps per revolution.

4. The _____ register is used for power control and baud rate selection.

5. The _____ bit is used to decide the baud rate in serial port operating modes.

6. In serial port mode-3, the _____ depends on the value of the SMOD bit in the PCON register.

7. The _____ bit of the SCON register can be used to enable or disable the serial reception.

8. When the crystal frequency is 11.0592 MHZ, the timer input clock frequency will be _____.

9. The _____ register is used to enable/disable the interrupts of an 8051.

10. The IP register can be programmed to make the _____ of any of the interrupt as highest.

11. The content of memory addressed by the stack pointer is considered _____.

Answers	
1. temperature sensor	7. REN
2. armature, field	8. 921.6 kHZ
3. 200	9. IE
4. PCON	10. Priority
5. SMOD	11. top of stack.
6. Baudrate	

II. State whether the following statements are True/False.

1. The stack memory of 8051 can be located anywhere in external RAM of 8051.
2. Upon reset, the SP of 8051 are cleared to zero.
3. The 8051 SP register is incremented by one after every write operation.
4. The 8051 SP register is decremented by one after every read operation.
5. In 8051, the oscillator frequency and microcontroller internal frequency are same.
6. In serial port mode-0, the serial port functions as a full-duplex serial port with a fixed baud rate.
7. In serial port mode-1, the serial port functions as a full-duplex serial port with variable baud rate.
8. In serial port mode-2, the serial port functions as a full-duplex serial port with baud rate of either 1/32 or 1/64 of the oscillator frequency.
9. The LCD module has six data pins and three control pins for interface with the microcontroller.
10. A key press is sensed by sending a low to a row and reading the columns.

Answers	
1. False	6. False
2. False	7. True
3. True	8. True
4. True	9. False
5. True	10. True

III. Choose the right answer for the following questions.

1. The _____ is used for precise incremental rotation in most of the computer peripherals.

   a) stepper motor   b) DC motor   c) a and b   d) neither a nor b

2. In stepper motor control system the free-wheeling diode is connected across each winding for _____.

   a) fast switchig   b) fast halt   c) both a and b   d) neither a nor b

3. The 8051 has two internal 16-bit timers/counters that can be programmed to work independently. They are called _____.

   a) timer-0/counter-0   b) timer-1/counter-1   c) both a and b   d) neither a nor b

4. The timer clock is internally derived by dividing the crystal frequency by _____.

   a) 12   b) 1/12   c) 1   d) 10

5. The TMOD register has separate fields _____ to program the operating mode of the timers.

   a) M0   b) M1   c) both a and b   d) neither a nor b

6. In mode-1, the _____ registers are cascaded to form a 16-bit timer register.

   a) TH   b) TL   c) both a and b   d) neither a nor b

7. In mode-0, the timer register is configured as a _____

   a) 13-bit register   b) 12-bit register   c) a and b   d) neither a nor b

8. In TCON register, TF flag is also used as an _____ to initiate the execution of a subroutine.

   a) interrupt signal   b) clock signal   c) neither a nor b   d) a and b

9. **In TCON register, IT bit is set to recognize _____ triggered external interrupt.**

   a) falling edge   b) rising edge   c) both a and b   d) neither a nor b

10. **When the 8-bit count register is selected, the maximum count value is___ and maximum possible time delay is _____ microseconds.**

    a) FFF_H, 128   b) $FFFF_H$, 256   c) FF_H, 128   d) FF_H, 256

Answers	
1. a	6. c
2. a	7. a
3. c	8. a
4. a	9. a
5. c	10. d

IV. Answer the following questions of 8051.

E5.1   Write an 8051 assembly language program to reset all the seven registers $(R_0$-$R_7)$ of register bank-3.

E5.2   Identify the operations performed by the following program

```
        MOV   A,#00H          MOV   @R0,A
        MOV   R0,08H          INC   R0
        MOV   R1,#05H         DJNZ  R1,BACK
BACK:   ADD   A,#02H    HALT: SJMP  HALT
```

E5.3   Write an 8051 program to exchange the contents of internal RAM 12_H and external data memory 0012_H.

E5.4   Write an 8051 program to exchange the contents of stack memory 50_H and the content of port-2.

E5.5   Write an 8051 program to perform the following simple task.

   (i)   Move a block of data (6 bytes) from internal RAM memory locations(60_H to 65_H) to internal RAM (70_H to 75_H)

   (ii)   Exchange the contents of 60_H to 65_H and 70_H to 75_H

   (iii)   Copy the content of 60H to 65_H at external data memory which starts at 2500_H.

E5.6   Write an 8051 assembly language program to add the data 12_H and 35_H. Also store the status of all flags of 8051.

E5.7   Write an 8051 program to add two 16-bit numbers.

E5.8   Write an 8051 program to add two BCD numbers.

E5.9   Identify the operation performed by the following 8051 program. Assume content of memory location $2000_H = 48_H$.

```
        MOV   DPTR,#2000H
        MOVX  A,@DPTR
        CPL   A
        INC   A
        INC   DPTR
        MOVX  @DPTR,A
HALT:   SJMP  HALT
```

E5.10   Write an 8051 program to calculate the distance (in km) travelled by a car when it travels at a speed of 50 kmph and reaches the destination in 2 hours.

E5.11 Write an 8051 program to count the number of 100's in the given data which is stored at 2300_H and store the result in 2301_H.

E5.12 Write an 8051 program to convert a packed BCD number into an unpacked BCD number. Assume the packed BCD number is at internal memory 20_H. Store the result in external memory location 3100_H and 3101_H.

E5.13 Write an 8051 program to find the square of a number.

E5.14 Write an 8051 program to evaluate the expression, $y = 4x + 6$, where x is an 8-bit hexadecimal number stored at 2500_H. Store the result at 3000_H and 3001_H.

E5.15 Write an 8051 program to perform the following tasks

a) Read the status of port-1 and if it is FF_H clear port-2

b) Set only 1,3,5 and 7th bit of port-1 to 1.

c) Reset only 1,3,5 and 7th bit of port-1 to 0.

d) Swap the lower nibble of port-0 and port-1.

e) To invert only the lower nibble of port-1

E5.16 Given an array of 6 data which is stored from 2501_H. Search for the data 25_H in the array, and if it is found store the value FF_H else 00_H in memory location 3000_H.

E5.17 Write an 8051 program to count the number of ones and zeros in the given data stored at 2000_H. store the number of zeros at 2001_H and the number of ones at 2002_H.

E5.18 Write an 8051 program to display 00_H at location 2001_H if the data available at memory location 2000_H is positive number else display 01_H.

E5.19 Check whether the data at location 2000_H is even or odd number. Display 00_H if even else 01_H at memory location 2001_H. Use 8051 microcontroller to program.

E5.20 Write an 8051 ALP to check whether the two numbers stored at memory locations 2000_H and 2001_H are same or not. If same display 00_H else 01_H at memory location 2002_H. Use logical instructions to compare.

E5.21 Write an 8051 ALP to convert a positive number into negative number. Assume positive data is stored at 3000_H.

E5.22 Write down the stepper motor switching sequence for full step rotation

E5.23 Write down the stepper motor switching sequence fo half-step rotation

APPENDIX I : LIST OF MICROPROCESSORS RELEASED BY INTEL

MICROPROCESSOR	DATE OF INTRODUCTION	NUMBER OF TRANSISTORS	CLOCK SPEED
4004	15th Nov, 1971	2,300	400 kHz
8008	Apr, 1972	3,500	500-800 kHz
8080	Apr, 1974	4,500	2 MHz
8085	Mar, 1976	6,500	5 MHz
8086	8th Jun, 1978	29,000	5/8/10 MHz
8088	Jun, 1979	29,000	5/8 MHz
80186	1982	....	10/12 MHz
80286	Feb, 1982	134,000	6/10/12 MHz
INTEL386 DX	17th Oct, 1985	275,000	16/20/25/33 MHz
INTEL386 SX	16th Jun, 1988	275,000	16/20/25/33 MHz
INTEL386 SL	15th Oct, 1990	855,000	20/25 MHz
INTEL486 DX	10th Apr, 1989	1.2 million	25/33/50 MHz
INTEL486 SX	16th Sep, 1991	900,000	16/20/25 MHz
INTEL486 SX	21st Sep, 1992	1.185 million	33 MHz
INTEL486 SL	4th Nov, 1992	1.4 million	20/25/33 MHz
INTELDX 2	3rd Mar, 1992	1.2 million	50/66 MHz
INTELDX 4	7th Mar, 1994	3.2 million	75/100 MHz
Pentium	22nd Mar, 1993	3.1 million	60/66 MHz
Pentium	7th Mar, 1994	3.2 million	75/90/100/120 MHz
Pentium	Jun, 1995	3.3 million	133/150/166/200 MHz
Pentium Pro	1st Nov, 1995	5.5 million	150/166/180/200 MHz
Pentium (MMX)	8th Jan, 1997	4.5 million	166/200/233 MHz
Mobile Pentium (MMX)	9th Sep, 1997	4.5 million	200/233/266/300 MHz
Pentium II	7th May, 1997	7.5 million	233/266/300/333/350/400/ 450 MHz
Mobile Pentium II	2nd Apr, 1998	7.5 million	233/266/300 MHz
Mobile Pentium II	25th Jan, 1999	27.4 million	333/366/400 MHz
Pentium II Xeon	29th Jun, 1998	7.5 million	400/450 MHz
Celeron	15th Apr, 1998	7.5 million	266/300 MHz
Celeron	24th Aug, 1998	19 million	333 MHZ to 2.7 GHz
Mobile Celeron	25th Jan, 1999	18.9 million	266 MHz to 2.4 GHz
Pentium III	26th Feb, 1999	9.5 million	450/500/550/600 MHZ

Appendix I continued...

MICROPROCESSOR	DATE OF INTRODUCTION	NUMBER OF TRANSISTORS	CLOCK SPEED
Pentium III	25th Oct, 1999	28 million	500 MHz to 1 GHz
Pentium III Xeon	17th Mar, 1999	9.5 million	500/550 MHz
Pentium III Xeon	25th Oct, 1999	28 million	600 to 900 MHz
Mobile Pentium III	25th Oct, 1999	28 million	400 MHz to 1 GHz
Mobile Pentium III	30th Jul, 2001	44 million	1/1.06/1.13/1.2/1.33 GHz
Pentium 4	20th Nov, 2000	42 million	1.4/1.5/1.6/1.7/1.8/1.9/2 GHz
Pentium 4	27th Aug, 2001	55 million	2 to 2.8 GHz
Pentium 4 (HT Technology)	14th Nov, 2002	55 million	2.4 to 3.3 GHz
Mobile Pentium 4	4th Mar, 2002	55 million	1.5 to 3.2 GHz
INTEL Xeon	21st May, 2001	42 million	1.4/1.5/1.7/2 GHz
INTEL Xeon	9th Jan, 2002	52 million	1.8/2/2.2/2.4/2.6/2.8 GHz
INTEL Xeon	18th Nov, 2002	108 million	1.4 to 3.2 GHz
INTEL Itanium	May, 2001	25 million	733/800 MHz
INTEL Itanium 2	8th Jul, 2002	220 million	900 MHz/1 GHz
INTEL Itanium 2	30th Jun, 2003	410 million	1/1.4/1.5 GHz
INTEL Pentium-M	12th Mar, 2003	77 million	900 MHz to 1.7 GHz

Note : The date mentioned here is the date of introduction of the lowest clock version of the processor. For the date of introduction of higher clock version of a processor please refer to INTEL website www.intel.com.

APPENDIX II : 8051 INSTRUCTIONS IN HEXADECIMAL ORDER

OPCODE IN HEX	MNEMONIC		OPCODE IN HEX	MNEMONIC		OPCODE IN HEX	MNEMONIC	
00	NOP		2B	ADD	A,R3	56	ANL	A,@R0
01	AJMP	addr11	2C	ADD	A,R4	57	ANL	A,@R1
02	LJMP	addr16	2D	ADD	A,R5	58	ANL	A,R0
03	RR	A	2E	ADD	A,R6	59	ANL	A,R1
04	INC	A	2F	ADD	A,R7	5A	ANL	A,R2
05	INC	DPTR	30	JNB	bit,offset	5B	ANL	A,R3
06	INC	@R0	31	ACALL	addr11	5C	ANL	A,R4
07	INC	@R1	32	RETI		5D	ANL	A,R5
08	INC	R0	33	RLC	A	5E	ANL	A,R6
09	INC	R1	34	ADDC	A,#data	5F	ANL	A,R7
0A	INC	R2	35	ADDC	A,direct	60	JZ	offset
0B	INC	R3	36	ADDC	A,@R0	61	AJMP	addr11
0C	INC	R4	37	ADDC	A,@R1	62	XRL	direct,A
0D	INC	R5	38	ADDC	A,R0	63	XRL	direct,#data
0E	INC	R6	39	ADDC	A,R1	64	XRL	A,#data
0F	INC	R7	3A	ADDC	A,R2	65	XRL	A,#direct
10	JBC	bit,offset	3B	ADDC	A,R3	66	XRL	A,@R0
11	ACALL	addr11	3C	ADDC	A,R4	67	XRL	A,@R1
12	LCALL	addr16	3D	ADDC	A,R5	68	XRL	A,R0
13	RRC	A	3E	ADDC	A,R6	69	XRL	A,R1
14	DEC	A	3F	ADDC	A,R7	6A	XRL	A,R2
15	DEC	direct	40	JC	offset	6B	XRL	A,R3
16	DEC	@R0	41	AJMP	addr11	6C	XRL	A,R4
17	DEC	@R1	42	ORL	direct,A	6D	XRL	A,R5
18	DEC	R0	43	ORL	direct,#data	6E	XRL	A,R6
19	DEC	R1	44	ORL	A,#data	6F	XRL	A,R7
1A	DEC	R2	45	ORL	A,direct	70	JNZ	offset
1B	DEC	R3	46	ORL	A,@R0	71	ACALL	addr11
1C	DEC	R4	47	ORL	A,@R1	72	ORL	C,bitaddr
1D	DEC	R5	48	ORL	A,R0	73	JMP	@A+DPTR
1E	DEC	R6	49	ORL	A,R1	74	MOV	A,#data
1F	DEC	R7	4A	ORL	A,R2	75	MOV	direct,#data
20	JB	bit,offset	4B	ORL	A,R3	76	MOV	@R0,#data
21	AJMP	addr11	4C	ORL	A,R4	77	MOV	@R1,#data
22	RET		4D	ORL	A,R5	78	MOV	R0,#data
23	RL	A	4E	ORL	A,R6	79	MOV	R1,#data
24	ADD	A,#data	4F	ORL	A,R7	7A	MOV	R2,#data
25	ADD	A,direct	50	JNC	offset	7B	MOV	R3,#data
26	ADD	A,@R0	51	ACALL	addr11	7C	MOV	R4,#data
27	ADD	A,@R1	52	ANL	direct,A	7D	MOV	R5,#data
28	ADD	A,R0	53	ANL	direct,#data	7E	MOV	R6,#data
29	ADD	A,R1	54	ANL	A,#data	7F	MOV	R7,#data
2A	ADD	A,R2	55	ANL	A,direct	80	SJMP	offset

Appendix II continued...

OPCODE IN HEX	MNEMONIC		OPCODE IN HEX	MNEMONIC		OPCODE IN HEX	MNEMONIC	
81	AJMP	addr11	AC	MOV	R4,direct	D7	XCHD	A,@R1
82	ANL	C,bit	AD	MOV	R5,direct	D8	DJNZ	R0,offset
83	MOVC	A,@A+PC	AE	MOV	R6,direct	D9	DJNZ	R1,offset
84	DIV	AB	AF	MOV	R7,direct	DA	DJNZ	R2,offset
85	MOV	direct,direct	B0	ANL	C,/bit	DB	DJNZ	R3,offset
86	MOV	direct,@R0	B1	ACALL	addr11	DC	DJNZ	R4,offset
87	MOV	direct,@R1	B2	CPL	bit	DD	DJNZ	R5,offset
88	MOV	direct,R0	B3	CPL	C	DE	DJNZ	R6,offset
89	MOV	direct,R1	B4	CJNE	A,#data,offset	DF	DJNZ	R7,offset
8A	MOV	direct,R2	B5	CJNE	A,direct,offset	E0	MOVX	A,@DPTR
8B	MOV	direct,R3	B6	CJNE	@R0,#data,offset	E1	AJMP	addr11
8C	MOV	direct,R4	B7	CJNE	@R1,#data,offset	E2	MOVX	A,@R0
8D	MOV	direct,R5	B8	CJNE	R0,#data,offset	E3	MOVX	A,@R1
8E	MOV	direct,R6	B9	CJNE	R1,#data,offset	E4	CLR	A
8F	MOV	direct,R7	BA	CJNE	R2,#data,offset	E5	MOV	A,direct
90	MOV	DPTR,#data16	BB	CJNE	R3,#data,offset	E6	MOV	A,@R0
91	ACALL	addr11	BC	CJNE	R4,#data,offset	E7	MOV	A,@R1
92	MOV	bit,C	BD	CJNE	R5,#data,offset	E8	MOV	A,R0
93	MOVC	A,@A+DPTR	BE	CJNE	R6,#data,offset	E9	MOV	A,R1
94	SUBB	A,#data	BF	CJNE	R7,#data,offset	EA	MOV	A,R2
95	SUBB	A,#direct	C0	PUSH	direct	EB	MOV	A,R3
96	SUBB	A,@R0	C1	AJMP	addr11	EC	MOV	A,R4
97	SUBB	A,@R1	C2	CLR	bit	ED	MOV	A,R5
98	SBBB	A,R0	C3	CLR	C	EE	MOV	A,R6
99	SBBB	A,R1	C4	SWAP	A	EF	MOV	A,R7
9A	SBBB	A,R2	C5	XCH	A,direct	F0	MOVX	@DPTR,A
9B	SBBB	A,R3	C6	XCH	A,@R0	F1	ACALL	addr11
9C	SBBB	A,R4	C7	XCH	A,@R1	F2	MOVX	@R0,A
9D	SBBB	A,R5	C8	XCH	A,R0	F3	MOVX	@R1,A
9E	SBBB	A,R6	C9	XCH	A,R1	F4	CPL	A
9F	SBBB	A,R7	CA	XCH	A,R2	F5	MOV	direct,A
A0	ORL	C,/bit	CB	XCH	A,R3	F6	MOV	@R0,A
A1	AJMP	addr11	CC	XCH	A,R4	F7	MOV	@R1,A
A2	MOV	C,bit	CD	XCH	A,R5	F8	MOV	R0,A
A3	INC	DPTR	CE	XCH	A,R6	F9	MOV	R1,A
A4	MUL	AB	CF	XCH	A,R7	FA	MOV	R2,A
A5	UNUSED		D0	POP	direct	FB	MOV	R3,A
A6	MOV	@R0,direct	D1	ACALL	addr11	FC	MOV	R4,A
A7	MOV	@R1,direct	D2	SETB	bit	FD	MOV	R5,A
A8	MOV	R0,direct	D3	SETB	C	FE	MOV	R6,A
A9	MOV	R1,direct	D4	DA	A	FF	MOV	R7,A
AA	MOV	R2,direct	D5	DJNZ	direct,offset			
AB	MOV	R3,direct	D6	XCHD	A,@R0			

APPENDIX III : 8051 INSTRUCTIONS IN ALPHABETICAL ORDER

HEX CODE	MNEMONIC		HEX CODE	MNEMONIC		HEX CODE	MNEMONIC	
11	ACALL	addr11	5F	ANL	A,R7	DE	DJNZ	R6,offset
31	ACALL	addr11	55	ANL	A,direct	DF	DJNZ	R7,offset
51	ACALL	addr11	56	ANL	A,@R0	D5	DJNZ	direct,offset
71	ACALL	addr11	57	ANL	A,@R1	04	INC	A
91	ACALL	addr11	54	ANL	A,#data	08	INC	R0
B1	ACALL	addr11	52	ANl	direct,A	09	INC	R1
D1	ACALL	addr11	53	ANl	direct,#data	0A	INC	R2
F1	ACALL	addr11	82	ANL	C,bit	0B	INC	R3
28	ADD	A,R0	B0	ANL	C,/bit	0C	INC	R4
29	ADD	A,R1	B5	CJNE	A,direct,offset	0D	INC	R5
2A	ADD	A,R2	B4	CJNE	A,#data,offset	0E	INC	R6
2B	ADD	A,R3	B8	CJNE	R0,#data,offset	0F	INC	R7
2C	ADD	A,R4	B9	CJNE	R1,#data,offset	05	INC	direct
2D	ADD	A,R5	BA	CJNE	R2,#data,offset	06	INC	@R0
2E	ADD	A,R6	BB	CJNE	R3,#data,offset	07	INC	@R1
2F	ADD	A,R7	BC	CJNE	R4,#data,offset	A3	INC	DPTR
25	ADD	A,direct	BD	CJNE	R5,#data,offset	20	JB	bit,offset
26	ADD	A,@R0	BE	CJNE	R6,#data,offset	10	JBC	bit,offset
27	ADD	A,@R1	BF	CJNE	R7,#data,offset	40	JC	offset
24	ADD	A,#data	B6	CJNE	@R0,#data,offset	73	JMP	@A+DPTR
38	ADDC	A,R0	B7	CJNE	@R1,#data,offset	30	JNB	bit,offset
39	ADDC	A,R1	E4	CLR	A	50	JNC	offset
3A	ADDC	A,R2	C3	CLR	C	70	JNZ	offset
3B	ADDC	A,R3	C2	CLR	bit	60	JZ	offset
3C	ADDC	A,R4	F4	CPL	A	12	LCALL	addr16
3D	ADDC	A,R5	B3	CPL	C	02	LJMP	addr16
3E	ADDC	A,R6	B2	CPL	bit	E8	MOV	A,R0
3F	ADDC	A,R7	D4	DA	A	E9	MOV	A,R1
35	ADDC	A,direct	14	DEC	A	EA	MOV	A,R2
36	ADDC	A,@R0	18	DEC	R0	EB	MOV	A,R3
37	ADDC	A,@R1	19	DEC	R1	EC	MOV	A,R4
34	ADDC	A,#data	1A	DEC	R2	ED	MOV	A,R5
01	AJMP	addr11	1B	DEC	R3	EE	MOV	A,R6
21	AJMP	addr11	1C	DEC	R4	EF	MOV	A,R7
41	AJMP	addr11	1D	DEC	R5	E5	MOV	A,direct
61	AJMP	addr11	1E	DEC	R6	E6	MOV	A,@R0
81	AJMP	addr11	1F	DEC	R7	E7	MOV	A,@R1
A1	AJMP	addr11	15	DEC	direct	74	MOV	A,#data
C1	AJMP	addr11	16	DEC	@R0	F8	MOV	R0,A
E1	AJMP	addr11	17	DEC	@R1	F9	MOV	R1,A
58	ANL	A,R0	84	DIV	AB	FA	MOV	R2,A
59	ANL	A,R1	D8	DJNZ	R0,offset	FB	MOV	R3,A
5A	ANL	A,R2	D9	DJNZ	R1,offset	FC	MOV	R4,A
5B	ANL	A,R3	DA	DJNZ	R2,offset	FD	MOV	R5,A
5C	ANL	A,R4	DB	DJNZ	R3,offset	FE	MOV	R6,A
5D	ANL	A,R5	DC	DJNZ	R4,offset	FF	MOV	R7,A
5E	ANL	A,R6	DD	DJNZ	R5,offset			

Appendix III continued...

HEX CODE	MNEMONIC		HEX CODE	MNEMONIC		HEX CODE	MNEMONIC	
A8	MOV	R0,direct	A4	MUL	AB	CC	XCH	A,R4
A9	MOV	R1,direct	00	NOP		CD	XCH	A,R5
AA	MOV	R2,direct	48	ORL	A,R0	CE	XCH	A,R6
AB	MOV	R3,direct	49	ORL	A,R1	CF	XCH	A,R7
AC	MOV	R4,direct	4A	ORL	A,R2	C5	XCH	A,direct
AD	MOV	R5,direct	4B	ORL	A,R3	C6	XCH	A,@R0
AE	MOV	R6,direct	4C	ORL	A,R4	C7	XCH	A,@R1
AF	MOV	R7,direct	4D	ORL	A,R5	D6	XCHD	A,@R0
78	MOV	R0,#data	4E	ORL	A,R6	D7	XCHD	A,@R1
79	MOV	R1,#data	4F	ORL	A,R7	68	XRL	A,R0
7A	MOV	R2,#data	45	ORL	A,direct	69	XRL	A,R1
7B	MOV	R3,#data	46	ORL	A,@R0	6A	XRL	A,R2
7C	MOV	R4,#data	47	ORL	A,@R1	6B	XRL	A,R3x
7D	MOV	R5,#data	44	ORL	A,#data	6C	XRL	A,R4
7E	MOV	R6,#data	42	ORL	direct,A	6D	XRL	A,R5
7F	MOV	R7,#data	43	ORL	direct,#data	6E	XRL	A,R6
F5	MOV	direct,A	72	ORL	C,bit	6F	XRL	A,R7
88	MOV	direct,R0	A0	ORL	C,/bit	65	XRL	A,direct
89	MOV	direct,R1	D0	POP	direct	66	XRL	A,@R0
8A	MOV	direct,R2	C0	PUSH	direct	67	XRL	A,@R1
8B	MOV	direct,R3	22	RET		64	XRL	A,#data
8C	MOV	direct,R4	32	RETI		62	XRL	direct,A
8D	MOV	direct,R5	23	RL	A	63	XRL	direct,#data
8E	MOV	direct,R6	33	RLC	A			
8F	MOV	direct,R7	03	RR	A			
85	MOV	direct,direct	13	RRC	A			
86	MOV	direct,@R0	D3	SETB	C			
87	MOV	direct,@R1	D2	SETB	bit			
75	MOV	direct,#data	80	SJMP	offset			
F6	MOV	@R0,A	98	SUBB	A,R0			
F7	MOV	@R1,A	99	SUBB	A,R1			
A6	MOV	@R0,direct	9A	SUBB	A,R2			
A7	MOV	@R1,direct	9B	SUBB	A,R3			
76	MOV	@R0,#data	9C	SUBB	A,R4			
77	MOV	@R1,#data	9D	SUBB	A,R5			
A2	MOV	C,bit	9E	SUBB	A,R6			
92	MOV	bit,C	9F	SUBB	A,R7			
90	MOV	DPTR,#data16	95	SUBB	A,direct			
93	MOVC	A, @A+DPTR	96	SUBB	A,@R0			
83	MOVC	A,@A+PC	97	SUBB	A,@R1			
E2	MOVX	A,@R0	94	SUBB	A,#data			
E3	MOVX	A,@R1	C4	SWAP	A			
E0	MOVX	A,@DPTR	C8	XCH	A,R0			
F2	MOVX	@R0,A	C9	XCH	A,R1			
F3	MOVX	@R1,A	CA	XCH	A,R2			
F0	MOVX	@DPTR,A	CB	XCH	A,R3			

APPENDIX IV : 8085 INSTRUCTIONS IN HEXADECIMAL ORDER

OPCODE IN HEX	MNEMONIC		OPCODE IN HEX	MNEMONIC		OPCODE IN HEX	MNEMONIC	
00	NOP		2B	DCX	H	56	MOV	D, M
01	LXI	B, d16	2C	INR	L	57	MOV	D, A
02	STAX	B	2D	DCR	L	58	MOV	E, B
03	INX	B	2E	MVI	L, d8	59	MOV	E, C
04	INR	B	2F	CMA		5A	MOV	E, D
05	DCR	B	30	SIM		5B	MOV	E, E
06	MVI	B, d8	31	LXI	SP, d16	5C	MOV	E, H
07	RLC		32	STA	addr16	5D	MOV	E, L
08	---		33	INX	SP	5E	MOV	E, M
09	DAD	B	34	INR	M	5F	MOV	E, A
0A	LDAX	B	35	DCR	M	60	MOV	H, B
0B	DCX	B	36	MVI	M, d8	61	MOV	H, C
0C	INR	C	37	STC		62	MOV	H, D
0D	DCR	C	38	---		63	MOV	H, E
0E	MVI	C, d8	39	DAD	SP	64	MOV	H, H
0F	RRC		3A	LDA	addr16	65	MOV	H, L
10	---		3B	DCX	SP	66	MOV	H, M
11	LXI	D, d16	3C	INR	A	67	MOV	H, A
12	STAX	D	3D	DCR	A	68	MOV	L, B
13	INX	D	3E	MVI	A, d8	69	MOV	L, C
14	INR	D	3F	CMC		6A	MOV	L, D
15	DCR	D	40	MOV	B, B	6B	MOV	L, E
16	MVI	D, d8	41	MOV	B, C	6C	MOV	L, H
17	RAL		42	MOV	B, D	6D	MOV	L, L
18	---		43	MOV	B, E	6E	MOV	L, M
19	DAD	D	44	MOV	B, H	6F	MOV	L, A
1A	LDAX	D	45	MOV	B, L	70	MOV	M, B
1B	DCX	D	46	MOV	B, M	71	MOV	M, C
1C	INR	E	47	MOV	B, A	72	MOV	M, D
1D	DCR	E	48	MOV	C, B	73	MOV	M, E
1E	MVI	E, d8	49	MOV	C, C	74	MOV	M, H
1F	RAR		4A	MOV	C, D	75	MOV	M, L
20	RIM		4B	MOV	C, E	76	HLT	
21	LXI	H, d16	4C	MOV	C, H	77	MOV	M, A
22	SHLD	addr16	4D	MOV	C, L	78	MOV	A, B
23	INX	H	4E	MOV	C, M	79	MOV	A, C
24	INR	H	4F	MOV	C, A	7A	MOV	A, D
25	DCR	H	50	MOV	D, B	7B	MOV	A, E
26	MVI	H, d8	51	MOV	D, C	7C	MOV	A, H
27	DAA		52	MOV	D, D	7D	MOV	A, L
28	---		53	MOV	D, E	7E	MOV	A, M
29	DAD	H	54	MOV	D, H	7F	MOV	A, A
2A	LHLD	addr16	55	MOV	D, L	80	ADD	B

Appendix IV continued...

OPCODE IN HEX	MNEMONIC		OPCODE IN HEX	MNEMONIC		OPCODE IN HEX	MNEMONIC	
81	ADD	C	AC	XRA	H	D7	RST	2
82	ADD	D	AD	XRA	L	D8	RC	
83	ADD	E	AE	XRA	M	D9	---	
84	ADD	H	AF	XRA	A	DA	JC	addr16
85	ADD	L	B0	ORA	B	DB	IN	addr8
86	ADD	M	B1	ORA	C	DC	CC	addr16
87	ADD	A	B2	ORA	D	DD	---	
88	ADC	B	B3	ORA	E	DE	SBI	d8
89	ADC	C	B4	ORA	H	DF	RST	3
8A	ADC	D	B5	ORA	L	E0	RPO	
8B	ADC	E	B6	ORA	M	E1	POP	H
8C	ADC	H	B7	ORA	A	E2	JPO	addr16
8D	ADC	L	B8	CMP	B	E3	XTHL	
8E	ADC	M	B9	CMP	C	E4	CPO	addr16
8F	ADC	A	BA	CMP	D	E5	PUSH	H
90	SUB	B	BB	CMP	E	E6	ANI	d8
91	SUB	C	BC	CMP	H	E7	RST	4
92	SUB	D	BD	CMP	L	E8	RPE	
93	SUB	E	BE	CMP	M	E9	PCHL	
94	SUB	H	BF	CMP	A	EA	JPE	addr16
95	SUB	L	C0	RNZ		EB	XCHG	
96	SUB	M	C1	POP	B	EC	CPE	addr16
97	SUB	A	C2	JNZ	addr16	ED	---	
98	SBB	B	C3	JMP	addr16	EE	XRI	d8
99	SBB	C	C4	CNZ	addr16	EF	RST	5
9A	SBB	D	C5	PUSH	B	F0	RP	
9B	SBB	E	C6	ADI	d8	F1	POP	PSW
9C	SBB	H	C7	RST	0	F2	JP	addr16
9D	SBB	L	C8	RZ		F3	DI	
9E	SBB	M	C9	RET		F4	CP	addr16
9F	SBB	A	CA	JZ	addr16	F5	PUSH	PSW
A0	ANA	B	CB	---		F6	ORI	d8
A1	ANA	C	CC	CZ	addr16	F7	RST	6
A2	ANA	D	CD	CALL	addr16	F8	RM	
A3	ANA	E	CE	ACI	d8	F9	SPHL	
A4	ANA	H	CF	RST	1	FA	JM	addr16
A5	ANA	L	D0	RNC		FB	EI	
A6	ANA	M	D1	POP	D	FC	CM	addr16
A7	ANA	A	D2	JNC	addr16	FD	---	
A8	XRA	B	D3	OUT	addr8	FE	CPI	d8
A9	XRA	C	D4	CNC	addr16	FF	RST	7
AA	XRA	D	D5	PUSH	D	--	---	
AB	XRA	E	D6	SUI	d8	--	---	

d8	→	8-bit data	addr16	→	16-bit address
d16	→	16-bit data	M	→	Memory
addr8	→	8-bit address	PSW	→	Program Status Word

APPENDIX V : 8085 INSTRUCTIONS IN ALPHABETICAL ORDER

OPCODE IN HEX	MNEMONIC		OPCODE IN HEX	MNEMONIC		OPCODE IN HEX	MNEMONIC	
CE	ACI	d8	E4	CPO	addr16	0A	LDAX	B
8F	ADC	A	CC	CZ	addr16	1A	LDAX	D
88	ADC	B	27	DAA		2A	LHLD	addr16
89	ADC	C	09	DAD	B	01	LXIB,d16	
8A	ADC	D	19	DAD	D	11	LXID,d16	
8B	ADC	E	29	DAD	H	21	LXIH,d16	
8C	ADC	H	39	DAD	SP	31	LXISP,d16	
8D	ADC	L	3D	DCR	A	7F	MOV	A,A
8E	ADC	M	05	DCR	B	78	MOV	A,B
87	ADD	A	0D	DCR	C	79	MOV	A,C
80	ADD	B	15	DCR	D	7A	MOV	A,D
81	ADD	C	1D	DCR	E	7B	MOV	A,E
82	ADD	D	25	DCR	H	7C	MOV	A,H
83	ADD	E	2D	DCR	L	7D	MOV	A,L
84	ADD	H	35	DCR	M	7E	MOV	A,M
85	ADD	L	0B	DCX	B	47	MOV	B,A
86	ADD	M	1B	DCX	D	40	MOV	B,B
C6	ADI	d8	2B	DCX	H	41	MOV	B,C
A7	ANA	A	3B	DCX	SP	42	MOV	B,D
A0	ANA	B	F3	DI		43	MOV	B,E
A1	ANA	C	FB	EI		44	MOV	B,H
A2	ANA	D	76	HLT		45	MOV	B,L
A3	ANA	E	DB	IN	addr8	46	MOV	B,M
A4	ANA	H	3C	INR	A	4F	MOV	C,A
A5	ANA	L	04	INR	B	48	MOV	C,B
A6	ANA	M	0C	INR	C	49	MOV	C,C
E6	ANI	d8	14	INR	D	4A	MOV	C,D
CD	CALL	addr16	1C	INR	E	4B	MOV	C,E
DC	CC	addr16	24	INR	H	4C	MOV	C,H
FC	CM	addr16	2C	INR	L	4D	MOV	C,L
2F	CMA		34	INR	M	4E	MOV	C,M
3F	CMC		03	INX	B	57	MOV	D,A
BF	CMP	A	13	INX	D	50	MOV	D,B
B8	CMP	B	23	INX	H	51	MOV	D,C
B9	CMP	C	33	INX	SP	52	MOV	D,D
BA	CMP	D	DA	JC	addr16	53	MOV	D,E
BB	CMP	E	FA	JM	addr16	54	MOV	D,H
BC	CMP	H	C3	JMP	addr16	55	MOV	D,L
BD	CMP	L	D2	JNC	addr16	56	MOV	D,M
BE	CMP	M	C2	JNZ	addr16	5F	MOV	E,A
D4	CNC	addr16	F2	JP	addr16	58	MOV	E,B
C4	CNZ	addr16	EA	JPE	addr16	59	MOV	E,C
F4	CP	addr16	E2	JPO	addr16	5A	MOV	E,D
EC	CPE	addr16	CA	JZ	addr16	5B	MOV	E,E
FE	CPI	d8	3A	LDA	addr16	5C	MOV	E,H

Appendix V continued...

OPCODE IN HEX	MNEMONIC		OPCODE IN HEX	MNEMONIC		OPCODE IN HEX	MNEMONIC	
5D	MOV	E,L	C1	POP	B	97	SUB	A
5E	MOV	E,M	D1	POP	D	90	SUB	B
67	MOV	H,A	E1	POP	H	91	SUB	C
60	MOV	H,B	F1	POP	PSW	92	SUB	D
61	MOV	H,C	C5	PUSH	B	93	SUB	E
62	MOV	H,D	D5	PUSH	D	94	SUB	H
63	MOV	H,E	E5	PUSH	H	95	SUB	L
64	MOV	H,H	F5	PUSH	PSW	96	SUB	M
65	MOV	H,L	17 RAL		D6	SUI	d8	
66	MOV	H,M	1F	RAR		EB	XCHG	
6F	MOV	L,A	D8	RC		AF	XRA	A
68	MOV	L,B	C9	RET		A8	XRA	B
69	MOV	L,C	20 RIM		A9	XRA	C	
6A	MOV	L,D	07 RLC		AA	XRA	D	
6B	MOV	L,E	F8	RM		AB	XRA	E
6C	MOV	L,H	D0	RNC		AC	XRA	H
6D	MOV	L,L	C0	RNZ		AD	XRA	L
6E	MOV	L,M	F0	RP		AE	XRA	M
77	MOV	M,A	E8	RPE		EE	XRI	d8
70	MOV	M,B	E0	RPO		E3	XTHL	
71	MOV	M,C	0F	RRC				
72	MOV	M,D	C7	RST	0			
73	MOV	M,E	CF	RST	1			
74	MOV	M,H	D7	RST	2			
75	MOV	M,L	DF	RST	3			
3E	MVI	A, d8	E7	RST	4			
06	MVI	B, d8	EF	RST	5			
0E	MVI	C, d8	F7	RST	6			
16	MVI	D, d8	FF	RST	7			
1E	MVI	E, d8	C8	RZ				
26	MVI	H, d8	98 SBB	B				
2E	MVI	L, d8	99 SBB	C				
36	MVI	M, d8	9A	SBB	D			
00	NOP		9B	SBB	E			
B7	ORA	A 9C	SBB	H				
B0	ORA	B 9D	SBB	L				
B1	ORA	C 9E	SBB	M				
B2	ORA	D	DE	SBI	d8			
B3	ORA	E 22	SHLD	addr16				
B4	ORA	H	30 SIM					
B5	ORA	L F9	SPHL					
B6	ORA	M	32 STA	addr16				
F6	ORI	d8	02 STAX	B				
D3	OUT	addr8	12 STAX	D				
E9	PCHL	37	STC					

d8	→	8-bit data	addr16	→	16-bit address
d16	→	16-bit data	M	→	Memory
addr8	→	8-bit address	PSW	→	Program Status Word

B.E/ B.Tech. DEGREE EXAMINATIONS, NOVEMBER/DECEMBER 2019

Fifth Semester

Electrical and Electronics Engineering

EE 8551 - MICROPROCESSORS AND MICROCONTROLLERS

(Common to Electronics and Instrumentation Engineering/ Instrumentation and Control Engineering)

(Regulation 2017)

Time: Three Hours

Maximum: 100 Marks

Answer ALL questions

PART-A

$(10 \times 2 = 20$ Marks)

1. List two major differences between INTR and the other hardware interrupts.

   Chapter 1, SA - Q1.128 **[Page No - 1.80]**

2. Does the 8085 support externally initiated operations. If yes how?

   Chapter 1, Section 1.6.2 **[Page No - 1.60]**

3. Illustrate the changes made to the content of registers during the execution of the instruction LXI B, 4000 H.

   Chapter 2, Section 2.4 **[Page No - 2.15]**

4. State the advantages of subtroutine.

   Chapter 2, SA - Q2.49 **[Page No - 2.109]**

5. Can single bit of a port be accessed in 8051? If yes, how? Give an example.

   Chapter 3, SA - Q3.27 **[Page No - 3.51]**

6. What are the flags supported by 8051 microcontroller?

   Chapter 3, Section 3.2.4 **[Page No - 3.11]**

7. Differentiate programmed I/O and interrupt driven I/O.

   Chapter 1, Section 1.4 **[Page No - 1.26]**

8. Why an interface is needed in between CPU and input-output devices?

   Chapter 1, SA - Q1.15 **[Page No - 1.67]**

9. Write a program to load the accumulator with the value 82H and complement the accumulator 700 times.

   Chapter 2, SA - Q2.76 **[Page No - 2.115]**

10. List any four applications of 8051 to automation systems.

    Chapter 5, Section **[Page No - 5.10]**

PART- B   (5 × 13 = 65 Marks)

11. a) With a functional block diagram, briefly discuss the architecture of the 8085 microprocessor.

           Chapter 1, Section 1.1.3 and 1.2 **[Page No - 1.6]**

(OR)

    b) Draw the timing diagram of the instruction MVI B, 45. Assume the memory address of the opcode and the data is 2000H and 2001H respectively.

           Chapter 1, Section 1.5 **[Page No - 1.47]**

12. a) i) Differentiate RAL and RLC instruction.   (3)

           Chapter 2, SA - Q2.77 **[Page No - 2.115]**

      ii) Write an assembly language program for 8085 microprocessor to count even numbers in series of 10 numbers.   (10)

Example:

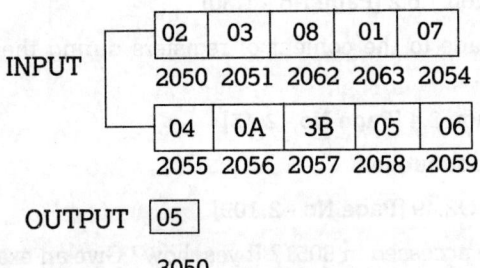

INPUT

02	03	08	01	07
2050	2051	2062	2063	2054

04	0A	3B	05	06
2055	2056	2057	2058	2059

OUTPUT   05

3050

           Chapter 2, SA - Q2.78 **[Page No - 2.115]**

    b) i) Briefly describe stack pointer register.

           Chapter 2, Section 2.10.5 **[Page No - 2.56]**

      ii) Briefly discuss the different types of addressing modes supported by the 8085 microprocessor with examples.

           Chapter 2, Section 2.2 **[Page No - 2.1]**

13. a) With a functional block diagram, briefly discuss the architecture of the 8051 microcontroller.

           Chapter 3, Section 3.1.2 and 3.2 **[Page No - 3.1]**

(OR)

b) (i) Summarise the similarities and difference between 8085 and 8051. (5)

Chapter 2, SA - Q2.77 **[Page No - 2.115]**

(ii) Discuss in detail the internal data memory organization of 8051 microcontroller. (8)

Chapter 3, Section 3.3 **[Page No - 3.16]**

14. a) (i) Interface 8255 with 8085 microprocessor and write an assembly language program to display 99 in port A, 1's complement of 99 in port B and 2's complement of 99 in port C. Assume the port addresses are 30_H, 31_H and 32_H for ports A, B and C respectively. (5)

Chapter 4, Section 4.2.1, SA - Q4.41 **[Page No - 4.72]**

(ii) Describe the operating modes and control words of 8255. (8)

Chapter 4, Section 4.2 and 4.2.3 **[Page No - 4.1 and 4.6]**

(OR)

b) With a functional block diagram, briefly discuss the architecture of the 8259 programmable interrupt controller.

Chapter 2, SA - Q2.77 **[Page No - 2.115]**

15. a) Show how to interface a stepper motor to 8051 micrcontroller. Also, write an assembly language program to demonstrate control of direction and speed of stepper motor rotation.

Chapter 5, Section 5.11 **[Page No - 5.73]**

(OR)

b) Show how to interface a servo motor to 8051 microcontroller. Also, explain the working principle to control a servo motor with angle rotations.

Chapter 5, Section 5.8 **[Page No - 5.62]**

PART- C (1 × 15 = 15 Marks)

16. a) Show how to interface a 8 × 8 matrix keyboard to the 8051 microcontroller and discuss in details the various stages for detection and identification of key activation by a microcontroller. Also, write an assembly language program to detect and identify the pressed key.

Chapter 5, Section 5.5 **[Page No - 5.52]**

(OR)

b) Show how to interface a Digital to Analog converter (DAC) with 8085 microprocessor and write an assembly language program to generate a square waveform. Also, discuss in detail the successive approximation technique for the process of conversion of analog signal to digital data.

Chapter 4, Section 4.6.2 **[Page No - 4.44]**

INDEX

CHIP INDEX